STRUCTURES

Fundamental Theory and Behavior

STRUCTURES

Fundamental Theory and Behavior

Second Edition

Richard M. Gutkowski, Ph.D., P.E.
Professor of Civil Engineering
Colorado State University
Fort Collins, Colorado

VAN NOSTRAND REINHOLD
New York

Printed in the United States of America

Van Nostrand Reinhold
115 Fifth Avenue
New York, New York 10003

Chapman & Hall
2-6 Boundary Row
London SE1 8HN, England

Thomas Nelson Australia
102 Dodds Street
South Melbourne, Victoria 3205, Australia

Nelson Canada
1120 Birchmount Road
Scarborough, Ontario M1K 5G4, Canada

16 15 14 13 12 11 10 9 8 7 6 5 4 3 2

Library of Congress Cataloging-in-Publication Data

Gutkowski, Richard M.
    Structures: fundamental theory and behavior / Richard M.
Gutkowski—2nd ed.
        p.   cm.
    ISBN 0-442-23012-5
    1. Structural analysis (Engineering)   I. Title.
TA645.G87   1990
624.1'71—dc20   89-21471
                                                     CIP

Contents

Preface to Second Edition   xi

Preface to First Edition   xvii

1 Introduction to Structural Analysis   1

    1.1   General Comments   1

    1.2   Structural Forms   1

    1.3   Idealized Models   2

    1.4   Types of Framed Structures   4

    1.5   Types of Continua   6

    1.6   Classification and Determination of Loads   9

    1.7   Load Paths   16

    1.8   Concepts of Work and Energy   19

    1.9   Objectives and Methods of Structural Analysis   21

    1.10 General Criteria for Structural Analysis   23

           Problems   24

2 Fundamental Mechanics of Members and Structures   26

    2.1   Introduction   26

    2.2   Elementary Definitions of Stress and Strain   26

    2.3   Basic Concepts in the Theory of Elasticity   28

    2.4   Nonlinear Material Behavior   31

    2.5   Inelastic Material Behavior   32

    2.6   Internal Stresses and Stress Resultants (Internal Forces)   33

    2.7   Actions   34

    2.8   Supports and External Reactions   35

    2.9   Equations of Equilibrium   36

    2.10 Free-Body Diagrams   37

    2.11 Stresses and Deformations in Simple Members   39

    2.12 Development of Idealized Models   45

    2.13 Stability of Structures   51

           Problems   59

3 Mechanics of Ordinary Beams   61

3.1   Statically Determinate Beams   61

3.2   Direct Analysis of Beams   68

3.3   Governing Equation for Flexural Displacement   78

3.4   Flexural Deformations by Methods of Integration   80

3.5   Deflection due to Shear   90

3.6   Assumptions in Ordinary Structural Analysis   92

3.7   Superposition   95

Problems   97

4 Moment-Area Method   103

4.1   Introduction   103

4.2   Moment-Area Method   103

4.3   Reference Tangents   107

4.4   Establishing a Sign Convention   113

4.5   Moment Diagrams By Parts   116

4.6   Example Problems on Displacement Computations   127

4.7   Effect of Chord Displacement   128

4.8   Closing Comments   132

Problems   132

5 Analysis of Statically Determinate Planar Trusses   136

5.1   General   136

5.2   Types of Trusses   136

5.3   Equilibrium of Planar Trusses   138

5.4   Forces in Statically Determinate Trusses   141

5.5   Geometrical Concepts   154

Problems   162

6 Analysis of Determinate Planar Frames   169

6.1   General Comments   169

6.2   Statically Determinate Frames   169

6.3   Mathematical Analysis of Flexural Behavior   176

6.4   Shear and Moment Diagrams   180

6.5   Displacements by the Moment-Area Method   181

6.6   Statical Indeterminacy   190

Problems   194

7 Angle Changes and Conjugate Structures   199

7.1   Introduction   199

7.2   Angle-Change Principle   199

7.3   The Conjugate-Grid Method   209

7.4   The Conjugate Beam Method   215

Problems   221

8 Work and Energy Principles in Structural Analysis   226

8.1   Introduction   226

8.2   Real Work and Strain Energy   226

8.3   Complementary Work and Strain Energy   229

8.4   Pseudo-Virtual Work and Pseudo-Virtual Strain Energy   230

8.5   Maxwell's Reciprocal Theorem   232

8.6   Betti's Reciprocal Energy Theorem   233

Problems   235

9 Unit-Load Method   237

9.1   Introductory Remarks   237

9.2   Development for Planar Trusses   238

9.3   Development for Beams   242

9.4   Extension to Planar Frames   249

9.5   General Structures   252

9.6   Unit-Load Method for Continua   256

Problems   257

10 Degrees of Freedom   260

10.1   Generalized Displacements and Deformations   260

10.2   Degrees of Freedom   262

10.3   Effect of Internal Hinges   265

10.4   Neglect of Axial Deformation   269

Problems   275

11 Flexibility Method   279

11.1   Fundamental Approach   279

11.2   Statical Indeterminacy   279

11.3   The Flexibility Method of Analysis   286

11.4   Symmetry of Flexibility Coefficients   303

11.5   Alternative Approach for Planar Trusses   305

11.6   Support Settlements   306

11.7   Initial Deformations   307

11.8   Compound Structures   308

11.9   Framed Structure Combined with A Continuum   310

11.10   Computer Programs   310

Problems   310

12 Slope-Deflection Method   316

12.1   Remarks   316

12.2   Slope-Deflection Equations for Prismatic Members   317

12.3   Application of Equilibrium Conditions   320

12.4   Examples   322

12.5   Nonprismatic Members   328

Problems   328

13 Moment-Distribution Method   331

13.1   Introduction   331

13.2   Basic Concept   332

13.3   Moment-Distribution Table   338

Problems   357

14 Basic Matrix Analysis of Trusses   360

14.1   Levels of Analysis   360

14.2   Basic Matrix Relationships   361

14.3   Example Problems   367

14.4   Alternative Matrix Analysis of Determinate Trusses   372
14.5   Treatment of Support Settlement   372
14.6   Treatment of Initial Member Deformations   376
14.7   Computer Programs   378
        Problems   378

15   **Direct Stiffness Analysis of Planar Structures   382**
15.1   Introduction   382
15.2   General Concepts   386
15.3   Direct Generation of [K]: Continuous Beams   388
15.4   Direct Stiffness Method: Beams   396
15.5   Direct Generation of [K]: Planar Trusses   405
15.6   Direct Stiffness Method: Planar Trusses   408
15.7   Generation of [K] by Physical Meaning: Planar Frames   415
15.8   Direct Stiffness Method: Planar Frames   418
15.9   Computer Programs   431
        Problems   431

16   **Transformation Matrices for Planar Structures   440**
16.1   Introduction   440
16.2   Member Stiffness Matrices   441
16.3   Assembly of the Structure Stiffness Matrix   451
16.4   Generation of the Load Matrix   452
16.5   General Solution Procedure   455
16.6   Example Problems   455
16.7   Special Conditions   463
16.8   Nonprismatic Members   471
16.9   Combined Systems   472
16.10   Remarks   474
        Problems   474

Appendix A Tables   477
Appendix B   Summary of Matrix Algebra   483
Appendix C   Computer Programs   492
Index   535

To my loving wife, Irene

The Story of My Life
Words and Music by Neil Diamond

The story of my life
Is very plain to read
It starts the day you came
And ends the day you leave

The story of my life
Begins and ends with you
The names are still the same
And the story's still the truth

I was alone
You found me waiting and made me your own
I was afraid
That somehow I never could be
The man that you wanted of me

You're the story of my life
And every word is true
Each chapter sings your name
Each page begins with you

It's the story of our times
And never letting go
And if I die today
I wanted you to know

Stay with me here
Share with me; care with me
Stay, and be near
And when it began
I'd lie awake every night
Just knowing somewhere deep inside
That our affair just might write

The story of my life
So very plain to read
It starts the day you came
And ends the day you leave

Preface to the Second Edition

About one decade has passed since the publication of the First Edition of this textbook. At that time desktop computer hardware for the individual user was emerging as a medium for structural engineering computations. Indeed, the author and most of his colleagues did not yet possess personal computers. Word processing was a capability in its infancy. Considerable change in the state of the art has occurred since that time.

Personal computer capability has become an essential need for students and practicing engineers and is readily available at low cost. External floppy diskettes with 360 kilobyte storage capacity have been virtually replaced by internal hard disks, commonly with 10 to 50 megabyte capacity, and much more in some computers. Furthermore, powerful workstations have become a dominant environment for computer based computation. User friendliness through-menu driven software, interactive formats, windowing techniques, etc. has simplified input, execution, output, and interpretation of results related to structural analysis performed using complex computer programs. Networking of hardware to allow in-place access to an extensive pool of computational and input/output devices and tools is commonplace. Computer-aided drafting (CAD) has enhanced and eased the world of engineering graphics. Expert systems are growing in importance as a format for allowing the everyday engineer to perform activities that traditionally would have to be done only by a specially skilled, knowledgeable, and well informed person, i.e., by an expert in the subject. In short, the contemporary structural engineer has a wealth of computing power at her/his hands.

In the First Edition of the present text, the author sought to provide balanced content, i.e., equal focus on traditional subject matter and on computer programming techniques. Given the ready availability of tools for computer-based computation, the Second Edition reflects a reduced emphasis on the latter need. The author believes that aspect of education has been affected by the advances made in recent years in making thorough, easily used commercial software available. Teaching undergraduate students to develop structural analysis software themselves has become much less common and less pertinent. Thus, students can readily become well skilled in structural analysis software usage, but potentially at the expense of time previously dedicated to forging knowledge of the basic concepts, theories, and limitations inherent in present-day structural analysis.

Structures: Fundamental Theory and Behavior, 2nd Edition, is a major restructuring of the original textbook from 12 chapters to 16 chapters. The objective has been to condense, reorganize, and fine-tune the material so as to provide a self-contained textbook for a single undergraduate course on the subject. The intention was to augment material on classical and computer methods of structural analysis, with an increased emphasis on providing a solid foundation in the fundamentals of the subject. In particular, the early chapters have been extended to include a stronger foundation on the basic mechanics of structural members.

Well documented commercial software packages exist for tutoring students on computer-based capabilities that are used in practice. Thus, one important aspect of this textbook is that a concentrated effort was made to enhance the reader's understanding of the make-up and applicability of techniques that are part of standard commercial software packages. The goal is to minimize the "black box" environment in which such packages might otherwise potentially be used. It is felt that this is best accomplished by combining use of manual methods to develop a "physical sense" of the principles and concepts with computer assignments that encourage the study of the influence of variables (e.g., geometry, dimensions, load position, member properties) on the structural response of a structure. Earlier-generation textbooks typically included extensive treatment of influence lines and criteria for determining maximum effects under moving and pattern loads. The author's experience is that contemporary curriculum demands only permit a nominal presentation of these interesting, valuable subjects. Some problem exercises are included on them, but reluctantly text on this subject is not included because of space limitations. Also, exercises to utilize computer analysis to illustrate the same points can be used to contribute to similar understanding of such behavior. Providing an understanding of the assumptions, versatility, and limitations of the methods (both classical and computer) described is equally important. However, the intricacies of actually developing software itself are reserved for studies at the graduate level of education.

The First Edition included several very lengthy chapters which encumbered readability and digestion of the material. These have been subdivided into shorter, self-standing chapters. The presentation of matrix structural analysis has been consolidated and enhanced. The direct stiffness method has been emphasized and extended. Content pertinent only to upper-level course work has been removed. The entire content of old Chapters 10 (Approximation Methods) and 11 (Introduction to the Finite-Element Method of Analysis) and content on space structures in the old Chapter 8 (Stiffness Analysis of Large Structures) have been removed. Rigorous work and energy definitions, related advanced theorems, and methods of analysis have been eliminated from the old Chapter 9 (Work and Energy Principles in Structural Analysis). At present it is the author's intention to develop a second volume of this edition of the book which would include the presently deleted material and other intermediate and advanced topics.

This book is a contemporary treatment of the subject of elementary structural analysis. In Chapter 1, various structural forms and systems used in buildings, bridges, etc. are introduced. General concepts for developing an appropriate mathematical model to represent any given framed structural system are presented. Determination of the applied loads and the manner (load paths) by which a structure transmits them through its skeleton (framework) to the supporting ground is detailed. Criteria for insuring equilibrium without failure or instability are explained.

Chapter 2 presents the fundamental laws of mechanics which a structure and its component members must obey. Principles learned in the prerequisite subjects of statics, properties of construction materials and mechanics of materials are reviewed and their interrelationship with the concepts of theory of elasticity illustrated. The different aspects of modeling complete three-dimensional structures, two-dimensional substructures and individual members are explained. Framing techniques necessary to the stabilization of a structural system against buckling and lateral wind loads are illustrated.

In the remaining chapters, well established analytical methods for determining the response of two-dimensional structures to statically appled loads are developed and their use illustrated. Determination of the displaced shape and internal forces of the members are the primary subject matter.

The principal classical methods are fully treated first. Chapter 3 focuses on the individual beam member. Assumptions in linear, elastic, first-order structural analysis are detailed. Procedures to calculate reactions, internal forces and shear and moment diagrams are developed. The governing differential equation for small deflection theory is derived and its integration illustrated, including use of binomial functions. Chapter 4 augments this material with the development and use of the moment-area theorems for deflection calculations. Techniques for using a moment diagram "drawn by parts" are emphasized.

Chapter 5 addresses statically determinate trusses, including simple, compound and complex trusses. The method of joints and the method of sections for calculating member forces are developed. Procedures for assessing geometrical stability are explained, including example problems. Chapter 6 provides similar treatment of planar frames. Methods are presented for calculating reactions, internal forces and shear and moment diagrams. Use of the equation of the elastic curve and the moment area method in calculation of displacements is presented. Chapter 7 covers the angle-change principle for simplifying the application of the moment area method. Use of conjugate grids and beams to convert a deflection geometry problem to an equivalent statics problem is explained.

In Chapter 8, work and energy principles that underlie all methods of structural analysis are presented. Concepts explained include real work and real strain energy, complementary work and complementary strain energy, pseudo-virtual work and pseudo-virtual strain energy. Maxwell's reciprocal theorem and Betti's reciprocal energy theorem are derived and applied to prove important aspects of the flexibility and stiffness methods of structural analysis. The unit-load method (virtual work method) is developed in Chapter 9, using the Law of Conservation of Work and Energy. A general equation is derived to calculate displacements in planar and space structures due to axial, flexural, shear, and torsional deformations.

Chapter 10 defines the concept of kinematic (discrete displacement) unknowns. Identification of a structure's "degrees of freedom" is detailed. The flexibility method for analyzing statically indeterminate structures is developed in Chapter 11. Application to beams, planar trusses, planar frames and compound structures is shown. Treatment of response to support settlements and initial member deformations is included. Chapters 12 and 13 contain succinct presentations of the classical slope-deflection method and the moment distribution method, respectively. These provide a backdrop to matrix structural analysis by introducing the two-stage stiffness method of structural analysis.

In Chapter 14, the basic relationships used in the matrix stiffness method of analysis are defined for statically indeterminate planar trusses. Example problems illustrate their use in calculating the complete structural response. Chapter 15 extends the matrix formulation to indeterminate beams and planar frames. The powerful direct stiffness method for automated assembly of the structural stiffness matrix is developed. In Chapter 16, fundamental techniques of computer-based matrix structural analysis are introduced. Use of transformation matrices for generating member stiffness matrices and extension of the automated assembly process to include equivalent joint loads is detailed. Efficient methods to account for support conditions, support motions, initial strains, internal hinges, inclined supports, nonprismatic members, and combined structures are included.

The order of the chapters retained from the First Edition has been changed and new content and chapters merged with them to improve the flow of the material. Additional example problems serve to fill voids identified in the First Edition. Problem exercises given at the end of chapters are predominantly new material. Computer algorithms presented in the First Edition are now obsolete and have been replaced by contemporary programs. Versions in both BASIC and FORTRAN are presented to extend the portability of the software. The programs are written in the simplest structure and form possible to help insure ease of understanding and of use. Extreme care has been taken to eliminate errata prevalent in the first printing of the original textbook. Classroom use for 9 years and a second printing served to minimize, if not eliminate, the carryover into the Second Edition. New text material has been classroom tested twice. The entire manuscript was read and edited by Mr. Thomas Purvis. Mr. Paul Marck confirmed the correctness of the BASIC algorithms and converted them to FORTRAN at the galley stage, to insure contemporary format and syntax. Both were graduate students in structural engineering at Colorado State University.

In the past decade, several other interesting textbooks have been published on the subject of structural analysis. Collectively, these all contribute complementary, supplementary, and alternative perspectives to the author's perception and knowledge of structrual analysis. My recent growth as a structural engineer has also been keenly influenced by a period of residency at the Institute for Wood Construction at the Swiss Federal Institute of Technology in Lausanne, Switzerland. Direct interaction with Professor Julius Natterer, Chair of that institute, has provided me the profound experience of working with a marvelously creative, uniquely accomplished structural engineer who is both outstanding and highly respected in his field. Mr. Wolfgang Winter of the same institute has very kindly shared thorough, detailed explanations of the architecture, framing, connection details, and load paths for many of actual timber structures in western Europe, including escorting me on extensive tours of numerous actual sites. These opportunities with these gifted practicing engineers provided me an important sense of balance vis-à-vis my dedication to understanding the theory and analytical techniques of structural engineering.

The First Edition of this book was motivated by a preceding, long period of encouragement by many successive groups of students who attended my courses. Otherwise, I would not have given any consideration to undertaking the activity. I tried to tell an interesting story and hope those who read it have enjoyed it. This Second Edition was initiated with the unselfish encouragement of my loving wife, Irene. This book is dedicated to her, knowing the strength, energy, enthusiasm and need she expressed for my work to be fulfilled may have delayed her own aspirations and desire for personal accomplishments. I do not know a kinder person and will be there for her as she achieves them. Nearly five years have passed in producing this edition and both her patience and bond have persisted despite the demands of my overhauling an extensive textbook at the expense of more time together with her and my very special daughter, Jody.

Richard Matthew Gutkowski
Ft. Collins, Colorado

To my teachers and my students

So it seems, and so it may be,
I can't know for sure.
But lately, seems I'm on a journey
To a place I've never been.
And baby, if I had the answers
I would lay them all before you.
Plainly it's a circle, one that ends
And then begins and begins again.

Yes I will

'deed I will

If I can

By NEIL DIAMOND

Preface to the First Edition

Contemporary structural engineering reflects the tremendous advances made in recent years in the area of computer-oriented methodology. About 15 years ago, highly sophisticated computers revolutionized the practical applications of engineering principles. Simultaneously, versatile electronic hand-held calculators replaced the slide rule as the engineer's major tool. Unlike their predecessors, contemporary structural analysts must possess extensive skills in digital as well as manual computation methods. Furthermore, these talents must be complemented with a keen perception and "feel" for the behavior of real structures.

Structures: Fundamental Theory and Behavior describes subjects the author considers fundamental background for the beginning structural analyst. A balanced context is achieved by emphasizing basic concepts, theories, and assumptions; including useful classical methods; addressing fundamental matrix (computer) methods; and extensively describing commonly used computer-programming techniques. A major aim has been to avoid a sequence of isolated topics and, instead, to link the chapters together in a somewhat narrative style, characterized by a smooth transition from one subject to the next. The order of presentation was chosen with that objective in mind. It is hoped that this treatment of elementary structural analysis unfolds like an interesting story.

Chapter 1 discusses the representation of real structures and the loads acting on them by idealized analytical models. Chapter 2, a key feature of the book, addresses the fundamental geometric concepts and subtle assumptions in structural analysis. Governing equations for the deformations of the individual member in space are developed in Chapter 3. Chapters 4 through 6 deal with the rudimentary analysis of planar trusses, beams, and planar frames, respectively. Well-known classical methods and a basic matrix method are presented with an emphasis on physical concepts.

Chapter 7 is a transitional chapter that serves as a bridge between the basic matrix formulation and the sophisticated semiautomated matrix computer methods detailed in Chapter 8. In Chapter 9, many important work and energy theorems are developed in a precise manner, using plain language and accompanied by numerical example problems. Variational principles are included as a prerequisite to the introductory presentation of the finite-element method in Chapter 11. Chapter 10 describes approximation methods commonly used to analyze irregularly shaped members. Chapter 12 assembles several well-known classical planar-frame analysis methods.

Several possibilities exist for covering the material. A single four-credit course might cover the entire text, but at an accelerated pace. Omission of parts of Chapters 9 and 10 (but not Section 10.4) would afford a comfortable pace. Alternatively, two consecutive three-credit courses would be most adequate, with the first covering Chapters 1 through 6 and Chapter 12, and the second covering the balance of the material. (At Colorado State University, some of the earlier material is covered in a prerequisite course—strength of materials. Chapters 1 through 7 and Chapter 12 are then completed in a three-credit course. Chapters 8 through 11 are reserved for the graduate level.) Some instructors may wish to select only certain topics in Chapter 12 to suit individual priorities or to omit it entirely to permit greater emphasis on matrix methods.

Because this text is introductory, it largely addresses established subject matter. My perspective of the material greatly reflects fortunate contacts with many superb educators and a continual effort to examine the writings of other scholars. I am particularly indebted to several inspirational teachers—Sister Mary Celine and Professors A. Fattah Chalabi, Allan Marcus, and Chu-Kia Wang—who, by example, encouraged me to be inquisitive, thorough, and unselfish. Special acknowledgment is also due the many fine students of my own who have come and gone—they have inspired this work.

The patient and sympathetic efforts of Debbie Harris, Dorothy Henderson, Cheryl Paget, Jane Rademacher, Janet Ragazzi, and Ruth Vickers, who typed the numerous class notes and manuscript drafts that preceded the book, are recalled with great appreciation. The continued moral support of my mother and father is impossible to repay. Finally, I am grateful to my loving wife, Irene, who gives me peace and comfort and care.

Learning awakens the mind. Teaching stirs the soul!

RICHARD MATTHEW GUTKOWSKI
Fort Collins, Colorado

STRUCTURES

Fundamental Theory and Behavior

1
Introduction to Structural Analysis

1.1 GENERAL COMMENTS

Throughout history, mankind has built various kinds of structures in order to achieve a desired degree of comfort, mobility, and convenience. In the earliest times, construction was a trial-and-error process. A structure was erected, and if it collapsed, it was modified in configuration and rebuilt until it stood successfully. As the centuries passed, building by experience evolved into engineering by basic physical laws. Scientific investigators began to recognize that natural and induced loads act on physical bodies, and they formulated techniques to predict the response of given structures to such loads. As a consequence, erection of a structural system today is normally preceded by an extensive phase of "structural analysis and design."

Structural analysis is the determination of the response of a structure to the loads that act upon it. *Design* is the creation and subsequent modification of the physical configuration of a structure to achieve a desired response (i.e., to cause it to function as intended). The design process is creative as well as technical and brings a conceived structure to realization. Structural analysis is an activity which is included in the design process. It is the means by which the response to all anticipated loadings is established for each potential framing configuration investigated by the structural engineer.

The availability of modern electronic computational systems has elevated this activity to a highly sophisticated level. Large structural systems are now routinely analyzed by very rigorous procedures.

Structural analysis and design both require a thorough understanding of the properties of construction materials and of the fundamental laws that govern material behavior. The application of the laws of statics and strength of materials, normal introductory subject matter for students of engineering science, is a small subset of structural analysis. Therefore, it is assumed that the reader has a proper understanding of statics and strength of materials. Extending this knowledge to the treatment of the common types of structural systems is the primary objective of this textbook.

1.2 STRUCTURAL FORMS

Engineering structures occur in our everyday surroundings in many different forms. They are usually distinguished by names that imply their general purpose—for example, bridges, buildings, pressure vessels, tanks, aircraft, and dams. Each of these has a specific function, and each is subjected to loads that are basically different. Although these complete assemblies are commonly called structures, from an analytical point of view there are only two fundamental categories of structures:

1

1. a framed structure
2. a continuum

A *framed structure* is an assemblage of individual *members*, which are either interconnected by mechanical fasteners (such as a residential home) or monolithically joined (such as a reinforced concrete building). The cross-sectional dimensions of a member are generally much smaller than its length, which may be either straight or curved.

A structure that cannot be distinguished as being comprised of individual members is classified as a *continuum*. Concrete slabs, domed roofs, and tanks are common examples. Generally, the thickness of a continuum is much smaller than its primary dimensions. In some instances, a continuum may be formed by mating individual flat or curved panels; however, often the resulting form can still be considered a continuum for analytical purposes.

1.3   IDEALIZED MODELS

Structures subjected to load are essentially energy systems. In the unloaded state, a structure has a particular configuration in space. When subjected to loads, the structure deforms and moves relative to its original position. As a consequence, the loads perform external work, and internal energy is stored in the structure. Neglecting the losses due to friction and other minor influences, the two quantities, external work and internal energy, must be in balance. This adherence to the Law of Conservation of Work and Energy is the fundamental concept that underlies structural analysis. A more detailed description of the roles of work and energy in structural analysis is given in Section 1.8.

Many methods of structural analysis have been developed in past years. Some methods rely upon the direct use of the basic concepts of work and energy and their interrelationship. Several of the more important work and energy theorems in structural analysis are developed and applied in Chapter 8. More typically, methods of structural analysis only reflect the work-energy balance indirectly, either in the

derivation stage or in the adopted material laws. Such methods, which are in predominant use, make up the major content of this book.

All methods of structural analysis rely upon the use of an *idealized model* to represent any given real structure. Development of a proper idealized model is a prerequisite to quantification of the behavior of the loaded structure. Essentially, one must establish a reliable means to *calculate* the expected response of a structure so that the prediction is in close agreement with the observed or *measured* behavior.

The basic features of idealized models are specified material properties and behavior (laws of material behavior), a simplified geometric representation (schematic diagram), and a sufficiently accurate mathematical analysis technique (mathematical model). Laws of material behavior and schematic diagrams will be discussed next, and the development of acceptable mathematical models will then constitute the remainder of this textbook.

1.3.1   Laws of Material Behavior

Laws developed in mechanics of materials are the means by which the engineer predicts the effects of internal energy—i.e., stress and strain. Structural analysis is the means by which the engineer predicts the integrated effects of stress and strain—namely, internal forces and external displacements. Because material laws are limited by assumptions made in the basic theory and by incomplete knowledge of the material properties, the predictions are approximate. Material and structural testing are the means of assessing the degree of approximation.

1.3.2   Schematic Diagrams

Conversion of a real structure to a schematic representation is depicted in Fig. 1.1. The rectangular framed structure shown in Fig. 1.1a has rigid joints at *B* and *C*. *Rigid* implies that the mechanical connection at these joints is such that, during deformation of the framework, the angles between the members at joints *B* and *C*

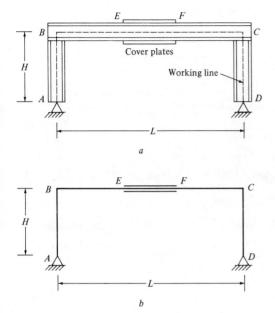

Fig. 1.1. Conversion to a schematic model. *a*, real structure; *b*, schematic model.

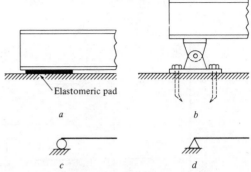

Fig. 1.3. Example idealization of support conditions. *a*, simple expansion bearing; *b*, double pin stand; *c*, schematic roller support, *d*, schematic pin support.

are maintained even if the joints themselves rotate as a whole.

Working lines shown in Fig. 1.1 are used to establish the dimensions of the schematic model depicted in Fig. 1.1b. Ideally, the working line of each member should be its centroidal axis, but this is not always so. For curved or tapered members, the working line is usually taken as some convenient straight line, as illustrated in Fig. 1.2. In schematic representations, the intersections of the various working lines are the joints of the structure.

Supports in the idealized model are representations of actual physical conditions. The particular representation chosen is dependent upon both the framing connections that exist and the judgment of the analyst. Figure 1.3 offers two situations commonly encountered in practice. The elastomeric bearing pad shown in Fig. 1.3a

Fig. 1.2. Working line for a double-tapered beam.

is used to permit thermal expansion to occur as freely as possible and is modeled as a roller support, as illustrated in Fig. 1.3c. Such elastomeric pads are usually made of neoprene, Teflon, or some other low-friction material. It should be noted that the roller symbol in Fig. 1.3c infers that resistance is provided against uplift as well as bearing. Pin supports are typified by the detail shown in Fig. 1.3b and the symbol shown in Fig. 1.3d. Such supports are anchored in place and permit free rotation of the member end with respect to the support itself.

Table 1.1 gives a summary of the possible support and joint conditions for planar structures. The idealized conditions listed in the table cannot be achieved exactly in real structures. However, standard support and connection details are available in standard design manuals to approximate these conditions in all building materials.

To convert real-world structures to idealized models is a difficult task for the structural analyst. Normally, actual support conditions and connections cannot be exactly modeled. Some continua have articulated shapes or openings. These are only two examples of the difficulties encountered in practice. Consequently, the modeling process is hardly automatic. Engineers must exercise care in defining a rational idealized structural model and should have the following objectives in mind:

Table 1.1.   Common Types of Supports and Joints.

Symbol	Resistance	Meaning
Fixed		Resists rotation and translations in two orthogonal directions.
Pinned		Allows rotations but resists translations in two orthogonal directions.
Roller		Allows rotation and longitudinal translation but resists transverse translation.
Free		Free to rotate and translate in all directions.
Internal hinge		Member free to rotate relative to others at the joint, but translational continuity is maintained.
Semirigid joint k	k $\theta = f(k)$	Rotation is a function of an elastic spring constant k.
Elastic support k	k $\Delta = f(k)$	Transverse displacement is a function of an elastic spring constant k.

1. Minimize the approximation that will result from the analytical solution
2. Simplify the mathematics as much as possible without compromising the first objective
3. Insure that, when finally constructed, the structure will respond to load in the manner assumed

1.4   TYPES OF FRAMED STRUCTURES

Framed structures are structures composed of an array of interconnected members. Historically, such structural systems have been classified into six basic types:

1. planar trusses

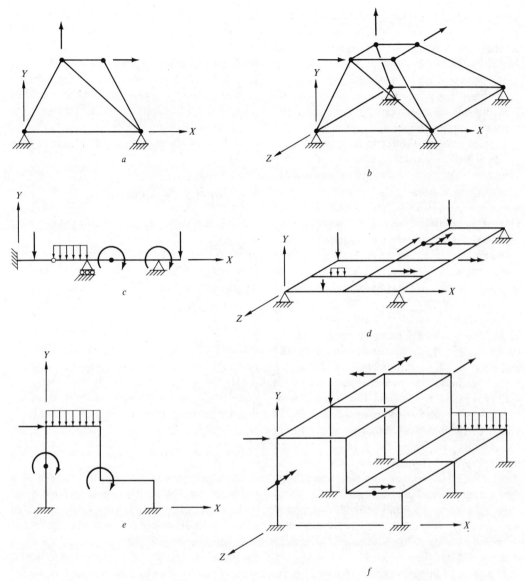

Fig. 1.4.   Types of framed structures. *a*, planar truss; *b*, space truss; *c*, beam; *d*, grid; *e*, planar frame; *f*, space frame.

2. space trusses
3. beams
4. grids
5. planar frames
6. space frames

A schematic example of each type is illustrated in Fig. 1.4.

A *space frame* is the most general of the framed structures; it is a three-dimensional array of members, can be subjected to loads acting in any direction, and can have all joints fully rigid. All other types of framed structures are a subset of the space frame. *Planar frames* lie in a single plane, and all loads are in that same plane. *Beams*, which consist of members that lie along a single axis, are subjected to loads that are transverse to that axis and lie in a common plane. Similarly, the members of a *grid* also lie in a common plane; and all loads act transverse to the plane that contains the grid. Normally, the joints in a grid are rigid. The *space truss* and *planar truss* are similar to the space frame and planar frame, respectively, except that all members are pin-connected and loads are applied only at the joints.

Framed structures differ in the general manner by which they respond to loads. It is sufficient here to distinguish the differences briefly. Members in trusses (either planar or space) transmit applied loads to their supports by either elongating or contracting. In a real sense, they act like unidirectional springs. Essentially, each member develops either a pure tensile or a pure compressive axial force. This behavior is a direct consequence of the truss being loaded only at its pinned joints.

By contrast, a member in either a beam or a planar frame can be transversely loaded between its ends. If so, the member flexes (bends) and internal "flexural moments" develop. In a beam, flexure is the primary manner by which load is resisted. In a planar frame, each member develops both flexural moments and an axial force. These two behaviors are additive, but one usually dominates the behavior of each member. Whether flexure or axial behavior dominates each member response depends on the configuration of the particular frame. Also, "second-order structural analysis" (defined subsequently) is needed to consider the interactive effect of axial force combined with flexural moment.

Members of grids and space frames also experience flexure when transversely loaded. In addition, they twist and develop internal torsional moments. Furthermore, by definition, each of these types of structures can be subjected to torsional moments as a loading. These loads also twist the members and produce internal torsional moments. Space frame members also transmit axial force, but grids do not do so.

Physically, beams and planar frame members can also experience torsion. Torsional moments are produced if either a torsional moment is directly applied to the member or a transverse load is not directed through a particular point in the member cross-section called the *shear center*. Customarily, and in this textbook, beams and planar frames are defined so as to exclude such possibilities. If they exist, it is usual to treat the structure as either a grid or a space frame, as appropriate, and use the corresponding analysis method. These special considerations are outside the scope of this textbook.

1.5   TYPES OF CONTINUA

All structures that do not fit into the category of framed structures are termed *continua*. Continua exist in many different physical shapes; however, analytically, a given continuum must be classified as one of three basic types:

1. flat plates
2. shells
3. membranes

Although the type of a particular continuum is usually suggested by its geometric shape, the distinction is properly made on the basis of the manner in which it resists applied loads.

1.5.1   Flat Plates

A staircase landing is a simple example of a flat-plate structure. Such structures have a flat surface, and their depth is very small compared to their plan dimensions. Loads are usually applied transverse to the flat surface. In plan, a flat plate is not restricted to a rectangular shape, and, in fact, a circular configuration is particularly common.

Figure 1.5 shows the schematic representation of a flat plate. The plate is rectangular; the dashes indicate that one edge is fixed. In this case, the loading consists of the downward concentrated load directly applied to the corner. Under this loading, the downward (transverse) displacement is the significant deformation that occurs. This is the case for most flat-plate structures. Displacements that occur are transverse to the loaded surface and are two-way. The displaced position of a point on the surface is mathematically defined by three-di-

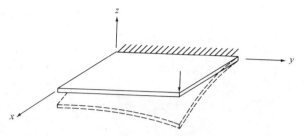

Fig. 1.5.   Displaced shape of a transversely loaded flat plate.

mensional geometry; i.e., the transverse displacement z is a function, $z(x, y)$, of the original position of the point and differs for each possible loading. This is unlike members in framed structures in which transverse displacement of a point is a function, $z(l)$, only of position l along the member length (i.e., one-way action occurs).

1.5.2   Shells

Flat plate structures are easy to construct but inefficient in structural performance, except for short spans. When using a continuum for intermediate- and long-span situations, use of either a *folded plate* or a *shell* is a far superior solution. A folded plate is accordion-like in shape. In ordinary terms, a shell is a flat plate curved (*developed*) into a new shape. A cylinder is an example of a developable shell. However, some shells are not developable from flat elements. A dome is an example of a nondevelopable shell.

In its early development, the cylindrical shell structure was used primarily to resist hydrostatic pressure. Barrels and tanks are the simplest examples. As time passed, many other applications of this structural form were developed. Pressurized boilers, submarines, pipe culverts, and similar structures adopted the cylindrical shape because of its effectiveness in resisting internal or external pressure. Major advances in the aircraft industry have resulted from the lightweight designs that have used a thin shell as the body of the aircraft.

In building construction, the classic form of shell structure is the domed roof. During the Renaissance, European architecture was

marked by the early predominance of the dome as a means of covering large areas of open space. Such structures are a reflection of daring engineering skill during a period when the science of mechanics had just been born. Unlike the earlier massive stone domes, contemporary applications (e.g., in stadiums, auditoriums, and arenas) are very economical. Furthermore, using modern technology and construction materials, today's engineers have refined their understanding of dome action and have thus erected structures of marvelous proportions. The Astrodome (504-ft diameter) in Houston, Texas, the Tacoma Dome (532-ft diameter) in Tacoma, Washington, and the Superdome (680-ft diameter) in New Orleans, Louisiana, are examples of such accomplishments.

On a smaller scale, there are a variety of shell forms in a common use. Several examples are depicted in Fig. 1.6. Such shell forms were usually developed for architectural appearance, and consequently each responds to load in a unique way. Although the shape and the analytical approach may differ for each type of shell, the internal stresses and strains are consistent with a set of basic laws of "shell analysis."

In many instances, a structure, such as the folded plate or the geodesic dome, is formed by molding or connecting panels to a supporting framework. The result is a *shell form* or *space form* that has the characteristics of both framed structures and continua. Complex analysis is required to properly assess the response of these formed shells.

Increased depth is the basic reason shells are more effective, structurally, than flat plates. Fig. 1.7a shows a flat plate with a given span.

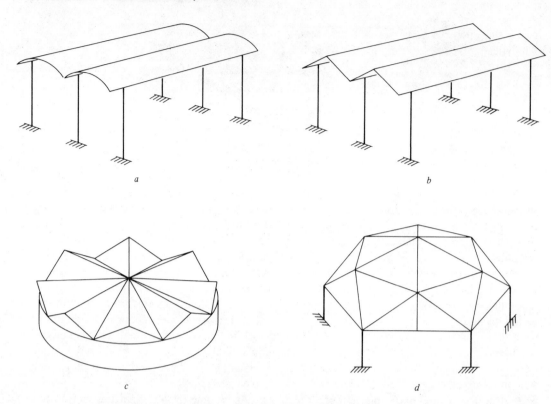

Fig. 1.6.   Common forms of shell structures. *a*, two-bay barrel vault; *b*, two-bay folded plate; *c*, circular folded plate; *d*, geodesic dome.

In Fig. 1.7b the same flat plate is converted to a folded plate without changing the amount of material. The folded plate tends to act like a very deep beam whose cross-sectional moment of inertia is dramatically increased relative to that of the flat plate. All shell forms have this same advantage. Additionally, curved shell forms tend to transmit loads either by internal tensile or internal compressive stresses instead of flexural stresses. This is a distinct advantage for most construction materials. However, when the internal stresses are compressive, the "buckling stability" of the continuum is a critical concern. Instability due to buckling can

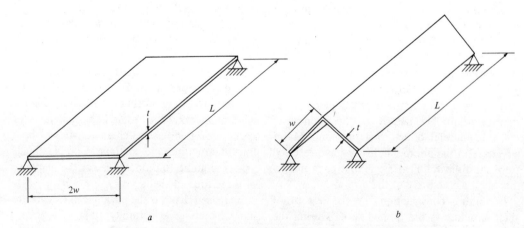

Fig. 1.7.   Effect of depth in a shell. *a*, flat plate; *b*, folded plate.

cause sudden, total collapse of the structure. Analytical procedures to examine this phenomenon do exist, but are a topic for advanced textbooks.

1.5.3   Membranes

Membrane structures are most easily exemplified by inflated tires or hot-air balloons. In a sense, a membrane is a "shell in reverse." Basically, a membrane resists applied internal pressure by developing tensile stresses (rather than the compressive stresses that develop in a shell) in all directions. Contemporary air-supported structures, such as the roof of Pontiac Stadium in Pontiac, Michigan, offer stunning structural examples of membranes.

Membranes are often made of cable-reinforced plastics. Because a membrane is not limited by the buckling tendencies inherent in compression members, the full strength of the materials is available for use. The result is an economical, lightweight structure. In practice, plate structures subjected to in-plane loads are sometimes referred to as membranes. This is improper terminology; it is preferable to say that a plate subjected to in-plane loads has developed *membrane-action*.

A type of structure similar to a membrane is the suspension structure. Cable-supported structures, suspension bridges, and cable-stayed bridges belong in this grouping. Madison Square Garden in New York City, is a notable example of a cable-supported roof system. This type of structure can be properly included in the framed structure category. The primary features are that, unlike general framed members, the cable members can only resist tensile action and are subject to significant *sag*.

1.6   CLASSIFICATION AND DETERMINATION OF LOADS

Loads that act upon a given structure are caused by many different effects. Principal categories of loads are:

1. *Dead loads*, which include the weight of the structural components and any attach-

ments permanently fixed in position. Floor slabs, roof deck, fixed walls and partitions, permanent equipment, ceilings, and flooring typify sources of dead load.

2. *Live loads*, which include most load effects that are not permanent in position or duration. Live loads are a reflection of the intended use and occupancy of the structure. Storage materials, people, traffic on a bridge, library books, and construction equipment are a few examples.

3. *Environmental loads*, which are the effects of nature that cause deformation of a structure. Snowfall, earthquake, wind, rainfall, temperature change, and ground settlement are the most significant types.

1.6.1   Sources of Load Data

Local building codes give requirements and guidelines for proper determination of most load values. In many cases, these codes rely on information given in national standards. Two common sources of load data for commercial structures are the American National Standards Institute (ANSI) and the Uniform Building Code (UBC). In addition, guidelines for design in a given material are available in specification codes produced by various trade organizations, such as the American Institute of Timber Construction (AITC), the American Institute of Steel Construction (AISC), and the American Concrete Institute (ACI).

Dead loads for common building materials are simply listed. With a knowledge of the appropriate unit-weight values, computation of the dead load for building materials is a simple matter. For ready usage, applicable values for common materials are generally listed in the format shown in Table 1.2. Values given are intended to be representative; if desired, more specific values can be obtained from manufacturers' literature.

The significant live load for highway bridges is vehicular traffic. Appropriate standard truck loads are specified by the American Association of State Highway and Transportation Officials (AASHTO). In buildings, the major live

Table 1.2. Example of Weights of Building Materials.

Material	Weight (psf)
Metal lathing	0.5
Gypsum block	
2-in.	9.5
4-in.	12.5
5-in.	14.0
Clay tile, 3-in.	17.0
Cement plaster, 1-in.	10.0

Table 1.3. Example of Recommended Occupancy Live Loads.

Occupancy	Live Load (psf)
Assembly-hall stage	100
Poolroom	75
Office	50
Library stack room	150
Restaurant	100
Garage (passenger cars)	50

load is due to occupancy. Such loads are specified by building codes and are probabilistic in nature. Generally, they are chosen to represent the maximum load one could *reasonably* expect to occur during the life of a structure. Higher loads could occur, but the probability is deemed much too low to justify the expense of designing for such load levels. Recommended values are generally listed as shown in Table 1.3. In warehouses, live load is more difficult to assess and requires engineering judgment because of the variability of the contents. As a design aid, some specifications tabulate *recommended live loads* for storage warehouses. A few typical listings are given in Table 1.4. It should be noted that recommended values are *not* standard in all specifications. Furthermore, listed values are intended for normal conditions. Unusual or special-purpose structures require considerable engineering judgment if proper live loads are to be assigned.

As live load is unlikely to act simultaneously throughout an entire building, most codes include *live load reduction criteria*. Such criteria allow lower values than usually recommended when the floor area to be carried by a member

Table 1.4. Example of Recommended Live Loads for Storage Warehouses.

Material	Weight (pcf)	Height of pile (ft)	Weight (psf)
Building Materials			
Asbestos	50	6	300
Bricks			
building	45	6	270
fire clay	75	6	450
Cement			
natural	59	5	354
portland	72 to 105	6	432 to 630
Gypsum	50	6	300
Lime and plaster	53	5	265
Tiles	50	6	300
Woods, bulk	45	6	270
Dry Goods, Cotton, Wood, Etc.			
Burlap, in bales	43	6	258
Carpets and rugs	30	6	180
Coir yarn, in bales	33	8	264
Cotton			
in bales, American	30	8	240
in bales, foreign	40	8	320
Cotton bleached goods, in cases	28	8	224

is large. The amount of reduction depends on the area supported and reflects the decreasing probability of a fully loaded situation occurring as the supported area increases.

Moving loads, such as traffic on a bridge, overhead cranes, and elevators cause dynamic effects both during motion and when motion begins or stops. These loads are termed *impact loads*. In design, *impact factors* are sometimes used to adjust the response calculated for static application of these loads. When the time-related response is desired a complete dynamic analysis must be performed.

Under normal conditions, the effects of environmental forces on a structure are the most significant analytical consideration. Likewise, for many situations, determination of the structural response requires complex analytical methods. For ordinary design, simplified approaches are permitted in most specification codes for consideration of wind, snow, and earthquake conditions. They are intended to reduce the analytical effort required of the engineer and are based upon many years of research, testing, field observations, and experience. Other environmental effects are generally accounted for through proper construction details, although unusual conditions may necessitate careful analysis. The remainder of this chapter discusses the basic concepts used in accounting for these effects of nature.

1.6.2 Wind Loads

Wind is air in motion. When the motion is obstructed by a building, some of the wind's kinetic energy is converted into pressure (or suction) on the building. The resulting pressure distribution depends upon the size and shape of the structure. A typical situation is depicted in Fig. 1.8. In reality, the pressure distribution and intensity fluctuate with time, and a proper determination of the building response is a complex aerodynamics problem. During approach to the structure, wind velocity is greatly affected by surrounding terrain, obstacles, and other sources of turbulence. Even if the ground is flat, wind velocity varies with elevation and time and can approach a structure from any di-

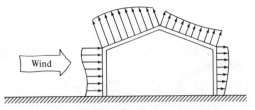

Fig. 1.8.   Wind pressures on a building.

rection (despite "prevailing" directions). Because structures primarily respond to short-time variations in wind speeds, or *gusts*, the fluid mechanics regarding the approaching air become extremely complex. Two approaches exist for a rigorous analysis of wind effects on a given structure:

1. Mathematical modeling using probabilistic methods and statistical data
2. Physical modeling and testing in a wind tunnel

The first alternative requires considerable computer time and suffers from its generality and the lack of abundant field data. On the other hand, wind-tunnel studies can be performed on any specific structure and terrain for a variety of conditions. However, physical modeling and testing are expensive.

Extensive wind studies are economically justified only for major structures. For ordinary design, wind-velocity recurrence maps, based upon statistical data accumulated by the U.S. Weather Bureau, are usually employed (see Fig. 1.9). One fundamental approach is to convert the dynamic pressure

$$q = \frac{1}{2}\rho v^2 \qquad (1.1)$$

where

q = dynamic pressure
ρ = mass density of air
v = wind velocity

into a pressure due to drag

$$p = 0.00256 C_s V^2 \qquad (1.2)$$

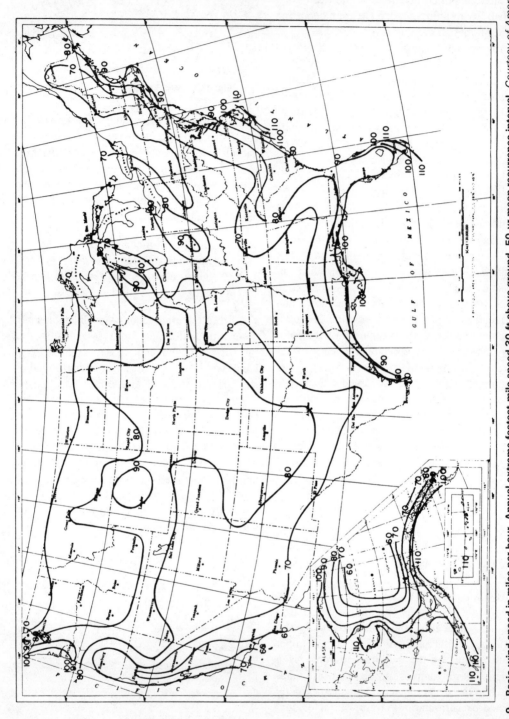

Fig. 1.9 Basic wind speed in miles per hour. Annual extreme fastest-mile speed 30 ft above ground, 50-yr mean recurrence interval. *Courtesy of American National Standards Institute.*

where

p = drag pressure

C_s = coefficient of drag or "shape factor" for a particular location on the structure

V = wind velocity in mph at a particular elevation

A design pressure is obtained by adjusting the drag pressure by multiplying by specified coefficients to account for gusts, terrain features, and other influences.

1.6.3   Snow Loads

The snow-load intensity to be used in design is generally specified in local building codes. For a particular location, the specified value is usually established on the basis of weather-bureau data, presented in a snow-load map such as that in Fig. 1.10. Values stated in the various codes are for flat surfaces, and the designer is usually permitted to make some allowance for reduced accumulations on pitched roofs when some snow is likely to fall off the roof.

Additional reductions can be made to account for different exterior exposure (terrain), thermal condition (heating), and the relative importance of the building. Special provisions exist to account for *drift loads* caused when snow builds up against vertical walls. Extra accumulation due either to snow falling off one building to another or to unusual roof constructions (e.g., sawtooth shape) must also be considered.

1.6.4   Earthquake Loading

Figure 1.11 illustrates the general effect of earthquake motion on a building. An earthquake is initiated by slippage along a fault line at the bedrock level below ground level. From this source, shock waves are transmitted through the bedrock and soil, eventually reaching the structure in the form of a base acceleration. A dynamic response results, causing accelerations at all points in the building and motion as depicted in Fig. 1.12. Rigorous investigation of the building's vibrations is the subject of dynamic analysis and probabilistic methods and is beyond the scope of introductory structural analysis. Furthermore, the expense of such analysis is justified only for major or special-purpose structures. The usual approach suggested in building codes involves computation of an equivalent static base shear force V, representative of the worst condition (see Fig. 1.13). Subsequently, the equivalent base shear is converted to a set of lateral loads statically applied at the various floor levels. The structure can then be analyzed by any of the methods of frame analysis described in this textbook.

Most codes also provide approximate methods to distribute the calculated floor loads into the resisting elements. The resisting elements may be either planar frames or "shear walls" (usually either reinforced concrete or reinforced masonry wall construction.) The approximate methods typically include consideration of the twisting that can occur about the vertical axis of the structure when the floor plan is unsymmetric about that axis.

1.6.5   Thermal Loads

When a change in temperature occurs, any solid matter experiences a change in dimensions. The magnitude of the change is directly related to the coefficient of thermal expansion of the particular material, the original dimensions of the body, and the degree of temperature change. For longitudinal bars, the change in length of the member, Δl, is given by

$$\Delta l = \alpha \Delta T l \qquad (1.3)$$

where

α = coefficient of thermal expansion

ΔT = a uniform change in temperature

l = original bar length

Free expansion of a material causes no stress in a member and is of little concern unless the magnitude of the deformation must be limited. Consequently, elastomeric supports, expansion joints, and other means are provided to allow a reasonable amount of free motion to occur in structural systems. Difficulties arise when such details are overlooked and unintentional re-

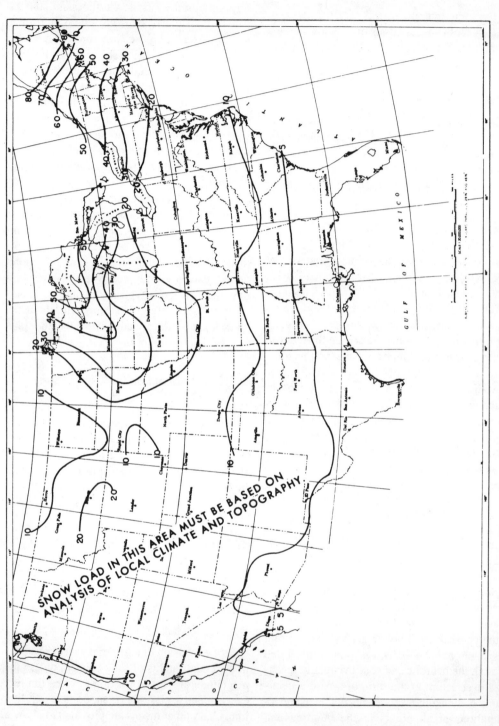

Fig. 1.10.   Snow load in pound force per square foot on the ground. 50-yr mean recurrence interval. *Courtesy of American National Standards Institute.*

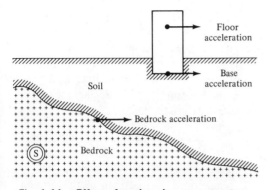

Fig. 1.11.   Effect of earthquakes on a structure.

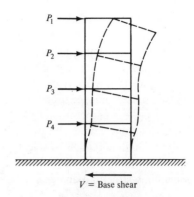

V = Base shear

Fig. 1.13.   Equivalent statical loading.

straint is present in a structural member. A simple example is given in Fig. 1.14. When a restrained bar is heated, the tendency for free expansion is resisted by the thrust that develops at the ends of the bar. Under conditions of extreme temperature change, the thrust value could be quite high, and the resulting compressive stress might cause failure.

A more serious condition exists when the thrusts developed have a natural eccentricity. Two examples are illustrated in Fig. 1.15. When the end supports are eccentric from the centroidal axis of a member by a distance e, the developed thrust R also produces end moments of magnitude $R \cdot e$. As a consequence, flexural stress and vertical displacement, which are usually unintended, occur in the member (Fig. 1.15a). A similar situation occurs in pitched structures as shown in Fig. 1.15b. The vertical eccentricity of the thrust from points on the structure causes a flexural deformation in the form of an upward "bowing" action.

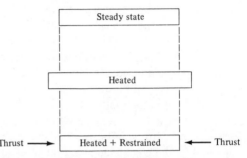

Fig. 1.14.   Effect of temperature change on a restrained bar.

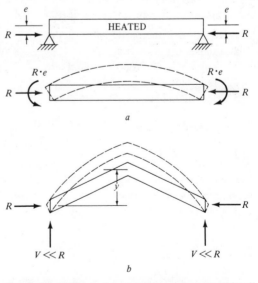

Fig. 1.15.   Effect of temperature when combined with eccentricity. a, heated bar with eccentric support; b, effect of heat on a pitched structure.

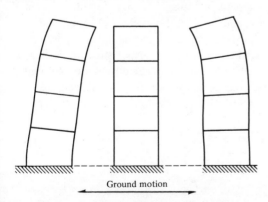

Fig. 1.12.   Motion due to an earthquake.

In the presence of thermal changes, structures can experience significant stress and deformation. These effects and the analytical difficulties are magnified by the existence of thermal gradients in the member, differential motion in members composed of more than one material (composite members), and daily fluctuation in temperature levels. Proper attenuation to support details and material selection can minimize these influences and eliminate the need for complex structural analysis.

1.6.6   Other Loads

In addition to the loads described previously, numerous other sources cause stress in structural members. Principal among them are hydrostatic pressure in tanks and culverts, soil pressure on retaining walls, differential motion due to support settlements, hydraulic effects in pressure vessels, and blast or impact loadings. A major source of unusual loadings is the erection process itself. In some instances, these loads can be more significant than those encountered by the completed structure. A notable example is the St. Louis Arch, which stood for days as two separate cantilevered "halves" prior to completion of the final connection at the top. To identify all significant load sources, the engineer must carefully consider:

1. function of the particular structure
2. intended service life
3. geographic and topographical site
4. climatological conditions
5. method and sequence of erection

1.7   LOAD PATHS

1.7.1   Concept

Various loads that can act on a structure have been described in the preceding section. A fundamental activity of the structural engineer is to create (devise and configure) a structure which can transmit all anticipated loads "to the ground" in an acceptable manner. Devising a means to transmit loads to the ground is commonly referred to as "establishing load paths within the structure." An acceptable load path has certain necessary features:

1. The devised structural arrangement is based on a clear perception of how the structural elements will convert the applied loads to reactions at the ground, i.e., at the supports of the overall structure.
2. The structure performs in a structurally sound (safe) manner.
3. The architectural and functional requirements of the owner are met.

1.7.2   Structural Foundation

Consider the requirement of providing a one-story structural framing system for the simple rectangular plan area shown in Fig. 1.16. Regardless of the framing system chosen, it must be supported by a *structural foundation* constructed at the ground level. The purpose of the foundation system is twofold: (1) to act as a bearing surface for the column and wall elements of the structure, and (2) to spread the reactions from these elements over a ground area large enough to avoid bearing pressure problems in the soil.

Various foundation systems are employed, including "slab-on-grade construction," "spread footings," "wall footings," "mat foundation," "piles," etc. Selection, analysis and design of a proper foundation system is outside the scope of usual structural analysis textbooks. For detailed study of the principles, the reader is referred to standard textbooks on foundation engineering. For the problem at

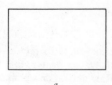

a

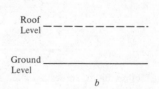

Roof Level

Ground Level

b

Fig. 1.16.   Example situation for devising a framing system. *a*, plan view; *b*, elevation view.

hand it is assumed an adequate foundation exists for any of the possible framing systems.

1.7.3   Primary and Secondary Load Paths

For a complete structural framework the load path typically consists of one *primary* and several *secondary* load paths. Framing elements are provided for each of these and are referred to as the *primary framing system* and the *secondary framing system*, respectively. In some cases, tertiary or even lower order load paths and corresponding framing systems might be used. Furthermore, in devising a load path, both the horizontal (plan) and vertical (elevation) directions must be considered. For simplicity, the discussion relating to Fig. 1.16 will be limited to providing a flat roof to be supported by individual vertical columns located wherever needed in the plan area. The roof framing itself is to be limited to "beam construction" i.e., entirely comprised of straight members lying within the plane of the roof. Thus, the columns provide the entire vertical support for the roof and constitute the primary *vertical* load path. The beam arrangement to be selected constitutes the *horizontal* load path.

Figure 1.17 illustrates alternative primary horizontal framing systems for the situation in Fig. 1.16, each being distinguished by its direction of load transmission. The broad arrows constitute the members and indicate the inherent horizontal load path. Based on the position of the beams and columns in plan, these six load paths are termed (1) *linear*, (2) *tangential*, (3) *radial-symmetric*, (4) *radial-unsymmetric*, (5) *orthogonal crossing*, and (6) *diagonal crossing*. It is usual to say that the members either "span linearly" or "span tangentially," etc. A common characteristic is that each beam is *column-connected* (or *column-supported*), i.e., its ends are directly supported by columns. The structural implication is that each member is capable of acting (of being loaded and responding to load) independently, i.e., acting without "load-sharing." In cases (e) and (f) this occurs if either the lowest of the two crossing members (member *CD*) alone is loaded or there is adequate clearance at the crossing point of the members.

Figure 1.18 illustrates six alternative load paths that are load-sharing systems. The common characteristic is that not all member ends are column-connected. Either some or all of the members are interconnected at one or more joints. In the particular cases illustrated all the members are tied together and act as a grid. In a grid, it is not possible for the members to act independently, i.e., a load applied at any location causes a response by all members. Thus the load-sharing characteristic is apparent. *Note:* If member *AB* is loaded at the same location in both Figs. 1.17e and 1.18e, the systems respond in somewhat (but not exactly) the same way. (What is the difference and how is

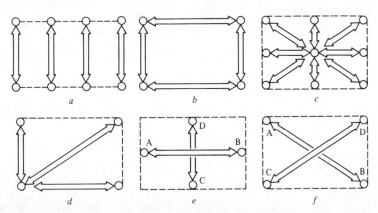

1.17.   Primary load paths without load sharing. *a*, linear; *b*, tangential; *c*, radial-symmetric; *d*, radial-asymmetric; *e*, orthogonal-crossing; *f*, diagonal-crossing. *(Adapted from J. Natterer and W. Winter, by private communication.)*

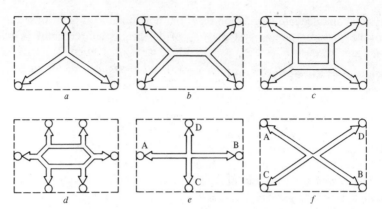

Fig. 1.18.   Primary load paths with load sharing. *(Adapted from J. Natterer and W. Winter, by private communication.)*

it significant?) The same observation and inquiry can be noted about Figs. 1.17f and 1.18f.

Of course, the primary framing systems in Figs. 1.17 and 1.18 would not be used alone. To complete any of the systems, a roof covering (decking) is usually placed atop shorter, smaller (shallower) beams that act to transmit loads directly to ("to span between") the primary framing system. These smaller beams are referred to as the *secondary framing system*. Figures 1.19a–d illustrate four complete framing systems. The narrow arrows constitute the spanning directions of the secondary beams. Secondary beams are used when the distance over which conventional decking can safely span is much less than the typical spacings of

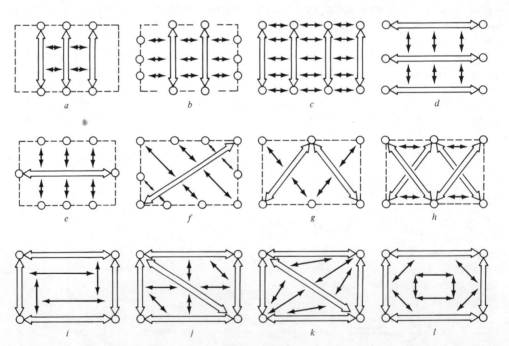

Fig. 1.19.   Example secondary load paths. *(Adapted from J. Natterer and W. Winter, by private communication.)*

members in the primary framing system. To reduce the needed span distance for the decking, a secondary framing system is incorporated. The narrow arrows in Figs. 1.19a–d represent both the smaller beams themselves and their inherent role of providing secondary load paths. In a real sense, the decking is a tertiary framing system since it spans between (transmits load to) the smaller beams. The important point is the way in which loads are made to pass to the ground, namely, from the decking to the secondary framing to the primary framing (comprised of the horizontal main beams and vertical columns.) Several other complete roof framing systems are shown in Figs. 1.19e–l.

Note that the secondary framing system is *not* an alternative to the primary framing system. "Secondary" here means "linked to it as in a chain." A concept which differs from this is that of an "alternative load path," which can be thought of as a "back-up system," i.e., one that comes into play only if a link in the principal load path ceases to function (fails). A simple example is a system of main cables combined with some slack secondary cables. The secondary cables would perform no function unless either excessive displacement or failure occurred in the main cables.

1.7.4 Composite Behavior

It is possible to connect decking to supporting beams in a manner that makes their behavior *interactive* or *composite*. Composite behavior means that two or more joined elements that would otherwise act independently are made to act together in resisting load. In the present case, the continuous decking, secondary beams, and primary beams could be converted to a composite floor system by appropriately connecting the decking to the tops of the beams. Composite behavior can be either *partial* or *complete*, depending on the nature of the interconnection of the elements. A completely composite system is interconnected so as to have no relative movement of the elements at the connected locations. A partially composite systems is subject to some degree of relative

movement (termed *slip*) at the connections. Structures exhibiting composite behavior are complex to analyze and considered a topic too advanced for an introductory textbook.

1.8 CONCEPTS OF WORK AND ENERGY

In the simplest of terms, a structure is no more than a collection of interconnected bodies in space. When subjected to an external force (a load in the context of structural analysis), a body must react (respond) in a manner consistent with basic laws of physics. The underlying condition is that the "closed system," consisting of the body and the applied forces, must obey the law of conservation of work and energy. Work performed by the forces within the closed system must be balanced by an equal amount of energy. The energy consists of the portion transferred to the body plus the portion that escapes the system ("energy losses"). Equation 1.4 is a mathematical statement of the work-energy balance:

$$W = E_S + E_L \qquad (1.4)$$

The work performed, W, is equal to the sum of the energy stored in the body, E_S, plus the energy losses, E_L. Heat lost to the atmosphere (considered to be outside the system) constitutes one type of energy loss.

In the context of a structure, the closed system consists of the structural members, all components of their connections and supports, and the applied loads. The combined work of all loads constitutes the external work of the system. Provided that the materials behave elastically, energy losses are insignificant in structural systems and can be ignored. Should the material be stressed beyond its elastic limit, some energy is dissipated as evidenced by the resulting *permanent set* (failure of the structure to fully return to its original shape after unloading). Systems that, ideally, experience no energy losses are termed *conservative systems*.

With energy losses ignored, all energy generated by the performance of work must be transferred directly to the structural components. The usual terminology for this stored energy is *elastic strain energy*, which has the symbolic designation U. Using this nomenclature, Eq. 1.4 becomes

$$W = U \qquad (1.5)$$

for a conservative structural system. It should be reiterated that U represents the combined strain energy stored in all the structural components. A number of assumptions are necessary if the system is to be considered conservative. A detailed discussion of the aspects of a closed, conservative structural system will be presented subsequently. Prior to the discussion of that concept, the physical and mathematical meaning of *work* will be reviewed.

The work performed by a force is the product of the magnitude of the force and the displacement that occurs along the line of action of the force. If the displacement is an infinitesimal amount, $d\delta$, the increment of work, dW, of the force, P, is expressed as

$$dW = Pd\delta \qquad (1.6)$$

The work performed by P moving through a finite distance, Δ, is obtained by integration of Eq. 1.6:

$$W = \int_0^\Delta Pd\delta \qquad (1.7)$$

Two conditions regarding the integration in Eq. 1.7 are of importance in structural analysis. If a force is "applied instantaneously," without causing any deformation, its magnitude P is assumed to be achieved prior to the occurrence of any motion. In this circumstance, the magnitude of the force would remain constant during the subsequent motion, Δ. Integration of Eq. 1.7 yields

$$W = P\Delta \qquad (1.8)$$

Work of this nature is normally termed *virtual work* because, strictly speaking, loads cannot

be applied instantaneously nor without causing deformation, and Eq. 1.8 describes a fictitious quantity. However, if the force P is allowed to reach a steady state and the displacement Δ is subsequently induced by another force P', Eq. 1.7 correctly defines the work performed by the force P during the displacement. The work is still termed *virtual* because it is created "by virtue" of a specific condition, namely, that the force P was present *prior* to the occurrence of Δ and had no part in creating it.

Realistically, loads must be "applied slowly" if dynamic effects are to be avoided. Under usual conditions, the load P is assumed to increase from zero value to its full magnitude during the time that the displacement Δ takes place. Furthermore, the magnitude of the force at any interim time is assumed to be proportional to the instantaneous displacement δ existing at that time, i.e.,

$$P = K\delta \qquad (1.9)$$

where K is a constant of proportionality called the *stiffness coefficient*. Equation 1.9 is recognized as having form identical to the familiar relationship ($F = KX$) relating the deformation of a spring to the applied force. Substitution of Eq. 1.9 into Eq. 1.6 and integrating produces

$$W = \tfrac{1}{2}K\Delta^2 \qquad (1.10)$$

When the displacement Δ has been attained, Eq. 1.9 gives

$$K = \frac{P}{\Delta} \qquad (1.11)$$

Substitution into Eq. 1.9 yields the desired work expression

$$W = \tfrac{1}{2}P\Delta \qquad (1.12)$$

Work created in the manner just described is termed *real work*. For the same magnitude of force and displacement, the real work is seen to be exactly half as great as the virtual work. Understanding the reason for the "$\frac{1}{2}$" in Eq. 1.12 is vital. In the virtual work expression, the displacement Δ is created by some outside force and is not due to P. Contrary to this, in the real work expression, the displacement Δ is created by the force P increasing from zero value to its full magnitude. The distinction between the two forms of work is a fundamental concept that must be well understood. Many important derivations presented in this text and others rely upon the reader's ability to make such a distinction.

With work adequately defined, the issue of a closed, conservative structural system can be addressed. Examine the structure depicted in Fig. 1.20. The two concentrated loads each contribute a certain amount of work to the system, which must be balanced by an equal amount of energy. Treating the structure as a closed, conservative system implies that all energy is transferred directly to the components of the system. Several conditions are necessary for this to be true:

1. Frictional resistance between the surfaces of the components and the surrounding air must be neglected. Air resistance does contribute a small amount of energy loss but can be safely ignored. (In dynamic analysis, air resistance and other "damping" effects inherent in the structure do play a significant role by eventually bringing a freely vibrating structure to a halt.)

2. Support components must be treated as rigid to remove the ability to transmit strain energy to the ground.

3. Supports must be treated as fixed in place—i.e., the reaction forces do not experience movement. In this way, they produce no work. If the supports do move (e.g., due to settlement), the reactions create an energy loss in the form of negative work. Such losses can be easily calculated and cause no analytical concern.

Viewed within these limits the structure is effectively sealed off from its surroundings in a mathematical sense. The pertinent fact is that all work performed by the loads is converted to elastic strain energy within the structure. This energy transfer materializes in the form of a deformed structure.

Work and elastic strain energy have been described in this section as a prelude to the broader treatment in Chapter 8. In that chapter, a number of energy theorems will be presented that provide the means for a variety of classical structural analysis procedures generally termed *work and energy methods*. They constitute an alternative to the geometric approaches such as the moment area method described in Chapter 4. For certain types of structures (e.g., curved members), energy methods are very powerful analytical tools.

1.9 OBJECTIVES AND METHODS OF STRUCTURAL ANALYSIS

Determination of the "response" of a given structure to the various loadings it will experience during its service life is the fundamental objective of the structural analyst. Prediction or estimation of the load sources and their intensity is, in itself, a major task in the analysis. "Response" connotes the end result of the application of loads to the structure. In the simplest enumeration, the consequences of any particular loading are:

1. A change in position and shape of the structural configuration

2. The internal forces that result in each

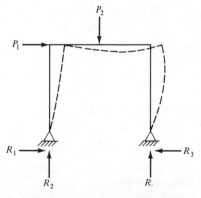

Fig. 1.20   Closed structural system.

member, due to the distortion of its original shape

In section 1.6, loads were classified by type (dead load, live load, etc.) according to the nature of the loading. From an analytical point of view, two major categories of applied loading exist

1. Static or slowly applied loads
2. Dynamic or time-dependent loads

1.9.1 Role of Statics and Dynamics

Static loading can produce both immediate and long-term effects. Immediate effects are those imparted by energy produced internally in the structure by the initial application of load. Long-term effects are usually due to the release of some of this energy in the form of creep or relaxation associated with some materials. Magnitude and duration of load are the primary variables affecting the creep observed in a given member. Dead load and sustained live load are the usual cause of these long-term deformations and associated stresses. Concepts introduced in the study of the laws of statics create a basis for the analysis of statically loaded structures. Long-term effects of static loads must be evaluated on the basis of experimentally obtained (''empirical'') relationships.

Dynamic loads are loads that cannot be categorized as being slowly applied to the structure. Earthquakes and wind are major dynamic loadings. Although their duration is relatively short, there are many instantaneous fluctuations of intensity while they are occurring. Response to dynamic loads is quantified by time histories obtained by application of the equations of motion. As structures generally have many component members, multiple equations of motion result, and solutions are complex in nature and time-consuming to obtain, even for simple structures. Such analysis is outside the scope of this textbook.

1.9.2 Role of Mechanics of Materials

Many methods of structural analysis have their origin in the area of study called the *theory of elasticity*. A truly ''exact'' analysis of the deformation of any loaded body is unattainable. The basic laws of elasticity provide the most precise formulation of material behavior available in contemporary times. Within the ''elastic limits'' of the material, all structural responses obey this theory. If the material is either ''inelastic'' or strained beyond its elastic limit, the theory of elasticity does not apply. Inelastic material considerations require knowledge of advanced concepts such as the ''theory of plasticity.'' However, development of either elasticity-solutions requires rigorous application of higher mathematics and is generally impractical for ordinary usage.

With appropriate simplifying assumptions (acceptable for normal conditions of material condition and structural geometry), one can reduce the degree of mathematical complexity required for analytical solutions. Mechanics of materials essentially accomplishes this task. Prior to the advent of computers as a major means of digital computation, laws of mechanics of materials were the source of most practical analysis methods. The analytical approaches developed before the computer age were, of necessity, amenable to manual computation and generally required such simplification. This is no longer necessary. However, within the elastic range of material behavior the use of mechanics of materials is sufficiently accurate for common framed structures.

1.9.3 Impact of Electronic Computation

Pre-computer age structural analysis procedures are commonly referred to as *classical methods of analysis*. Contemporary computer systems are very sophisticated machines with enormous computational capability and efficiency. During the sixties and seventies, structural analysis procedures intended for computer application developed at a rapid pace. Two fundamental approaches are the source of most computer methods of analysis:

1. Matrix methods of analysis for framed structures
2. Finite-element methods of analysis for continua

Matrix methods offer the structural engineer the advantage of automated structural analysis and theoretical refinements not normally attainable by classical approaches. Large, complex frames that defy expedient analysis by classical methods are quickly and easily treated by matrix methods. Finite-element methods represent a very powerful means of applying the theory of elasticity solutions to complex continuum problems. Elasticity solutions are generally limited to problems of regular geometry and simple loadings. The governing mathematical equations generally prove to be intractable for continua with irregular geometries. Finite-element techniques, which are not limited by such complexities, permit approximate but highly accurate elasticity solutions for all types of continua.

The combination of computer methods of analysis and the high speed of contemporary computers enables engineers to effect inelastic solutions much more readily than in the past. Methods to do this belong in the category of "nonlinear" methods of structural analysis. Nonlinear analysis is a cyclic process involving repeated elastic analysis of the structure. Modification of the material properties in each cycle is necessary when the behavior is inelastic.

In the past, some space structures were treated, when rational, as a set of isolated planar structures in order to make manual solutions possible. With the advent of computers, any framed structure can now be readily analyzed in its actual configuration (whether 2-D or 3-D). As the number of members and the complexity of the geometry increases, the number of unknown quantities increases as does the computational effort needed to determine them. This poses little difficulty because current-generation computers have great storage capacity, and efficient computer algorithms for execution of the structural analysis.

Both classical and computer methods of structural analysis are the subjects of this textbook. Beyond the pure historical value, classical approaches still have useful practical application for simpler structures. In addition, they give the structural engineer a basic understanding of the physical behavior of structural elements. Computer methods offer the capability of rapid analysis of large structures and the opportunity for parametric investigations. However, such analysis can be costly for very large systems, and so it is usually preceded by a series of smaller-scale preliminary studies, sometimes performed manually. Furthermore, the correctness of the output of a computer solution requires verification. Physical understanding of structural behavior is mandatory during this phase, and manual spot-checks create confidence in the solution.

1.10 GENERAL CRITERIA FOR STRUCTURAL ANALYSIS

All structural response must satisfy the criteria associated with three basic concepts commonly referred to as: (1) equilibrium, (2) laws of material behavior, and (3) compatibility. Implementation of these concepts *for a chosen schematic model* constitutes the fundamentals of structural analysis.

1.10.1 Criteria for Continua

Structures classified as continua are analyzed by application of the theory of elasticity. Equilibrium and compatibility are represented by differential equations, which are derived from a consideration of statics and geometry, respectively. Material behavior is represented by constitutive laws that describe the relationships between stress and strain for the component material. By invoking equilibrium, material laws, and compatibility for a particular form of continuum, the so-called governing differential equation describing its behavior is obtained. Because continua are three-dimensional, this equation is generally a partial differential equation. Mathematical solution of the governing differential equation and satisfaction of the existing boundary conditions constitute the structural analysis.

1.10.2 Criteria for Framed Structures

In the context of framed structures, the theory of elasticity is replaced by its simplified integrated form—mechanics of materials. In this form, the three criteria of structural analysis are incorporated as follows:

- *Equilibrium.* A free-body diagram of a structure or any of its components must satisfy the laws of statics or dynamics. Applied loads, external reactions, and internal forces exposed in the free-body diagram must be in equilibrium. If the loads are suddenly applied or have magnitudes that are a function of time (e.g., wind), the equations of motion must be obeyed.
- *Laws of material behavior.* Each structural member must respond to forces in a manner consistent with the observed behavior of its component material. Theory of elasticity provides the most precise formulation of material behavior. For most common materials used in framed structures, some strains may have negligible effect on structural response. In this circumstance, mechanics of materials is a sufficient base for modeling the material behavior.
- *Compatibility.* External displacements must be consistent with the internal member deformations. Pieces of the structure, which were continuous originally, must remain continuous after load is applied. Members connected to a joint must remain connected to the joint, and, if rigidly connected, their ends must experience the same displacements as the joint itself. In simple terms, compatibility means that geometry must be obeyed.

The result of applying these criteria to an individual member produces a governing equation. The form of the equation depends on the type of defamations caused by the loads. Also, because mechanics of materials are used (in place of theory of elasticity) the governing equation for static loads is an ordinary differential equation.

1.10.3   Sources of Error

Predicted response and true response for a given structure are never in exact agreement. Differ-

ences observed between the two must be attributed to one of two basic sources: (1) discrepancies inherent in the selected idealized model, or (2) inaccuracies introduced in applying the three criteria of structural analysis. Some of the more apparent sources of error warrant comment.

Equilibrium and compatibility are affected by mathematical and physical influences. In generating the governing equations, a number of simplifying assumptions are necessary either to render the resulting equations mathematically tractable or to reduce them to a lower level of mathematics. However, even when given the most precise mathematical model possible, the numerical data that are used when applying the equations are subject to some error. Load values are seldom known accurately, and structural dimensions on drawings and in the field are certain to differ slightly.

Laws of material behavior also induce error into the solution process. Unlike equilibrium and compatibility, the errors in these laws are not imbedded in any chosen mathematical theory. Material laws are observed phenomena and reflect "errors of measurement." Relationships between stress and strain and the material constants embodied in them are the product of laboratory experimentation and testing. For this reason and because of the natural variability of material properties, such relationships can never be exact models.

Errors produced in performing a structural analysis are dependent upon all the previously mentioned factors, as well as others. The major aim of the structural analyst is to select a mathematical and schematic model that produces, as well as possible, an acceptable degree of accuracy. Usually, extreme accuracy is not necessary in the analysis of ordinary structures. The nature and limitations of such "ordinary structural analysis" are treated in Section 3.6.

PROBLEMS

1-1. Examine the content of recent issues of several engineering magazines (e.g., *Civil Engineer-*

ing, Engineering News Record). Submit xerox copies of photographs and/or schematic drawings representing actual examples of each of the six basic types of framed structures (as defined in Section 1.4). Identify each example by type.

1-2. Repeat Problem 1-1, but identify structures which exemplify each of the following as the principal form used in the design: (a) flat plate, (b) shell, and (c) membrane.

1-3. Research the published literature to determine the existing structures which constitute each of the following:

(a) world's tallest building
(b) world's longest clear-span bridge
(c) world's largest dome based on diameter
(d) world's tallest shell form
(e) world's longest clear-span timber bridge
(f) world's longest clear-span steel or concrete bridge not employing cable suspension.

1-4. Repeat Problem 1-3 except replace "world's" with the state in which you either live or attend college.

1-5. Examine a copy of the building code for the community where your college or university campus is located. Briefly summarize the provisions for examining snow, wind, and earthquake loading.

1-6. Comment on the significant differences in the $P - \Delta$ effect when a tension force acts instead of a compression force.

1-7. A certain spring has a linear spring constant; i.e., the force in the spring is directly proportional to the stretch of the spring. If the load applied to the spring is $2P$ instead of P, will the work performed be doubled?

1-8. The force in a certain spring is known to be directly proportional to the square of its stretch. Show that the work performed by a load P, in creating a displacement Δ, is $\frac{1}{3}P\Delta$. How much work is performed if the load applied is $2P$?

2

Fundamental Mechanics of Members and Structures

2.1 INTRODUCTION

Chapter 1 includes a description of the three general criteria of structural analysis, namely, (1) equilibrium, (2) laws of material behavior, and (3) compatibility. Fundamental concepts of and procedures for implementing the first two criteria are treated in the subjects of *statics* and *mechanics of materials*. Complete coursework in these subjects is prerequisite to undertaking a study of structural analysis at the level treated in this textbook. The purpose of this chapter is to present an elementary review of those subjects and to relate them to the first two criteria of structural analysis. The major part of the remainder of the text treats their application to the various types of structures as well as the presentation of compatibility concepts. In the succeeding chapters, various "methods of structural analysis" are derived and implemented. In each case, the three general criteria of structural analysis are used as the bases of development.

2.2 ELEMENTARY DEFINITIONS OF STRESS AND STRAIN

At the rudimentary level of understanding, *stress* and *strain* are defined in the context of an axial tensile test of a longitudinal specimen

(Fig. 2.1). If F_x is the axial tensile force applied through the centroid of a specimen of measured cross-sectional area A_x, the corresponding *axial stress* σ_x is defined as

$$\sigma_x = \frac{F_x}{A_x} \qquad (2.1)$$

The subscript x indicates both the force and the member cross section face in the direction of the longitudinal axis of the member.

Because the member elongates when acted on by the tensile force, the concept of axial strain is used as a measure of the change in length. The *axial tensile strain* ϵ_x is defined as

$$\epsilon_x = \frac{\Delta L}{L_0} \qquad (2.2)$$

in which ΔL is the experimentally measured extension that occurs in the preestablished reference length L_0, termed the *gage length*.

The axial stress σ_x and axial strain ϵ_x, as defined in Eqs. 2.1 and 2.2 are average values. The stress is averaged over the cross-sectional area and the strain is averaged over the gage length. In this sense, the stress is considered to be "uniformly distributed over the cross section" and the strain is considered to be "con-

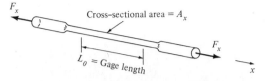

Fig. 2.1.   Typical tensile test specimen.

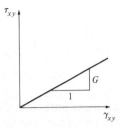

Fig. 2.3.   Linear, elastic stress-strain diagram for shear.

stant over the gage length." For general situations of loading, stress and strain vary throughout the volume of a body and are properly defined at the particle level. Each particle of the body has an associated stress and strain created by the loading. For the particular case of a concentrated axial tensile force applied to a straight member of constant cross section within the gage length, the average stress and strain values are theoretically correct. Except in regions very close to the concentrated load, all particles do experience the same magnitude of stress and strain. For nearly all other loading situations, the use of average stress and strain values gives only approximate results.

The results of an axial tensile test are generally presented as a *stress-strain diagram*. A stress-strain diagram is a plot of the relationship between stress and strain observed as the tensile force on an axially loaded specimen is increased until fracture occurs. For materials that are linear, elastic until failure, the diagram appears as shown in Fig. 2.2. A straight line results because of the linearity of the material. The slope, E of the stress-strain diagram is termed the *modulus of elasticity*. "Elastic" refers to the fact that the loading and unloading occur along reverse paths. Specifically, during loading the stress-strain behavior follows path OA in Fig. 2.2. During unloading the stress-strain behavior follows path AO.

If the stress-strain behavior in the x-y plane of a linear, elastic material subjected to a pure shear force is observed, the stress-strain diagram of Fig. 2.3 is the result. τ_{xy} is the *shearing stress* and γ_{xy} is the *shearing strain*. The slope of the diagram, G, is referred to as the *shearing modulus of elasticity*.

From the preceding paragraphs it is evident that the functional relationships between the stress and strain in Figs. 2.2 and 2.3 can be expressed by

$$\sigma_x = E\epsilon_x \qquad (2.3)$$

$$\tau_{xy} = G\gamma_{xy} \qquad (2.4)$$

When a uniaxial tensile force is applied to a body, its elongation is not the only deformation that occurs. As shown in Fig. 2.4, a member will also shorten in the y-direction, i.e., perpendicular to the direction of loading. The change in transverse dimension is referred to as the *Poisson effect* and is quantified by use of *Poisson's ratio* v_{yx}:

$$v_{yx} = -\frac{\epsilon_y}{\epsilon_x} \qquad (2.5)$$

ϵ_y is the absolute value of the strain caused in the y-direction by introducing ϵ_x in the x-direction. A similar effect occurs in the z-direction (perpendicular to the page) and is quantified by

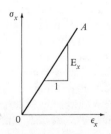

Fig. 2.2.   Linear, elastic stress-strain diagram for axial tension.

Fig. 2.4.   Poisson effect in a tension member.

$$v_{zx} = -\frac{\epsilon_z}{\epsilon_x} \qquad (2.6)$$

The negative signs in Eqs. 2.5 and 2.6 reflect the fact that the sense of the Poisson effect is reverse to that of the applied load. A tensile load along the longitudinal axis of the member reduces the dimensions of the cross section. Conversely, a compressive contraction along the member increases the dimensions of the cross section.

For an isotropic* material the Poisson's ratios v_{yx} and v_{zx} are equal, i.e.,

$$v_{yx} = v_{zx} = v \qquad (2.7)$$

and the three material constants are related according to

$$G = \frac{E}{2(1 + v)} \qquad (2.8)$$

Note that for an isotropic material the strains due to the Poisson effect represented by Eqs. 2.5 and 2.6 can be expressed as follows:

$$\epsilon_y = -v\epsilon_x = -v\frac{\sigma_x}{E} \qquad (2.9)$$

$$\epsilon_z = -v\epsilon_x = -v\frac{\sigma_x}{E} \qquad (2.10)$$

2.3   BASIC CONCEPTS IN THE THEORY OF ELASTICITY

2.3.1   Field Problems

Consider the arbitrary three-dimensional (3-D) continuum shown in Fig. 2.5. When subjected to applied loads (or other actions) the continuum displaces into a new position with a "deformed shape" such as that indicated by the dashed lines. To assume this deformed shape the particles of the body experience displacements $u(x, y, z)$, $v(x, y, z)$, and $w(x, y, z)$ in

___
*An *isotropic* material is such that all particles have the same material constants E, G, and v and these properties are independent of the direction of loading. Otherwise the material is termed *anisotropic*.

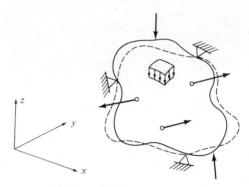

Fig. 2.5.   Deformed three-dimensional continuum.

the x-, y-, and z-coordinate directions, respectively. Collectively, these displacement functions are termed the *displacement field*. The displacement field defines the position of all particles in the body.

Stresses and strains that develop throughout the body are related to the displacement field. Collectively, the stresses and strains are referred to as the *stress field* and *strain field*, respectively. The complete response of a continuum can be determined if either the stress field or the strain field or the displacement field is established. Determination of the response of the body is accomplished by use of the theory of elasticity and is referred to as a *field problem*.

2.3.2   Strain-Stress and Stress-Strain Relationships

The displacements of a given body are the integrated effects of the strains created in the body by the application of load. Displacements are not the sole consequence of inducing strains in a body. Various types of stress are a direct result of induced strains. Simple computation of stress and strain is the subject matter of mechanics of materials. However, all laws of materials are fundamentally derived from relationships formulated in the theory of elasticity. Stress and strain are *tensor* quantities, and exhibit a relationship to one another. In the theory of elasticity, the relationship is described by the *constitutive laws of materials*. In strength of materials, the constitutive laws are reduced to simplified forms for common types of materials and are simply referred to as "strain-stress relationships" or their inverse, the "stress-strain relationships."

Consider an elemental volume of a linear, elastic, 3-D continuum subjected to the general stresses shown in Fig. 2.6. For an isotropic material, it is evident the strains created by the combined stresses are

$$\epsilon_x = \frac{1}{E}\left(\sigma_x - \nu\sigma_y - \nu\sigma_z\right)$$

$$\epsilon_y = \frac{1}{E}\left(-\nu\sigma_x + \sigma_y - \nu\sigma_z\right)$$

$$\epsilon_z = \frac{1}{E}\left(-\nu\sigma_x - \nu\sigma_y + \sigma_z\right)$$

$$\gamma_{xy} = \tau_{xy}/G$$

$$\gamma_{yz} = \tau_{yz}/G$$

$$\gamma_{zx} = \tau_{zx}/G \qquad (2.11)$$

The symbols σ_x, σ_y, and σ_z are the normal stresses in three orthogonal directions (x, y, and z, respectively), and the symbols ϵ_x, ϵ_y, and ϵ_z are the corresponding normal strains. Symbols τ_{xy}, τ_{yz}, and τ_{zx} are the independent shear stresses on the faces of a three-dimensional differential element, and γ_{xy}, γ_{yz}, and γ_{zx} are the corresponding shear strains. Knowledge of the material constants E (the modulus of elasticity), ν (Poisson's ratio), and G (the shearing modulus of elasticity) is required in order to invoke the relationships in Eq. 2.11. For anisotropic materials, the number of material constants needed exceeds three, and the strain-stress relationships become more complex.

Note in Eqs. 2.11 that the normal strains relate directly to the normal stresses and the shearing strains relate directly to the shearing stresses. No interrelationship exists between the two effects, normal and shearing, since the first three of Eqs. 2.11 are not coupled to the last three. Also, Eqs. 2.11 could be rearranged to express the six stresses in terms of the six strains i.e., the *stress-strain relationships*. This is left as an exercise for the reader.

Equations 2.11 describe a 3-D strain-stress relationship. In many situations, certain stress components or strain components either have zero magnitude or can be neglected. Two specific cases, the *plane-stress* and the *plane-strain* states, are prevalent.

2.3.2.1 Plane-Stress State

In a plane-stress state, the stress components σ_z, τ_{yz} and τ_{zx} are of zero magnitude. Imposing these conditions reduce the strain-stress relationships to

$$\epsilon_x = \frac{\sigma_x}{E} - \nu\frac{\sigma_y}{E}$$

$$\epsilon_y = \frac{\sigma_y}{E} - \nu\frac{\sigma_x}{E}$$

$$\gamma_{xy} = G\tau_{xy} \qquad (2.12)$$

and

$$\epsilon_z = -\frac{\nu}{E}\sigma_x - \frac{\nu}{E}\sigma_y \qquad (2.13)$$

The distinguishing feature is that all stress components lie in the same plane, hence the nomenclature "plane stress." Note that ϵ_z does not have zero value and deformations do occur in all three orthogonal directions. Strain ϵ_z occurs freely because there is no normal stress present to resist it. For most plane-stress problems, it is not important to know ϵ_z.

Equations 2.12 can be solved simultaneously

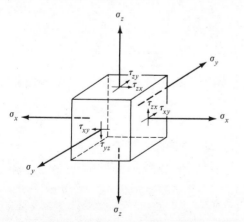

Fig. 2.6. General stresses on an elemental volume of a three-dimensional continuum.

for the stress values, producing

$$\sigma_x = \frac{E}{1 - \nu^2} (\epsilon_x + \nu\epsilon_y)$$

$$\sigma_y = \frac{E}{1 - \nu^2} (\epsilon_y + \nu\epsilon_x)$$

$$\tau_{xy} = G\gamma_{xy} \qquad (2.14)$$

2.3.2.2   Plane-Strain State

In contrast to a plane-stress state, the plane-strain state is defined as a condition in which σ_z, τ_{xz}, and τ_{yz} have zero value. For this condition

$$\epsilon_x = \frac{1}{E} (\sigma_x - \nu\sigma_y - \nu\sigma_z)$$

$$\epsilon_y = \frac{1}{E} (\sigma_y - \nu\sigma_x - \nu\sigma_z)$$

$$0 = \frac{1}{E} (\sigma_z - \nu\sigma_x - \nu\sigma_y) \qquad (2.15)$$

$$\gamma_{xy} = \frac{1}{G} \tau_{xy}$$

The third relationship gives

$$\sigma_z = \nu\sigma_x + \nu\sigma_y \qquad (2.16)$$

Substitution of Eq. 2.16 into Eqs. 2.15 gives

$$\epsilon_x = \frac{1}{E} [(1 - \nu^2) \sigma_x - \nu(1 + \nu) \sigma_y]$$

$$\epsilon_y = \frac{1}{E} [(1 - \nu^2) \sigma_y - \nu(1 + \nu) \sigma_x]$$

$$\gamma_{xy} = \frac{1}{G} \tau_{xy} \qquad (2.17)$$

Solving Eqs. 2.17 for the stresses yields

$$\sigma_x = \frac{E}{(1 + \nu)(1 - 2\nu)} [(1 - \nu) \epsilon_x + \nu\epsilon_y]$$

$$\sigma_y = \frac{E}{(1 + \nu)(1 - 2\nu)} [(1 - \nu) \epsilon_y + \nu\epsilon_x]$$

$$\gamma_{xy} = \frac{1}{G} \tau_{xy} \qquad (2.18)$$

Eqs. 2.14 and 2.18 are the stress-strain relationships for plane-stress and plane-strain states, respectively.

2.3.3   Strain-Displacement Relationships

Consider a small element of a body in a two-dimensional field (Fig. 2.7). In the undeformed state, the element has dimensions dx and dy. In the deformed state, each corner of the body displaces, and the differential motion each corner experiences is reflected in the creation of strain. Displacements in the orthogonal x- and y-coordinate directions are designated by u and v, respectively. Examination of Fig. 2.7 indicates that the approximate strains are

$$\epsilon_x = \frac{u + \dfrac{\partial u}{\partial x} dx - u}{dx} = \frac{\partial u}{\partial x} \qquad (2.19)$$

$$\epsilon_y = \frac{v + \dfrac{\partial v}{\partial y} dy - v}{dy} = \frac{\partial v}{\partial y} \qquad (2.20)$$

and

$$\gamma_{xy} = \gamma_1 + \gamma_2$$

$$= \frac{v + \dfrac{\partial v}{\partial x} dx - v}{dx} + \frac{u + \dfrac{\partial u}{\partial x} dy - u}{dy}$$

$$\gamma_{xy} = \frac{\partial v}{\partial x} + \frac{\partial u}{\partial y} \qquad (2.21)$$

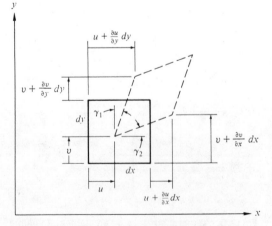

Fig. 2.7.   Displacements in a two-dimensional field.

Equations 2.19 through 2.21 are the strain-displacement relationships for a two-dimensional body. Because these relationships are derived from a consideration of geometry only, they apply to both plane-stress and plane-strain problems.

Expanding the above development to 3-D continua involves only a similar examination of the displacements in the x-z and y-z planes. In this way, the six strain-displacement relationships are established as

$$\epsilon_x = \frac{\partial u}{\partial x} \qquad \epsilon_y = \frac{\partial v}{\partial y} \qquad \epsilon_z = \frac{\partial w}{\partial z}$$

$$\gamma_{xy} = \frac{\partial v}{\partial x} + \frac{\partial u}{\partial y} \qquad \gamma_{yz} = \frac{\partial v}{\partial z} + \frac{\partial w}{\partial y}$$

$$\gamma_{zx} = \frac{\partial w}{\partial x} + \frac{\partial u}{\partial z} \qquad (2.22)$$

The displacements u, v, and w are collectively referred to as a *displacement field* because the magnitudes of these displacements usually vary throughout the body. In implicit terms they would be written in functional notation as $u(x, y, z)$, $v(x, y, z)$, and $w(x, y, z)$. In this text the functional notation is omitted for brevity.

2.3.4   Problem Solution Process

Sections 2.3.1 to 2.3.3 focus on some of the fundamental relationships developed in theory of elasticity. It is evident from the material included in those sections that the response of a continuum is completely defined if the displacement field is known. If the functions u, v, and w are known, the strain field can be determined from the strain-displacement relationships (Eqs. 2.22).

The appropriate strain-stress relationships, either Eqs. 2.14 or Eqs. 2.18 can then be applied to establish the stress field.

Exact displacement fields are difficult to establish and only a limited number of field problems in theory of elasticity have been obtained in closed form. In contrast, approximate solutions are readily obtained by using approximate displacement fields and analytical methods such

as the Rayleigh-Ritz method or its equivalent in numerical methods, the "finite element method." The concepts and equations presented in the preceding sections of this chapter are pertinent to the development of the finite element method. Detailed presentation of methods used to obtain solutions to field problems is treated in numerous textbooks on the theory of elasticity and the finite element method.

2.4   NONLINEAR MATERIAL BEHAVIOR

Standard laboratory tests are the means used for establishing the mechanical properties of materials. Material constants needed for structural analysis of frames and continua comprised of isotropic materials are E, G, and v. As described in Section 2.2, each of these constants is obtained from an appropriate stress-strain diagram.

Normally, E and v are obtained by conducting either an axial tension or axial compression test and G is obtained by conducting a pure torsion test. The latter test is employed because it produces pure shearing stresses in the standard specimen. For steel, the value of E (and v) is essentially the same for both tension and compression. For plain concrete, E is obtained from the results of a compression test conducted on a standard cylindrical test specimen of 6-inch diameter and 12-inch height. Cross-sectional properties of a reinforced concrete member are based on the use of a "transformed section." Computation of the properties involves the use of the modular ratio, i.e., the ratio of the E values for the reinforcing steel and the concrete. In structural analysis, E for concrete is considered to be equal for both tension and compression.

Standard tests also exist for other properties of concrete and for all other construction materials (wood, plastics, aluminum etc.) The American Society for Testing and Materials publishes details for the conduct of all standard test procedures used in the U.S.

In linear structural analysis the stress-strain behavior of materials is assumed to be linear, i.e., the material properties are constant over the full stress range. When linearity either exists or is assumed, parameters such as E, G, and ν are referred to as *material constants*. For most materials linear behavior is not the case. The stress-strain diagram for a mild structural steel is shown in Fig. 2.8a. Linearity exists only up to the "proportional limit," above which the stress is not proportional to the strain. When member stresses exceed the proportional limit "nonlinear structural analysis" must be employed.

Nonlinear structural analysis takes into account the variation in material behavior over the full range of stress. In theory, for reinforced concrete nonlinear structural analysis is necessary throughout its entire stress-strain range. This is because the observed stress-strain behavior is as shown in Fig. 2.8b. Practically, the curve is nearly linear up to approximately 50% of the ultimate stress level. During service concrete members are often stressed below this stress level and the use of a constant E value is reasonable. Similar approximations are made for wood members.

With the exception of some topics in Chapter 8, all methods of structural analysis described in this textbook are based on the assumption of linear, elastic material behavior. Work and energy methods are introduced in Chapter 8. Some of these apply to both linear and nonlinear materials.

For steel, a particular type of nonlinear analysis, termed "plastic analysis," is commonly used when behavior beyond the proportional limit is to be taken into account. In plastic analysis the stress-strain behavior is approximated as ideally "elasto-plastic." Elasto-plastic behavior is depicted in Fig. 2.8c and consists of an initial region of linearly, elastic behavior up to the "yield stress" level followed immediately by a plateau of inelastic or "plastic" behavior.

Plastic behavior infers a condition in which large increases in strain occur with no corresponding increase in stress, i.e., the material exhibits "ductility." In other words, during plastic behavior the stress intensity remains constant with a magnitude equal to the yield stress of the material. Complete textbooks have been written on the subject of plastic analysis methods. Since plastic analysis is limited to elasto-plastic behavior the topic is not treated in this textbook.

2.5 INELASTIC MATERIAL BEHAVIOR

A material whose stress-strain behavior during unloading does not follow the same path as during loading is termed *inelastic*. Figure 2.9 depicts inelastic behavior in a qualitative manner. A material that follows path $OABO$ experiences "full recovery," i.e., it returns to its original state. Inelastic structural materials do

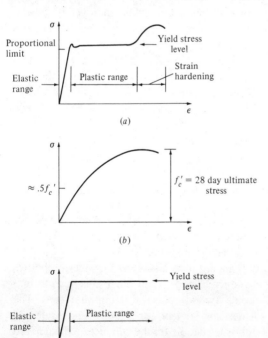

Fig. 2.8. Stress-strain behavior for (a) mild steal, (b) concrete, and (c) elasto-plastic material.

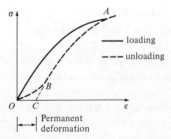

Fig. 2.9. Inelastic material behavior.

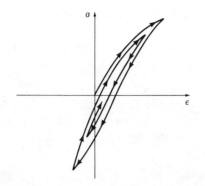

Fig. 2.10. Hysteretic behavior of an inelastic material.

not fully recover and some "permanent deformation" remains after unloading. Path *OABC* in Fig. 2.9 represents such behavior.

Permanent deformation occurs because of yielding of some fibers, fracture of some fibers (e.g., cracking of concrete) or other degradation. Under "cyclic loading" (repetitions of loading and unloading) alternating into the tensile and compressive stress ranges, inelastic, ductile materials exhibit an "hysteretic effect." Hysteresis is illustrated in Fig. 2.10. The diminishing nature of the resistance reflects energy dissipation that occurs during the loading and unloading cycles. Hysteretic behavior is particularly important in the structural analysis of systems subjected to seismic loadings.

2.6 INTERNAL STRESSES AND STRESS RESULTANTS (INTERNAL FORCES)

Framed members subjected to load must develop internal stresses to resist the loads and prevent a material failure. The integrated effects of stresses are force quantities of a particular magnitude and direction. A frequent terminology used to refer to these force quantities is *stress resultants*. "Internal forces" is an alternative terminology. Internal forces that can be created in framed members are categorized into four types:

1. axial force
2. shear force
3. flexural moment
4. torsional moment

The determination of these force quantities is one of the basic aims of structural analysis. The particular force quantities created in a given structure depend upon its behavior. For a planar frame member, the significant internal forces are the shear and axial direct forces and the flexural moment. Figure 2.11 depicts the manner in which these generalized forces are developed by the internal stresses. The resultants of the shear stress distribution τ, and the axial stress distribution σ_1, when integrated over the member cross section, are the transverse shear force V and the longitudinal (or axial) force P, respectively. Flexural stresses are the source of *two* internal forces. Tensile and compressive regions of this stress distribution, σ_2, produce the compressive force C and tensile force T, respectively. Unlike the resultant P, which acts at the centroid, the resultants T and C are separated by a distance, and neither of their lines of action coincides with the centroid. Consequently, the net effect of T and C is the creation of an internal resisting moment, M.

Shear and moment resultants that develop in an individual framed member exhibit a relationship to each other and to the transverse loads applied to the member. These fundamental interrelationships will be formulated in Chapter 3. Flexural moment is also a major cause of deflections, and this subject is also discussed, as is deflection due to shear. Internal forces developed in each type of framed structure are presented in Table 2.1.

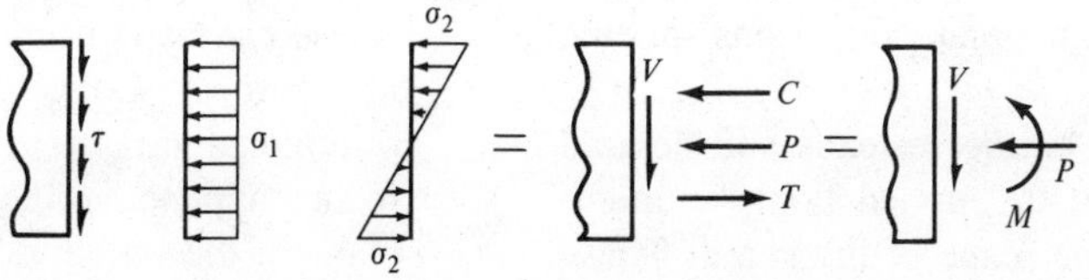

Fig. 2.11. Internal stresses and corresponding stress resultants.

Table 2.1.   Internal Forces (Stress Resultants) for the Basic Structural Types.

Structural Type	Schematic of Assumed Member Forces	Description
Plane truss and space truss		Only axial force, F_1, is significant.
Beam and plane frame		Only in-plane shear force F_1, flexural moment F_2, and axial force F_3, are significant. For beams, axial force generally does not exist.
Grid		Only shear force F_1 and flexural moment F_2 (both transverse to the plane of the grid) and the torsional moment F_3 are significant.
Space frame		All stress resultants are significant: axial force F_1, shear forces F_2 and F_3, torsional moments F_4, and flexural moments F_5 and F_6.

2.7 ACTIONS

It is common to refer to loads as "forces that are applied to a structure." "Applied" is taken to mean "having an identifiable location or point of application." Each of the six types of framed structures were illustrated in Fig. 1.4. Inspection of that figure indicates the applied load types that are included in each of the six idealized models.

1. *Planar truss.* Concentrated loads applied in two orthogonal directions at the joints.
2. *Space truss.* Concentrated loads applied in three orthogonal directions at the joints.
3. *Beams.* a, Transverse loads and flexural moments that are applied along the member and in the plane of the beam. These can be concentrated or distributed in na-

ture. b, Concentrated moments applied at the joints and acting in the plane of the beam.

4. *Grid.* a, Concentrated or distributed loads (transverse to the plane of the grid) and flexural moments applied along the member length. b, Concentrated or distributed torsional moments acting along the member length. c, Concentrated loads (transverse to the plane of the frame) and out-of-plane moments applied at the joints.
5. *Planar frame.* a, Transverse loads and flexural moments that are applied along the member and in the plane of the frame. These can be concentrated or distributed in nature. b, Concentrated loads in two orthogonal directions and in-plane moments applied at the joints and in the plane of the frame.
6. *Space frame.* a, Transverse loads and

flexural moments that are applied along the member length. These may be concentrated or distributed in nature and act in any plane passing through the entire member. *b*, Concentrated or distributed torsional moments acting along the member length. *c*, Concentrated loads and moments applied at the joints and in any of the three orthogonal planes.

Various load categories were described in Section 1.6. The weight of objects (either dead or live load) and the hydrostatic pressure of water are two examples of applied loads. In each case there is contact between the load source and the loaded structure.

In a strict technical sense, loads are not always applied to the structure. Frequently, structures are subjected to phenomena not commonly referred to as loads. Temperature change and shrinkage of material are two examples. Each of these phenomena cause a structure to experience strain and stress and, consequently, to deform. These are the same kinds of effects as caused by applied loads. When such effects are included, it is conventional to refer to the general category of loads as "actions." In this text the terms "load" and "action" are treated as synonomous, both meaning any effect that causes stress and/or strain in a structure.

Any action must satisfy Newton's first law,

i.e., it must cause a reaction. The fundamental concepts of how a structure reacts to an action are developed in the balance of this chapter.

2.8 SUPPORTS AND EXTERNAL REACTIONS

In structural analysis terminology, the consequence of subjecting a structure to an action is termed the "response" of the structure. A structure subjected to an action (whether static or dynamic) will move (translate and/or rotate) indefinitely unless the action is resisted in some way. "Supports" are provided as the means of preventing free movement of the structure. A statically loaded structure will deform into a new shape but remain attached to its supports and at rest in its new position. A dynamically loaded structure will remain attached to its supports but its deformed shape will change as a function of time. "Damping" will eventually bring it to rest but a finite length of time is required for this to occur. A statically loaded structure is assumed to be subjected to simultaneously applied loads and come to rest instantaneously.

Supports prevent free motion of the structure as a whole by developing forces to counteract the load actions. The counteracting forces are commonly termed "external reactions" or simply "reactions." The reader should note the

Table 2.2   Reaction Forces for Various Supports.

Type of Support	Reactions	Explanation
Fixed	R_3 R_2 R_1	Reactions resist rotation and orthogonal translations
Pinned	R_1 R_2	Reactions resist orthogonal translations
Roller	R_1	Reactions resist translations transverse to roller surface
Free end	———	No reactions

distinction between the terms "response" and "reaction." Sometimes these terms are, erroneously, considered to be equivalent. Response refers to the collective consequences of subjecting a structure to an external action. Reactions are a subset of the response.

Schematic representations of the common types of supports and joints for planar structures and the restraint each provides were illustrated in Table 1.1. Reactions corresponding to fixed, pin, and roller supports are shown in Table 2.2. Internal hinges, semi-rigid joints, and supports for space structures are described in subsequent chapters.

2.9 EQUATIONS OF EQUILIBRIUM

When static loads act on a properly supported structure, the structure deforms in some way but comes to rest. A body at rest is said to be in "equilibrium." "External equilibrium" exists when the support reactions acting on the structure "balance" the loads. "Internal equilibrium" refers to the fact that when part of a structure is detached from the whole, the loads and reactions acting on the part removed are balanced by the exposed internal forces (stress resultants). A balanced state exists when the resultant of the load actions, internal forces and reactions acting on the body is neither a force nor a couple. Quantitatively, a body is in equilibrium when the equations of statics are satisfied.

2.9.1 Space Structures

Space structures are three-dimensional framed assemblages. In 3-D Cartesian space, the equations of statics are

$$\Sigma F_x = 0 \quad \Sigma F_y = 0 \quad \Sigma F_z = 0$$
$$\Sigma M_x = 0 \quad \Sigma M_y = 0 \quad \Sigma M_z = 0$$

$$(2.23)$$

The former three equations state that the net force in each of three orthogonal directions (x, y, and z) caused by all actions and reactions is zero. The latter three equations state that the net moment about each of the corresponding orthogonal axes caused by all actions and reactions is zero. The orthogonal directions are arbitrary. Thus Eqs. 2.23 indicate the net force in any direction and the net moment about any axis must be equal to zero.

When all forces have lines of action passing through a common point they are referred to as "concurrent" forces. The common point is called the "point of concurrency." For concurrent forces no net moment is created in any plane passing through the point of concurrency. Because this condition of net zero moment is automatic, the equations of statics for concurrent forces are reduced to

$$\Sigma F_x = 0 \quad \Sigma F_y = 0 \quad \Sigma F_z = 0 \quad (2.24)$$

Concurrent forces are an important feature in the analysis of trusses.

Direct application of scalar algebra to effect the equations of statics is often complicated for space structures. Usually, this is because of the inherent 3-D geometry. Computation of distances, moment arms, etc. can require tedious effort. In conventional applications and in many statics textbooks, 3-D analysis is performed by the use of vector algebra. In contemporary structural analysis, space structures are most readily analyzed using matrix structural analysis and electronic computation.

2.9.2 Planar Structures

Planar structures are 2-D assemblages. In two-dimensional Cartesian space, the equations of statics are reduced to

$$\Sigma F_x = 0 \quad \Sigma F_y = 0 \quad \Sigma M_z = 0 \quad (2.25)$$

The use of these equations requires that the loads and reactions be "coplanar," i.e., that their lines of action all lie in a common plane. The z-axis is orthogonal to the x-y plane. The specific meaning of Eqs. 2.25 is that the net force in the x and y directions and net moment about the z-axis are zero. The effective meaning of Eqs. 2.25 is that the net force in any direction within the plane of the forces and net moment about any axis perpendicular to the plane of the loads must be zero. For concurrent force systems Eqs. 2.25 reduce to

$$\Sigma F_x = 0 \qquad \Sigma F_y = 0 \qquad (2.26)$$

Equations 2.26 are useful in the analysis of planar trusses.

Equations 2.25 are a literal statement of the equations of planar statics, but execution of the inherent "conditions of statics" may differ from these literal equations. Often it is useful to execute the two conditions

$$\Sigma M_1 = 0 \qquad \Sigma M_2 = 0 \qquad (2.27)$$

where the subscripts 1 and 2 refer to axes through two selected points on the structure. Afterward, forces perpendicular to the line joining points 1 and 2 are summed and equated to zero. This sequence is useful when a structure has only two supports.

In Eqs. 2.25, the x- and y-axes can be any orthogonal axes but are often taken as the horizontal and vertical directions. If so, the equations of statics are sometimes expressed as

$$\Sigma H = 0 \qquad \Sigma V = 0 \qquad M_z = 0 \qquad (2.28)$$

where H and V refer to "in the horizontal direction" and "in the vertical direction," respectively.

Finally, when a set of planar forces are parallel, equilibrium is automatically satisfied in the direction perpendicular to the forces and the number of equations of statics is reduced to two.

2.10 FREE-BODY DIAGRAMS

An essential aspect of the structural analysis process is the isolation of the force quantities that act upon the given structure. For a single member, these consist of the applied loads, the reaction forces that provide external support, and the internal forces that maintain the continuity of the member. For a multimember structure, the forces transmitted by the connections between the members are additional actions. In statics, the *free-body diagram* was introduced as the mode for accomplishing this task. A proper free-body diagram is a drawing of a body or assemblage of bodies illustrating all forces acting upon it. Although the body in full size could be employed in a free-body diagram, it is usually sufficient to use the appropriate schematic diagram. For framed structures, a single member, group of members, or the entire structure could be shown in a free-body diagram. For continua, the free-body diagram consists of the complete structure or a portion "sliced out of the whole." The former would generally include only applied loads and support reactions, while the latter would also include the generalized internal forces or, alternatively, the stresses acting upon the cut surface.

As a review of the concept of free-body diagrams, consider the truss given in Fig. 2.12a. The idealized representations shown in Fig. 2.12b and c serve as free-body diagrams of the complete structure and the assemblage of members, respectively. The load P, applied at joint a, and the three support reactions constitute the only forces acting on the system (the gravity effect of the member weights is neglected). Figure 2.12c differs from Fig. 2.12b only in the removal of the supports from the free-body diagram. This is depicted by indicating that the reaction forces are transmitted directly to joints b and c.

Free-body diagrams of the truss joints, isolated by broken lines in Fig. 2.12c, are drawn in Fig. 2.12d. Each of the three resulting diagrams consists of the intact joint, the pieces of attached members, the stress resultants at the cut portion of each member, and the applied loads and/or reactive forces acting at the particular joint. The free-body diagram is simplified by the inclusion of the intact joint. It should be remembered that the actual transfer of forces at the joints is accomplished in the manner detailed in Fig. 2.12e: free-body diagrams are shown for each of the members by itself and for the pin that connects the two members. By application of statics to the free-body diagram of the pin, it is clear that H_1 and H_2 are equal in magnitude, as are V_1 and V_2. However, the direction of the forces is reversed by transmission through the pin.

The ability to visualize the transfer of forces at a joint is vital to the construction of correct

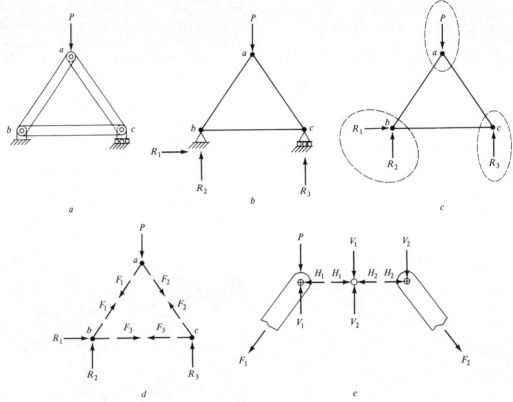

Fig. 2.12.   Truss free-body diagrams.

free-body diagrams of framed structures. Equally important is the recognition that it is necessary to assume a direction for all unknown forces and to draw all associated free-body diagrams in a consistent manner. As an example, the member forces in Fig. 2.12d are consistently assumed to be in tension at all joints. Furthermore, should subsequent computations indicate that a particular force is, in fact, compressive, no adjustment of the free-body diagrams is necessary. A simple notation indicating that the sense is reversed can be made when listing the computed member forces. In subsequent computations involving the use of the force quantity, a negative sign can be given to its numerical value when substitution into equations is required.

Free-body diagrams become more complex for structures other than a planar truss because of the increased number of internal forces associated with each member. As an example, consider the planar frame in Fig. 2.13a. In the analysis of such structures, it is usually necessary to isolate each member and joint (including supported joints) as a separate free-body diagram. The diagrams necessary for the structure in Fig. 2.13a have been constructed in Fig. 2.13b. Directions for the internal forces at the end of each member are unknown and assumed to act as shown. In this case, all end moments, M_1-M_6, have been shown to act in the clockwise direction at the member ends. For consistency, the free-body diagrams of the joints *must show these moments acting in the reverse sense*, i.e., in the counterclockwise direction. For members 1 and 2, the internal direct forces have been indicated as an orthogonal pair of axial and shear forces and are labeled as A_1-A_4 and S_1-S_4, respectively. As in the case of the moments, the direction of these forces are reversed when shown on the free-body diagrams of the joints. No requirement exists for depicting direct internal forces as orthogonal axial and shear components. Statically equivalent com-

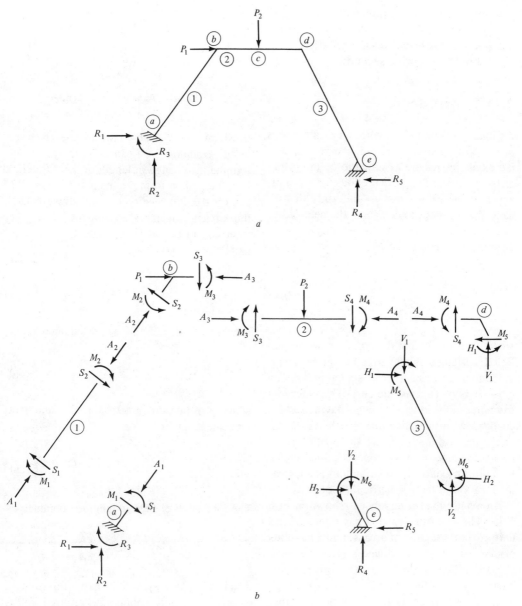

Fig. 2.13. Free-body diagrams for a planar frame. *a*, planar frame structure; *b*, free-body diagrams of members and joints.

ponents in any other orthogonal directions are acceptable and are sometimes computationally advantageous representations. As an example, the direct internal forces for member 3 have been shown as horizontal and vertical components on the member ends and on the associated joints, *c* and *e*. Usually, the internal forces are shown as axial and shear force components.

2.11 STRESSES AND DEFORMATIONS IN SIMPLE MEMBERS

Direct computation of the "internal forces" that exist throughout a structure is one objective of structural analysis. Internal forces are the resultants of the various internal stresses that are developed over a member's cross section. Determination of the corresponding stress distributions for linear, elastic materials is reviewed in the following sections.

2.11.1   Axial Stress

As stated in Section 2.2, when an axial force is directed through the centroid, the corresponding axial stresses are assumed to be uniformly distributed over the cross section. Stated in reverse, if the axial stress is uniformly distributed, the resultant force acts at the centroid of the cross section. The axial stress due to an axial force F_x is computed as

$$\sigma_x = \frac{F_x}{A_x} \tag{2.1}$$

where A_x is the area of the cross section. Equation 2.1 is applicable to straight members having either a constant or a smoothly varying cross section. Tapered and curved members are examples of the latter. Members having sudden changes in cross section due to notches, holes, or change of dimensions are subject to "stress concentrations." Stress concentrations create nonuniform axial stress distributions at and near the location of the geometric discontinuity.

An axially loaded member is shown in Fig. 2.14. The member depicted has a length L, a uniform cross section of area A, and a modulus of elasticity E. A coordinate system has been chosen, which aligns the x-axis with the longitudinal axis of the member. Inspection of the material laws given by Eq. 2.11 indicates that the principal strains produced are

$$\epsilon_x = \frac{\sigma_x}{E} \tag{2.29}$$

$$\epsilon_y = -\nu \frac{\sigma_x}{E} \tag{2.30}$$

$$\epsilon_z = -\nu \frac{\sigma_x}{E} \tag{2.31}$$

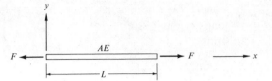

Fig. 2.14.   Axially loaded member.

in which ν is Poisson's ratio for the given material. Equation 2.29 can be employed to determine the change in length of the differential length dx taken from the member. Figure 2.15 depicts the deformation of such an element. If the left end displaces an amount, Δ, and the right end displaces an amount, $\Delta + d\Delta$, the axial strain is given by

$$\epsilon_x = \frac{d\Delta}{dx} \tag{2.32}$$

Equations 2.29 and 2.32 can be equated to give

$$\frac{d\Delta}{d_x} = \frac{\sigma_x}{E} \tag{2.33}$$

which, by introduction of Eq. 2.1 and rearrangement, gives

$$d\Delta = \frac{F}{EA} d_x \tag{2.34}$$

The total axial change in length is obtained as

$$\Delta = \int_0^L d\Delta = \frac{FL}{EA} \tag{2.35}$$

which is a familiar relationship used in strength of materials and commonly referred to as *Hooke's law*.

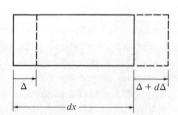

Fig. 2.15.   Deformation of a differential element subjected to axial force.

If Δ and F are considered functions of position along the axis of the member, Eq. 2.34 can be rewritten as

$$\frac{d\Delta(x)}{dx} = \frac{F(x)}{AE} \qquad (2.36)$$

Integration of Eq. 2.36 yields

$$\Delta(x) = \int \frac{F(x)}{EA}\, dx + C_1$$

Introduction of the boundary condition

$$@x = 0 \qquad \Delta = 0$$

gives $C_1 = 0$ and

$$\Delta(x) = \int \frac{F(x)}{EA}\, dx \qquad (2.37)$$

Equation 2.37 has little practical value, but serves as a comparison with the development of the equation of the elastic curve presented in Chapter 3. The equation also demonstrates that axial displacement varies linearly with distance along the member when the axial force is constant.

The strains ϵ_y and ϵ_z, given by Eqs. 2.30 and 2.31, respectively, can also be used to determine displacements in the remaining principal directions. Integrations similar to Eq. 2.35 result, but each of the upper limits of integration are a cross-sectional dimension. As a consequence, the two transverse displacements are very small and can be safely neglected.

Example 2.1. Determine the reactions at each end of the member shown in Fig. 2.16a. The cross-sectional areas of the different segments are A_1, A_2, and A_3 as shown. E is constant over the entire length.

Solution: The problem can be solved by superposition of the structures shown in Figs. 2.16b and c and requiring the net axial displacement to be zero. For Fig. 2.16b, Eq. 2.37 gives

$$\Delta = \int_0^L \frac{F(x)}{EA}\, dx = \int_0^{L_3} \frac{P}{EA_3}\, dx = \frac{PL_3}{EA_3} \rightarrow$$

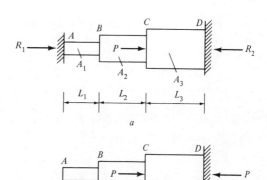

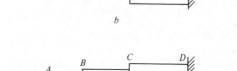

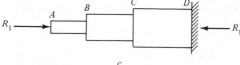

Fig. 2.16. Member for Example 2.1.

For Fig. 2.16c, Eq. 2.37 gives

$$\Delta' = \int_0^L \frac{F(x)}{EA}\, dx = \int_0^{L_1} \frac{R_1}{EA_1}\, dx + \int_{L_1}^{L_1+L_2} \frac{R}{EA_2}\, dx$$

$$+ \int_{L_1+L_2}^{L_1+L_2+L_3} \frac{R}{EA_3}\, dx$$

$$= \frac{R_1 L_1}{EA_1} + \frac{R_1 L_2}{EA_2} + \frac{R_1 L_3}{EA_3} = \frac{R_1}{E}$$

$$\cdot \left(\frac{L_1}{A_1} + \frac{L_2}{A_2} + \frac{L_3}{A_3} \right) \rightarrow$$

For a net displacement of zero

$$\Delta + \Delta' = 0$$

$$\frac{PL_3}{EA_3} + \frac{R_1}{E}\left(\frac{L_1}{A_1} + \frac{L_2}{A_2} + \frac{L_3}{A_3} \right) = 0$$

$$\therefore R_1 = -P \left(\frac{\dfrac{L_3}{A_3}}{\dfrac{L_1}{A_1} + \dfrac{L_2}{A_2} + \dfrac{L_3}{A_3}} \right)$$

For horizontal equilibrium

$$R_2 - R_1 = P$$

$$\therefore R_2 = P + R_1 = P \left(\frac{\dfrac{L_1}{A_1} + \dfrac{L_2}{A_2}}{\dfrac{L_1}{A_1} + \dfrac{L_2}{A_2} + \dfrac{L_3}{A_3}} \right)$$

Note: If all three segments have the same area ($A_1 = A_2 = A_3 = A$) and L is the total member length ($L_1 + L_2 + L_3 = L$), then

$$R_1 = P \left(\frac{\frac{L_3}{A}}{\frac{L_1}{A} + \frac{L_2}{A} + \frac{L_3}{A}} \right)$$

$$= P \left(\frac{L_3}{L_1 + L_2 + L_3} \right) = P \frac{L_3}{L}.$$

$$R_2 = P \left(\frac{L_1 + L_2}{L} \right)$$

Thus, for a constant cross-section the reactions are inversely proportional to their distances from the location of the axial load.

2.11.2   Flexural Stress

2.11.2.1   Uniaxial Bending

The cross section shown in Fig. 2.17 is subjected to a uniaxial bending moment, M_z. If "uniaxial bending" deformation results, normal stresses on the cross-section are computed from the simple "flexure formula"

$$\sigma_x = \frac{M_z y}{I_z} \qquad (2.38)$$

where I_z is the moment of inertia about the centroidal z-axis and y is a coordinate on the cross section measured from the neutral axis. Eq. 2.38 indicates the stress is normal to the cross section and varies linearly with distance y. Uniaxial bending occurs when moment is present about only one axis and the member displaces in a direction perpendicular to that axis and does not twist. It is further required that the

cross section be symmetric about the centroidal y-axis. It need not be symmetric about the centroidal x-axis. Also the load on the member that creates M_z must act in the y-direction and be directed through the centroid of the cross section.

2.11.2.2   Biaxial Bending

The member cross-section shown in Fig. 2.18 is subjected to "biaxial moments," i.e., the vectors M_y and M_z are orthogonal. If the moments are aligned with the principal axes of a cross section which is symmetric about both the y- and z-axes, "biaxial bending" results. Biaxial bending further refers to the fact that when bending occurs about both the x- and y-axes, the net displacement is the vector sum of the deflections in the x- and y-directions. By corollary, if either moment acts alone, uniaxial bending results. In biaxial bending, the cross-section does not twist about the x-axis. Under the above conditions, the normal stress σ_x is given by

$$\sigma_x = \frac{M_z y}{I_z} + \frac{M_y z}{I_y} \qquad (2.39)$$

where I_y is the moment of inertia about the centroidal y-axis and z is a coordinate on the cross section and is measured from the neutral axis. For Eq. 2.39 to be valid, the line of action of the loads must pass through a point in the cross-section referred to as the "shear center." Determination of the location of the shear center is a subject in advanced mechanics of materials and the reader is referred to the numerous available textbooks. For many members the shear center is coincident with the centroid of the cross section.

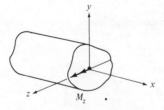

Fig. 2.17.   Uniaxial moment.

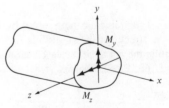

Fig. 2.18.   Biaxial moments.

2.11.2.3 Unsymmetrical Bending

If either the member cross-section in Fig. 2.17 is not symmetric about the y-axis or the member cross section in Fig. 2.18 is unsymmetric about either the y-axis or the z-axis, unsymmetrical bending results. Qualitatively, the member deflects in a direction that is not aligned with either the y-axis or the z-axis. The bending stress σ_x which results at any point (y, z) on the cross section is given by

$$\sigma_x = \left(\frac{M_zI_y - M_yI_{yz}}{I_yI_z - I_{yz}^2}\right)y + \left(\frac{M_yI_z - M_zI_y}{I_yI_z - I_{yz}^2}\right)$$

(2.40)

where I_{yz} is the "product of inertia" for the cross section.

Derivation of Eq. 2.40 is a topic in advanced mechanics of materials and the reader is not expected to be familiar with its usage. However, a cross section that is symmetric about either of its principal axes has a zero value for I_{yz}. Hence, when $I_{yz} = 0$, Eq. 2.40 reduces to Eq. 2.39 as expected.

2.11.2.4 Flexural Deformation

Deformation due to flexure is reflected in the member into a curved shape, sometimes accompanied by twisting. This was described briefly in the preceding three subsections. Computation of the displaced shape of a member subjected to uniaxial bending is treated extensively in Chapter 3. The deformations due to biaxial bending are flexure in each of the principal axis directions. The net displacement of any point on the neutral axis is the resultant of the two components of flexural motion. Deformations due to unsymmetrical bending is beyond the scope of this textbook.

2.11.3 Shearing Stresses

When the uniaxial bending stresses vary in magnitude along the length of a beam, as in Fig. 2.19a, "horizontal shearing stresses" are developed. Figure 2.19b shows horizontal

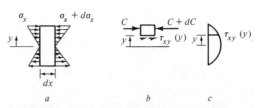

Fig. 2.19. a, differential bending stresses; b, c, horizontal shear stress distribution.

shearing stresses must exist to maintain internal equilibrium of the member. Due to the varying bending moment, the flexural stresses differ in magnitude on opposite faces of a differential element taken from the length of the beam. The differential difference dC in the normal stress resultant C, which exists when the beam element is cut at any level y, necessitates formation of the shearing stresses τ_{xy} to satisfy statics. The magnitude of the shearing stresses varies through the depth of the element according to

$$\tau_{xy}(y) = \frac{1}{I_zt}\frac{dM_z(x)}{dz}\int_y^{c_t} y\, dA \quad (2.41)$$

where t is the member width at level y and c_t is the distance from the neutral axis to the top of the member. Equation 2.41 is usually written as

$$\tau_{xy} = \frac{V_yQ_z}{I_zt} \quad (2.42)$$

where Q_z is defined as the statical moment about the neutral axis of the portion of the cross section that lies above (or below) the level y at which the shearing stress is to be calculated. For a rectangular cross section the variation in τ_{xy} is parabolic with a maximum value at the neutral axis. This distribution of shearing stresses is depicted in Fig. 2.19c.

Equation 2.42 is based on the assumption that the transverse loads on the member act through the "shear center." If symmetry about the y-axis exists for the cross section, the centroid is the shear center.

For biaxial bending there is an additional shearing stress

$$\tau_{xz} = \frac{V_z Q_y}{I_y t'} \qquad (2.43)$$

acting in the z-direction, i.e., orthogonal to τ_{xy}. For Eqs. 2.42 and 2.43 to be valid in biaxial bending, both components of the transverse load must act through the shear center. For doubly symmetric cross sections the shear center coincides with the centroid.

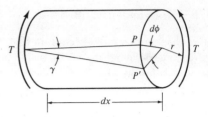

Fig. 2.21. Torsional deformation of a differential element.

2.11.4 Torsional Shearing Stresses and Deformations

Torsion is a phenomenon which occurs when either torque is applied about the longitudinal x-axis of a member or applied loads do no pass through the shear center. Members subjected to torsion will twist and develop internal torsional shearing stresses.

Torsional behavior of simple members is treated in most elementary textbooks on mechanics of materials. At that level, the members are generally taken to be circular. For the circular member, (see Fig. 2.20) it is valid to assume the twisting action causes the cross section to rotate about the longitudinal axis while remaining in its original plane. This rotational effect is termed *pure torsion*. A member whose cross section is not circular can experience out-of-plane distortion in addition to the rotational motion. The additional distortion is generally termed *warping*. In ordinary structural analysis, the longitudinal deformations due to warping torsion are considered to be negligible and not included in any formulations. Structural analysis, which includes warping deformations, is a complex undertaking and extends beyond the elementary level addressed herein. Only the twisting deformation due to pure torsion is of interest on this textbook. Governing relationships for pure torsion only are developed in the following paragraphs.

Circular members (either solid or hollow) subjected to pure torsion are assumed to deform as depicted in Fig. 2.21. Twisting caused by the pair of torsional moments of magnitude T produces shearing strains throughout the body. In the differential element drawn in Fig. 2.21, the strain on the outer longitudinal surface is reflected by the strain angle, γ, defined by the relative rotation of the two end cross sections. The relative movement of point P from its original position to the location P' can be defined in two ways. Recognizing that the movement is related either to γ or to $d\phi$, the relative rotation of the cross section, one can write

$$\gamma \, dx = r \, d\phi$$

or

$$\frac{d\phi}{dx} = \frac{\gamma}{r} \qquad (2.44)$$

Generalizing Eq. 2.4 converts Eq. 2.44 to

$$\frac{d\phi}{dx} = \frac{\tau_r}{Gr} \qquad (2.45)$$

where τ_r is the shear stress due to the applied torque. τ_r acts perpendicular to the radial distance r and is given by

$$\tau_r = \frac{T_x r}{J} \qquad (2.46)$$

which is developed in mechanics of materials. J is the polar moment of inertia, which, for a circular section, is

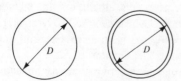

Fig. 2.20. Circular cross sections.

$$J = \int_0^A r^2 \, dA = \frac{\pi r^4}{2} \qquad (2.47)$$

Substitution of Eq. 2.46 into Eq. 2.45 produces the governing equation for pure torsion

$$\frac{d\phi}{dx} = \frac{T_x}{GJ} \qquad (2.48)$$

If T varies over the finite length of the member

$$\frac{d\phi}{dx} = \frac{T_x(x)}{GJ} \qquad (2.49)$$

where $T_x(x)$ is the variation in the torsional moment. For a given torsional moment variation, Eq. 2.49 is easily integrated to obtain the rotation of the member.

For noncircular solid sections, the governing equation for pure torsion is

$$\frac{d\phi}{dx} = \frac{T_x(x)}{GK} \qquad (2.50)$$

K is a torsional constant that describes the torsional resistance of the particular cross section. For structural shapes composed of n rectangular pieces, the common procedure is to use

$$K = \sum_{i=1}^{n} \frac{1}{3} b_i t_i^3 \qquad (2.51)$$

where b_i and t_i are the width and thickness, respectively, of the ith part. Improved K values can be found in advanced texts on strength of materials.

Example 2.2. Determine the relative rotation of the ends of a member subjected to constant torsion of magnitude T. The member has constant GK over its entire length.

Solution: From Eq. 2.50

$$\frac{d\phi}{dx} = \frac{T}{GK}$$

By integration

$$\phi = \int_0^L \frac{T}{GK} \, dx = \frac{TL}{GK}$$

Noncircular cross sections have more complex torsional shearing stress distributions than given by Eq. 2.46. This is particularly true for "open" cross sections such as shown in Fig. 2.22. In addition to more complicated twisting effects, "warping deformation" occurs if the cross-section is not longitudinally restrained. When the cross-section is restrained "warping stresses" are produced. The reader is directed to textbooks on advanced mechanics of materials for detailed presentations about torsion of complex cross sections.

2.12   DEVELOPMENT OF IDEALIZED MODELS

2.12.1   3-D Structural Models

On the exterior, buildings can have a variety of appearances. Architectural desires and a wide range of decorative building products have lead to great creativity in modern building construction. Despite the array of visul forms that exist, however, the skeletal makeup of common commercial buildings is quite simple.

In most low-rise and high-rise buildings, the supporting framework is a three-dimensional framework of interconnected members. For ease of construction, the framework generally has a multi-tiered rectangular configuration. The nature of the physical connections at the joints is dependent upon the basic building ma-

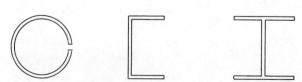

Fig. 2.22.   Open cross sections.

terial selected for the structural members. Reinforced concrete structures are generally monolithic—that is, the members and joints are essentially cast as a unit and behave as rigidly connected. In timber construction, virtually all connections are pinned in nature. Steel-framed buildings are generally classified into three types by the American Institute of Steel Construction. These types and their characteristics are:

1. *Type 1.* Rigid frame construction in which all beam-to-column joints virtually maintain the original angles between the connected members during deformation. This type of framework is generally called a *continuous frame.*

2. *Type 2.* Simple frame construction in which all joint connections offer little or no restraint against rotation. Connected members transmit shear only and are free to rotate relative to the joint itself.

3. *Type 3.* Semirigid frame construction in which the connections possess a known degree of partial restraint against rotation.

Regardless of the nature of the connections, all buildings are, as stated earlier, three-dimensional in configuration. The proper schematic model for structural analysis is the space frame. Figure 2.23 depicts several examples of three-dimensional frameworks. If one considers the possibility of wind or earthquake loading from

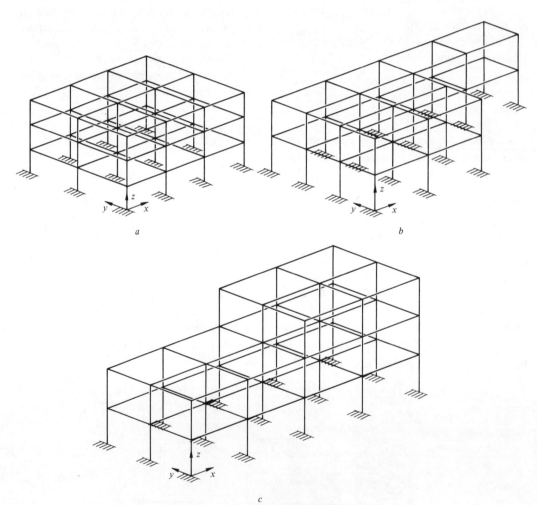

Fig. 2.23. Three-dimensional frameworks.

any direction in addition to the normal gravity loadings, the analytical model becomes quite complex. However, within the limits of the simplifying assumptions of strength of materials, a rigorous analysis of the three-dimensional behavior is possible. Because the number of unknown force and displacement quantities is large, direct manual solution methods are impractical. If structural engineers desire to make an analysis of the complete space frame, they generally employ one of a number of available computer-based analysis procedures. However, obtaining computer solutions for large space frames can be expensive, and so it is sometimes necessary to reduce the size of the problem. The usual approach is to view the three-dimensional structure as an array of component planar frames and to make separate two-dimensional analyses of each of the components. Indeed, for reinforced concrete buildings, the American Concrete Institute has formulated the *equivalent frame method* for accomplishing such an analysis.* Prior to the computer age, such substructuring techniques were a necessity.

2.12.2 Simplification to 2-D Planar Frames

Multistory frameworks commonly are configured in a regular, orthogonal pattern of beams and columns. Examples of possible arrangements are depicted schematically in Fig. 2.23. When any of these frameworks is subjected to load, all component members interact in some way to produce a response. For any given loading, all joints translate and rotate, and each member develops internal forces. Consequently, determination of the complete response of the space frame acting as a unit involves a large number of unknown quantities. Depending upon the regularity of the structural geometry, it is sometimes feasible to reduce the computational effort by a suitable substructuring of the system.

Box-shaped frames, such as shown in Fig.

2.23a, exhibit the lattice arrangement most suitable for substructuring. This particular structure may be visualized as being composed of a series of interlocking planar frames lying in the orthogonal x and y directions. Examination of Fig. 2.24 demonstrates this more clearly. All planar frames lying in a particular direction have an identical geometrical arrangement. Frames in the x-z plane have the configuration illustrated in Fig. 2.25a, and those in the y-z plane have the configuration shown in Fig. 2.25b. Under gravity loading or wind loading parallel to the plane of the frames, each interior planar frame can be expected to have nearly the same response. The same observation can be made about the exterior planar frames. Consequently, it is a common practice to apportion the total load effect among the component frames in a rational manner and analyze each as an isolated planar frame. For wind load acting in the direction of the y-axis,

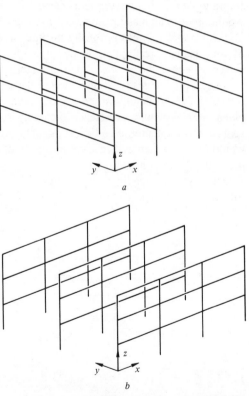

Fig. 2.24.   Two-dimensional frames comprising the space frame of Fig. 2.23a.

---

*See: Building Code Requirements for Reinforced Concrete (ACI 318-83), American Concrete Institute, 1983.

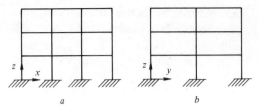

Fig. 2.25.   Substructures for space frame of Fig. 2.23a.

one would employ the substructuring shown in Fig. 2.24a. The substructures shown in Fig. 2.24b would be selected for consideration of wind loading acting in the direction of the x-axis. Both subsystems would be used for investigation of gravity loads.

Care must be taken to avoid a casual decision to divide a space frame into substructures. A primary feature of the space frame of Fig. 2.23a is the double symmetry that exists when the structure is viewed in plan. The space frame depicted in Fig. 2.23b does not have this particular property. Because of the unsymmetrical framing plan, a subdivision into planar frames would produce an inadequate modeling. Figure 2.26 illustrates the reason for the insufficiency of this analysis technique. Under wind load acting in either the x or y direction, lack of symmetry produces a twisting action in the x-y plane. Under this condition, no two component planar frames would experience the same response. Consequently, a vertical substructuring would be unjustified and lead to an inaccurate prediction of response.

Under gravity loading, application of the substructuring technique to a space frame of the

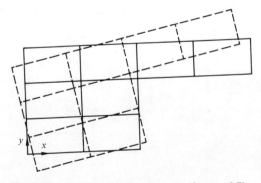

Fig. 2.26.   Twisting motion in space frame of Fig. 2.23b (as seen in a plan view of the structure).

type shown in Fig. 2.23c would normally provide a reasonably accurate result. For this particular structure, the three basic configurations shown in Fig. 2.27 could be employed with some confidence. However, abrupt changes in the regularity of the tiered pattern or the loading in adjacent bays will inherently induce torsional effects into the response. Caution is warranted to insure that substructuring is used only when the torsional response can be safely neglected in the analysis.

2.12.3   Simplification to Grids

One can visually conceive of substructuring a space frame into a stack of horizontally oriented planar frames. This concept is illustrated in Fig. 2.28. It is reasonable to assume that the flexibility of the columns is sufficient to permit the use of simple supports at the joints. While this modeling normally would be superfluous to the analysis described in the previous section, it is nonetheless useful to introduce the concept. A primary feature of this horizontal subdivision is that under gravity loading, the component frames are subjected to loads that are transverse to the plane of the structure itself. In effect, the substructures are grids, as defined in Chapter 1. If simple connections (Type 2 construction) are used, the horizontal floor framing at each story is indeed a grid supported by columns. However, it is common in the design process to further subdivide the grid into individual beams wherever possible.

Subdivision of a space frame into a series of horizontal frameworks is not a common analytical procedure. One exception is the analysis for membrane action due to lateral loading on a building. It is common practice to provide *shear walls* in a building as a means of resisting lateral loads acting on the sides of the building. These are generally incorporated by designing particular walls in the structure to behave as tall, vertically cantilevered beams. In such cases, it is necessary to insure that the lateral loads can be transmitted to the shear walls by the in-place or membrane action of the floor system. However, in this circumstance, the loads of interest do not act in a direction trans-

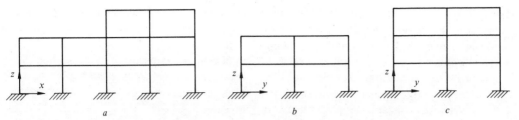

Fig. 2.27.   Substructures for space frame of Fig. 2.23c.

verse to the isolated structure, and the term *grid* is inappropriate. However, true grid structures do occur in practice—for example, *bridge flooring, lifting rigs* and *grillage*. The presence of two-way action (tendency to span in two directions) complicates the structural analysis of grids. The transverse (out-of-plane) loads generally produce torsional moments in addition to the expected flexural effects. In addition, the torsional stiffness of the members can be significant and must be considered in the determination of the joint displacement quantities.

2.12.4   The Individual Member

Structural analysis has been defined as the use of an idealized mathematical model to predict the manner in which a structural system responds to loading. In framed structures, one is confronted with an array of individual members that behave in a complex way to create the deformed geometry or the framework. Internal forces and equilibrium of a structure were discussed in Section 2.6. Stresses associated with internal forces are reviewed in Section 2.11.

Fundamentally, the analyst seeks to determine the displaced configuration of the structure and the internal member forces caused by this distortion of the structure's original shape.

Basically, the internal effect of loading a structure is the creation of stress and strain in the component members. Internal member forces are the integrated effects of stress, and these forces maintain equilibrium of the structure.

The integrated effects of the various strains constitute the deformations that a member exhibits. *Change in member length* is the result of uniaxial strain, *curvature* is the consequence of flexural strain, and *twist* is the outcome of torsional shear strain. A free member in space simply adjusts to these strains by assuming a new shape. However, the individual members do not deform independently; they also affect the other members that frame into them. This multiple interaction leads to the overall change in position of the entire framework. Consequently, to properly understand the behavior of the complete structure, it is necessary to investigate the performance of the individual mem-

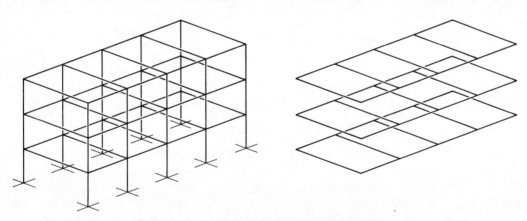

Fig. 2.28.   Substructuring into a series of grids.

ber. Prediction of the response of a single member to imposed loads is the essence of the analysis of an entire framed structural system.

The idealized model of a member consists of a schematic representation of its physical configuration, properly selected material laws, and the mathematical concepts adopted for use in predicting its response. In the process of choosing a mathematical base for application to structural analysis, it is necessary to make and accept certain assumptions regarding material behavior and the geometry of the deformations. Generally, the introduction of assumptions reduces the level of mathematics necessary for a solution. Basic assumptions of a first-order structural analysis are enumerated and discussed in Section 3.6. Under normal conditions and within the constraint of the present knowledge of material behavior, this level of analysis can be expected to result in accurate solutions.

2.12.5   Selecting an Idealized Model

2.12.5.1   Neglecting Minor Stress Resultants

It is the function of the designer to configure a structure to behave in a desired fashion. Each of the basic structural types defined in Chapter 1 differs with respect to the type of stress resultants created by the applied loads. Determination of the appropriate idealized model, plane truss, plane frame, and so on, for prediction of the response is a task for a structural analyst. In each idealized model, some of the stress and strain quantities are inferred to be

negligible. The further inference is that the integrated effects they cause—member forces and deformation—are also negligible. These tacit implications are discussed in this section. Stress resultants that are assumed to be significant for each of the six structural types were described in Table 2.1.

2.12.5.2   Structural Analysis

Space frames represent the most general structural system. When subjected to loads, they displace in all three coordinate directions. Deformations induced in each member are such that all possible types of internal forces are created to offer resistance to the movements of the structure. These forces consist of the axial force A, torsional moment T, flexural moments M and M', and shear forces V and V', directed as shown in Fig. 2.29a. A complete force analysis of a space frame consists of determining these six quantities at critical points for each member in the frame.

Under certain conditions, a space frame can be substructured in one of the two ways described in Section 2.12.2. Vertical substructuring infers the isolation of a portion of the structure that lies in a vertical plane. Figure 2.29b depicts the vertical substructure of the partial space frame of Fig. 2.29a obtained by isolating the members contained in the x-y plane. Two comments regarding the implication of this action are pertinent. First, the substructured system would be analyzed as a planar frame, and each member would develop only three internal

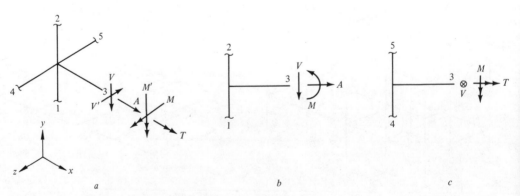

Fig. 2.29.   Comparison of three-dimensional and two-dimensional structures. *a*, space frame member—internal forces; *b*, substructured planar frame; *c*, substructured grid.

resisting forces, namely, the axial force, flexural moment, and shear force. This circumstance infers one condition for the validity of vertical substructuring. Unless the torsional moment and "out-of-plane" moment M' and shear V' are negligible in each member of the space frame, an improper modeling is obtained. Second, the magnitude of the force quantities in Fig. 2.29b would differ from the corresponding quantities shown in Fig. 2.29a because only a portion of the structure remains. The requirement that these differences be small (in a practical sense) is a second condition for vertical substructuring.

Horizontal substructuring of the member system shown in Fig. 2.29a produces the result shown in Fig. 2.29c. When analyzed as a grid, the three forces shown in this figure are the unknown force quantities for each member. This differs from vertical substructuring in three ways: (1) axial forces are absent, (2) torsional moments are present, and (3) the flexural moment and shear force would differ in magnitude from those depicted in either Fig. 2.29a or b. By nature of this circumstance, two conditions are inferred for a permissible horizontal substructuring. First, the axial force and "in-plane" (where "plane" refers to the plane of the resulting grid) flexural moment M' and shear force V' must be negligible in the original structure. Second, the difference in the magnitudes of the forces in Fig. 2.29c and the corresponding forces in Fig. 2.29a must be small.

Figure 2.29 and its implications discussed in the preceding paragraph illustrate significant differences between three- and two-dimensional structural analysis. Substructuring greatly reduces the complexity of the analysis problem but cannot be employed casually. Clearly, the analyst must make an a priori judgment concerning the suitability of substructuring a particular space frame. This requires some experience as well as careful consideration of the three-dimensional geometry of the space frame and its loading. Generally, substructuring is applicable only to the lattice-type systems described earlier in this chapter and only when the loading is symmetrical and has a regular pattern.

Explicit guidelines for choosing between three- and two-dimensional structural analysis are beyond the scope of this textbook. In the remainder of the book, each of these types of structures will be discussed separately. Proficiency in choosing a particular type as the proper schematic model for a given real-world structure requires practical experience not expected of the reader.

2.13 STABILITY OF STRUCTURES

2.13.1 Statical and Geometrical Stability

A structure which is configured and supported so as to be in equilibrium under all possible loadings is termed "stable." To be stable both a minimum number and a proper orientation of reactions are necessary. If, for any single loading, equilibrium is not possible the structure is "unstable." An unstable structure can be either "statically unstable" or "geometrically unstable." If a structure is statically unstable, it can be either "statically unstable externally" or "statically unstable internally" or both.

A structure which is statically unstable externally has an insufficient number of reaction forces, NRF, to satisfy the applicable number of global statics conditions. For planar structures, three statics conditions apply ($\Sigma F = 0$, $\Sigma F = 0$, $\Sigma M = 0$) and at least three reactions are needed. For space structures, six static conditions apply and at least six reactions are needed. If a structure has at least the minimum NRF needed for external statical stability but is still unstable externally, it is termed "geometrically unstable." For planar structures, geometric instability occurs if the reaction forces in the plane of the structure are either parallel or concurrent. A space structure is geometrically unstable if the reactions in any plane are either parallel or concurrent.

A set of parallel reactions (no matter how many in number) can only resist loads that are applied in the direction of parallelism. In Fig. 2.30 the body would be stable for the given loads, but a load in any other direction (e.g., horizontally applied) would cause the body to roll freely. A set of concurrent reaction forces

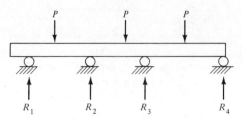

Fig. 2.30. System of parallel reaction forces.

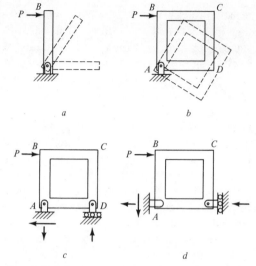

a

b

c

d

Fig. 2.32. External instability considerations in a structure.

can resist any load that passes through the point of concurrency (see Fig. 2.31). However, a load that is not directed through this point creates a moment about the point of concurrency. Since the resultant of the reaction forces passes through the point of concurrency, the moment created by the load cannot be resisted. Consequently, the body would rotate freely about the point of concurrency for the reaction forces.

Figure 2.32a illustrates the physical nature of statical instability. In the figure a single member stands in vertical position, being pinned to the ground only at its base. Clearly, in this condition stability in the presence of any horizontal load is not possible because of the inability to satisfy needed moment equilibrium about the base. Application of a load that is not directed through the base causes the member to rotate about the base and fall to the ground.

In Fig. 2.32b, the single member has been replaced by a rectangular frame with continuous joints. Despite the added members, the framework (taken as a whole) experiences rigid body rotation about the pinned support. Adding a roller support, in the manner shown in Fig. 2.32c, creates a stable configuration because

motion is not possible unless the members deform.

Figure 2.32d emphasizes the importance of the orientation of the supports. The structure is the same as the frame in Fig. 2.32c, which is stable, except the orientation of the roller support is modified. Despite the same number of reactions, the modified structure is geometrically unstable. Summing moments about the pin support demonstrates this fact. Note that under a symmetric vertical loading, the situation in Fig. 2.32a is stable. However, the stability is tenuous. Any slight perturbation (e.g., wind) or imperfection (e.g., nonuniform weight distribution through the width) upsets the balance and would immediately cause collapse. It is often said that a statically unstable structure "cannot carry its own weight." While not a proper technical definition, the statement is literally true for most unstable structures.

In summary, the following is evident for planar structures:

if NRF $\leq$ 3:   statically unstable externally
if NRF = 3 (properly arranged):   statically and geometrically stable externally
if NRF > 3 (properly arranged):   statically and geometrically stable but statically indeterminate externally

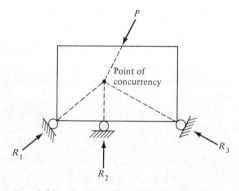

Fig. 2.31. System of concurrent reactions.

and for space structures:

if NRF < 6:   statically unstable externally

if NRF = 6 (properly ar-   statically and geometrically
ranged):   stable externally

if NRF > 6 (properly ar-   statically and geometrically
ranged):   stable but statically indeterminate externally

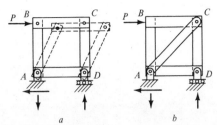

Fig. 2.34. Internal statical instability considerations.

For a planar set of reactions to be "properly arranged" it must be possible that their resultant be a 2-D force vector and a moment vector (or a couple). The reaction force and moment must equilibrate the applied loads in accordance with the three conditions of statics. For a space structure, it must be possible that their resultant be a 3-D force vector and a 3-D moment vector. The reaction force and moment must equilibrate the applied loads in accordance with the six conditions of statics.

Of course, it is possible that the nature of a particular loading is such that stability exists even with less than the minimum NRF needed for general stability. This is possible if the proper support orientation requirement is met. For example, the members shown in Fig. 2.33 are stable for the given loads, but cannot resist horizontal loads. To reiterate, to be classified as stable, a structure must exhibit stability for all possible loadings.

A structure can also be "statically unstable internally." Figure 2.34a illustrates a rectangular truss which is properly supported at two locations. As no rotational resistance exists between pin-connected members, collapse occurs in the manner shown when horizontal load is applied. Each vertical member experiences rigid body rotation about its base. The top member "goes along for the ride." Inserting a diagonal member, Fig. 2.34b, prevents independent motion of the members and a stable structure results. For example, joint C cannot move as in Fig. 2.34a without straining the di-

agonal member. Internal instability is discussed in greater detail in subsequent chapters.

Figure 2.35 demonstrates that a statically unstable structure does not necessarily collapse. The simply supported beam in Fig. 2.35a is obviously statically stable. However, the introduction of an internal hinge, Fig. 2.35b, produces statical instability and the illustrated collapse. In contrast, if the roller support is converted to a pin support, Fig. 2.35c, then collapse is not possible. This is because horizontal movement is prevented at joint B. The point is that the displaced shape cannot exist without stretching (straining) the members. Despite this observation, the structure is clearly unstable, as is evident from a free-body diagram of member BC (Fig. 2.35d). The shear force V transferred at the internal hinge causes a moment about point B. Since the moment cannot be equilibrated, the member starts to rotate. Once a finite rotation occurs the structure assumes a stable configuration (dashed lines in Fig. 2.35c). The developed horizontal force H in Fig. 2.35e has a moment about B that equilibrates that due to V. This is irrelevant to the basic issue. Specifically, it is the statical stability of the structure in its original position that is in question. This is the important matter in first-order structural analysis. Stated simply, an unstable structure cannot maintain its *original* configuration when subjected to load. It either

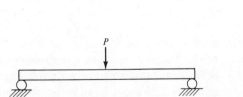

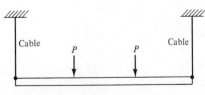

Fig. 2.33. Unstable structures exhibiting stability for a particular loading.

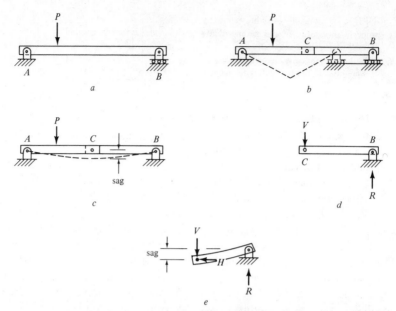

Fig. 2.35.   Physical nature of statical instability.

collapses or displaces into the range where second-order effects must be taken into account.

2.13.2   Buckling Stability

Slender members stressed in concentric axial compression are subject to possible "buckling." Buckling is the sudden change from a straight configuration to one with a large lateral bending deformation. Buckling of this type is depicted in Fig. 2.36. The load at which buckling occurs is the "critical buckling load," P_{cr}.

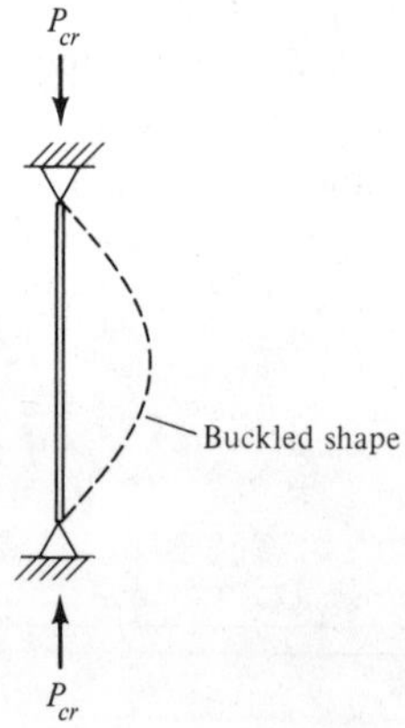

Buckling stability exists when the compressive force does not exceed P_{cr}.

Buckling can occur as a result of either "elastic instability" or "inelastic instability." Elastic instability means the stress at which buckling of the member initiates is within the linear, elastic range of material behavior. Otherwise, the instability is inelastic.

The critical load for elastic buckling, derived by Leonhard Euler in 1757, is given by

$$P_{cr} = \frac{\pi^2 EI}{(kL)^2} \qquad (2.52)$$

The corresponding critical stress is

$$\sigma_{cr} = = \frac{\pi^2 E}{(kL/r)^2} \qquad (2.53)$$

E is the modulus of elasticity, I and r are the moment of inertia and radius of gyration about the minor axis (axis of least resistance), respectively, L is the member length, and kL is the "effective length" of the member. The effective length (specifically, the value of k) depends on the end support conditions. Equations 2.52 and 2.53 apply to all materials exhibiting linear, elastic behavior. For inelastic instabil-

P_{cr}

Buckled shape

P_{cr}

Fig. 2.36.   Buckling of a slender member.

ity, the critical buckling load is given by empirical equations that differ for the various construction materials.

When the compressive load is equal to P_{cr}, a "state of bifurcation" exists. Bifurcation refers to the fact that if a load greater than P_{cr} is applied to a member, two displaced states are possible. In the first state axial shortening of the member takes place but no lateral (sideways) motion accompanies it. The second state will exist if the member is laterally disturbed by the slightest influence. Should a disturbance occur that creates any eccentricity of the axial load, the member will bend into a curved shape. Even at very small increments of axial load (say 15% of P_{cr}) above P_{cr}, the lateral deformations are much in excess of those sustainable by usual construction materials. In brief, at bifurcation a compressive member remains straight but is unstable against any diturbance. In contrast, at loads less than P_{cr}, a disturbance would cause a momentary slightly bent shape, with the member quickly returning to a straight configuration.

2.13.3   Lateral Buckling Stability

Because of the compressive portion of the bending stress distribution, buckling is also a problem in flexural members. This buckling behavior (termed "lateral instability") is depicted in Fig. 2.37 and occurs when a critical moment M_{cr} is attained at the location of maximum moment. The magnitude of M_{cr} depends on the member length, loading, support conditions, and cross section. Derivation of M_{cr} is complex and not treated in this textbook. How-

ever, M_{cr} decreases as either the member length or its depth increase.

2.13.4   Bracing for Stabilization and Wind Loads

To insure the overall stability of a structure, an adequate "bracing system" must be provided. A bracing system is a set of structural members that are provided primarily to prevent either buckling instability or excessive lateral displacements under wind loading. Bracing of individual members and of the structure as a whole must be accomplished. Typically, bracing members are framed laterally into the member that is to be braced.

2.13.4.1   Stabilization

Figure 2.38 illustrates a simple example of the usual considerations involved in bracing a structural system for stabilization. If the vertical columns in Fig. 2.38a are pin-connected at their ends, they would topple as shown under the action of slightest nonconcentric load. To prevent this inherent statical instability a "diagonal bracing system" could be configured, e.g., as in Fig. 2.38b. Unfortunately, this framing alone is inadequate for totally stabilizing the structural system. Being pinned at the bottom of both ends only, the beams have no

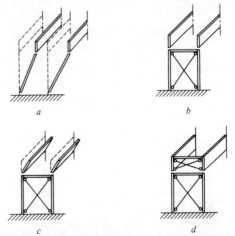

Fig. 2.38.   Bracing against: *a, b,* column statical instability; *c, d,* beam statical instability. *(Adapted from J. Natterer and W. Winter, by private communication.)*

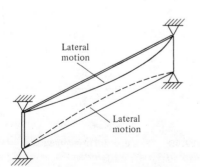

Fig. 2.37.   Lateral buckling of a beam.

means of resisting overturning motion. Despite stabilization of the columns, the beams would topple as shown in Fig. 2.32c under the slightest nonconcentric load. Consequently, a second diagonal bracing system, as shown in Fig. 2.38d, is needed to remove the inherent statical instability of the beams.

With the sources of statical instability removed, buckling instability is the next consideration. If the anticipated compressive loads in the columns exceed their buckling capacities (e.g., as given by Eq. 2.52), instability occurs as shown in Fig. 2.39a. Column buckling can be prevented by either increasing the member's cross-sectional area or decreasing its effective length. By altering the diagonal bracing as illustrated in Fig. 2.39b, the length over which buckling could occur is reduced. In other words, if lateral displacement is prevented at mid-height, the effective length is half the column height. Examination of Eq. 2.52 indicates the consequence of halving kL is the buckling capacity is quadrupled. Note that the mid-height bracing shown in Fig. 2.39b prevents only the motion in the plane of the bracing system. Additional mid-height bracing might also be needed in the orthogonal plane (parallel to the beams) depending on the shape of the column cross-section (why just ''might''?) Fig. 2.39c depicts the lateral buckling instability that

might exist if either the beam length or depth is excessive. To prevent lateral buckling, the beam could require periodic lateral bracing such as provided by the diagonal members illustrated in Fig. 2.39c.

2.13.4.2   Lateral Wind Loading

In the preceding section, bracing against possible buckling effects was described. Lateral wind loading requires additional considerations. A simple example will serve to introduce the basic factors to be considered in providing ''wind bracing.''

Consider the action of a concentrated lateral load acting on a one-story building framework. Figure 2.40 depicts several ways of resisting such a load. In Fig. 2.40a, the single column can be designed to resist the loading but a ''moment-resisting connection detail'' is needed at the foundation level. Such a detail is needed to provide resistance to both the translation and overturning tendencies caused by the horizontal load. The remainder of the structural framing would then serve to resist gravity loadings. A moment-resisting connection detail can be avoided by additional framing, namely by a ''vertical bracing system.'' By providing either a diagonal member or the truss construction shown in Figs. 2.40b and 2.40c, respectively. The vertical reactions (separated by a distance) create a resultant couple needed to resist overturning. Such bracing details are often economical alternatives when the moment to be developed is too large for practical moment-resisting connection details.

For large lateral loads, options such as shown in Fig. 2.41 can be employed to brace a structure vertically. In Fig. 2.41a, the full width of

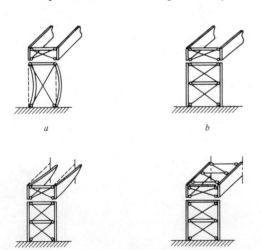

Fig. 2.39.   Bracing against: *a, b,* column buckling, *c, d,* beam lateral buckling. *(Adapted from J. Natterer and W. Winter, by private communication.)*

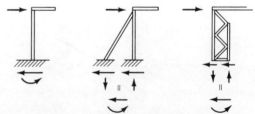

Fig. 2.40.   Concepts for bracing against lateral load. *(Adapted from J. Natterer and W. Winter, by private communication.)*

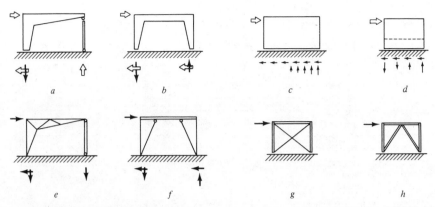

Fig. 2.41.   Concepts for vertical bracing to resist lateral load. *(Adapted from J. Natterer and W. Winter, by private communication.)*

the structure is used to develop the resisting couple. The use of "continuous construction" at the top of the windward column provides additional stiffness in the structure. If both joints are made continuous, as shown in Fig. 2.41b, the frame itself may be stiff enough to prevent large displacements, without need of bracing. The internal couple created by the column reactions serves to satisfy statics (stabilize the structure.) Figures 2.41c and d each illustrate the use of a "shear wall" to resist lateral load. In the former case the in-plane shear resistance of the solid masonry wall serves as the primary mode for stiffening the structure and transmitting the applied load to the ground level. Soil bearing pressure and shear strength serve to stabilize the overall structure. In light-frame wood construction (Fig. 2.41d), shear walls are provided by "sheathing" (usually plywood paneling) nailed to common stud-wall construction.

Each of the vertical bracing systems depicted in Figs. 2.41a–d can be realized by truss construction. Corresponding alternatives to each are shown directly below in Figs. 2.41e–h. The framing techniques depicted in Figs. 2.41g and h are commonly referred to as "X-bracing" and "K-bracing," respectively.

Under lateral load one member of a diagonal bracing system develops tensile force and the other develops compressive force. This is necessary to satisfy horizontal and vertical equilibrium, as pictured in Fig. 2.42 for an X-bracing system. The diagonal members are normally

designed to behave as "counters." A counter has a sufficiently large L/r ratio (called the "slenderness ratio") so that when stressed in even moderate compression it buckles and forces the tension diagonal to act alone. This state renders the analysis under lateral wind load statically determinate. Note, because the lateral wind load can (and usually does) reverse its direction, both members are needed with the counter behavior alternating from one to the other. In a "fully effective" X-bracing system, the compression member has a small enough slenderness ratio to avoid buckling and is included in the analysis. Fully effective X-bracing systems are statically indeterminate. K-bracing acts in a manner similar to X-bracing but can more easily accommodate door and window openings. K-bracing also reduces the span of the beam above it. One disadvantage is that each member of a K-brace develops an additional force due to the gravity load reaction from the beam above it.

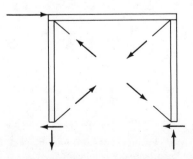

Fig. 2.42.   Resistance to lateral loading by X-bracing.

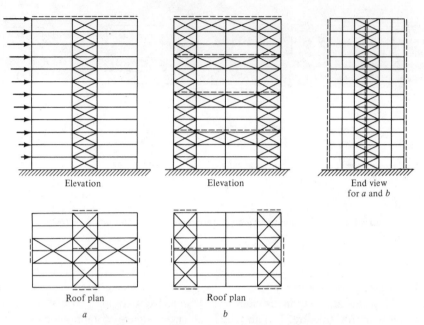

Fig. 2.43.   **Examples of diagonal bracing systems.**

In multistory buildings, diagonal bracing is usually configured as vertically cantilevered trusses. Some examples are shown in the elevation views in Figs. 2.43a and b. (Alternatively, reinforced masonry or concrete walls—vertically cantilevered deep beams—can be used as shear wall construction.) Dashed lines in each view represent end views of bracing systems that do not lie in the plane of the figure. In the roof and selected floor levels, horizontal diagonal bracing may be used to tie together the vertical bracing systems. When horizontal bracing systems are used, they are usually designed to behave as large, simply supported deep beams. Figure 2.44 shows a variety of possible horizontal bracing systems. Dashed lines in each figure represent vertical framing systems depended upon for end support of the horizontal bracing systems. Observe in Figs. 2.43 and 2.44 that for rectangular buildings, bracing for lateral loads might be used in each of the orthogonal primary framing directions. Also, the bracing concepts presented in Section 2.13.4 apply to all structures exposed to wind, particularly bridges and towers.

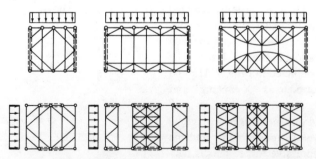

Fig. 2.44.   **Examples of horizontal bracing systems for lateral loading.** *(Adapted from J. Natterer and W. Winter, by private communication.)*

PROBLEMS

In the following problems, use $E = 30,000$ ksi (200 $\times 10^3$ MPa), $G = 12,000$ ksi (80 $\times 10^3$ MPa), and $\mu = 0.30$. These are properties for steel.

2-1. Convert Eqs. 2.12 to Eqs. 2.14.

2-2. The two-dimensional body shown is analyzed by a method based on theory of elasticity. Its displaced shape is established as

$$u(x, y) = .001 + .0005x + .0015y + .0007xy$$

$$v(x, y) = .0005 + .00004x + .0002y + .00006xy$$

where the displacements are in units of inches (cm). Determine the normal and shear stresses at the center of the body. Plot the variation in normal force along each edge. Assume a plane stress situation.

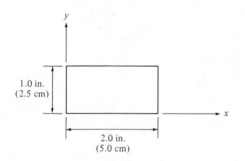

2-3. Repeat Problem 2-2 but assume a plane-strain state.

2-4. A flat plate with a thickness of .75 inches (20 mm) is stretched as shown. Calculate the stresses σ_x, σ_y, and τ_{xy} at the upper right-hand corner. Assume a plane stress state.

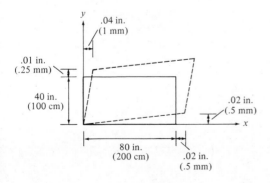

2-5. The square plate shown is subjected to $\sigma_x = \sigma_y = 20$ ksi and $\tau_{xy} = 10$ ksi. Calculate the change in length of both its diagonals.

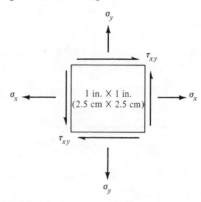

2-6. A solid 2-inch (50-mm) diameter steel cylinder is bonded inside a hollow 4-inch (100-mm) diameter (outer) copper tube having a wall thickness of 0.20 inch (5 mm). A rigid block weighing 5 kips is placed atop the assembly (the cylinder and tube are of equal length). Determine the load carried by each circular member. Would either member buckle elastically? For copper, use $E = 15 \times 10^3$ ksi (100 $\times 10^3$ MPa).

2-7. Given the system described in Problem 2-6, what temperature change would cause the load on the steel member to be totally relieved? Repeat for the copper member. Use $\alpha = 6.5 \times 10^{-6}$ in./in./°F (12 $\times 10^{-6}$ cm/cm/°C) for steel and 9×10^{-6} in./in./°F (16 $\times 10^{-6}$ cm/cm/°C) for copper.

2-8. A tapered member is 20 ft (6.5 m) long. The ends have diameters of 1.0 in. (2.5 cm) and 4.0 in. (10 cm). Calculate the change in length caused by an applied axial compressive load of 8 kips.

2-9. Calculate the horizontal and vertical movement of joint A in the given planar truss. Structural analysis indicates the force in member AB to be 26 kips (5.2 kN) tension and the force in member AC to be 24 kips (4.8 kN). The areas are $AB = 6$ in.^{2} (100 mm^2) and $AC = 9$ in.^{2} (150 mm^2).

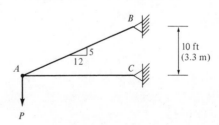

2-10. A torque T is applied 6 ft (2 m) from end C of the assemblage of solid cylinders shown in the accompanying figure. Calculate the reactions at the fixed ends. Use $G = 4,000$ ksi $(28 \times 10^3$ MPa$)$ for aluminum.

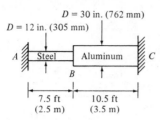

2-11. A torque T is applied at the free end of the given assemblage. Determine the end rotation.

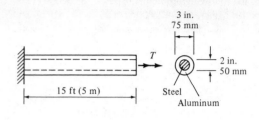

3

Mechanics of Ordinary Beams

3.1 STATICALLY DETERMINATE BEAMS

3.1.1 Definition

A beam has been defined as a member that lies along a single axis through the centroid of its cross section and subjected to loads which act transverse to that axis and lie in a common plane. Figure 3.1 illustrates some common beam configurations and their external support reactions. The beams are classified by type as being simple, cantilevered, fixed-end, etc. according to the manner of support.

Stable structures are classified as either "statically determinate" or "statically indeterminate" depending on whether the equations of statics are either sufficient or insufficient to determine all its external and internal forces. If all its reactions can be calculated by use of statics alone, a structure is "statically determinate externally." If all internal forces throughout the structure can be determined by use of statics alone, it is "statically determinate internally." When either the reactions or the internal forces cannot be determined by statics alone, a structure is termed "statically indeterminate externally" or "statically indeterminate internally," respectively. When a structure is statically determinate both internally and externally, it is termed "statically determinate."

Structures that are statically indeterminate are said to have a particular "degree of statical indeterminacy" which is the sum of the degrees of external and internal indeterminacy. Detailed explanations of external and internal statical indeterminacy of beams and how to determine the degree of indeterminacy of beams are presented in Section 3.1.3. It will be shown that all beams are statically determinate internally. Thus, the degree of statical indeterminacy for a beam equals the degree of external statical indeterminacy, i.e. it equals the number of external reactions beyond those that can be calculated by use of statics alone. Once the reactions of a beam are known it is always possible to determine the internal forces by continued use of statics alone.

3.1.2 Calculation of Reactions

In Fig. 3.1, it is evident that the beams in cases a, b and c are all statically determinate. Each beam has three reactions and the three available equations of statics are sufficient in number to compute them. The following example problem demonstrates how to calculate reactions for each of these beams.

Example 3.1. Calculate the reactions for the simple beam of Fig. 3.2a.

Solution: For the purpose of executing statics the distributed loading is replaced by its resultant acting

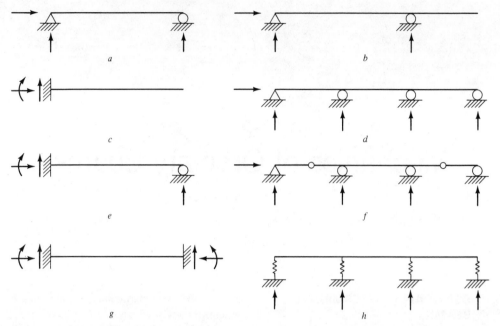

Fig. 3.1.   Example beams: *a*, simple beam; *b*, simple beam with an overhang; *c*, cantilever beam; *d*, continuous beam; *e*, propped cantilever; *f*, continuous beam with internal hinges; *g*, fixed-end beam; *h*, continuous beam on elastic supports.

through the centroid of the actual loading. In this case, the resultant of the uniform (rectangular) loading is 4 kips/ft times 10 ft = 40 kips located at the midpoint of the rectangular loading as shown in Fig. 3.2b. The three reactions are determined as follows:

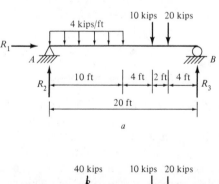

a

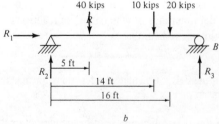

b

Fig. 3.2.   Beam for Example 3.1.

$$\Sigma F_x = R_1 = 0$$

$$\therefore R_1 = 0$$

$$\Sigma M_A = -R_3 (20) + 40 (5) + 10 (14)$$
$$+ 20 (16) = 0$$

$$\therefore R_3 = \frac{5}{20} (40) + \frac{14}{20} (10) + \frac{16}{20} (20)$$

$$\therefore R_3 = 10 + 7 + 16 = 33 \text{ kips} \uparrow$$

$$M_B = R_2(20) - 40 (15) - 10 (6)$$
$$- 20 (4) = 0$$

$$\therefore R_2 = \frac{15}{20} (40) + \frac{6}{20} (10) + \frac{4}{20} (20)$$

$$\therefore R_2 = 30 + 3 + 4 = 37 \text{ kips} \uparrow$$

Note the contribution of a concentrated load to the left and right reactions can be obtained by proportioning in accordance with Fig. 3.3. For example, in computing R_2, the 40 kip concentrated resultant is located one-quarter (5/20ths) of the span length from the support at A. Thus, three-quarters (15/20ths) of the load will be contributed to the reaction R_2. The remaining one-quarter (5/20ths) of

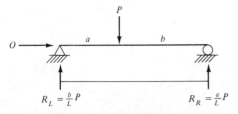

Fig. 3.3.   Simple beam reactions for a concentrated load.

the load is contributed to the reaction R_2. Thus the first and third lines of computation preceded by $\therefore$ can be deduced directly from study of Fig. 3.2b. This is particularly useful if one begins the solution by replacing all distributed loads by their resultants as was done in Fig. 3.2b.

Also note that the use of two moment equilibrium conditions to calculate R_2 and R_3 was done to illustrate the validity of proportioning loads to the simple beam reactions. Once R_3 was known, R_2 could have been calculated using vertical equilibrium:

$$\Sigma F_y = R_2 + R_3 - 40 - 10 - 20 = 0$$

$$R_2 + 33 - 40 - 10 - 20 = 0$$

$$\therefore R_2 = 37 \text{ kips} \uparrow$$

Example 3.2.   Calculate the rections for the simple beam with an overhang, loaded as shown in Fig. 3.4a.

Solution: The linearly varying distributed loads are replaced by their resultants as shown in Fig. 3.4b.

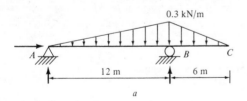

a

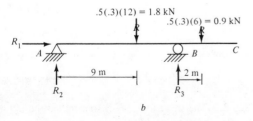

b

Fig. 3.4.   Beam for Example 3.2.

Hence

$$\Sigma F_x = R_1 = 0$$

$$\therefore R_1$$

$$\Sigma M_A = -R_3 (12) + 1.8(9) + 0.9$$

$$\cdot (12 + 2) = 0$$

$$\therefore R_3 = 2.40 \text{ kN} \uparrow$$

$$\Sigma F_y = R_2 + R_3 - 1.8 - 0.9 = 0$$

$$\therefore R_2 + 2.4 - 2.7 = 0$$

$$\therefore R_2 = 0.3 \text{ kN} \uparrow$$

Example 3.3.   Calculate the reactions for the beam shown in Fig. 3.5a.

Solution: Distributed loads are replaced by the resultants as shown in Fig. 3.5b. Next, the horizontal reaction is determined as

$$\Sigma F_x = R_1 = 0 \qquad \therefore R_1 = 0$$

The two remaining statics conditions are insufficient to calculate any of the remaining reactions. Because internal hinges cannot develop moment resistance, the internal moment is known to be of zero magnitude at points B and E. Thus, the five free-body diagrams shown in Fig. 3.5c can be isolated. Note the internal moment is properly omitted at the internal hinge locations. The shear forces V_B and V_F and horizontal forces H_B and H_F at locations B and F are also transmitted from one beam segment to the next via the internal hinges.

Step 1.   Using segment A-B:

$$M_A = 8 (2) + V_B (4)$$

$$\therefore V_B = -4 \text{ kN} = 4 \text{ kN}$$

directed opposite

to direction shown

$$F_y = R_2 - 8 - V_B = 0$$

$$R_2 - 8 - (-4) = 0$$

$$\therefore R_2 = 4 \text{ kN} \uparrow$$

It should be recognized that if V_B is shown acting downward on segment AB, it must be shown to be of equal magnitude but opposite direction on the left

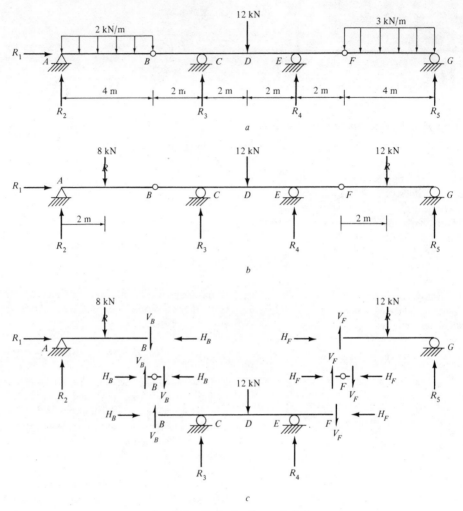

Fig. 3.5.   Beam for Example 3.3.

side of the hinge at point *B*. This is consistent with the general convention shown in Fig. 3.6 wherein the equal and opposite forces on the two adjacent faces tend to move the left body downward relative to the right body.

Vertical equilibrium requires an equal but oppositely directed shear force to be developed on the

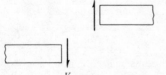

Fig. 3.6.   Convention for positive shear.

right side of the hinge at *B*. Again, consistent with Fig. 3.6 an upward shear force V_B is then transmitted to segment *BF*.

Step 2.   Using segment F-G, statics requires

$$M_F = 12\,(2) - R_5\,(4) = 0$$

$$\therefore R_5 = 6 \text{ kN} \uparrow$$

$$\Sigma F_y = V_F + R_5 - 12 = 0$$

$$V_F + 6 - 12 = 0$$

$$\therefore V_F = 6 \text{ kN directed as shown}$$

Step 3.   The shear force V_F is transmitted through the hinge to segment *BF*.

Step 4.   Using segment BF

$$M_c = V_B\,(2) + 12\,(2) - R_4\,(4) + V_F\,(6) = 0$$

$$-4(2) + 24 - R_4(4) + (6)\,(6) = 0$$

$$\therefore R_4 = 13 \text{ kN} \uparrow$$

$$F_y = V_B + R_3 - 12 + R_4 - V_F = 0$$

$$-4 + R_3 - 12 + 13 - 6 = 0$$

$$\therefore R_3 = 9 \text{ kN} \uparrow$$

3.1.3   Degrees of Statical Indeterminacy

3.1.3.1   General Equation

Example 3.3 demonstrates that, for the purpose of calculating reactions, the presence of internal hinges creates known conditions that can be used to augment the equations of statics. In that example the free-body diagram of the complete beam contained five reactions that could not be calculated by applying the the three statics conditions to that body alone. However, by removing each hinge as a free-body diagram and then isolating the adjacent beam segments, it was possible to calculate all five reactions. Although not proven in general, the example confirms the validity of the following equation for establishing the external degree of statical indeterminacy I_e of any beam.

$$I_e = \text{NRF} - \text{NIH} - 3 \qquad (3.1)$$

in which

  NRF = number of external reactions (including moments)
  NIH = number of internal hinges

As all beams are statically determinate internally, the total degree of statical indeterminacy I is equal to I_e. Thus

$$I = \text{NRF} - \text{NIH} - 3 \qquad (3.2)$$

In words, if the number of external reactions is equal to the total number of internal hinges plus the number of available statics equations, a beam is statically determinate. If

$$\text{NRF} > \text{NIH} + 3$$

Eq. 3.2 gives the degree of statical indeterminacy. If

$$\text{NRF} < \text{NIH} + 3$$

a beam is usually (some unusual exceptions exist) statically unstable. In the case of Example 3.3, Eq. 3.2 gives

$$I = 5 - 2 - 3 = 0$$

as expected.

It is important to recognize why the hinges render the beam in Example 3.3 statically determinate. Note that the set of free-body diagrams for the three beam segments (Fig. 3.5c) include a total of nine unknowns, namely R_1 through R_5 plus V_B, V_F, H_B, and H_F. Applying the three statics conditions to each segment would produce nine independent equations involving these nine unknowns. Simultaneous solution would yield the values of the unknowns. In the actual solution, H_B and H_F were not calculated as they were not needed. Also, R_1 was determined from the free-body diagram of the complete beam. (Note that the three equilibrium conditions that are applied to the complete beam are not independent of the nine equations obtained from its separate parts!) Instead, applying the conditions $F_x = 0$ (omitted throughout the solution) to segments FG, BF, and AB in that order yields the following

$$\text{Segment } FG: \ \Sigma\, F_x = H_F = 0$$

$$\therefore H_F = 0$$

$$\text{Segment } BF: \ \Sigma\, F_x = H_B - H_F = 0$$

$$H_B - 0 = 0$$

$$\therefore H_B = 0$$

$$\text{Segment } AB: \ \Sigma\, F_x = R_1 - H_B = 0$$

$$R_1 - 0 = 0$$

$$\therefore R_1 = 0$$

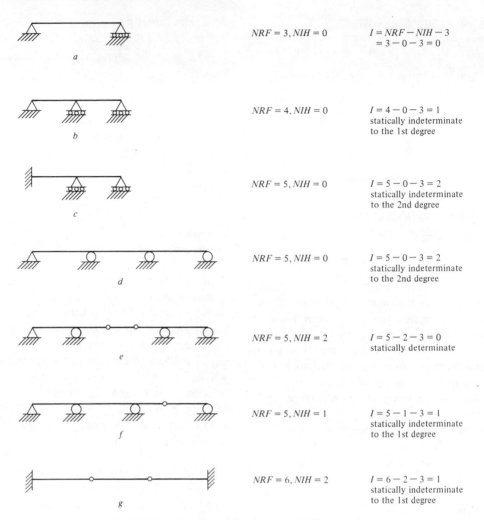

$$NRF = 3, NIH = 0 \qquad \begin{aligned} I &= NRF - NIH - 3 \\ &= 3 - 0 - 3 = 0 \end{aligned}$$

a

$NRF = 4, NIH = 0 \qquad I = 4 - 0 - 3 = 1$
statically indeterminate
to the 1st degree

b

$NRF = 5, NIH = 0 \qquad I = 5 - 0 - 3 = 2$
statically indeterminate
to the 2nd degree

c

$NRF = 5, NIH = 0 \qquad I = 5 - 0 - 3 = 2$
statically indeterminate
to the 2nd degree

d

$NRF = 5, NIH = 2 \qquad I = 5 - 2 - 3 = 0$
statically determinate

e

$NRF = 5, NIH = 1 \qquad I = 5 - 1 - 3 = 1$
statically indeterminate
to the 1st degree

f

$NRF = 6, NIH = 2 \qquad I = 6 - 2 - 3 = 1$
statically indeterminate
to the 1st degree

g

Fig. 3.7.   Determination of degree of statical indeterminacy.

Figure 3.7 illustrates the determination of the degree of statical indeterminacy for a number of beam configurations. Case (g) requires some discussion. This particular beam is statically indeterminate to the first degree. However, it is possible to calculate all reactions except the horizontal reactions. These are interrelated by the condition $F_x = 0$ applied to the complete beam. If no horizontal load is applied, both have zero magnitude. If a nonzero horizontal load is applied the proportion resisted by each horizontal reaction can be determined either by the procedure or the equations developed in Example 2.1.

Also, when no axial loads are applied to a beam, there is no need to distinguish between a horizontal roller support and a pin support. If in such cases a pin support is shown, there will be no horizontal reaction. However showing a pin support instead of rollers will increase the calculated degree of statical indeterminacy because the zero-valued horizontal reactions are included in the counting. From this point forward in the text beams will be considered to be transversely loaded only. If axially loaded, the beam will be considered to be a degenerate form of a planar frame, i.e., a frame having one member.

3.1.3.2 Stability of Beams

Types of statical instability (external, internal and geometric) were defined and illustrated in Chapter 2. Assessment of the stability of beams in particular is the subject matter of this subsection. For external statical stability and geometric stability to exist, it is necessary to have enough properly oriented reaction forces to satisfy statics. For beams, the minimum number is three. For internal statical stability to exist, it must be possible to resist all possible loadings by flexural action (bending) without collapse either of the beam or of any of its separate segments.

A negative degree of statical indeterminacy is an indicator of instability, i.e.,

$$I \geq 0, \quad \text{beam is likely stable}$$
$$I < 0, \quad \text{beam is unstable}$$

The condition $I \geq 0$ is a necessary condition for stability but is not a sufficient condition. Based on this condition, each beam in Fig. 3.7 appears to be stable and they indeed are stable. Application of the above conditions to several unstable beams is illustrated in Fig. 3.8.

Case (a) obviously lacks the minimum of three reaction forces needed for statical stability externally. The member is free to rotate about the support, as depicted by the dashed line. By contrast, case (b) has four external reactions and $I = 1$, which suggests that stability exists. However, the beam is geometrically unstable, being free to roll horizontally. Case (c) contains an internal hinge which renders the beam unstable. By computation, $I = -1$. The computation is correct because any loading will cause the members to freely rotate (without bending) in the manner shown. The rationale

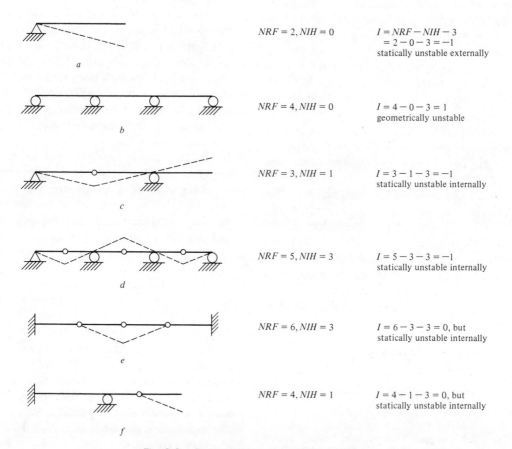

$NRF = 2, NIH = 0$ 　　　$I = NRF - NIH - 3$
$= 2 - 0 - 3 = -1$
statically unstable externally

$NRF = 4, NIH = 0$ 　　　$I = 4 - 0 - 3 = 1$
geometrically unstable

$NRF = 3, NIH = 1$ 　　　$I = 3 - 1 - 3 = -1$
statically unstable internally

$NRF = 5, NIH = 3$ 　　　$I = 5 - 3 - 3 = -1$
statically unstable internally

$NRF = 6, NIH = 3$ 　　　$I = 6 - 3 - 3 = 0$, but
statically unstable internally

$NRF = 4, NIH = 1$ 　　　$I = 4 - 1 - 3 = 0$, but
statically unstable internally

Fig. 3.8. Determination of instability of beams.

for the instability comes from examination of Eq. 3.2. For $I_e = 0$,

$$\text{NRF} - \text{NIH} - 3 = 0$$

$$\text{NRF} = 3 + \text{NIH} \qquad (3.3)$$

Equation 3.3 implies the number of reaction forces required for stability of a beam containing internal hinges equals three plus one per hinge. This condition explains the stability of case (e) in Fig. 3.8. Extra reactions render the structure statically indeterminate. This explains the indeterminacy of case (f) in Fig. 3.7 and why case (g) in that figure is statically determinate except for the horizontal reactions.

Continuing with Fig. 3.8, case (d) calculates to be statically unstable internally, as expected (six reactions are needed and only five exist.) Case (e) calculates to be stable but is statically *unstable* internally. The tri-hinged segment of the beam can freely deform as shown. Note that for loads outside this segment the beam is stable. Case (f) has a similar outcome, due to the local instability of the segment overhanging to the right of the hinge. For loads to the left of this hinge, the beam is stable.

Equation 3.2 has been developed with the term NIH defined as the number of internal hinges. Internal hinges reduce the degree of statical indeterminacy of beams. Other construction features sometimes exist that also reduce the degree of statical indeterminacy. Two examples are sleeved and slotted member ends, such as depicted in Fig. 3.9. Neglecting friction, these connection features eliminate the transfer of either axial force, case a or shear force, case (b), and, thus, reduce the degree of statical indeterminacy in a manner similar to an internal hinge. For more general application, Eq. 3.2 can be written as

$$I = \text{NRF} - \text{NIF} - 3 \qquad (3.4)$$

a   b

Fig. 3.9. Internal connections with special features. a, sleeved member end; b, slotted member end.

where NIF is the number of internal forces removed by the connection features. If features such as those in Fig. 3.9 exist at the exterior supports of a beam, NRF is simply calculated accordingly and NIF is not increased due to their presence. This is synonymous with an exterior pin connection, whose hinge is not to be included in NIH either (why?).

3.2 DIRECT ANALYSIS OF BEAMS

3.2.1 Reactions, Shear, and Moment

By definition, the reactions and internal forces of a statically determinate beam can be computed by application of statics alone. Calculation of reaction forces has just been described. Generally, internal shear and moment vary as a function of distance x along the span. Shear and moment values, V_{x*} and M_{x*}, at any specified location, $x*$, are determined by conditions of internal equilibrium. If the beam is severed at the specified location, the result is the pair of free-body diagrams shown in Fig. 3.10. Both of the bodies must satisfy statics. Thus, if the reaction and loads on each body create an imbalance in either vertical equilibrium or moment equilibrium or both, internal forces are needed to prevent translation and rotation. V_{x*} and M_{x*} accomplish this.

V_{x*} and M_{x*} at the chosen location are calculated by application of statics to either segment of the member. Conceptually, this procedure could be repeated for a number of locations, and the computed shear and moment values plotted to provide approximate shear and moment diagrams. This rudimentary approach is valid but impractical and unnecessary. Instead, the precise variation in shear $V(x)$ and moment $M(x)$ can be determined by letting the location of the cut in Fig. 3.10 be considered a variable instead of a specified value $x*$. The resulting functions are then plotted to yield the shear and moment diagrams. Normally, it is also a tedious process to generate shear and moment diagrams by this method, so the approach is not undertaken in practice. An alternative technique is described in Section 3.2.3. The ability to write shear and moment equations in the manner just described, however, is

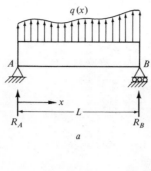

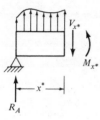

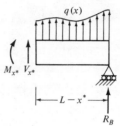

Fig. 3.10.   Equilibrium of a simple beam.

vital to the unit-load method of analysis as described in Section 9.3 and will be demonstrated
by example.

Example 3.4.   Determine the shear and moment
equations for the beam in Fig. 3.11.

Solution: Reactions are obtained by enforcing
statics for the beam as a whole.

$$\Sigma M_D = R_B (10) - 2.5 (4) (12) - 20(5) = 0$$
+ cw (clockwise)

$$\therefore R_B = 22 \text{ kN} \uparrow$$

$$\Sigma F_y = R_B + R_D - 2.5 (4) - 20 = 0$$
+ up

$$\therefore R_D = 8 \text{ kN} \uparrow$$

Free-body diagrams given in Fig. 3.11b through d
demonstrate that the makeup of the diagram is a
function of the distance x to the selected cut in the
member. Because three distinct free-body diagrams
are evident in this beam, separate shear and moment
equations are needed to describe each different region of the span:
For $0 < x < 4$ (Fig. 3.11b):

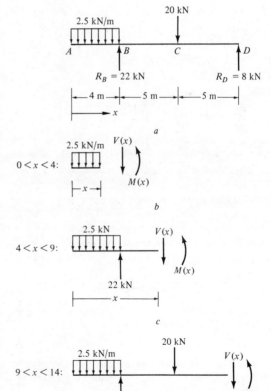

Fig. 3.11.   Beam and free-body diagrams for
Example 3.4.

$$\Sigma F_y = V(x) + 2.5(x) = 0$$
+ down

$$\Sigma M_{cut} = M(x) + 2.5(x)(x/2) = 0$$
+ ccw (counterclockwise)

$$\therefore V(x) = -2.5x \text{ kN}$$

$$M(x) = -1.25x^2 \text{ kN} \cdot \text{m}$$

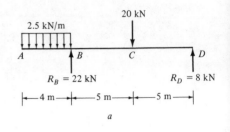

For $4 < x < 9$ (Fig. 3.11c):

$$\Sigma F_y = V(x) + 2.5(4) - 22 = 0$$
+ down

$$\Sigma M_{cut} = M(x) + 2.5(4)(x - 2)$$
$$+ \text{ ccw} - 22(x - 4) = 0$$

$$\therefore V(x) = 12 \text{ kN}$$

$$\therefore M(x) = 12x - 68 \text{ kN} \cdot \text{m}$$

For $9 < x < 14$ (Fig. 3.11d):

$$\Sigma F_y = V(x) + 2.5(4) - 22 + 20 = 0$$
+ down

$$\Sigma M_{cut} = M(x) + 2.5(4)(x - 2) - 22(x - 4)$$
+ ccw
$$- 20(x - 9) = 0$$

$$\therefore V(x) = -8 \text{ kN}$$

$$\therefore M(x) = -8x + 112 \text{ kN} \cdot \text{m}$$

Concentrated forces create a discontinuity in the shear equations. This is indicated by expressing the limitations on the location x, as inequalities. Also, the free-body diagrams to the left of the cut were employed in this solution. With a proper recognition of the origin of x being at the left end of the beam, the right-side free-body diagrams would produce the same results. The shear and moment equations are plotted in Fig. 3.12 to produce the shear and moment diagrams.

Example 3.5. Calculate the shear and moment equations for the beam shown in Fig. 3.13a.

Solution: The reactions are calculated by proportioning the resultant load $\frac{1}{2}qL$, which is located two-thirds of the span distance from A.

Cutting the member at a distance x from the left support isolates the free-body diagram shown in Fig. 3.13b. In Fig. 3.13c, the partial loading has been replaced by its resultant. Thus

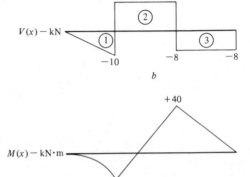

Fig. 3.12. Beam used in Examples 3.4 and 3.9.

$$\Sigma F_y = V(x) - \frac{qL}{6} + \frac{qx^2}{2L} = 0$$

$$V(x) = \frac{qL}{6} - \frac{qx^2}{2L}$$

$$\Sigma M_{cut} = M(x) - \frac{qL}{6}(x) + \frac{qx^2}{2L}\left(\frac{x}{3}\right) = 0$$
+ ccw

$$M(x) = \frac{qLx}{6} - \frac{qx^3}{6L} = \frac{q\,x}{6L}(L^2 - x^2)$$

Earlier in this chapter it was stated that all beams are statically determinate internally. This can now be validated. Whether statically determinate externally or not, it is possible to calculate the external reactions by methods described in this textbook. If the reactions are known, any external indeterminacy has been removed. Furthermore, the internal shear force, axial force, and moment can be calculated at any location along the span. This is done by cutting the beam at the desired location and applying statics in the manner just described in Example 3.5. Since any location can be so

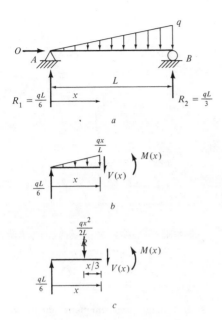

Fig. 3.13.   Beam for Example 3.5.

treated, the internal forces can always be determined throughout the entire beam. Methods to do so efficiently are the subject of the following subsections.

3.2.2   General Relationship Among Load, Shear, and Flexural Moment

Internal forces in a member usually vary in magnitude along its length. A free-body diagram of a differential element taken from a loaded member is drawn in Fig. 3.14. Internal forces acting in the directions shown will be considered as positive values. These sign conventions and loading can be stated as follows:

1. Positive moment produces compression in the top fibers and tension in the bottom

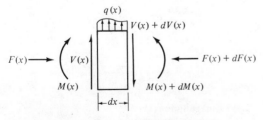

Fig. 3.14.   A loaded, differential element.

fibers (usually termed the *designer's convention*).
2. Positive axial force is in compression.
3. Positive shears act upward on the left-side faces and downward on the right-side faces.
4. Positive loads act in the upward direction.

Although axial forces are shown in Fig. 3.14 they are not a consideration at this point, since their magnitudes are normally zero for a beam.

Because the differential element is subjected to transverse load, the shear and moment forces on opposite faces are not equal. The differences in values can be established by applying statics to the body shown in Fig. 3.14. Differential length dx is assumed to be sufficiently small to permit neglecting the variation in load along the element. Thus, $q(x)$ is considered to be constant over the length dx. By summing all forces in the direction of the shear resultants,

$$V(x) - [V(x) + dV(x)] + q(x)\,dx = 0$$

which can be simplified to

$$dV(x) = q(x)\,dx \qquad (3.5)$$

or

$$\frac{dV(x)}{dx} = q(x) \qquad (3.6)$$

Moment equilibrium about the lower right corner of the element establishes

$$M(x) - [M(x) + dM(x)] \\ + V(x)\,dx + q(x)\,dx\,(dx/2) = 0$$

By neglecting the term $(dx)^2$ in comparison with dx

$$dM(x) = V(x)\,dx \qquad (3.7)$$

or

$$\frac{dM(x)}{dx} = V(x) \qquad (3.8)$$

Equations 3.5 and 3.7 are the desired relationships and, simply stated, show that if the loading function is known, the shear and moment functions can be determined by direct integration over the length of the finite member. Conversely, Eqs. 3.6 and 3.8 demonstrate that a known moment function can be differentiated to establish the shear and loading expressions. It is further evident from these differentiations that the maximum (and minimum) values of shear and moment within a loaded segment occur at locations where the load and shear values, respectively, are zero.

The normal sequence for establishing the shear and moment expressions is to employ a known loading function in the following sequence

$$V(x) = \int dV(x) = \int q(x)\, dx \quad (3.9)$$

$$M(x) = \int dM(x) = \int V(x)\, dx \quad (3.10)$$

Functions $V(x)$ and $M(x)$ are commonly referred to as *shear and moment equations*. Direct integration is the fundamental method for establishing these functions. However, a more direct approach will be discussed in the next section.

––––––––––––

Example 3.6.   Use the basic load-shear-moment relationships to determine the shear and moment equations for the beam shown in Fig. 3.15. Determine the maximum shear and moment.

Solution: The reactions shown were calculated in Example 3.5. By use of Eqs. 3.9 and 3.10,

$$V(x) = \int \left(-\frac{qx}{L}\right) dx = -\frac{qx^2}{2L} + C_1$$

$$M(x) = \int \left(-\frac{qx^2}{2L} + C_1\right) dx$$

$$= -\frac{qx^3}{6L} + C_1 x + C_2$$

Constants of integration are determined from known conditions

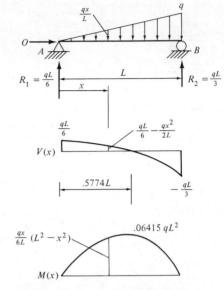

Fig. 3.15. Shear and moment diagrams for Example 3.6.

$$@x = 0 \quad M(x) = C_2 = 0$$

$$\therefore C_2 = 0$$

$$@x = L \quad M(x) = -\frac{qL^2}{6} + C_1 L$$

$$\therefore C_1 = \frac{qL}{6}$$

Consequently,

$$V(x) = -\frac{qx^2}{2L} + \frac{qL}{6} = \frac{qL}{6} - \frac{qx^2}{2L}$$

$$M(x) = -\frac{qx^3}{6L} + \frac{qLx}{6} = \frac{qx}{6L}(L^2 - x^2)$$

which agree with the results obtained in Example 3.5.

The maximum moment occurs where $V(x) = 0$

$$\therefore \quad -\frac{qx^2}{2L} + \frac{qL}{6} = 0$$

$$\therefore x = \frac{L}{\sqrt{3}} = .5774L$$

and its magnitude is

$$M_{max} = -\frac{q\,(L/\sqrt{3})^3}{6L} + \frac{qL\,(L/\sqrt{3})}{6}$$

$$= \frac{qL^2}{9\sqrt{3}} = .06415\ qL^2$$

The preceding results are plotted below the beam in Fig. 3.15.

Example 3.7. Use the basis load-shear-moment relationships to determine the shear and moment equations for the beam shown in Fig. 3.16.

Solution: By use of Eqs. 3.9 and 3.10

$$V(x) = \int (-q)\,dx = -qx + C_1$$

$$M(x) = \int (-qx + C_1)\,dx = -\frac{qx^2}{2} + C_1 x + C_2$$

Constants of integration C_1 and C_2 are determined from known conditions:

$$@x = 0 \quad V(x) = C_1 = 0$$

$$\therefore C_1 = 0$$

$$@x = 0 \quad M(x) = C_2 = 0$$

$$\therefore C_2 = 0$$

Consequently,

$$V(x) = -qx$$

$$M(x) = -\frac{qx^2}{2}$$

Plots of these two functions are shown below the beam in Fig. 3.16. When plotted, in this fashion,

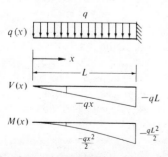

Fig. 3.16. Shear and moment in a uniformly loaded cantilever beam (Example 3.7).

the graphical results are commonly referred to as the *shear diagram V(x)* and the *moment diagram M(x)*. Shear and moment at any point along the beam are determined by substitution of its x-coordinate into the functional relationships, e.g., at the fixed end,

$$V(L) = -qL$$

$$M(L) = \frac{-qL^2}{2}$$

which are the absolute maximum values. In this case the maximum moment does not occur where $V(x) = 0$. Why?

Example 3.8. Repeat Example 3.7 for the beam shown in Fig. 3.17. Determine the maximum shear and moment.

Solution: Using subscripts 1 and 2 for segments AB and BC, respectively, Eqs. 3.9 and 3.10 yield

$$V_1(x) = \int (-q)dx = -qx + C_1$$

$$M_1(x) = \int (-qx + C_1)dx = -\frac{qx^2}{2} + C_1 x + C_2$$

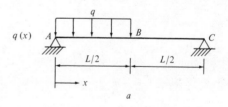

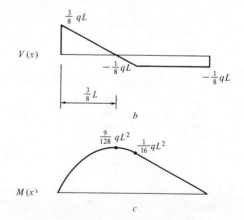

Fig. 3.17. Beam for Examples 3.8 and 3.13.

and

$$V_2(x) = \int (0)dx = C_3$$

$$M_2(x) = \int (C_3)dx = C_3 x + C_4$$

for segments AB and BC, respectively.

It is known that the moment is zero at the support locations, and the shear and moment where the segments join must be equal.

@ $x = 0$ $M_1(x) = C_2 = 0$

@ $x = L$ $M_2(x) = C_3 L + C_4 = 0$

@ $x = \dfrac{L}{2}$ $V_1(x) = V_2(x)$

$$-\tfrac{1}{2}qL + C_1 = C_3$$

@ $x = \dfrac{L}{2}$ $M_1(x) = M_2(x)$

$$-\tfrac{1}{2}qL^2 + C_1\frac{L}{2} + C_2 = C_3\frac{L}{2} + C_4$$

By simultaneous solution

$$C_1 = \tfrac{3}{8}qL \qquad C_3 = -\tfrac{1}{8}qL$$

$$C_2 = 0 \qquad O_4 = \tfrac{1}{8}qL^2$$

Consequently

$$V_1(x) = -qx + \tfrac{3}{8}qL$$

$$M_1(x) = -\tfrac{1}{2}qx^2 + \tfrac{3}{8}qLx$$

$$V_2(x) = -\tfrac{1}{8}qL$$

$$M_2(x) = -\tfrac{1}{8}qLx + \tfrac{1}{8}qL^2$$

The shear and moment diagrams corresponding to these equations are plotted below the beam in Fig. 3.17. The location of the maximum shear is evident. From Eq. 3.8, the maximum value of M_1 is where

$$\frac{dM_1(x)}{dx} = V_1(x) = 0$$

Thus

$$x = \tfrac{3}{8}L$$

and

$$M_1\left(\tfrac{3}{8}L\right) = \frac{9}{128}qL^2$$

which is the maximum moment in the span.

Two examples have been presented to illustrate the use of Eqs. 3.9 and 3.10. The reader should fully examine the validity of Eqs. 3.6 and 3.8 for these examples, particularly in regard to the determination of maximum (and minimum) values of shear and moment. Note the mathematical maxima (and minima) sometimes occur outside the length of the beam segment. In such cases the highest (and lowest) values within the segment must be deduced from the plotted shear and moment diagrams. Such is the case for Example 3.7.

3.2.3 Shear and Moment Diagrams

Shear and moment diagrams serve many useful purposes in structural analysis and design. Knowledge of maximum shear and moment values that occur for given loadings permit a proper selection of member size to avoid an overstress of the materials. Economical placement of steel reinforcement in concrete beams requires information regarding the variation in shear and moment values. In the context of this text, the use of shear and moment diagrams for the determination of the displaced shape of a structural member constitutes their most useful purpose. Integration of the shear and moment equations to produce the equation of the elastic curve of a beam is described in Section 3.2.2. Additional methods for computing displacement values for framed members exist and will be developed in subsequent chapters. Many of the classical computational methods rely upon shear and moment equations or the corresponding graphical diagrams as an integral part of the computational process. Thus, it is valuable to develop an expeditious method for obtaining these important diagrams.

In Section 3.2.2, statics was applied to a differential element of a loaded beam to produce the relationships

$$\frac{dV(x)}{dx} = q(x) \qquad (3.6)$$

$$\frac{dM(x)}{dx} = V(x) \qquad (3.8)$$

or equivalently,

$$dV(x) = q(x)\, dx \qquad (3.5)$$

$$dM(x) = V(x)\, dx \qquad (3.7)$$

Equations 3.5 and 3.7 can be integrated over a finite-beam length to produce

$$V(x_1 - x_2) = \int_{x_1}^{x_2} q(x)\, dx \qquad (3.11)$$

$$M(x_1 - x_2) = \int_{x_1}^{x_2} V(x)\, dx \qquad (3.12)$$

Symbols $V(x_1 - x_2)$ and $M(x_1 - x_2)$ are used to represent the change in shear and the change in moment, respectively, between two locations x_1 and x_2 located along the member length. Integrals that appear in Eqs. 3.11 and 3.12 have physical interpretations. The integral in Eq. 3.11 is recognized as the area under the loading function between points x_1 and x_2. Similarly, the right side of Eq. 3.12 is recognized as the area under the shear function between points x_1 and x_2. Two important conclusions are drawn from the preceding discussion:

1. The difference in shear values between two points on a loaded beam element is numerically equal to the area under the loading function (*loading diagram*) for the segment of the beam that lies between the two points.
2. The difference in moment values between two points on a loaded beam element is numerically equal to the area under the shear function (*shear diagram*) for the segment of the beam that lies between the two points.

The major importance of the preceding statements lies in the observation that shear and moment diagrams can be plotted without resort to mathematical integration. If the shear and moment values are known at any single point in the beam, the shear and moment values at any other point are determined by application of Eqs. 3.11 and 3.12. If V_{x_1} and M_{x_1} are the known shear and moment at location x_1 then V_{x_2} and M_{x_2}, the shear and moment at location x_2 are given by

$$V_{x_2} = V_{x_1} + V(x_1 - x_2) \qquad (3.13)$$

$$M_{x_2} = M_{x_1} + M(x_1 - x_2) \qquad (3.14)$$

If the change in either shear or moment occurs over an infinitesimal distance due to either a concentrated load, concentrated moment, or a reaction, the integrals in Eqs. 3.11 and 3.12 are replaced by the magnitude of the concentrated effect and are rewritten

$$V_{x_1^+} = V_{x_1^-} + V\,(x_1^- - x_1^+) \qquad (3.15)$$

$$M_{x_1^+} = M_{x_1^-} + M\,(x_1^- - x_1^+) \qquad (3.16)$$

Locations x_1^- and x_1^+ are an infinitesimal distance to the left and right, respectively, of location x_1. The primary step is determining the areas under the $q(x)$ and $V(x)$ diagrams between the point of known shear and moment values and the point of interest. Because the loading and shear diagrams are generally composed of simple shapes, the areas under these curves can be calculated by known formulas instead of integration. Geometric properties of shapes commonly encountered in the loading and shear diagrams are presented in Appendix A, Table A.1, as an aid in computations. This table is also useful in the moment-area method, which is discussed in Chapter 4.

Equations 3.6 and 3.8 relate the slope at any given point of the shear and moment diagrams to the load value and shear value, respectively, that exist at the point under consideration. This information is valuable in determining the shape of the shear and moment diagrams. Equations 3.11 and 3.12 have merit only if the shear and moment are known at a point in the member. Because the loading is always known, it is a simple matter to apply statics to determine the shear and moment values at a convenient point in any statically determinate beam.

Example 3.9.   Determine the shear and moment diagrams for the beam of Fig. 3.12a.

Solution: Plotting of the shear diagram begins with the known value of zero shear at point A, i.e., $V_A = 0$. The area under the loading diagram between points A^+ and B^- is

$$V(A^+ - B^-) = -2.5(4) = -10 \text{ kN}$$

$$\therefore V_B^- = V_A^+ + V(A^- - B^-)$$

$$= 0 - 10 = -10 \text{ kN}$$

Point A^+ is an infinitesmal distance to the right of A. Point B^- is an infinitesimal distance to the left of B. A negative area results because the loading is negative (acts downward). The shape of the shear diagram between A and B is linear. This follows from the recognition that the loading diagram is integrated to produce changes in shear. In this case, the uniform load integrates to a linear decrease in shear. The slope of the line is negative and has a value of -2 kN/m based on Eq. 3.6.

A concentrated jump in shear occurs at point B, because of the upward reaction of 22 kN.

$$\therefore V_{B^+} = V_{B^-} + V(B^- - B^+)$$

$$= -10 + 22 = +12 \text{ kN}$$

No load is encountered between point B and C. As a result, the value of shear remains constant has zero slope in this region. At point C, the concentrated load of 20 kN acting downward causes a sudden change in shear value

$$V_{C^+} = V_{C^-} - V(C^- - C^+)$$

$$= 12 - 20 = -8 \text{ kN}$$

Between points C and D, no load is applied, and the shear remains constant. The complete shear diagram is given in Fig. 3.12.

The moment diagram also begins with a known value of moment (zero) at point A. The area under the shear diagram between points A and B is

$$M(A - B) = \tfrac{1}{2}(-10)(4) = -20$$

$$\therefore M_B = M_A + M(A - B) = 0 + (-20)$$

$$= -20 \text{ kN} \cdot \text{m}$$

Because the shear variation between A and B is linear, it integrates to a second-order parabolic moment relationship. The moment diagram plots as concave downward. This is evident because the shear diagram indicates the slope of the moment equation is zero at A and has an increasingly negative value from A to B. At B the slope is -10 kN/m. Remaining ordinates are obtained as follows:

$$M(B - C) = +12(5) = +60 \text{ kN} \cdot \text{m}$$

$$\therefore M_C = M_B + M(B - C) = +40 \text{ kN} \cdot \text{m}$$

Between B and C the shear is constant. Thus the moment variation is linear with a positive slope of 12 kN/m.

$$M(C - D) = -8(5) = -40 \text{ kN} \cdot \text{m}$$

$$\therefore M_D = M_C + M(C - D) = 0$$

The shear diagram indicates the moment variation is linear between C and D with a slope of -8 kN/m.

Example 3.10.   Determine the shear and moment diagrams for the beam in Fig. 3.18a.

Solution: Reactions shown were calculated in Example 3.3. The shear diagram plots as shown in Fig. 3.18b, according to the following sequence. For equilibrium with the reaction R_2, the shear just to the right of A must be +4 kN. Subsequently,

$V(A - B) = -2(4) = -8 \text{ kN}$	$V_{B^-} = -4 \text{ kN}$
$V(B^- - B^+) = 0$	$V_{B^+} = -4 \text{ kN}$
$V(B - C) = 0$	$V_{C^-} = -4 \text{ kN}$
$V(C^- - C^+) = +9 \text{ kN}$	$V_{C^+} = +5 \text{ kN}$
$V(C - D) = 0$	$V_{D^-} = +5 \text{ kN}$
$V(D^- - D^+) = -12 \text{ kN}$	$V_{D^+} = -7 \text{ kN}$
$V(D - E) = 0$	$V_{E^-} = -7 \text{ kN}$
$V(E^- - E^+) = +13 \text{ kN}$	$V_{E^+} = +6 \text{ kN}$
$V(E - F) = 0$	$V_{F^-} = +6 \text{ kN}$
$V(F^- - F^+) = 0$	$V_{F^+} = +6 \text{ kN}$
$V(F - G) = -3(4)$	$V_{G^-} = -6 \text{ kN}$
$\quad = -12 \text{ kN}$	

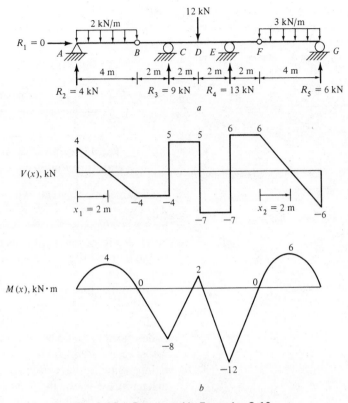

Fig. 3.18. Beam used in Examples 3.10.

Since V_{G^-} is negative, it acts downward on the cross section just to the left of G and equilibrates the upward 6 kN reaction, R_5.

For moment equilibrium at joint A, the internal moment must be zero, since no external moment is applied at that joint. The remaining ordinates are found as follows.

The distance x_1 to the point of zero shear is found by recognizing the slope of the shear diagram in the segment AB equals the loading, -2 kN/m. Thus,

$$x_1 = \frac{4 \text{ kN}}{2 \text{ kN/m}} = 2 \text{ m}$$

Similarly the distance x_2 is

$$x_2 = \frac{6 \text{ kN}}{3 \text{ kN/m}} = 2 \text{ m}$$

Integrating the V(x) diagram produces

$$M(A - x_1) = (4)(2) = 4 \text{ kN} \cdot \text{m}$$

$$\therefore M_{x_1} = 4 \text{ kN} \cdot \text{m}$$

$$M(x_1 - B) = \tfrac{1}{2}(-4)(2) = -8 \text{ kN} \cdot \text{m}$$

$$\therefore M_B = 0$$

$$M(B - C) = -4(2) = -8 \text{ kN} \cdot \text{m}$$

$$\therefore M_C = -8 \text{ kN} \cdot \text{m}$$

$$M(C - D) = 5(2) = 10 \text{ kN} \cdot \text{m}$$

$$\therefore M_D = 2 \text{ kN} \cdot \text{m}$$

$$M(D - E) = -7(2) = -14 \text{ kN} \cdot \text{m}$$

$$\therefore M_E = -12 \text{ kN} \cdot \text{m}$$

$$M(E - F) = 6(2) = 12 \text{ kN} \cdot \text{m}$$

$$\therefore M_F = 0$$

$$M(F - x_2) = \tfrac{1}{2}(6)(2) = 6 \text{ kN} \cdot \text{m}$$

$$\therefore M_{x_2} = 6 \text{ kN} \cdot \text{m}$$

$$M(x_2 - G) = \tfrac{1}{2}(-6)(2) = -6 \text{ kN} \cdot \text{m}$$

$$\therefore M_G = 0$$

The resulting $V(x)$ and $M(x)$ diagrams are plotted in Fig. 3.18b. The shape of each segment of these diagrams is deduced by observing Eqs. 3.6 and 3.8. Note that the shear values at the internal hinges agree with the values calculated in Example 3.3. As expected, the moment is zero at both internal hinges.

Example 3.11. Redo Example 3.6 by the method used in Example 3.10.

Solution: From Fig. 3.15 the reactions are known. Thus $V_{A^+} = ql/6$ and

$$V_{B^-} = V_{A^+} + V(A - B)$$

$$= \frac{qL}{6} + [\tfrac{1}{2}(-q)(L)] = -\frac{qL}{3}$$

At a distance x to the right of A

$$V(A - x) = \frac{1}{2}\left(-\frac{qx}{L}\right)(x) = -\frac{qx^2}{2L}$$

and zero shear occurs at the location $x = x_0$ where

$$0 = V_{A^+} + V(A - x) = \frac{qL}{6} + \left(-\frac{qx_0}{2L}\right)$$

$$x_0 = \frac{L}{\sqrt{3}} = .5774L$$

Therefore the maximum moment is

$$M(x_0) = M_A + M(A - x_0)$$

$$= 0 + \frac{2}{3}\left(\frac{qL}{6}\right)\left(\frac{L}{\sqrt{3}}\right) = .06415qL$$

where $M(A - x_0)$ is obtained as the area under the parabolic shear diagram between points A and x_0 (See Table A.1 in Appendix A and note that the slope of $V(x)$ is zero at A!)

A general formula for $M(x_0 - B)$ is not available. Thus,

$$M(x_0 - B) = \int_{.5774L}^{L}\left(\frac{qL}{6} - \frac{qx^2}{2L}\right) dx$$

$$= \left(\frac{qLx}{6} - \frac{qx^3}{6L}\right)\Bigg|_{.5774L}^{L}$$

$$M(x_0 - B) = -.06415L$$

and

$$M_B = M_{x_0} + M(x_1 - B)$$

$$= .06415L - .06415L = 0$$

The results confirm the outcome of Example 3.6.

3.2.4 Total Degree of Statical Indeterminacy

If all the reactions are known for a given beam, it is always possible to draw complete shear and moment diagrams. In the context of Section 3.2.1, the three internal forces (moment, shear and axial force) can be calculated at any location by cutting the beam at that location and applying the three conditions of statics. Alternatively, with the reactions established, the loading diagram $q(x)$ can always be integrated as described in Sections 3.2.2 and 3.2.3 to yield the $V(x)$ and $M(x)$ diagrams. Consequently, it is impossible to have a beam that is statically indeterminate internally. This logic is the basis for earlier statements that the total degree of statical indeterminacy for a beam is always equal to the external degree of statical indeterminacy.

3.3 GOVERNING EQUATION FOR FLEXURAL DISPLACEMENT

The schematic model for a solitary framed member in the undeformed position is a straight line. Under the action of loads, the element will change shape and move to a new position. With a knowledge of the elastic material behavior under flexure, analytical geometry, and calculus, the *governing differential equation* for the deflection of the member due to bending moment can be derived. Preliminary to this, it is necessary again to view a different element, as shown in Fig. 3.19, taken from the full-size member. In the deformed position, the element of original length dx assumes a curved shape, but the length of the element remains unchanged. This fundamental geometric approximation is discussed in Section 3.6 and essentially means the displaced arc length can be

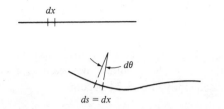

Fig. 3.19. Effect of flexure on a differential element.

taken to be the same as the corresponding chord length. In a *first-order analysis* (explained in Section 3.6) longitudinal deformation due to internal axial force is considered to be negligible in comparison to longitudinal deformation due to flexure and is generally ignored. Under this circumstance, the centroidal axis length would not change. For this reason it is termed the "neutral axis." Consequently, the inference is that the line diagram shown in Fig. 3.19 represents the neutral axis of the member. Locating the displaced position of the centroidal axis is the subject matter of this section.

An expanded view of the differential element is drawn in Fig. 3.20a. The effect of the flexural moment M acting on the section is to deform it into the shape shown in Fig. 3.20b. It can be seen that the angular deformation $d\theta$

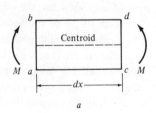

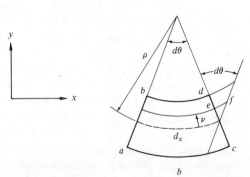

Fig. 3.20. Differential element in a deformed position.

causes the two faces $a\text{-}b$ and $c\text{-}d$ to rotate relative to each other and introduce strain in the longitudinal direction of the member. Geometrical considerations permit quantification of the strain effects.

At a distance v measured perpendicular to the centroidal axis, the strain effect is evident by the change in length, distance $e\text{-}f$, at that location. Numerically, this distance is equal to $vd\theta$. Consequently, the strain ϵ_v at that level is

$$\epsilon_v = \frac{vd\theta}{dx} = \frac{vd\theta}{\rho d\theta} = \frac{v}{\rho} \qquad (3.17)$$

where ρ is the radius of curvature to the centroidal axis. The assumption of a circular curved shape is an approximation that is normally acceptable.

The strain ϵ_v can be quantified by consideration of basic material laws. The flexural stress at a distance v from the centroidal axis is

$$\sigma_v = \frac{Mv}{I}$$

and the corresponding strain is

$$\epsilon_v = \frac{\sigma_v}{E} = \frac{Mv}{EI} \qquad (3.18)$$

By equating the expressions 3.17 and 3.18, one obtains

$$\frac{1}{\rho} = \frac{M}{EI} \qquad (3.19)$$

The radius of a curvature at a point in a line is a geometrical characteristic, which is defined mathematically as

$$\frac{1}{\rho} = \frac{d^2y/dx^2}{\left[1 + \left(\dfrac{dy}{dx}\right)^2\right]^{3/2}} \qquad (3.20)$$

where x and y are the coordinates of the point on the curved line. The right-hand side of Eq. 3.20 is positive because the curvature in Fig. 3.19 is positive in a mathematical sense (concave upward). In ordinary structural analysis, the rotations that occur are quite small, and the squared slope term $(dy/dx)^2$ can be safely ignored in Eq. 20 when compared to unity. This approximation constitutes a fundamental assumption in *small-deflection theory* (discussed in Sect. 3.6.3) and permits a less complicated mathematical model. Equating Eqs. 3.19 and 3.20 and eliminating the squared slope term gives

$$\frac{d^2y}{dx^2} = \frac{M}{EI} \qquad (3.21)$$

In any given member, the value of the flexural moment will vary throughout its length. If the variation in moment is expressed as a function $M(x)$, as described in Section 3.2.2, Eq. 3.21 becomes

$$\frac{d^2y}{dx^2} = \frac{M(x)}{EI} \qquad (3.22)$$

Equation 3.22 is a basic relationship in small-deflection theory of structural analysis. Transverse displacement y is shown to have a definite relationship to the internal moments that exist in a beam segment. An implication of this equation is that the deflection is in a single plane and the applied loads lie in that same plane. In space frames, biaxial bending of a member occurs, and the deflecton can be predicted by considering the relationship defined by Eq. 3.22 as separately valid in two orthogonal directions. For this approach to have validity, the line of action of the loads must pass through the shear center.

Equation 3.22 can be differentiated twice according to the following sequence:

$$\frac{d^3y}{dx^3} = \frac{d}{dx}\left[\frac{M(x)}{EI}\right] \qquad (3.23)$$

$$\frac{d^4y}{dx^4} = \frac{d^2}{dx^2}\left[\frac{M(x)}{EI}\right] \qquad (3.24)$$

For members for which EI is constant over the entire length, Eq. 3.23 degenerates to

$$\frac{d^3y}{dx^3} = \frac{1}{EI}\frac{dM(x)}{dx} = \frac{1}{EI}V(x) \qquad (3.25)$$

where $V(x)$ is the equation that expresses the variation in shear within the beam segment. Consequently, Eq. 3.24 becomes

$$\frac{d^4y}{dx^4} = \frac{1}{EI}\frac{dV(x)}{dx} = \frac{1}{EI}q(x) \qquad (3.26)$$

This relationship is the governing differential equation for the deflection of a flexural element of constant cross section and material properties and is generally written in the form:

$$EI\frac{d^4y}{dx^4} = q(x) \qquad (3.27)$$

Transverse displacement y is shown to be a function of the loading $q(x)$, applied to the segment. This, of course, is an expected result.

3.4   FLEXURAL DEFORMATIONS BY METHODS OF INTEGRATION

Differential equations governing the behavior of an individual framed member were developed in Sections 3.2.2 and 3.3.3. In particular, Eqs. 3.22 and 3.27 constitute the bases for a pair of methods for the determination of the displaced shape of individual beam elements. When properly integrated, either equation yields the *equation of the elastic curve*. The two options available for performing the necessary computations are presented in Table 3.1. Use of Eq. 3.22 is termed the *double integration method*, and use of Eq. 3.27 is termed the *quadruple integration method*. It should be noted that, by either method, it is necessary to evaluate the constants of integration. This is accomplished by recognition of boundary conditions inherent in any given problem. Indeed, the number and nature of the known boundary conditions establish which of the integration methods can be employed to achieve a solution. Double integration requires two known conditions regarding slope and deflection, in addition to the function $M(x)$. The quadruple

Table 3.1. Flexural Displacements by Direct Integration.

Double Integration	Quadruple Integration
	$EI\dfrac{d^4y}{dx^4} = q(x)$
	$EI\dfrac{d^3y}{dx^3} = \displaystyle\int q(x) + C_1$
$EI\dfrac{d^2y}{dx^2} = M(x)$	$EI\dfrac{d^2y}{dx^2} = \displaystyle\iint q(x) + C_1x + C_2$
$EI\dfrac{dy}{dx} = \displaystyle\int M(x) + K_1$	$EI\dfrac{dy}{dx} = \displaystyle\iiint q(x) + \dfrac{C_1x^2}{2} + C_2x + C_3$
$EIy = \displaystyle\iint M(x) + K_1x + K_2$	$EIy = \displaystyle\iiiint q(x) + \dfrac{C_1x^3}{6} + C_2\dfrac{x^2}{2} + C_3x + C_4$

integration procedure requires knowledge of four boundary conditions regarding displacement, slope, and higher derivatives of y, in addition to the function $q(x)$. Boundary conditions applicable to each of the common support conditions are enumerated in Table 3.2. It should be noted that the sign convention associated with the various force quantities are identical with those established for plotting shear and moment diagrams.

3.4.1 Example Problems

Example 3.12. Determine the equation of the elastic curve and the maximum displacement at point B for the beam shown in Fig. 3.21. Assume EI is constant over the entire span. Determine and plot the shear and moment equations.

Solution: By the double integration method (Eq. 3.22)

$$EI\frac{d^2y}{dx^2} = M = \frac{qL}{2}x - \frac{q}{2}x^2$$

$$EI\frac{dy}{dx} = \frac{qL}{4}x^2 - \frac{q}{6}x^3 + K_1$$

$$EIy = \frac{qL}{12}x^3 - \frac{q}{24}x^4 + K_1x + K_2$$

Boundary conditions consist of the known zero displacement at each support.

@x = 0 $Ely = K_2 = 0$ ∴$K_2 = 0$

@x = L $Ely = \dfrac{qL^4}{12} - \dfrac{qL^4}{24} + K_1(L)$

$$\therefore K_1 = -\frac{qL^3}{24}$$

$$\therefore y(x) = \frac{1}{EI}\left(-\frac{q}{24}x^4 + \frac{qL}{12}x^3 - \frac{qL^3}{24}x\right)$$

$$\therefore y\left(\frac{L}{2}\right) = -\frac{5qL^4}{384EI} = \frac{5qL^4}{384EI}\ \text{downward}$$

Alternate solution: By quadruple integration (Eq. 3.27)

$$EI\frac{d^4y}{dx^4} = -q\ \text{(the load is downward!)}$$

$$EI\frac{d^3y}{dx^3} = -qx + C_1$$

$$EI\frac{d^2y}{dx^2} = -\frac{q}{2}x^2 + C_1x + C_2$$

$$EI\frac{dy}{dx} = -\frac{q}{6}x^3 + \frac{C_1}{2}x^2 + C_2x + C_3$$

$$EIy = -\frac{q}{24}x^4 + \frac{C_1}{6}x^3 + \frac{C_2}{2}x^2 + C_3x + C_4$$

Four boundary conditions are available and employed as follows:

Table 3.2. Common Boundary Conditions.

Support	Boundary Conditions	
Fixed	$y = 0$	$\dfrac{dy}{dx} = 0$
Exterior pin	$y = 0$	$EI\dfrac{d^2y}{dx^2} = M^*$
Roller	$y = 0$	$EI\dfrac{d^2y}{dx^2} = M^*$
Free	$EI\dfrac{d^3y}{dx} = V$ Note: V is nonzero only if a concentrated force is applied at the free end.	$EI\dfrac{d^2y}{dx^2} = M^*$
Interior pin	$\left(\dfrac{dy}{dx}\right)_a = \left(\dfrac{dy}{dx}\right)_b$	$y = 0$
Internal hinge	$(y)_a = (y)_b$	$EI\dfrac{d^2y}{dx^2} = 0$

*Usually the applied moment, M, is zero.

$@x = 0 \quad EIy = C_4 = 0 \qquad \therefore C_4 = 0$

$@x = 0 \quad EI\dfrac{d^2y}{dx^2} = M = C_2 = 0 \quad \therefore C_2 = 0$

$@x = L \quad EI\dfrac{d^2y}{dx^2} = M = -\dfrac{qL^2}{2} + C_1 L = 0$

$$\therefore C_1 = \dfrac{qL}{2}$$

$@x = L \quad EIy = -\dfrac{qL^4}{24} + \dfrac{qL^4}{12} + C_3 L = 0$

$$\therefore C_3 = \dfrac{qL^3}{24}$$

$$\therefore y(x) = \dfrac{1}{EI}\left(-\dfrac{q}{24}x^4 + \dfrac{qL}{12}x^3 - \dfrac{qL^3}{24}x\right)$$

$$\dfrac{dy}{dx} = \dfrac{1}{EI}\left(-\dfrac{q}{6}x^3 + \dfrac{qL}{4}x^2 - \dfrac{qL^3}{24}\right)$$

$$M(x) = EI\dfrac{d^2y}{dx^2} = -\dfrac{q}{2}x^2 + \dfrac{qL}{2}x$$

$$V(x) = EI\dfrac{d^3y}{dx^3} = -qx + \dfrac{qL}{2}$$

$$q(x) = EI\dfrac{d^4y}{dx^4} = -q \quad \text{(as expected)}$$

The shear and moment equations are plotted in Fig. 3.21b. The maximum moment is at midspan as de-

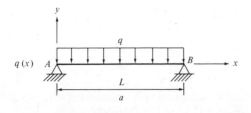

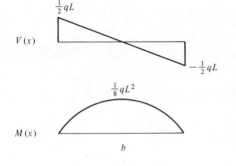

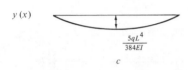

Fig. 3.21.   Beam for Example 3.12.

termined by setting $V(x) = 0$. By symmetry, the maximum displacement occurs at midspan (where the slope is zero) and is given by

$$y_{max} = y\left(\frac{L}{2}\right) = \frac{1}{EI}\left[-\frac{q}{24}\left(\frac{L}{2}\right)^4 + \frac{qL}{12}\left(\frac{L}{2}\right)^3\right.$$

$$\left. - \frac{qL^3}{24}\left(\frac{L}{2}\right)\right]$$

$$y_{max} = -\frac{5qL^4}{384EI} = \frac{5qL^4}{384EI} \downarrow$$

which is shown in Fig. 3.21c.

Example 3.13. Determine the equation of the elastic curve for the beam in Example 3.8. Assume EI is constant over the entire span.

Solution: Using the double-integration method requires treating segments AB and BC separately. Let y_1 and y_2 describe displacements in segments AB and BC, respectively, and measure x from the left end (joint A).

Segment AB:

$$EI\frac{d^2y_1}{dx^2} = M = \frac{3qL}{8}x - \frac{q}{2}x^2$$

$$EI\frac{dy_1}{dx} = \frac{3qL}{16}x^2 - \frac{q}{6}x^3 + K_1$$

$$EIy_1 = \frac{qL}{16}x^3 - \frac{q}{24}x^4 + K_1x + K_2$$

Segment BC:

$$EI\frac{d^2y_2}{dx^2} = M = \frac{qL}{8}(L - x) = \frac{qL^2}{8} - \frac{qL}{8}x$$

$$EI\frac{dy_2}{dx} = \frac{qL^2}{8}x - \frac{qL}{16}x^2 + K_3$$

$$EIy_2 = \frac{qL^2}{16}x^2 - \frac{qL}{48}x^3 + K_3x + K_4$$

In addition to the zero displacement values required at each support, compatibility of slope and displacement must be enforced at point B.

Segment AB:

$$@x = 0 \quad EIy_1 = K_2 = 0 \quad \therefore K_2 = 0$$

At point B:

$$EI\frac{dy_1}{dx} = EI\frac{dy_2}{dx}$$

$$\frac{3qL}{16}\left(\frac{L}{2}\right)^2 - \frac{q}{2}\left(\frac{L}{2}\right)^3 + K_1$$

$$= \frac{qL^2}{8}\left(\frac{L}{2}\right) - \frac{qL}{16}\left(\frac{L}{2}\right)^2 + K_3$$

or

$$K_1 - K_3 = \frac{qL^3}{48} \qquad (a)$$

and

$$EIy_1 = EIy_2$$

$$\frac{qL}{16}\left(\frac{L}{2}\right)^3 - \frac{q}{24}\left(\frac{L}{2}\right)^4 + K_1\left(\frac{L}{2}\right)$$

$$= \frac{qL^2}{16}\left(\frac{L}{2}\right)^2 - \frac{qL}{48}\left(\frac{L}{2}\right)^3 + K_3\left(\frac{L}{2}\right) + K_4$$

or

$$K_1\left(\frac{L}{2}\right) - K_3\left(\frac{L}{2}\right) - K_4 = \frac{qL^4}{128} \qquad \text{(b)}$$

Segment BC:

$$@x = L \quad EIy_2 = \frac{qL^2}{16}(L)^2 - \frac{qL}{48}(L)^3$$

$$+ K_3(L) + K_4 = 0$$

or

$$K_3L + K_4 = -\frac{qL^4}{24} \qquad \text{(c)}$$

Evaluating a, b, and c simultaneously yields

$$K_1 = -\frac{9qL^3}{384}$$

$$K_3 = -\frac{17qL^3}{384}$$

$$K_4 = +\frac{qL^4}{384}$$

$$\therefore EIy_1 = \frac{qL}{16}x^3 - \frac{q}{24}x^4 - \frac{9qL^3}{384}x$$

and

$$EIy_2 = \frac{qL^2}{16}x^2 - \frac{qL}{48}x^3 - \frac{17qL^3}{384}x + \frac{qL^4}{384}$$

Example 3.14. Determine the equation for the elastic curve and the midspan displacement for the beam in Fig. 3.22. Assume EI is constant over the entire span.

Solution: Because the beam is statically indeterminate, the reactions and the moment equation cannot be determined. This prevents the use of the double-integration method. By quadruple integration

$$EI\frac{d^4y}{dx^4} = -q$$

$$EI\frac{d^3y}{dx^3} = -qx + C_1$$

$$EI\frac{d^2y}{dx^2} = -\frac{q}{2}x^2 + C_1x + C_2$$

$$EI\frac{dy}{dx} = -\frac{q}{6}x^3 + \frac{C_1}{2}x^2 + C_2x + C_3$$

$$EIy = -\frac{q}{24}x^4 + C_1\frac{x^3}{6} + \frac{C_2}{2}x^2 + C_3x + C_4$$

These are identical to the equations developed in Example 3.12 except the boundary conditions are

$$@x = 0 \quad EI\frac{dy}{dx} = C_3 = 0 \qquad \therefore C_3 = 0$$

$$@x = 0 \quad EIy = C_4 = 0 \qquad \therefore C_4 = 0$$

$$@x = L \quad EI\frac{dy}{dx} = -\frac{qL^3}{6} + C_1\frac{L^2}{2} + C_2L = 0$$

$$@x = L \quad EIy = -\frac{qL^4}{24} + C_1\frac{L^3}{6} + C_2\frac{L^2}{2} = 0$$

Solving the latter two equations simultaneously yields

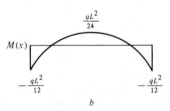

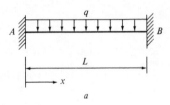

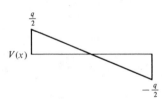

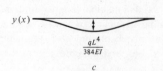

Fig. 3.22.   Beam for Example 3.14.

$$C_1 = \tfrac{1}{2}qL$$

$$C_2 = -\tfrac{1}{12}qL^2$$

$$\therefore y(x) = \frac{1}{EI}\left(-\frac{q}{24}x^4 + \frac{qL}{12}x^3 - \frac{qL^2}{24}x^2\right)$$

$$M(x) = EI\frac{d^2y}{dx^2} = -\frac{q}{2}x^2 + \frac{qL}{2}x - \frac{qL^2}{12}$$

$$V(x) = EI\frac{d^3y}{dx^3} = -qx + \frac{qL}{2}$$

$$q(x) = EI\frac{d^4y}{dx^4} = -q \quad \text{(as expected)}$$

These results are plotted in Fig. 3.22b and c. Finally,

$$y_{\text{max}} = y\left(\frac{L}{2}\right) = -\frac{qL^4}{384EI} = \frac{qL^4}{384EI} \downarrow$$

With some exceptions, the maximum (and minimum) displacement occurs where dy/dx is equal to zero. Physically, this is equivalent to stating "where the slope is zero." For some cases, e.g., a uniformly loaded cantilever beam, the maximum displacement occurs at a location where the slope is nonzero. Usually, this is because the function $y(x)$ does not have both a maxima and a minima.

Finally, it should be understood that although the requirement of the double-integration method that $M(x)$ be known is an inconvenience, it does not restrict the method to statically determinate beams. For statically indeterminate beams, determination of $M(x)$ must be preceded by application of any of the indeterminate structural analysis methods presented in subsequent chapters of this text.

3.4.2   Use of Binomial Functions

3.4.2.1   Double Integration Method

Because of the particular loading considered in Example 3.13, determining the deflected shape of the beam was tedious. The function $M(x)$ differed for segments AB and BC. Thus, separate double integrations were executed for each segment. Continuity of the segments had to be established by enforcing joint boundary conditions at their junction. This involved solution process can be simplified by the use of bino-

mial functions. Use of binomial functions allow development of a single moment function applicable over the entire length of the beam. While some care must be taken, the resulting integration operations are typically reduced to about half the ordinary effort. An example problem will be used to illustrate the concept and serve as a base for further description.

Example 3.15.   Determine the displaced shape and maximum displacement of the beam given in Fig. 3.23. Use the double-integration method.

Solution: The usual moment equations for the beam are

$$0 \le x \le \frac{L}{3} \qquad M(x) = Px$$

$$\frac{L}{3} \le x < \frac{2L}{3} \qquad M(x) = Px - P\left(x - \frac{L}{3}\right)$$

$$\frac{2L}{3} \le x \le L \qquad M(x) = Px - P\left(x - \frac{L}{3}\right)$$
$$- P\left(x - \frac{2L}{3}\right)$$

By use of binomial functions a single expression is written to replace the above, as follows:

$$M(x) = P\langle x\rangle - P\left\langle x - \frac{L}{3}\right\rangle - P\left\langle x - \frac{2L}{3}\right\rangle$$

The general term $\langle x - a \rangle$ implies the term is non-existent if $x < a$. Otherwise, it can be operated on like $(x - a)$.

Continuing the solution

$$EI\frac{d^2y}{dx^2} = P\langle x\rangle - P\left\langle x - \frac{L}{3}\right\rangle - P\left\langle x - \frac{2L}{3}\right\rangle \quad \text{(d)}$$

$$EI\frac{dy}{dx} = \frac{P}{2}\langle x\rangle^2 - \frac{P}{2}\left\langle x - \frac{L}{3}\right\rangle^2$$
$$- \frac{P}{2}\left\langle x - \frac{2L}{3}\right\rangle^2 + C_1 \quad \text{(e)}$$

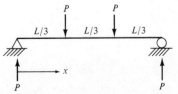

Fig. 3.23.   Beam for Example 3.15.

$$Ely = \frac{P}{6}\langle x\rangle^3 - \frac{P}{6}\left\langle x - \frac{L}{3}\right\rangle^3$$

$$- \frac{P}{6}\left\langle x - \frac{2L}{-3}\right\rangle^3 + C_1\langle x\rangle + C_2 \quad (f)$$

Note the term $\langle x - a\rangle$ is integrated like the term $(x - a)$ except all resulting terms involving constants of integration are collected together in the expression C_1 (for the first integration) and $C_1\langle x\rangle + C_2$ (for the second integration.) The reader should directly integrate expression c and validate its rearrangement into the form of expression f.

Application of boundary conditions establishes C_1 and C_2.

$$@x = 0 \;\; Ely = \frac{P}{6}(0)^3 + C_1(0) + C_2 = 0$$

$$C_2 = 0$$

$$@x = L \;\; Ely = \frac{P}{6}(L)^3 - \frac{P}{6}\left(L - \frac{L}{3}\right)^3$$

$$- \frac{P}{6}\left(L - \frac{2L}{3}\right)^3 + C_1(L) = 0$$

$$C_1 = -\frac{PL^2}{9}$$

$$Ely = \frac{P}{6}\langle x\rangle^3 - \frac{P}{6}\left\langle x - \frac{L}{3}\right\rangle^3$$

$$- \frac{P}{6}\left\langle x - \frac{2L}{3}\right\rangle^3 - \frac{PL^2}{9}\langle x\rangle$$

If desired, the result can be rewritten in ordinary form as

$$0 < x < \frac{L}{3} \quad Ely = \frac{P}{6}x^3 - \frac{PL^2}{9}x$$

$$\frac{L}{3} < x < \frac{2L}{3} \quad Ely = \frac{P}{6}x^3 - \frac{P}{6}\left(x - \frac{L}{3}\right)^3$$

$$- \frac{PL^2}{9}x$$

$$Ely = \frac{PL}{6}x^2 - \frac{PL^2}{6}x + \frac{PL^3}{162}$$

$$\frac{2L}{3} < x < L \quad Ely = \frac{P}{6}x^3 - \frac{P}{6}\left(x - \frac{L}{3}\right)^3$$

$$- \frac{P}{6}\left(x - \frac{2L}{3}\right)^3$$

$$- \frac{PL^2}{9}x$$

$$Ely = -\frac{P}{6}x^3 + \frac{PL}{2}x^2$$

$$- \frac{7PL}{18}x + \frac{PL^3}{18}$$

The maximum displacement, which occurs at midspan, is

$$Ely\left(\frac{L}{2}\right) = \frac{PL}{6}\left(\frac{L}{2}\right)^2 - \frac{PL^2}{6}\left(\frac{L}{2}\right) + \frac{PL^3}{162}$$

$$y_{max} = \frac{-23PL^3}{648\,EI} = \frac{23}{648}\frac{PL^3}{EI}$$

Example 3.16:   Repeat Example 3.13, but use binomial functions.

Solution: The usual moment equations are

$$0 < x < \frac{L}{2} \quad M(x) = \frac{3qL}{8}x - \frac{qx^2}{2}$$

$$\frac{L}{2} < x < L \quad M(x) = \frac{3qL}{8}x - q\left(\frac{L}{2}\right)\left(x - \frac{L}{2}\right)^2$$

In contrast to Example 3.13, these cannot be directly accumulated into a single function. However this becomes possible if the loading is replaced by its statical equivalent shown in Fig. 3.24 and the moment equations rewritten as

$$0 < x < \frac{L}{2} \quad M(x) = \frac{3qL}{8}x - \frac{qx^2}{2}$$

$$\frac{L}{2} < x < L \quad M(x) = \frac{3qL}{8}x - \frac{qx^2}{2}$$

$$+ \frac{q}{2}\left(x - \frac{L}{2}\right)^2$$

The preceding expressions are replaced by

$$M(x) = \frac{3qL}{8}\langle x\rangle - \frac{q}{2}\langle x\rangle^2 + \frac{q}{2}\left\langle x - \frac{L}{2}\right\rangle^2$$

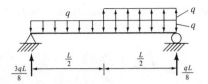

Fig. 3.24.   Statically equivalent loading for beam in Example 3.16.

Thus,

$$EI\frac{d^2y}{dx^2} = \frac{3qL}{8}\langle x\rangle - \frac{q}{2}\langle x\rangle^2 + \frac{q}{2}\left\langle x - \frac{L}{2}\right\rangle^2$$

$$EI\frac{dy}{dx} = \frac{3qL}{16}\langle x\rangle^2 - \frac{q}{6}\langle x\rangle^3$$

$$+ \frac{q}{6}\left\langle x - \frac{L}{2}\right\rangle^3 + C_1$$

$$EIy = \frac{qL}{16}\langle x\rangle^3 - \frac{q}{24}\langle x\rangle^4 + \frac{q}{24}\left\langle x - \frac{L}{2}\right\rangle^4$$

$$+ C_1x + C_2$$

From boundary conditions,

$$@x = 0 \quad EIy = \frac{qL}{16}(0)^3 - \frac{q}{24}(0)^4 + C_1(0)$$

$$+ C_2 = 0$$

$$C_2 = 0$$

$$@x = L \quad EIy = \frac{qL}{16}(L)^3 - \frac{q}{24}(L)^4 + \frac{q}{24}(L)^4$$

$$+ C_1(L) = 0$$

$$C_1 = -\frac{9}{384}qL^3$$

$$\therefore Ely = \frac{qL}{16}\langle x\rangle^3 - \frac{q}{24}\langle x\rangle^4 + \frac{q}{24}\left\langle x - \frac{L}{2}\right\rangle^4$$

$$- \frac{9}{384}qL^3\langle x\rangle$$

Conversion to ordinary form yields the same expressions for Ely_1 and Ely_2 obtained in Example 3.13.

A binomial function has the form

$$(x + a)^n, \quad n > 0$$

where a is any positive or negative constant. In the preceding examples, the form

$$\langle x - a\rangle^n, \quad n > 0$$

has been used, with the pointed brackets inferring the term exists only if $x > a$. In particular, the moment equation for a concentrated load, P located at $x = a$ is

$$M(x) = P\langle x - a\rangle \qquad (3.28)$$

For a uniform load of intensity q, beginning at $x = a$, the moment equation is

$$M(x) = q \cdot \tfrac{1}{2}\langle x - a\rangle^2 \qquad (3.29)$$

Note the term $\tfrac{1}{2}\langle x - a\rangle^2$ is the integral of $\langle x - a\rangle$, according to the rule

$$\int \langle x - a\rangle^n\, dx = \frac{\langle x - a\rangle^{n+1}}{n + 1} \qquad (3.30)$$

This rule has been used throughout Examples 3.15 and 3.16. In these examples it is also evident that multiple integration of the term $\langle x - a\rangle$ is a fundamental operation. The validity of the integration rule for this term is confirmed in Table 3.3. In the

Table 3.3.

$f(x)$	$\langle x - a\rangle = \langle x - a\rangle$	$x - a$
$\int f(x)dx$	$\tfrac{1}{2}\langle x - a\rangle^2 = \tfrac{1}{2}\langle x^2 - 2ax + a^2\rangle + K_1$	$\dfrac{x^2}{2} - ax + C_1$
$\int\int f(x)dx$	$\tfrac{1}{6}\langle x - a\rangle^3 = \tfrac{1}{6}\langle x^3 - 3ax^2 + 3a^2x - a^3\rangle + K_1x + K_2$	$\dfrac{x^3}{6} + ax^2 + C_1 + C_2$
$\int\int\int f(x)dx$	$\tfrac{1}{24}\langle x - a\rangle = \tfrac{1}{24}\langle x^4 - 4ax^3 + 6a^2x^2 - 4a^3x + a^4\rangle + K_1\dfrac{x^2}{2} + K_2x + K_3$	$\dfrac{x^4}{24} - ax^3 + C_1\dfrac{x^2}{2} + C_2x + C_3$

second column the term $(x - a)$ is integrated according to Eq. 3.30 and then expanded. In the third column the expression $(x - a)$ is integrated directly (in the usual way.) By collecting like terms on the right side of the equation, in column 2, each expression can be shown to be identical to the result of the usual integration. For example, consider the second derivative.

$$\frac{1}{2}\langle x^2 - 2ax + a^2 \rangle + K_1 = \frac{x^2}{2} - ax + C_1$$

$$\frac{x^2}{2} - ax + \left(\frac{a^2}{2} + K_1\right) = \frac{x^2}{2} - ax + C_1$$

$$\frac{x^2}{2} - ax + C_1 = \frac{x^2}{2} - ax + C_1$$

3.4.2.2   Quadruple Integration Method

Binomial functions can be used in the quadruple integration method provided the right side of the expression

$$EI\frac{d^4y}{dx^4} = q(x)$$

can be written in binomial form.

For uniform loading of intensity q beginning at the location $x = a$, $q(x)$ is written in the symbolic form

$$q(x) = q\langle x - a \rangle^0 \qquad (3.31)$$

which implies, by integration,

$$V(x) = EI\frac{d^3y}{dx^4} = q\langle x - a \rangle^1 \qquad (3.32)$$

$$M(x) = EI\frac{d^2y}{dx^2} = q \cdot \tfrac{1}{2}\langle x - a \rangle^2 \qquad (3.33)$$

Equations 3.32 and 3.33 are the correct terms for the contributions made to shear and moment, respectively. Higher integrals for slope and displacement are proper, too. Equation 3.31 simply implies that whenever $x > a$

$$q(x) = q\langle x - a \rangle^0 = q$$

Similarly, for a concentrated load, P, located at $x = a$, the symbolic form

$$q(x) = P\langle x - a \rangle^{-1} \qquad (3.34)$$

is used to describe it. By integration,

$$V(x) = EI\frac{d^3y}{dx^2} = P\langle x - a \rangle^0 \qquad (3.35)$$

$$M(x) = EI\frac{d^2y}{dx^2} = P\langle x - a \rangle^1 \qquad (3.36)$$

and the correct higher integrals follow.

Note, the expression 3.34 is symbolic only, as it does not fit the form of a binomial function. However, it does produce the correct results upon multiple integration. Indeed, if the inverse is taken literally,

$$q(x) = \frac{P}{\langle x - a \rangle} \qquad (3.37)$$

and the load approaches infinity as x approaches a and diminishes as x increases beyond a. This is shown qualitatively in Fig. 3.25a. However, the integration produces the desired results for computational purposes. The area under the function is concentrated very close to the location of the singularity (i.e., it accumulates mostly where the abscissa approaches infinity.) Thus, it can be approximated by the distributed load shown in Fig. 3.25b of intensity

$$q(x) = \frac{P}{\Delta x} \qquad (3.38)$$

Thus,

$$\lim_{\Delta x \to 0} \int_a^{a+\Delta x} \frac{P}{\Delta x}\,dx = \lim_{\Delta x \to 0} \left(\frac{P}{\Delta x}x\Big|_a^{a+\Delta x}\right)$$

$$= \lim_{\Delta x \to 0}\left(\frac{P\Delta x}{\Delta x}\right)$$

$$\lim_{\Delta x \to 0} \int_a^{a+\Delta x} \frac{P}{\Delta x}\,dx = P$$

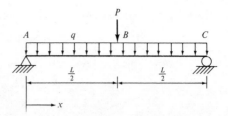

Fig. 3.26.   Beam for Example 3.17.

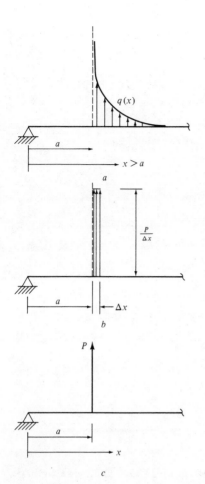

Fig. 3.25.   Concentrated load represented as a singularity.

By use of L'Hospital's Rule and the desired result, Fig. 3.25c, is achieved. The symbolic expression 3.34 is intended to symbolize the physical concepts depicted in Fig. 3.25

Example 3.17.   Determine the maximum displacement for the beam in Fig. 3.26 by quadruple integration. EI is constant.

Solution:

$$q(x) = EI \frac{d^4y}{dx^4} = -q\langle x \rangle^0 - P\left\langle x - \frac{L}{2} \right\rangle^{-1}$$

$$V(x) = EI \frac{d^3y}{dx^3} = -q\langle x \rangle - P\left\langle x - \frac{L}{2} \right\rangle^0$$

$$+ C_1$$

$$M(x) = EI \frac{d^2y}{dx^2} = -\frac{q}{2}\langle x \rangle^2 - P\left\langle x - \frac{L}{2} \right\rangle$$

$$+ C_1 x + C_2$$

$$EI \frac{dy}{dx} = -\frac{q}{6}\langle x \rangle^3 - \frac{P}{2}\left\langle x - \frac{L}{2} \right\rangle^2$$

$$+ C_1 \frac{x^2}{2} + C_2 x + C_3$$

$$EIy = -\frac{q}{24}\langle x \rangle^4 - \frac{P}{6}\left\langle x - \frac{L}{2} \right\rangle^3$$

$$+ C_1 \frac{x^3}{6} + C_2 \frac{x^2}{2} + C_3 x + C_4$$

Boundary conditions require

$$@x = 0 \ y = 0 \qquad\qquad \therefore C_4 = 0$$

$$@x = 0 \ M(x) = 0 \qquad\qquad \therefore C_2 = 0$$

$$@x = L \ EIy = 0 = -\frac{qL^4}{24}$$

$$-\frac{PL^3}{48} + C_1 \frac{L^3}{6} + C_3 L$$

$$@x = L \ M(x) = 0 = -\frac{qL^2}{2}$$

$$-\frac{PL}{2} + C_1 L$$

Therefore

$$C_1 = \frac{qL}{2} + \frac{P}{2}$$

$$C_3 = -\frac{qL^3}{24} - \frac{PL^2}{16}$$

and

$$EIy = -\frac{q}{24}\langle x \rangle^4 - \frac{P}{6}\left\langle x - \frac{L}{2} \right\rangle^3 + \frac{qLx^3}{12}$$

$$-\frac{qL^3 x}{24} + \frac{Px^3}{12} - \frac{PL^2 x}{16}$$

For segment AB

$$Ely = -\frac{qx^4}{24} + \frac{qLx^3}{12} - \frac{qL^3x}{24} + \frac{Px^3}{12} - \frac{PL^2x}{16}$$

For segment BC

$$Ely = -\frac{qx^4}{24} + \frac{qLx^3}{12} - \frac{qL^3x}{24} - \frac{Px^3}{12}$$

$$+ \frac{PLx^2}{4} - \frac{3PL^2x}{16} + \frac{PL^3}{48}$$

The maximum displacement is at midspan. Substitution of $x = \frac{L}{2}$ into either displacement equation yields

$$y_{max} = -\frac{5qL^4}{384EI} - \frac{PL^3}{48EI}$$

$$y_{max} = \left(\frac{5qL^4}{384EI} + \frac{PL^3}{48EI}\right)_{downward}$$

Example 3.18. Determine the equation of the elastic curve for the beam in Example Fig. 3.11a. Calculate the displacement at A.

Solution: Choosing a reference at A and using the known reactions

$$q(x) = EI\frac{d^4y}{dx^4} = -\frac{5}{2}\langle x\rangle^0 + \frac{5}{2}\langle x - 4\rangle^0$$

$$+ 22\langle x - 4\rangle^{-1} - 20\langle x - 9\rangle^{-1}$$

$$V(x) = EI\frac{d^3y}{dx^3} = -\frac{5}{2}\langle x\rangle + \frac{5}{2}\langle x - 4\rangle$$

$$+ 22\langle x - 4\rangle^0 - 20\langle x - 9\rangle^0 + C_1$$

$$M(x) = EI\frac{d^2y}{dx^2} = -\frac{5}{4}\langle x\rangle^2 + \frac{5}{4}\langle x - 4\rangle^2$$

$$+ 22\langle x - 4\rangle - 20\langle x - 9\rangle$$

$$+ C_1x + C_2$$

$$EI\frac{dy}{dx} = -\frac{5}{12}\langle x\rangle^3 + \frac{5}{12}\langle x - 4\rangle^3 + 11\langle x - 4\rangle^2$$

$$- 10\langle x - 9\rangle^2 + C_1\frac{x^2}{2} + C_2x + C_3$$

$$Ely = -\frac{5}{48}\langle x\rangle^4 + \frac{5}{48}\langle x - 4\rangle^4 + \frac{11}{3}\langle x - 4\rangle^3$$

$$- \frac{10}{3}\langle x - 9\rangle^3 + C_1\frac{x^3}{6} + C_2\frac{x^2}{2}$$

$$+ C_3x + C_4$$

Boundary conditions require

@$x = 0$	$V(x) = 0$	$C_1 = 0$
@$x = 0$	$M(x) = 0$	$C_2 = 0$
@$x = 4$	$Ely = 0 = -26.67 + C_3(4)$	
@$x = 14$	$Ely = 0 = 290 + C_3(14) + C_4$	

$$C_3 = -31.7 \qquad C_4 = 153$$

$$Ely = -\frac{5}{48}\langle x\rangle^4 + \frac{5}{48}\langle x - 4\rangle^4 + \frac{11}{3}\langle x - 4\rangle^3$$

$$- \frac{10}{3}\langle x - 9\rangle^3 - 31.7x + 153$$

$$x = 0 \qquad Ely = EI\Delta_A = 153$$

$$\Delta_A = \frac{153}{EI} \uparrow$$

3.5 DEFLECTION DUE TO SHEAR

A differential element of a frame member subjected to pure shear is depicted in Fig. 3.27a. Under the action of the pair of shear forces of magnitude V, the element experiences an angular shear strain, γ, and deforms as shown in Fig. 3.27b. The small vertical displacement, dy, is related to the strain angle according to the relationship

$$dy = -\gamma dx \qquad (3.39)$$

The negative sign indicates that positive shear produces a downward (negative) displacement. From laws of material behavior (Eq. 2.11), the shear stress-strain relationship is

$$\gamma = \frac{\tau}{G} \qquad (3.40)$$

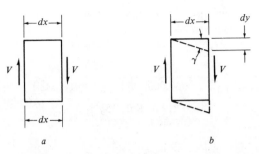

Fig. 3.27. Shear deformation of a differential element. *a*, shear force; *b*, shear deformation.

Substitution of Eq. 3.40 into Eq. 3.39 and rearranging terms gives

$$\frac{dy}{dx} = -\frac{\tau}{G} \qquad (3.41)$$

The shear stress τ varies throughout the cross section, but an average value is generally used in structural analysis. It is common to express the average value as

$$\tau = \frac{nV}{A} \qquad (3.42)$$

where A is the full cross-sectional area and n is a constant, to account for the use of an average stress. The numerical value of n depends upon the shape of the cross section. Substitution of the average shear stress into Eq. 3.41 yields

$$\frac{dy}{dx} = -\frac{nV}{GA} \qquad (3.43)$$

Recognizing that the shear force varies throughout the length of the member and can be expressed as a function $V(x)$, Eq. 3.43 becomes

$$\frac{dy}{dx} = -\left[\frac{nV(x)}{GA}\right] \qquad (3.44)$$

Equation 3.44 permits calculation of the slope at a point in a member due to the internal shear. If $V(x)$ is known, the equation for the deflection due to shear is easily determined by direct integration. This deflected shape must be superimposed upon that which is produced by any flexural moments that exist in the member.

Equation 3.44 can be differentiated once to produce the relationship

$$\frac{d^2y}{dx^2} = -\frac{n}{GA}\frac{dV(x)}{dx} \qquad (3.45)$$

With the substitution of Eq. 3.8, it gives

$$\frac{d^2y}{dx^2} = -\frac{n}{GA}\frac{d^2M(x)}{dx^2} \qquad (3.46)$$

Combining Eq. 3.46 with Eq. 3.22 yields

$$\frac{d^2y}{dx^2} = \frac{M(x)}{EI} - \frac{n}{GA}\frac{d^2M(x)}{dx^2} \qquad (3.47)$$

Solving this differential equation for a given moment variation yields the deflected shape with shear deformations included.

Example 3.19. Determine the relative displacement of the ends of a member of length, L, subjected to constant shear of magnitude V.

Solution: From Eq. 3.44

$$\frac{dy}{dx} = -\frac{nV}{GA}$$

by integration

$$\Delta = y = \int_0^L \left(-\frac{nV}{GA}\right) dx = -\frac{nVL}{GA}$$

The negative shear indicates that positive shear produces a downward displacement.

Example 3.20. Determine the elastic displacement of the free end of the cantilever beam pictured in Fig. 3.28. Include shear deformations and use $E = 30{,}000$ kips/in.^{2} and $G = 12{,}000$ kips/in.^{2}. For a rectangular cross section, n can be taken as 1.2. Compare results for (a) $L = 8$ ft and (b) $L = 1$ ft.

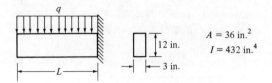

Fig. 3.28.   Beam for Example 3.20.

Solution: Using a reference at the free end

$$M(x) = -\tfrac{1}{2}qx^2$$

and

$$\frac{d^2M(x)}{dx^2} = -q$$

Substitution into Eq. 3.47 gives

$$\frac{d^2y}{dx^2} = -\frac{1}{2}\frac{qx^2}{EI} - \frac{n}{GA}(-q)$$

and by integrating twice

$$y = -\frac{qx^4}{24EI} + \frac{nq\,x^2}{GA\,2} + C_1 x + C_2$$

Applying the boundary conditions

$$@x = L \qquad y = \left(\frac{dy}{dx}\right)_f^* = 0$$

gives

$$C_1 = -\frac{qL^3}{6EI}$$

$$C_2 = \frac{qL^4}{8EI} + \frac{nqL^2}{2GA}$$

Therefore, the deflected shape of the beam is

$$y = -\frac{qx^4}{24EI} + \frac{nqx^2}{2GA} + \frac{qL^3}{6EI}x$$

$$+ \frac{qL^4}{8EI} + \frac{nqL^2}{2GA}$$

At the free end $(x = 0)$

$$y_{max} = \frac{qL^4}{8EI} + \frac{nqL^2}{2GA}$$

*Due to flexure only

A comparison of the flexural and shear displacements is variable.

For $L = 8' = 96$ in.

$$y_{max} = \frac{q(96)^4}{8(30,000)(432)} + \frac{1.2q(96)^2}{2(36)(12,000)}$$

$$= 0.819\ q + 0.013q = 0.832q \text{ in.}$$
$$\quad\ \text{(flexure)} \qquad \text{(shear)}$$

For $L = 1' = 12$ in.

$$y_{max} = \frac{q(12)^4}{8(30,000)(432)} + \frac{1.2q(12)^2}{2(36)(12,000)}$$

$$= 0.0002q + 0.0002q = 0.0004q \text{ in.}$$
$$\quad\ \text{(flexure)} \qquad \text{(shear)}$$

For a span of 8 ft, the displacement due to shear is 1.6% of the total displacement and can be neglected. However, for a span of 1 ft, the percentage of displacement due to shear increased to 50%! This observation concerning short spans is general. In addition, shear displacement is inversely proportional to square of the beam depth, and flexural deflection is inversely proportional to the cube of the beam depth. As the beam depth increases, the relative importance of shear rises. In conclusion, a deep beam having a short span is likely to exhibit measurable shear displacement, and its effect should be included in the computations. However, under normal circumstances, the geometry is such that only flexural displacements are significant.

3.6   ASSUMPTIONS IN ORDINARY STRUCTURAL ANALYSIS

At the introductory level, the analytical techniques of structural analysis are confined to a ''linear, elastic, first-order analysis of statically loaded structures.'' Furthermore, the resulting displacements are considered to be ''small.'' Many limiting assumptions are implied in such an analysis and will be discussed in detail. Large deflections and other conditions, which necessitate a more complex mathematical model, are outside the scope of this textbook.

3.6.1   Linear, Elastic Analysis

Linear, elastic structural analysis incorporates several conditions. Foremost is the condition that the geometry is such that all movements of a point in a structure can be described by straight-line relationships, i.e., linear equations. A clearer picture of the intent of this statement will be given in a subsequent discussion of small-deflection theory.

Material properties must also exhibit a linear behavior. Basically, this means that the material properties E, G, and ν must remain constant in value throughout the entire response of the structure. In this way, simple mechanics of material relationships are applicable and suffice to predict the deformations of any particular member. A linear stress-strain curve is a prerequisite for such performance. For a steel, this is a true condition provided the proportional limit is not exceeded. Conversely, reinforced concrete does not exhibit a linear stress-strain behavior.

An elastic material has the ability to return to its original position after unloading. In the process, it retraces the path followed during loading. A linear, elastic material is further restricted to linear behavior during loading or unloading. Such material performance allows the use of the *principle of superposition* in structural analysis. This vital principle will be discussed later.

3.6.2 First-Order Analysis

First-order analysis has three significant implications. First, the equations of equilibrium are written using *undeformed geometry*—i.e., the coordinates of the structure in its original position. After deformation, the line of action of loads applied to a structure has been slightly adjusted due to movement of the joints, and to curvature and change of length in the members. Recalculation of reactions and internal forces to account for these motions requires an iterative procedure due to the nonlinearity of the relationships. However, the error introduced by omitting this tedious refinement is insignificant in normal situations.

Second, the line of action of any applied axial load or internal axial force is assumed to act through the centroid of a member's cross section. In reality, the eccentricity of an axial force

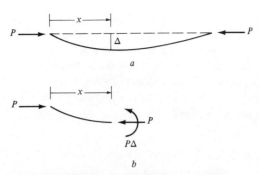

Fig. 3.29.   Development of a $P - \Delta$ moment. _a,_ deformed member; _b,_ member cut at _x._

from the displaced member axis produces a moment, $P - \Delta$, as depicted in Fig. 3.29. In first-order structural analysis, this moment is ignored. To include its effect would also require an iterative method, but normally it is an unnecessary refinement.

The third implication of a first-order analysis is the use of uncoupled *member stiffnesses*. Although *stiffness* has not yet been defined, the meaning of the preceding statement can be demonstrated by a simple example. Figure 3.30a and b show two identical simple beams. The only difference in the two structures is the presence of an axial load in Fig. 3.30b. If the member stiffnesses are "uncoupled," both spans will experience identical end rotations. Presumably, the presence of an axial force, whether tension or compression, has no influence on the end rotations. In reality, this is not true, but inclusion of its effects normally has only minor influence. Member stiffnesses are treated in several chapters of this textbook and are assumed in all cases to be uncoupled.

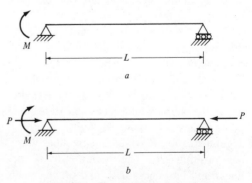

Fig. 3.30.   Simple flexural member.

Three conditions normally neglected in first-order structural analysis have been described. When any of the three stated refinements are included, the analysis is termed a *second-order analysis*. Such methods are a topic for advanced structural analysis and are not treated in this textbook.

3.6.3   Small-Deflection Theory

Two obvious conditions would justify the neglect of the $P - \Delta$ moment in structural analysis. If either the axial force P or the deflection Δ (or both) is very small, this effect would be negligible. When the latter circumstance is considered to be true, *small-deflection theory* is said to apply. In simple words, deflections are considered "small" when the external displacements have magnitudes much less than cross-sectional dimensions. Also, rotation at any point of the member is assumed to be sufficiently small to consider the cosine of the angle defining the motion to equal unity. Likewise, the sine and tangent are taken as equal to the value of the angle in radians.

Adoption of the small-deflection concept also permits a relaxation of some geometric conditions that would otherwise create great mathematical difficulty. A subtle, but very important, approximation is employed in all ordinary structural analysis on the basis of this theory. This approximation involves the strain created by transverse rotation and flexure of a member. Figure 3.31 serves to allow discussion of this concept. Recognition of the effects depicted in this figure is paramount to fundamental understanding of the structural analysis of framed systems.

Consider a member that is subjected to loads that produce an "axial deformation" and a *rigid body* rotation, as shown in Fig. 3.31a. The relative deformations can be viewed as occurring in two stages. With point A held in place, the member undergoes an extension and point B moves to point B_1. Subsequently, the angular rotation R swings the member into the new position AB'. The motion B_1B' is curvilinear and introduces a second-order equation

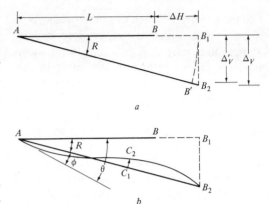

Fig. 3.31.   Concepts of small-deflection theory, *a*, axial deformation + chord rotation; *b*, axial deformation + chord rotation + flexure.

into the computations. The vertical distance to point B' can be calculated as

$$\Delta_V' = (L + \Delta_H) \sin R \qquad (3.48)$$

which is a nonlinear relationship (e.g., doubling R does not double Δ_V'), which will later be shown to be undesirable. However, for small angles Eq. 3.48 becomes

$$\Delta_V' = (L + \Delta_H)R$$

If instead, the transverse motion is assumed to move point B_1 to point B_2, the vertical movement is computed as

$$\Delta_V = (L + \Delta_H) \tan R \qquad (3.49)$$

which for small angles also becomes

$$\Delta_V = (L + \Delta_H)R \qquad (3.50)$$

Clearly, if R and Δ_V are small, the distance $B'B_2$ is also small and the difference between Δ_V' and Δ_V is insignificant. Under usual conditions, the deformations are indeed small enough to justify this approximation. Furthermore, the use of Eq. 3.50 has the advantage of being a linear function of R. It is interesting to note that Eq. 3.50 can be obtained by alternative reasoning. Lines AB_1 and AB' are of equal length. If the motion B_1B' is circular, the arc length is given by the product of the radius of the circle, $L + \Delta_H$, and the enclosed angle R. This computation is recognized as the right-hand side of Eq. 3.50. For small values of R, Δ_V' and the arc length differ by a negligible amount and

$$\Delta_V' = B_1B' = (L + \Delta_H)R \quad (3.51)$$

For small deflections, the distance $B'B_2$ is also quite small; the difference between Δ_V' and Δ_V can be safely ignored, and Eq. 3.51 converts to Eq. 3.50.

Employment of the approximations implied in Eq. 3.50 is a trademark of small-deflection theory and can be stated as follows:

In small-deflection theory, a member is assumed to deform in two stages. Axial elongation or contraction causes any point to move in the direction of the original member axis. Angular rotation causes the point to move perpendicular to the original member axis.

In the context of Fig. 3.31a, point B undergoes two motions, Δ_H and Δ_V. The former constitutes the *axial deformation* and the latter is the *transverse deformation*. The new position of the member as defined by these deformations approximates the true position. Negligible error results if the deformations are very small.

A second feature of small-deflection theory can be discussed with reference to Fig. 3.31b. In addition to the axial deformation and rigid-body rotation shown in Fig. 3.31a, the member has been subjected to "flexural deformation" causing the curved shape AB_2. Clearly, the member would have to stretch (experience additional axial deformation) if point B_2 is to remain in position during the flexural deformation. In small-deflection theory, this axial deformation is considered to be very small compared to the distance BB_1 and is neglected. To calculate its effect would require nonlinear equations. Neglecting this additional axial deformation is acceptable when the transverse motion due to flexure (e.g., C_1C_2) is small. Normally, this is the case.

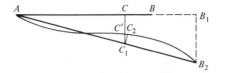

Fig. 3.32.   Flexure in small-deflection theory.

A further implication of small-deflection theory is depicted in Fig. 3.32. In the numerical work presented in subsequent chapters, it will prove to be easier and more desirable to consider the distance $C'C_1$ equal to the transverse distance C_1C_2. In effect, the new position of point C will be taken as C' instead of the true position C_2. This will not lead to significant error if the distance C_1C_2 and the rotation R are small.

By now, it is clear that structural analysis, at the level addressed herein, embodies a number of geometric simplifications. In the case of Fig. 3.32, point B is considered to move to point B_2 by longitudinal and transverse displacements (relative to the original position of the member). The flexure of the member is considered to have no effect on the position of point B_2. All approximations discussed in the preceding paragraphs are encompassed in a single statement: If, in Fig. 3.32, the straight-line distances AB_1 and AB_2 and the curvilinear distance AB_2 can all be treated as essentially equal, then the deformations that are depicted in the figure can be safely treated as "small." The value of the use of small-deflection theory is that the inherent mathematical analysis is linear. This is a necessary condition if one is to employ the important concept of superposition.

3.7 SUPERPOSITION

Limiting oneself to linear, elastic, first-order structural analysis and small-deflection theory is highly advantageous in the computational stage. Acceptance of the assumptions inherent in such an analysis permits the use of the very powerful *Principle of Superposition. To superpose* means "to superimpose"; the principle of superposition implies the validity of combining

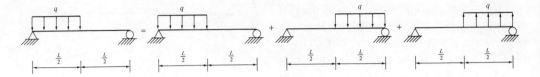

Fig. 3.33　Superposition of loadings used in Ex. 3.16.

one or more structural responses to obtain another response. The principle can be stated as follows:

When limited to a linear, elastic, first-order analysis, the response of a structure to some static loading can be linearly combined with the response to any other static loading. The combined response is the same as would be obtained if the structure were subjected to the combined loading itself.

Implementation of this principle is mathematically represented by

$$R' = \sum_{i=1}^{n} \alpha_1 R_i \qquad (3.52)$$

in which R_i is the response of the structure to a loading L_i, α_i is a constant of any magnitude, and n is the number of separate loading cases. The response R' obtained by this summation would be identical with that of the same structure if subjected to a loading L' as described by

$$L' = \sum_{i=1}^{n} \alpha_i L_i \qquad (3.53)$$

Constant coefficients α_i through α_n have the same values in Eq. 3.52 and 3.53.

Equation 3.52 signifies the validity of combining entire responses. In practice, the Principle of Superposition is actually applied to individual response quantities. As an example, the displaced shape of a beam subjected to a series of concentrated loads could be obtained by analyzing the beam for each load separately and combining the individual displaced shapes. In this case, the α constants in Eqs. 3.52 and 3.53 would all have a value of unity.

Example 3.16 illustrates the use of superposition. To permit the use of binomial functions the superposition of loadings shown in Fig. 3.33 was implied in the solution.

In a subsequent chapter, a distinction will be made between determinate and indeterminate structures. The principle of superposition applies to both, but is more important to the latter category. A qualitative example of how statically determinate responses might be superposed to produce the response of a statically indeterminate structure is illustrated in Fig. 3.34. Freedom to linearly combine individual loadings is the very foundation of indeterminate structural analysis.

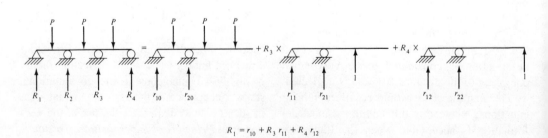

$$R_1 = r_{10} + R_3 \, r_{11} + R_4 \, r_{12}$$
$$R_2 = r_{20} + R_3 \, r_{21} + R_4 \, r_{22}$$

Fig. 3.34　Statically indeterminate beam analysis by superposition.

PROBLEMS

Note: Unless stated otherwise, neglect shear deformations in each of the following problems.

3-1. Are each of the given beams stable? If yes, determine the degree of statical indeterminacy. If no, show why each is unstable. Use a pictorial illustration of a loading which the beam cannot support and explain the instability. Show how to stabilize the beam.

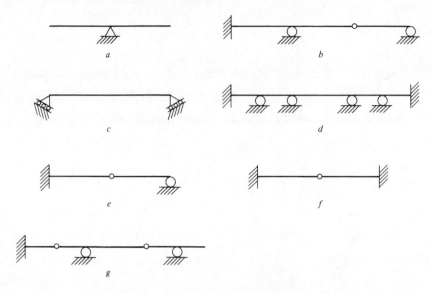

3-2. Calculate the external reactions for each of the statically determinate beams shown.

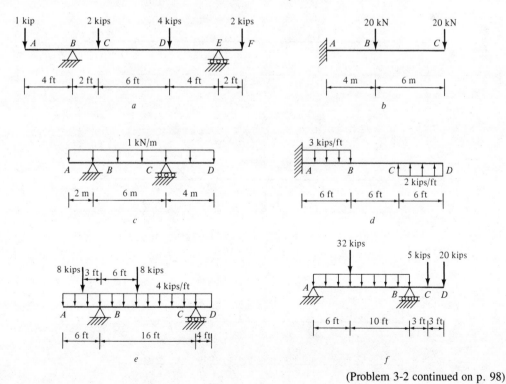

(Problem 3-2 continued on p. 98)

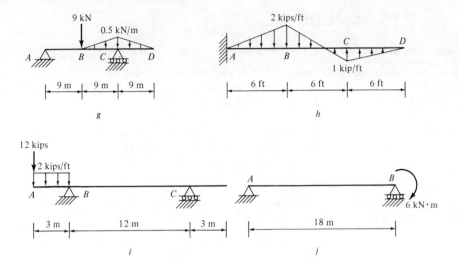

g

h

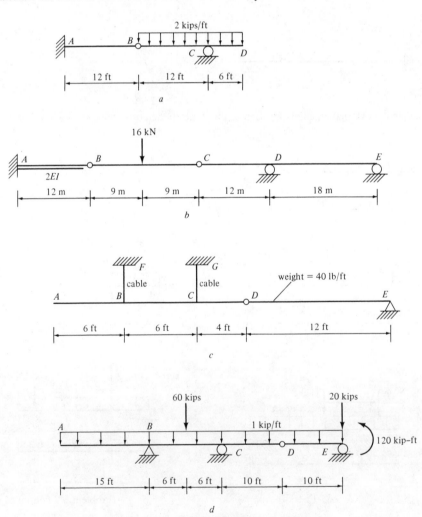

i

j

3-3. Calculate the external reactions for each of the statically determinate beams shown.

a

b

c

d

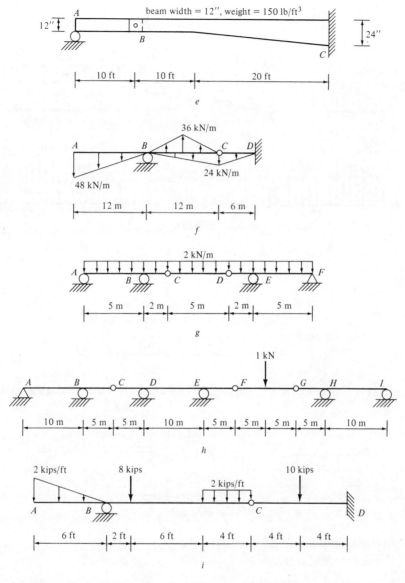

3-4. Determine shear and moment equations for each beam. Use the left end as a reference for the x-axis. Plot the resulting equations. See Section 3.2.1.

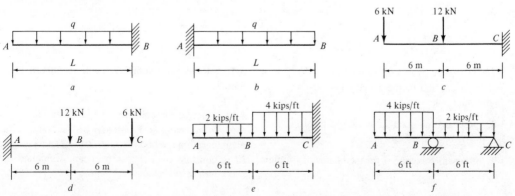

3-5. Repeat Problem 3-4 for the given beams.

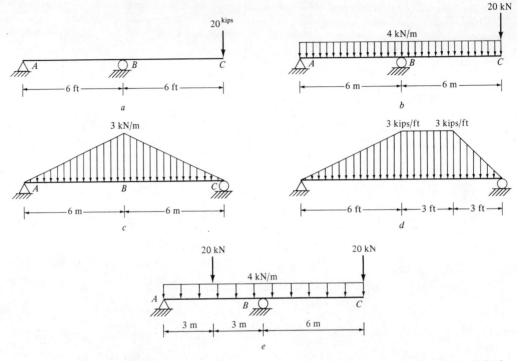

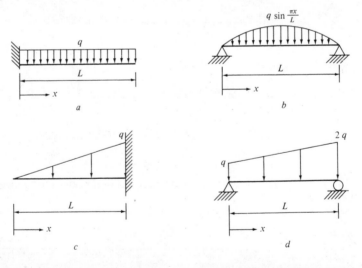

3-6. Determine expressions for shear and moment in each of the given beams using Eqs. 3.9 and 3.10.

3-7. Using the method described in Section 3.2.3, plot the shear and moment diagrams for the beams of Problem 3-4.

3-8. Repeat Problem 3-7 for the beams of Problem 3-5.

3-9. Repeat Problem 3-7 for the beams of Problem 3-2.

3-10. Repeat Problem 3-7 for the beams of Problem 3-3.

3-11. Derive the equation of the elastic curve for each of the beams in Problem 3-6. Use the quadruple integration method (Section 3.4.2.2) and assume EI is constant over the entire span. Calculate the maximum displacement.

3-12. Derive the equations that describe the elastic curve for each of the given beams. Indicate the region of the beam for which each equation is applicable. *EI* is constant over the entire span. Use either double or quadruple integration (Section 3.4).

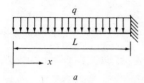

a

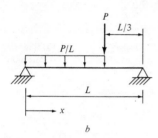

b

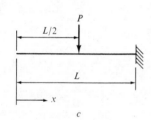

c

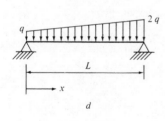

d

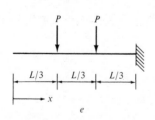

e

3-13. A beam rests on soil that produces an upward force $F(x)$ which is proportional to the deflection of the beam, i.e., $F(x) = ky$. Derive the equation of the elastic curve. Assume the beam is simply sup-

ported at its ends and *EI* is constant. Calculate the maximum deflection.

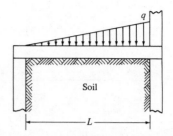

Soil

3-14. Repeat Problem 3-13 for a uniform loading.

3-15. Calculate the midspan displacement for the given beam by (*a*) double integration and (*b*) quadruple integration.

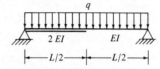

3-16. Change the supports in Problem 3-15 to fixed ends. Calculate the midspan displacement by quadruple integration.

3-17. An earthquake occurs in which the horizontal acceleration at a given instant is given by

$$a(x) = A \cos \frac{5\pi}{h_t} x$$

in which A is the amplitude and h_t is the building height. Assuming Newton's second law ($F = ma$) holds, what is the amplitude, A, if the horizontal displacement at the top of the building is 0.0025 h_t? The building weighs 20 kips per foot of height.

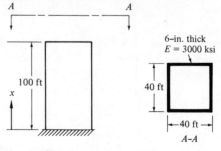

3-18. Calculate the slope and displacement at the free end of the given beam.

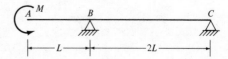

3-19. A simply supported beam is subjected to unequal end moments. Calculate the resulting slope at each end of the member. Convert the results to a pair of equations that give the end moments as functions of the end slopes. Assume EI is constant over the beam length.

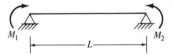

3-20. Repeat Problem 3-19 for the beam geometry of Problem 3-15.

3-21. Determine the equation of the displaced shape for each beam in Problem 3-4. Use double integration and binomial functions (Section 3.4.2.1). Reduce the result to normal algebraic form, by segments. Calculate the displacement at the free end.

beam subjected to uniform load over the entire span. Use the cross-section properties from Example 3-20. Do this for spans of:

(a) $L = 2$ ft
(b) $L = 8$ ft
(c) $L = 16$ ft

3-26. Repeat Problem 3-6 except include shear deformation. If the cross-sectional properties are as in Example 3.20, what percent of the displacement is due to shear for

(a) $L = 2$ ft
(b) $L = 8$ ft
(c) $L = 16$ ft

3-27. Use the principle of superposition, as shown in the accompanying figure. To calculate the reaction R_B.

Prob. 3-27

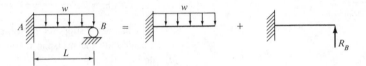

3-22. Repeat Problem 3-21 using quadruple integration and binomial functions (Section 3.4.2.2).

3-23. Determine the equation of the displaced shape for the beam in Problem 3-5 using double integration and binomial functions. Reduce the result to normal algebraic form, by segments.

3-24. Repeat Problem 3-23 using quadruple integration.

3-25. Compare the deflection due to flexure with the deflection due to shear in a simply supported

3-28. The building in Problem 3-17 is subjected to

$$a(x) = 17636 \cos \frac{5\pi}{100} x - 4000 \cos \frac{7\pi}{100} x$$

Calculate the displacement at the top of the building by superposing the two modes.

3-29. Determine the equation of the elastic curve, if the pinned end of the given beam settles downward by an amount Δ. Calculate the moment reaction at A.

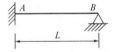

4
Moment-Area Method

4.1 INTRODUCTION

In Section 3.4, it was shown that the displaced shape of a beam can be determined by direct integration. Either the double or the quadruple integration method is used to accomplish the computation. These methods are suitable only if either the loading function $q(x)$ or the moment function $M(x)$ is not too complex, preferably describing the function over the entire beam length with a single expression. This is seldom the case, which makes either method unwieldy for practical usage. In addition, knowledge of the complete elastic curve is an uncommon need. The typical situation confronting the analyst is the desire to know displacements only at particular locations along the member. An alternative approach to mathematical integration, which satisfies this need, is discussed in the next section.

4.2 MOMENT-AREA METHOD

4.2.1 Derivation

The interrelationship of moment, stiffness EI, and curvature provides a means for deriving two conditions referred to as the *moment-area theorems*. Use of these theorems to calculate beam displacements is called the "moment-area method." Principles underlying the moment area method are developed in this section. A framed member is shown in Fig. 4.1a in its original straight-line position $A'B'$ and its curvilinear displaced and deformed position AB. The original position is intended to be general and represents a member located at some arbitrary inclination θ in space, where θ is any angle, large or small. Consider the problem of computing the difference in slope, t_{AB}, between points A and B. The angle t_{AB} is the angle between the tangent to A (line AE) and the tangent to B (line BF) and is commonly referred to as an "angle change." The desired angle change can be established by considering the effect of curvature in a small element of length dl located at a distance l from point B'. For this piece, a small difference in slope, dt, exists between the tangents to its endpoints. The relationship between dt and the moment $M(l)$ at the point can be quantified by consideration of Fig. 4.1b. In this figure, the differential element is expanded in scale. Due to the internal flexural moment, the surface 3-4 rotates relative to surface 1-2. Because plane sections are assumed to remain plane, the magnitude of this relative rotation is equal to the change in slope, dt. The angular movement causes the extreme fiber 1-3 to compress by an amount equal to l_{3-6}, as indicated by the broken line 3-6. Point 6 is located by extending a line from the centroid (point 5) in a direction parallel to surface 1-2. For small deformations, the lengths 5-6 and 5-3 can be considered equal without intro-

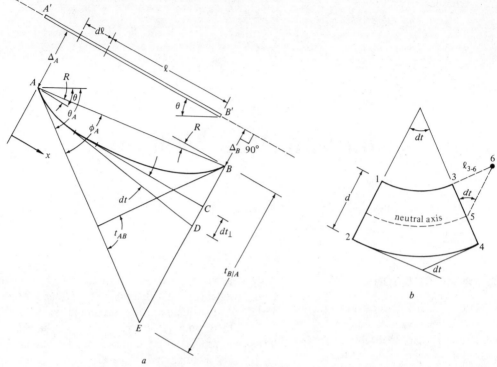

Fig. 4.1.   Geometry of a displaced member. *a*, member displaced in space; *b*, differential element.

ducing significant error and

$$l_{3-6} = dt(d/2) \qquad (4.1)$$

Compression of the outer fiber is a direct result of the flexural strain at the location caused by the flexural moment, and l_{3-6} can be expressed alternatively as

$$l_{3-6} = \epsilon \, dl \qquad (4.2)$$

where ϵ is the unit flexural strain at the outer fiber. Flexural strain, ϵ, is given by

$$\epsilon = \frac{M(l)\,d/2}{EI} \qquad (4.3)$$

which allows conversion of Eq. 4.2 to the form

$$l_{3-6} = \frac{M(l)\,d}{2EI}\,dl \qquad (4.4)$$

By equating the two expressions (4.1) and

(4.4), the desired relationship

$$dt = \frac{M(l)\,dl}{EI} \qquad (4.5)$$

is obtained. The total change in angle between points A and B in Fig. 4.1a is obtained by integrating Eq. 4.5 over the entire length of the member.

$$t_{AB} = \int_{0}^{L} \frac{M(l)}{EI}\,dl \qquad (4.6)$$

Mathematical integration is usually unnecessary because the result of the integration is recognized as the area under the M/EI diagram for the member. Under normal circumstances, this area can be computed by simple formulas because the M/EI diagram is composed of simple shapes. Equation 4.6 is the mathematical statement of the first moment-area theorem.

First moment-area theorem: The difference in slope between two points A and B that lie on a continuous, elastic curve is equal to the area under the M/EI diagram that exists between the two points.

The M/EI diagram has an additional role, which is useful in displacement computations. In Fig. 4.1a, it can be observed that distance $dt_\perp$ between points C and D is geometrically related to the angle change dt. Because dt and dl are considered to be small,

$$dt_\perp = l\, dt \qquad (4.7)$$

The quantity $dt_\perp$ represents a segment of the total distance, $t_{B/A}$, between points B and E and is measured perpendicular to the original member axis. By integration over the length of the member,

$$t_{B/A} = \int_0^L l\, dt = \int_0^L \frac{M(l)}{EI}\, l\, dl \qquad (4.8)$$

which is the desired quantity. As with the angular rotation, mathematical integration is normally unnecessary because the integral in Eq. 4.8 also has a physical meaning. This physical interpretation constitutes the second moment-area theorem.

Second moment-area theorem: The length of a line measured perpendicular to the original axis of a member, that extends from a point B on its continuous elastic curve to a tangent line extended from a point A on the same elastic curve is equal to the first moment of the M/EI diagram between points A and B taken about point B.

It is a convenient and common practice to refer to the quantity $t_{B/A}$ as a *tangential deviation*.

A key phrase in the moment-area theorems is "on a continuous elastic curve." An internal hinge in a member creates a slope discontinuity in the member, which was not considered in the derivation of the theorems. Note that the integration inherent in the method cannot be performed if a hinge is present between the two points under consideration.

4.2.2 Geometrical Concepts

Examination of Eq. 3.22 indicates that a moment diagram, when divided by EI, is an indicator of the displaced shape of a member. Indeed, Eq. 3.19 states that the M/EI diagram is a plot of the curvature, $1/\rho$, as a function of position along the member. These relationships are the basis for establishing a well-known procedure, which is based upon *geometrical* concepts, for determining the elastic curve of a framed member. This analytical process, the *moment-area method*, relies upon two fundamental theorems regarding the geometry of a member that has displaced into a curved position. The two theorems will be developed after a general discussion of flexural displacements. It is important to recall that Eqs. 3.19 and 3.22 infer an assumption of small deflections. Consequently, the moment-area method is a first-order analysis technique.

The direct meaning of Eq. 3.19 was depicted in Fig. 3.20. In essence, moment causes curvature. Positive moment causes positive curvature (concave upward), and negative moment causes negative curvature (concave downward) in a member. This simple statement permits the analyst to approximate the curved shape of a member from examination of its M/EI diagram. An ability to discern the general shape of a displaced beam is a distinct advantage in the moment-area method. As an introduction to the application of the relationships given in Eqs. 3.19 and 3.22, consider the beam in Fig. 4.2a. For the loading depicted, the M/EI diagram is likely to have a shape similar to that shown in Fig. 4.2b. Numerical values for M_1 and M_2 would depend upon the magnitude of loads P_1 and P_2. For presentation purposes, it is assumed that P_1 is not large enough to cause M_2 to become negative. The point of zero moment in segment BD is labeled F in Fig. 4.2b. As indicated below the M/EI diagram, the negative M/EI region to the left of point F causes segment AF to assume a negative curvature (downward concavity). Conversely, segment FD is subjected to positive M/EI, and its deformation produces positive curvature (upward concavity).

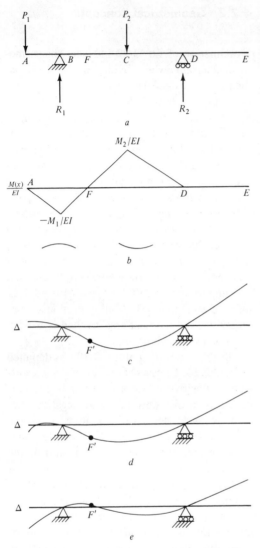

Fig. 4.2.   Example of flexural displacements in a beam.

duced by recognizing both the sense of the curvatures and the need to maintain contact with the supports. Several comments in regard to this shape follow.

Point F is the location of a change in the sense of the curvature (a point of zero moment) and is identified as such by the use of a solid dot. A location corresponding to a change in curvature is normally called an *inflection point*, and several may exist in any given beam. It is also significant to realize that no moment exists in the right overhang of the structure, and consequently the curvature is zero. The inference is that segment DE undergoes a rotation at D and continues from that point as a straight line to the right and upward.

The upward movement of point A is a presumption, and the displaced shape drawn in Fig. 4.2d is an alternative possibility. The true position can only be established through computation. Finally, the nature of the M/EI diagram suggests that the shape depicted in Fig. 4.2e is also possible. In this displaced shape, point F' is assumed to be above the original axis of the member. The reader should realize that if point F moves upward, then point A must move downward. The fact that upward motion of point A is inadmissible in Fig. 4.2e is inferred by the negative curvature that exists in segment AB and the need to remain attached to the support at point B.

As a more general comment regarding the displaced shape of a beam, it can be stated that if the M/EI diagram and the support conditions are given, the probable deformed shapes can be drawn with no knowledge of the loading itself. In fact, it is advisable to ignore the loads entirely when drawing the displaced shape. An example of the motivation for this suggestion is the fact that in Fig. 4.2a point A can move *upward* even though the load P_1 strongly suggests a *downward* motion. The point is that curvature, not loading, suggests the displaced shape! Finally, because several possible shapes can exist in a given situation, the analyst is forced to assume one prior to initiating computations. Subsequent computations may sometimes indicate errors in the assumption, and the chosen trial shape can be modified as the solution process progresses.

After recognition of these curvature conditions, a reasonable representation of the displaced beam configuration can be sketched by considering the compatibility conditions imposed by the supports. An exact displaced shape for the beam in Fig. 4.2a can be established only by numerical computation. However, depicting a possible shape is usually a vital part of the analytical solution. Several possible configurations exist for the beam under discussion. Figure 4.2c illustrates one of the possibilities. This displaced shape is pro-

4.3 REFERENCE TANGENTS

4.3.1 Definition

Concepts illustrated in Fig. 4.2 are important to proper implementation of the moment-area theorems. Combined with a known moment diagram and an ability to establish an appropriate *reference tangent*, the rotation and displacement can be calculated at any desired location in a given beam. A reference tangent is a tangent to the displaced curve of a beam at a location where the rotation and deflection of the beam is known. The succeeding paragraphs detail the determination of a proper reference tangent and its use in the moment-area method.

4.3.2 Cantilever Beams

The cantilever beam is the most direct example for initially describing a reference tangent. Fig. 4.3a shows a cantilever beam subject to a con-

centrated load at the free end. Because the rotation θ_A and vertical deflection Δ_A are zero by definition of a fixed end, point A is a proper location for a reference tangent. Being tangent to A, the reference tangent is the horizontal line AB shown in Fig. 4.3b. In this case the reference tangent coincides with the original position of the member. The displaced shape of the member is a deviation from the reference tangent. The deviation is due to the curvature created by the $M(x)/EI$ variation shown in Fig. 4.3c. Because the moment is negative, causing compression in the bottom fibers of the member over the entire length of the beam, the entire deflected shape is concave downward. Furthermore, since the reference tangent is horizontal, all points on the beam deflect downward and rotate clockwise relative to it.

The clockwise rotation θ_x and downward displacement Δ_x at any location x are established from the moment area theorems. In general,

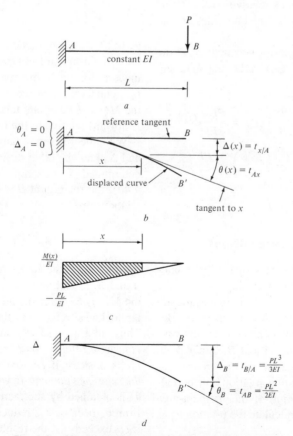

Fig. 4.3. Moment-area concepts for a cantilever beam.

$$\theta(x) = \theta_A + t_{Ax} \qquad (4.9)$$

$$\Delta(x) = \Delta_A + t_{x/A} \qquad (4.10)$$

but the fixed end reduces these to

$$\theta(x) = t_{Ax} \qquad (4.11)$$

$$\Delta(x) = t_{x/A} \qquad (4.12)$$

The quantity t_{Ax} is the angle change between points A and x, and, quantitatively, equals the shaded area shown in Fig. 4.3c. The quantity $t_{x/A}$ is the tangential deviation of point x with respect to point A. Quantitatively, $t_{x/A}$ equals the first moment of the shaded area in Fig. 4.3c taken about point x. In other words, $t_{x/A}$ is the first moment of t_{Ax} taken about point x.

The various terms in relationships 4.9 to 4.12 have an implied sign convention, i.e., they are consistent with Fig. 4.3b. Any quantity substituted into the equations is assigned a positive value if it is either a clockwise rotation or a downward displacement relative to the reference tangent. Otherwise, the quantities are assigned negative values.

If the rotation θ_B and displacement Δ_B at the free end are desired, Eqs. 4.11 and 4.12 become

$$\theta_B = t_{AB} = \frac{1}{2}\left(\frac{PL}{EI}\right)(L) = \frac{PL^2}{2EI} \text{ cw}$$

$$\Delta_B = t_{B/A} = \left(\frac{PL^2}{2EI}\right)\left(\frac{2L}{3}\right) = \frac{PL^3}{3EI} \downarrow$$

as indicated in Fig. 4.3d.

4.3.3 Simply-Supported Beams

A simply supported beam subjected to a concentrated load is shown in Fig. 4.4a. The $M(x)/EI$ diagram, Fig. 4.4b, indicates the deflected beam will have upward concavity. Because the displacement is known (in this case equal to zero) at points A and B, either point could serve as the location of a reference tangent if the corresponding rotation can be calculated. Choosing point A, the moment-area concepts are used to calculate the rotation θ_A as follows.

In Fig. 4.4c, a reference tangent is shown at A and is extended to B' (located directly below B). In the first-order structural analysis, point B' would be the displaced position of the support at B, if the rigid body rotation θ_A was imposed on the member. Thus, the reference tangent from a point on the member constitutes the position of the member due to a rigid body rotation about that point. Since the support at B does not displace under the applied loading, the flexure of the member must be such that the curved member deviates upward from the reference tangent, by an amount $t_{B/A}$, such that B has a zero displacement.

From the triangle ABB' it is evident that

$$\tan \theta_A = \frac{BB'}{L} = \frac{t_{B/A}}{L} \qquad (4.13)$$

or, for small displacements,

$$\theta_A = \frac{t_{B/A}}{L} \qquad (4.14)$$

θ_A has been shown clockwise, consistent with the known upward curvature. Positive $t_{B/A}$ is upward for the same reason. Quantitatively, $t_{B/A}$ equals the first moment of the area under the $M(x)/EI$ diagram taken about point B and is upward relative to the reference tangent.

With the reference tangent established (θ_A and Δ_A known) other desired quantities can be determined. For example (as depicted in Fig. 4.4d), the rotation at B is established from the relationship

$$\theta_B = \theta_A - t_{AB} \qquad (4.15)$$

in which t_{AB} is the angle change between points A and B. Quantitatively, t_{AB} is the area under the $M(x)/EI$ diagram and is, obviously, counterclockwise relative to the reference tangent. Thus, in Eq. 4.15, θ_A and θ_B are positive if *clockwise relative to the reference tangent* and t_{AB} is positive if *counterclockwise relative to the reference tangent*. In words, Eq. 4.15 states θ_B is obtained by subtracting the angle change from θ_A because t_{AB} reduces the clockwise θ_A. If t_{AB} exceeds θ_A, the result is a negative θ_B and

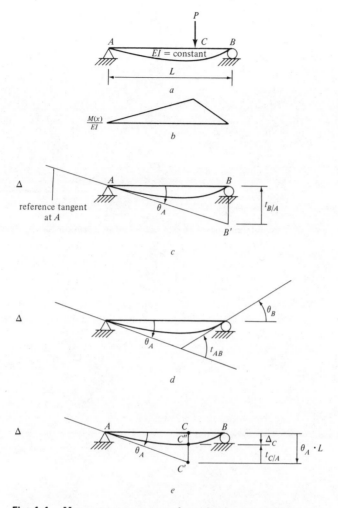

Fig. 4.4. Moment-area concepts for a simply supported beam.

indicates θ_B is counterclockwise. Fig. 4.4d infers this is the expected outcome.

Similarly (from Fig. 4.4e), the *downward* displacement of point C is given by

$$\Delta_C = \theta_A \cdot L - t_{C/A} \qquad (4.16)$$

where positive θ_A is *clockwise relative to the horizontal* and $t_{C/A}$ is positive if *upward relative to the reference tangent*. Although not shown in Fig. 4.4e, the clockwise rotation θ_C can be obtained in a manner similar to θ_B, namely

$$\theta_C = \theta_A - t_{AC} \qquad (4.17)$$

where positive t_{AC} is *counterclockwise relative to the reference tangent*. In the preceding, t_{AC} is the area under the $M(x)/EI$ diagram between A and C and $t_{C/A}$ is the first moment of t_{AC} about C.

Example 4.1. Determine the slope and vertical displacement at point B and the slope at pont C of the beam shown in Fig. 4.5. Use the moment-area theorems.

Solution: The moment diagram and the M/EI diagram are constructed in Fig. 4.5b and c, respectively. Using the schematics of the displaced shape given in Fig. 4.5d and e, the moment-area theorems can be employed as follows

$$\theta_A = \frac{t_{C/A}}{24}$$

$$\theta_B = \theta_A - t_{AB}$$

$$\theta_C = \theta_A - t_{AC}$$

$$\Delta_B = 12\theta_A - t_{B/A}$$

Using the moment diagram, the numerical values are computed as follows:

t_{AB} = area under the M/EI diagram between points A and B

$$= \frac{1}{2} \times \frac{30}{EI} \times 12 = \frac{180}{EI} \text{ ccw} = \frac{180}{EI}$$

$t_{B/A}$ = first moment of t_{AB} taken about point B

$$= \frac{180}{EI} \times 4 = \frac{720}{EI} \uparrow = \frac{720}{EI}$$

t_{AC} = area under the M/EI diagram between points A and C

$$= \frac{180}{EI} + \frac{1}{2} \times \frac{60}{EI} \times 12 = \frac{540}{EI} \text{ ccw} = \frac{540}{EI}$$

$t_{C/A}$ = first moment of t_{AC} taken about point C

$$= \frac{180}{EI} \times 16 + \frac{360}{EI} \times 8 = \frac{5760}{EI} \uparrow$$

Thus,

$$\theta_A = \frac{5760/EI}{24} = \frac{240}{EI} \text{ cw}$$

$$\theta_B = \frac{240}{EI} - \frac{180}{EI} = \frac{60}{EI} \text{ cw}$$

$$\theta_C = \frac{240}{EI} - \frac{540}{EI} = -\frac{300}{EI} = \frac{300}{EI} \text{ ccw}$$

$$\Delta_B = 12 \times \frac{240}{EI} - \frac{720}{EI} = \frac{2160}{EI} \downarrow$$

Example 4.2. Determine the maximum displacement and its location for the beam of Fig. 4.5a.

Solution: The maximum displacment occurs at the location of a zero value of slope. Moment-area principles can be applied to Fig. 4.5f and g to locate the point and compute Δ_{max}. Noting from Fig. 4.5c that the moment at a distance x from point C is $5x$.

$$t_{CX} = \frac{1}{2}\left(\frac{5x}{EI}\right)(x) = \frac{2.5x^2}{EI} \text{ ccw} = \frac{2.5x^2}{EI}$$

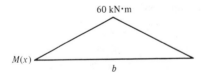

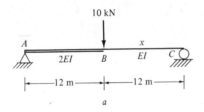

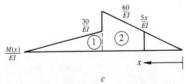

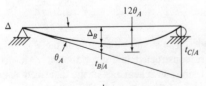

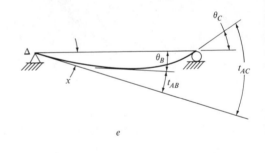

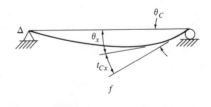

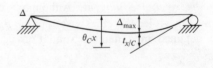

Fig. 4.5.   Displacements by the moment-area method—Examples 4.1, 4.2.

$t_{X/C}$ = first moment of t_{CX} about point X

$$= \frac{2.5x^2}{EI}\left(\frac{x}{3}\right) = \frac{2.5x^3}{3EI} \uparrow = \frac{2.5x^3}{3EI}$$

$$\theta_X = \theta_C + t_{CX} = 0$$

$$= \frac{300}{EI} - \frac{2.5x^2}{EI} = 0$$

$$\therefore x = 10.95 \text{ m} < 12 \text{ m}$$

$$\therefore t_{X/C} = \frac{2.5(10.95)^3}{3EI} = \frac{1094}{EI}$$

$$\Delta_{\max} = \theta_C x - t_{X/C}$$

$$= \frac{300}{EI}(10.95) - \frac{1094}{EI}$$

$$= \frac{2191}{EI}$$

$$\therefore \Delta_{\max} = \frac{2191}{EI} \downarrow$$

4.3.4 Overhangs

A simply supported beam with an overhang (or overhangs) is basically treated by an extension of the concepts presented in the preceding section. An example problem will be used to demonstrate that extension.

Example 4.3. Calculate the rotation and displacement of point C in the given beam (Fig. 4.6a.) Assume EI is constant from A to C.

Solution: The $M(x)/EI$ is easily established and is shown in Fig. 4.6b without supporting computations. The negative magnitude throughout indicates downward concavity and the displaced shape shown in Fig. 4.6a is obvious. A reference tangent at point A is selected and the rotation θ_A is calculated as follows.

$$t_{B/A} = \frac{1}{2}\left(\frac{100}{EI}\right)(25)\left(\frac{2}{3} \times 25\right) = \frac{31250}{3EI} \uparrow$$

$$\theta_A = \frac{t_{B/A}}{25} = \frac{1250}{3EI} = \frac{416.67}{EI} \text{ ccw}$$

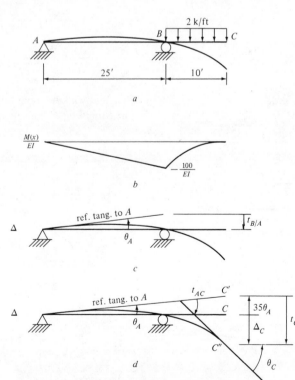

Fig. 4.6. Beam for Example 4.3.

In Fig. 4.6d, the reference tangent is extended to C', directly above C and C''. The rotation and displacement at C are established as follows:

$$\theta_C = \theta_A - t_{AC}$$

It is presumed (properly) that positive θ_A and θ_C are counterclockwise relative to the horizontal and positive t_{AC} is clockwise relative to the reference tangent.

$$\Delta_C = t_{C/A} - 35\theta_A$$

Positive Δ_C is downward relative to the horizontal (original position of the member) and $t_{C/A}$ is downward relative to the reference tangent.

Using Appendix A, Table A2.

$$t_{AC} = \frac{1}{2}\left(\frac{100}{EI}\right)(25) + \frac{1}{3}\left(\frac{100}{EI}\right)(10)$$

$$= \frac{1250}{EI} + \frac{1000}{3EI} = \frac{4750}{3EI} \text{ cw}$$

$$t_{C/A} = \frac{1250}{EI}\left(\frac{25}{3} + 10\right) + \frac{1000}{3EI}\left(\frac{3}{4}\right)(10)$$

$$= \frac{76250}{3EI}$$

$$\theta_C = \frac{1250}{3EI} - \frac{4750}{3EI} = -\frac{3500}{3EI}$$

$$= \frac{3500}{3EI} = \frac{1166.67}{EI} \text{ cw}$$

$$\Delta_C = \frac{76250}{3EI} - 35\left(\frac{1250}{3EI}\right) = \frac{32500}{3EI} \downarrow$$

$$= \frac{32500}{3EI} \downarrow = \frac{10833.33}{EI} \downarrow$$

Example 4.4. Calculate Δ_C for the beam in Example 4.3, but use a reference tangent at B.

Solution: From Fig. 4.7

$$\theta_B = \theta_A - t_{AB}$$

where positive θ_A and θ_B are counterclockwise relative to the horizontal and positive t_{AB} is clockwise relative to the reference tangent. Thus

$$t_{AB} = \frac{1}{2}\left(\frac{100}{EI}\right)(25) = \frac{1250}{EI} = \frac{1250}{EI} \text{ cw}$$

$$\theta_B = \frac{1250}{3EI} - \frac{1250}{EI} = -\frac{2500}{3EI} = \frac{833.33}{EI} \text{ ccw}$$

At the overhang

$$\Delta_C = 10\theta_B + t_{C/B}$$

where positive Δ_C is downward relative to the horizontal (original position of the member) and positive $t_{C/B}$ is downward relative to the reference tangent:

$$t_{C/B} = \frac{1}{3}\left(\frac{100}{EI}\right)(10)\left(\frac{3}{4}\right)(10) = \frac{2500}{EI} \downarrow$$

$$\Delta_C = 10\left(\frac{2500}{3EI}\right) + \frac{7500}{3EI} = \frac{32500}{3EI}$$

$$= \frac{32500}{3EI} = \frac{10833.33}{EI} \downarrow$$

which agrees with the result of Example 4.7.

Note, from Fig. 4.7, an alternative calculation of θ_B is

$$\theta_B = \frac{t_{A/B}}{25} = \frac{\dfrac{1}{2}\left(\dfrac{100}{EI}\right)(25)\left(\dfrac{50}{3}\right)}{25}$$

$$= \frac{2500}{3EI} \text{ ccw}$$

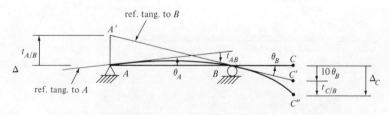

Fig. 4.7. Alternative relationships for the beam in Example 4.3.

The reader should deliberate about the sign conventions inferred in this alternative computation.

4.4 ESTABLISHING A SIGN CONVENTION

4.4.1 Rotation

An all-encompassing sign convention for the moment area method is awkward to develop and enforce. Instead, a convention is established on a common sense, albeit problem-dependent, basis, i.e., the sign convention used is deduced from the form of the feasible displaced shape used in any particular problem. Many of the common possibilities are discussed in the following paragraphs.

If the reference tangent has been established at a location A with a known rotation θ_A, then the rotation θ_B at another location is always given by

$$\theta_B = \theta_A \pm t_{AB}$$

The rotation θ_A and θ_B are always measured relative to the horizontal axis of the member in its original position. The positive sense of the quantities depends on the direction θ_A. The known direction of θ_A for any given situation is always taken as the positive sense for other rotations. The sign preceding t_{AB} depends on whether the angle change is either additive (in the same direction) or subtractive (in the opposite direction) to θ_A.

Figure 4.8 illustrates six general possibilities. In Fig. 4.8a, b, and c, a segment of a deflected member is shown. In each case the ro-

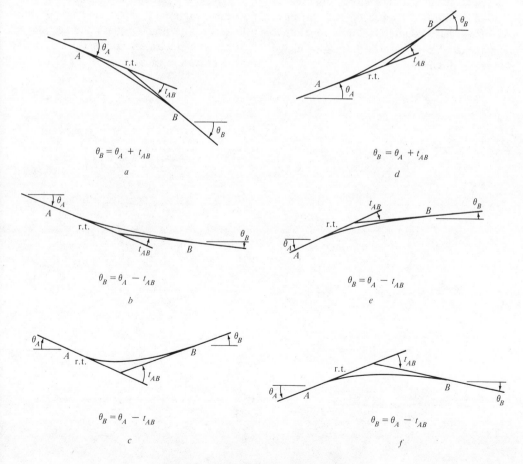

Fig. 4.8. Moment-area rotation concepts.

tation θ_A of the reference tangent (r.t.) is clockwise. In case a, the clockwise angle change t_{AB} is additive to θ_A, producing

$$\theta_B = \theta_A + t_{AB} \qquad (4.19)$$

as shown. In case b, t_{AB} is counterclockwise and thus subtractive to θ_A, producing

$$\theta_B = \theta_A - t_{AB} \qquad (4.20)$$

as shown. Figure 4.8b indicates that the calculated θ_B will be a positive value (clockwise.) Case c in Fig. 4.8c is similar to case b. Therefore, θ_B is also given by Eq. 4.20. However, Fig. 4.8c indicates the calculated θ_B will be a negative value (counterclockwise.)

In Figs. 4.8d, e, and f, a segment of a beam is shown with a counterclockwise rotation θ_A for a reference tangent. Therefore, any rotation equation established by use of this reference tangent will have counterclockwise rotation θ, taken as positive. In case d, the angle change is additive, producing

$$\theta_B = \theta_A + t_{AB} \qquad (4.21)$$

which appears identical to Eq. 4.19, but is not. The difference is the sign convention for θ_A and θ_B. In cases e and f, the relationship is

$$\theta_B = \theta_A - t_{AB} \qquad (4.22)$$

which appears identical to Eq. 4.20, but is not. Again, the difference is due to the inherent sign convention for θ_A and θ_B. Also note, for case e (case f), Eq. 4.22 would produce a positive (negative) value for θ_B.

4.4.2   Displacement

The situation of an overhanging portion of a beam is useful to illustrate how to establish a sign convention for use in displacement computations. Figure 4.9 shows six cases of interest. In cases a, b, and c, the reference tangent at B has a clockwise orientation θ_B. Depending on the curvature of segment BC and the dis-

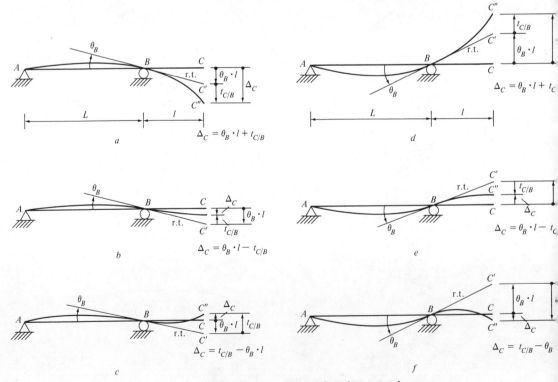

Fig. 4.9.   Moment-area concepts for a beam overhang.

placed position of point C, completely different relationships evolve for computing Δ_C.

In case a, Δ_C is obtained from

$$\Delta_C = \theta_B \cdot l + t_{C/B} \qquad (4.23)$$

with the arrowheads in Fig. 4.9a serving to illustrate the positive sense of each quantity. In this case, the quantities Δ_C, $\theta_B \cdot l$, and $t_{C/B}$ are positive if downward. Note that the first two quantities are measured relative to the horizontal and always are. The tangential deviation $t_{C/B}$ is always measured relative to the extended reference tangent.

In case b, Δ_C is obtained from

$$\Delta_C = \theta_B \cdot l - t_{C/B} \qquad (4.24)$$

In contrast to case a, $t_{C/B}$ is positive if upward because the curvature of BC in case b is opposite to that in case a. If the curvature is sharp enough, point C could deflect to a position above the horizontal. This condition is shown in case c. In this case

$$\Delta_C = t_{C/B} - \theta_B \cdot l \qquad (4.25)$$

where Δ_C and $t_{C/B}$ are positive if upward and $\theta_B \cdot l$ is positive if downward.

Figures 4.9d, e, and f are similar to the three cases just described. However, the rotation θ_B of the reference tangent is assumed to be clockwise. Again, depending on the curvature of segment BC and displaced position of point C, completely different relationships evolve for computing Δ_C. The results for each case are shown in the corresponding figures. The reader is urged to deliberate on these and confirm the implied sign conventions as a special exercise.

Figures 4.8 and 4.9 served to illustrate the development of relationships for computing rotation and displacement by the moment-area method. An overhang was used as an illustration, but the concepts presented are general. Figures 4.10a, b, and c show the application of the concepts to a simply supported portion of a beam. If the dashed portion of the displaced shape is ignored and a reference tangent at B is used, it is clear the segment BC will always fit one of the situations described for an overhang. The reader should confirm the correctness of the different relationships for computing Δ_C shown in the various cases in Fig. 4.10.

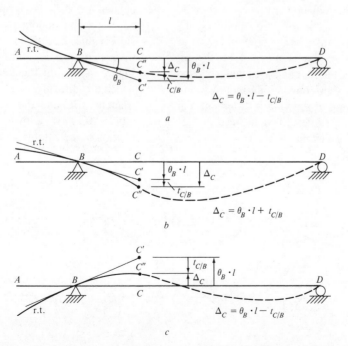

Fig. 4.10. Moment-area concepts for simply supported beam segments.

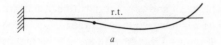

Fig. 4.11. Examples of reverse curvature.

When the sense of the curvature is constant (either upward or downward) for segment BC in Figs. 4.9 and 4.10, the development of relationships for computing rotation and displacement is usually straightforward. The sign conventions also are usually readily apparent, if the deflected shape is properly labeled. In executing the computations a proper labeling of the displaced shape will usually produce a situation in which the sign of all substituted numerical values is positive.

An inflection point is a location along the elastic curve where the sense of the curvature reverses (either from upward to downward or vice-versa.) When an inflection point occurs between the ends of a beam segment, such as those shown by solid dots in Fig. 4.11, the implementation of the moment-area concepts and designation of sign conventions requires caution. The concept of drawing a "moment diagram by parts," as described in Section 4.5, considerably reduces the difficulty.

4.5 MOMENT DIAGRAMS BY PARTS

4.5.1 Complex M/EI Diagrams

Employment of the moment-area method explicitly involves the computation of areas under the M/EI diagram and their corresponding first moments about some selected point along the span. In some instances, establishing sign convention and the execution of the associated numerical work is complicated by the nature of the given moment diagram. Some examples of segments of moment diagrams that cause difficulty are illustrated in Fig. 4.12. Some computational problems associated with each of the diagrams can be enumerated as follows:

- *Figure 4.12.* Initially, the M/EI diagram must be divided into shapes suitable for computation. Six individual areas must be computed and their centroids located before proceeding with further analysis. In addition, the distances x_1 and x_2 to the locations of a zero moment value must be determined in the process.

- *Figure 4.12.* In addition to a need to locate the points of zero moment, the two cross-hatched areas cannot be directly calculated by a single formula because the apex of each parabola does not have a zero slope value. Location of the centroids for these two areas is a further complication. Figure 4.13 illustrates a procedure by which this difficulty can be overcome. The triangle ABC and the portion of the parabola AC have known properties. Subtraction of the latter from the former would

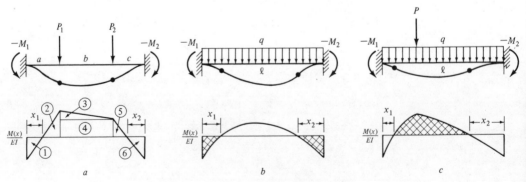

Fig. 4.12. Example moment diagrams.

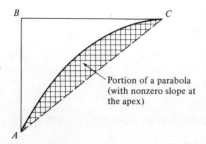

Fig. 4.13.   Subdivision of an *M/EI* area.

yield the desired area. Although valid, this procedure is time-consuming and doubles the number of *M/EI* areas to be computed.

• *Figure 4.12.* All the difficulties illustrated in Fig. 4.12b are also present in Fig. 4.12c. As a further difficulty, the center cross-hatched area does not have a zero slope at its apex. Evaluation of its area in the manner illustrated in Fig. 4.14 is possible but is a tedious undertaking.

Situations depicted in Fig. 4.12 are common occurrences. From these cases, it is evident that undesirable computational difficulties will be a frequent aspect of the moment-area method. By introducing the concept of a *moment diagram by parts*, it is possible to avoid the complications inherent in *actual* moment diagrams and simplify the numerical work.

An actual moment diagram is the diagram that depicts the true variation in moment in the beam segments and gives a direct indication of the member's curvature. The displaced shapes in Fig. 4.12 are deduced because positive (negative) moment corresponds to upward (downward) curvature. Because two curvatures exist a decision must be made as to which would be

considered positive in establishing sign conventions in the moment-area method. This is discussed subsequently.

A moment diagram by parts is a modified version of the actual moment diagram, which facilitates the moment-area computations. Because it is drawn differently, the moment diagram by parts does not resemble the actual moment diagram. Indeed, while the actual moment diagram is unique, the moment diagram by parts can be drawn in many different ways. Although it is useful in the computation of specific displacements, the moment diagram by parts is of no assistance in interpreting the general characteristics of the displaced shape of the member.

Complexity in a moment diagram is due to the complexity of the loading itself. For single loads, whether concentrated or uniform in type, the moment diagram is relatively simple in form. Construction of a moment diagram by parts is based upon a recognition of this fact. Basically, the phrase, "by parts," implies that the moment diagram can be drawn in a disassembled fashion. Commonly, this is done in one of two ways.

4.5.2   Method 1: Cantilever Method (CM)

A beam can be visualized as a body that is in equilibrium in space. By statics, moment equilibrium must exist at any point along the span. Viewed in this manner, each force (whether a load or a reaction) contributes to the moment, which is created internally at that point. On this basis, a procedure for drawing the moment diagram by parts can be stated.

1. Select an appropriate point in the span as a reference point.
2. Draw a separate moment diagram for each individual force acting on the beam as if it were a force cantilevered from the chosen reference point. When the reference point is in the interior of the beam, the resulting diagram is two-sided, i.e., it consists of separate diagrams to the left and to the right of the reference point.

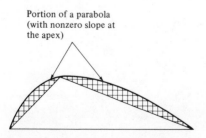

Fig. 4.14.   Subdivision of an *M/EI* area.

The CM is best introduced by use of example problems.

Example 4.5. For the beam given in Fig. 4.15, draw the following diagrams:

a. Shear and actual moment.
b. Moment diagram by parts using the CM and point A as a reference point.
c. Moment diagram by parts using the CM and point B as a reference point.

Solution:

Part a: Support reactions are given in Fig. 4.15a and can be verified by statics. Shear and moment diagrams are determined by conventional numerical integrations and are drawn in Figs. 4.15b and c. Because the location of zero moment is tedious to locate, the actual moment diagram is not readily usable in the calculations of the moment-area method.

Part b: Proceeding across the span from left to right, each of the forces will be treated separately. For each force, the moment value created internally at point A will be calculated and the shape of the diagram noted. Each load is treated as though it was cantilevered from point A.

Force	M_A	Shape
6.72 kN↑	0	None
1.6 kN/m ↓	$-1.6(14)(7)$ $= -156.8$ kN · m	Parabolic, downward curvature
15.68 kN ↑	$15.68(10)$ $= 156.8$ kN · m	Linear

The corresponding moment diagram by parts is given in Fig. 4.15d. If the parts were combined, the actual moment diagram (Fig. 5-13c) would result.

Part c: The cantilever value at point B and shape of curve created by each force are tabulated as follows.

Force	M_B	Shape
6.72 kN ↑	$6.27(10)$ $= 67.2$ kN · m	Linear
1.6 kN/m ↓	$-1.6(10)(5)$ $= -80$ kN · m	Parabolic, downward curvature
15.68 kN ↑	0	None
1.6 kN/m ↓	$-1.6(4)(2)$ $= -12.8$ kN · m	Parabolic, downward curvature

In treating the uniform load, the portions to the left of and to the right of the reference point must be treated individually. As a check on the solution, it should be observed that the moments at B caused by forces to the left of this reference point sum to

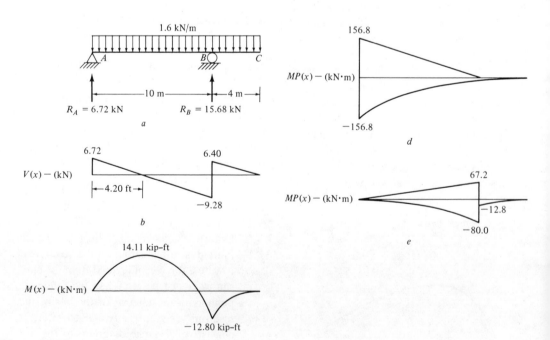

Fig. 4.15.   Moment diagrams by parts—Examples 4.5 and 4.13.

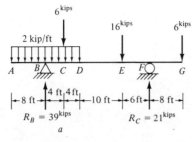

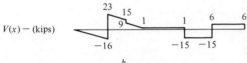

a

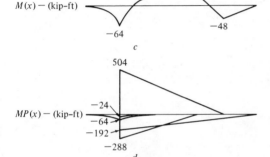

$V(x) - (\text{kips})$

b

$M(x) - (\text{kip-ft})$

c

$MP(x) - (\text{kip-ft})$

d

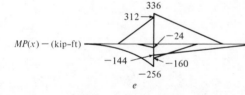

$MP(x) - (\text{kip-ft})$

e

Fig. 4.16. Moment diagrams by parts—Examples 4.5 and 4.13.

−12.80 kN-m as do the moments caused by forces to the right of the reference point (in this example, only one such force exists). Figure 4.15 illustrates the moment diagram by parts corresponding to the computed values.

Example 4.6. A simply supported beam with two overhanging ends is subjected to the loads shown in Fig. 4.16a. Draw each of the following diagrams:

a. Shear and moment diagrams.
b. Moment diagram by parts, using the CM and point *B* as a reference point.

c. Moment diagram by parts, using the CM and point *D* as a reference point.

Solution:
Part a: Reaction forces are calculated by application of statics to the free-body diagram of the beam as a whole. Shear and moment diagrams are given in Fig. 4.16b and c.

Part b: Proceeding across the span from left to right:

Force	M_B	Shape
2.0 kip/ft ↓	$-2(8)(4)$ $= -64$ kip-ft	Parabolic, downward curvature
39 kips ↑	0	None
2.0 kip/ft ↓	$-2(8)(4)$ $= -64$ kip-ft	Parabolic, downward curvature
6 kips ↓	$-6(4) = -24$ kip-ft	Linear
16 kips ↓	$-16(18) = -288$ kip-ft	Linear
21 kips ↑	$21(24) = 504$ kip-ft	Linear
6 kips ↓	$6(32) = -192$ kip-ft	Linear

The MP diagram is shown in Fig. 4.16d.
Part c: Proceeding across the span from left to right:

Force	M_D	Shape
2.0 kip/ft ↓	$-2(16)(8)$ $= -256$ kip-ft	Parabolic, concave downward
39 kips ↑	$39(8) = 312$ kip-ft	Linear
6 kips ↓	$-6(4) = -24$ kip-ft	Linear
10 kips ↓	$-16(10) = -160$ kip-ft	Linear
21 kips ↑	$21(16) = 336$ kip-ft	Linear
6 kips ↓	$-6(24) = -144$ kip-ft	Linear

The MP diagram is shown in Fig. 4.16e.

Example 4.7. Drawn an MP/EI diagram for the beam given in Fig. 4.17a using the CM.

Solution: The natural fixed end will be taken as the reference point. Each load is taken separately as follows

Force	M_A	Shape
60 kN/m	$-60(7)(3.5) = -1470$ kN · m	Parabolic, downward curvature
50 kN/m	$-50(3) = -150$ kN · m	Linear

The results are shown in Fig. 4.17b. Division by the *EI* values produces the diagram shown in Fig. 4.17c. The moment at *C* is a prerequisite to drawing the MP/EI diagram and is given by

$$M_C = -60(3)(1.5) = -270 \text{ kN · m}$$

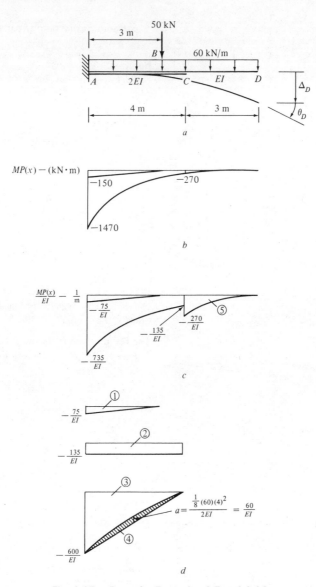

Fig. 4.17.   Beam for Examples 4.7 and 4.14.

4.5.3   Method 2: Simple Beam Method (SBM)

An alternative procedure for constructing a moment diagram by parts is the so-called "simple beam method" (SBM). Development of the method begins with an understanding of the internal equilibrium of an isolated segment of a beam and its identity with an appropriately loaded simply supported beam. Figure 4.18a depicts an arbitrary beam with a general load-

ing. The nature of the beam geometry (spans, supports, etc.) is irrelevant and the particular beam shown has no bearing on the development. Qualitative shear and moment diagrams for the beam are also shown (Figs. 4.18b and c.)

Consider the equilibrium of a segment between locations A and B. Extracting values from the $V(x)$ and $M(x)$ diagrams, a free-body diagram of the loaded segment is as shown in Fig. 4.18d. As statics must be satisfied, the in-

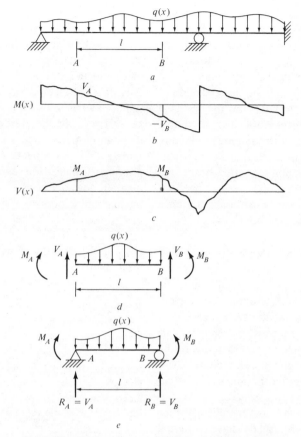

Fig. 4.18.   Basic concept of the simple beam method.

ternal forces V_A, V_B, M_A and M_B maintain the equilibrium of the structure. In particular, the summation of moments about each end must be zero in each case.

Now consider the simply supported beam shown in Fig. 4.18e. It is loaded with the same $q(x)$ as the beam segment just considered. In addition, applied end moments M_A and M_B have the same magnitude and direction as the internal moments of the segment AB. Application

of statics (by summing moments about each end) will confirm the reactions of the simply supported beam must be equal in magnitude to the internal shears V_A and V_B. Consequently, the shear and moment diagrams for the beam in Fig. 4.18e must be identical to those of the beam segment in Fig. 4.18d.

Next, consider the superposition of the effects shown in Fig. 4.19, wherein the loading in Fig. 4.18e has been divided into parts. The

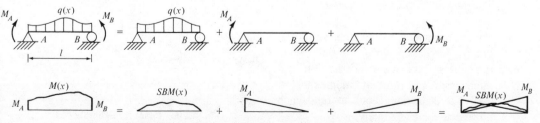

Fig. 4.19.   Moment diagram by parts by superposition of simple beam loadings.

first part is the "simple beam loading," wherein the transverse loading $q(x)$ has been isolated. The second and third parts isolate the applied end moments M_A and M_B, respectively. This subdivision permits the replacement of the actual moment diagram by the superposition of moment diagrams indicated below the beams in Fig. 4.19. When collected together, as shown at the far right, the result is a moment diagram by parts. As each part is obtained from a particular load on a simply supported beam, the method is termed the simple beam method. Note that if several transverse loads act on a beam segment, a separate $SBM(x)$ diagram can be drawn for each load.

Table A.2 in Appendix A gives moment-diagram shapes and maximum values for common simple-beam loadings and is a useful aid in the superposition method of separating the moment diagram into parts. A subtle advantage of SBM diagrams is that the points of zero moment values need not be located as they will always be at the ends of the segment. This eliminates a troublesome aspect often present in the actual moment diagram. When a separate SBM diagram is drawn for each transverse load, the resulting diagrams will usually have common shapes for which the area and centroid location are easily determined. In light of these observations, a second procedure for drawing a moment diagram by parts can be enumerated.

1. Visualize the beam to be comprised of individual segments. Normally, a segment is delineated as a piece of member that lies between two supports (including internal hinges) or is an overhanging element.
2. Draw the actual moment diagram for the entire beam.
3. For each designated segment, extract the end moment values from the actual moment diagram.
4. For each segment, draw the SBM diagrams caused by each of the three separate effects: transverse loads, left end momentum, and right end moment. Normally, it is advantageous to treat each transverse load separately.

The distinction between the actual *moment diagram* (*M* diagram) and the *moment diagram by parts* (*MP* diagram) must be reemphasized. The actual moment diagram is a plot of the actual moment function $M(x)$. As such, it is the only pictorial method for determining the maximum moment within the span or assessing the general form of the elastic curve. A moment diagram by parts divides the moment diagram into several components for purposes of simplifying displacement computations. Such a diagram does not permit one to directly locate the maximum moment or indicate the curvature of the member. Furthermore, such diagrams can be drawn in many ways and are not unique in form. Whether drawn by the CM or the SBM these distinctions do not change.

Example 4.8.   Use the SBM to construct the moment diagram by parts for the beam given in Fig. 4.20a.

Solution: Fig. 4.20b shows the free-body diagrams for segments *AB* and *BC*. End shears and moments were extracted from Fig. 4.15b of Example 4.5. Figure 4.21 shows, in complete detail, the sub-

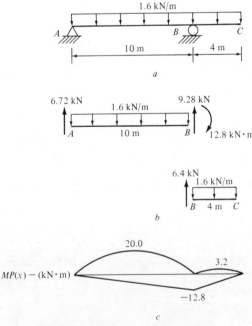

Fig. 4.20.   Beam for Example 4.8.

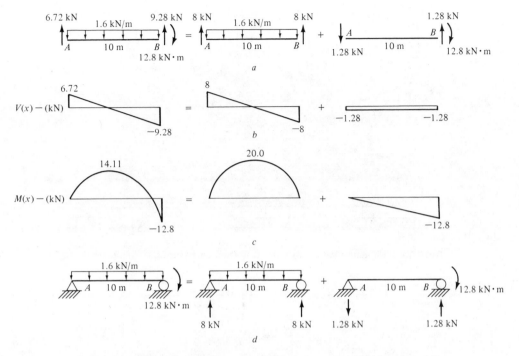

Fig. 4.21.   Moment diagram by parts (SBM) for segment AB**—Example 4.8.**

division of segment AB into simple beam loadings and development of the moment diagram by parts. To emphasize the fact that the use of simple beams is merely a superposition, Fig. 4.21a shows a subdivision into individual loaded *segments*. In this case, the subdivision is the addition of a segment loaded with the 1.6 kN/m transverse load to a segment loaded with the 12.8 kN · m right end moment. End shears for each can be calculated by statics.

The shear and moment diagrams for each segment are drawn below them in Figs. 4.21b and c. Addition of the separate diagrams to produce the actual shear and moment diagrams (to the left of the equality signs) is also shown. This is unnecesasry but confirms the correctness of the superposition by comparing with Fig. 4.15. Figure 4.22 completes the computations by examining segment BC in similar manner to segment AB.

The lengthy preceding presentation is executed for in depth understanding. Normally, the solution begins by visualizing the simple beam superposition shown in Fig. 4.21d and 4.22d. It proceeds as follows:

Treating AB and BC as segments of the beam, the moment diagram by parts is determined as follows.

AB

SBM	$\frac{1}{8}(1.6)(10)^2 = 20.0$ kN · m
M_L	0
M_R	-12.8 kN · m

BC

SBM	$\frac{1}{8}(1.6)(4)^2 = 3.2$ kN · m
M_L	-12.8 kN · m
M_R	0

Formulas for simple beam moments are obtained from Table A.2. M_L and M_R are the internal moments at the left and right ends of the segment, respectively. The moment diagram by parts is given in Fig. 4.20c. Shapes for the SBM diagrams are also taken from Table A.2.

Example 4.9.   Repeat Example 4.6 by the SBM.

Solution: The beam is shown again in Fig. 4.23a. Segments AB, BF and FG will be used. End moments at locations B and F are either extracted from the actual moment diagram, Fig. 4.16c, or calculated by isolating each overhang as a free body diagram and applying statics.

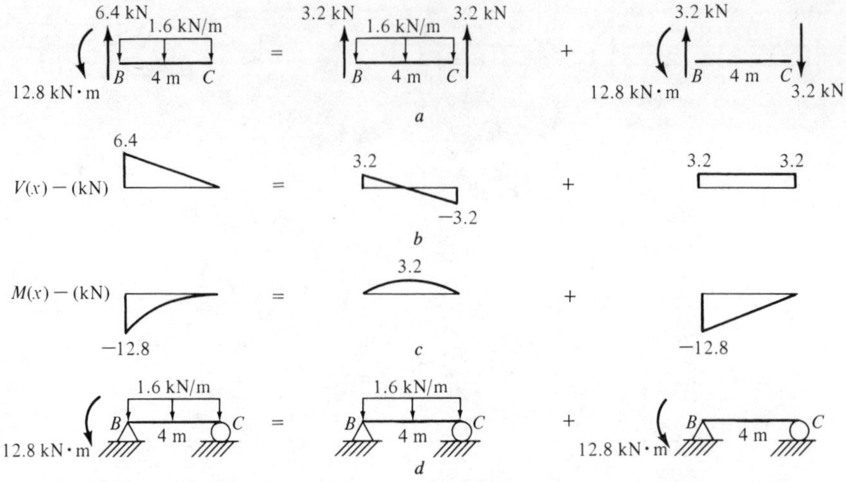

$V(x) - (kN)$

$M(x) - (kN)$

Fig. 4.22.   Moment diagram by parts (SBM) for segment BC—Example 4.8.

Using Table A.2

	AB	BF	FG
SBM	Actual moment diagram	$\dfrac{6(4)(20)}{24} = 20$ kip-ft	Actual moment diagram
		$\dfrac{16(18)(6)}{24} = 72$ kip-ft	
		$\dfrac{2(16)(8)^2}{2(24)} = 42.67$ kip-ft	
M_L		-64 kip-ft	
M_R		-48 kip-ft	

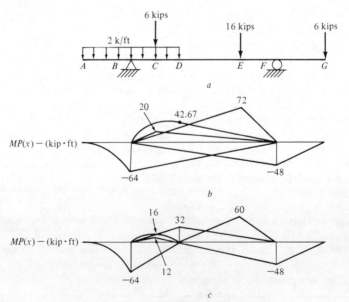

Fig. 4.23.   Beam for Examples 4.9 and 4.10.

The moment diagram by parts is shown in Fig. 4.23b. In this case, the actual moment diagram has been used for segments *AB* and *CD*. Normally, this is a desirable format for overhanging elements and leads to less computational effort.

Example 4.10. Repeat Example 4.9, but divide the beam into four segments, namely *AB*, *BD*, *DF*, and *FG*.

Solution: In the previous solution, the partial uniform load on segment *BF* produced an awkward moment diagram for computational purposes. Specifically, the technique illustrated in Fig. 4.14 must be used in the ensuing moment area computations. The diagram requested in this example problem eliminates that need. The end moments are extracted from the actual moment diagram. Using Table A.2:

	AB and FG	BD	DF
SBM	Actual Moment Diagram	$\dfrac{6(8)}{4} - 12$ kip-ft	$\dfrac{16(10)(6)}{16} = 60$ kip-ft
		$\dfrac{1}{8}(2)(8)^2 = 16$ kip-ft	
M_L		-64 kip-ft	32 kip-ft
M_R		32 kip-ft	-48 kip-ft

The results are plotted in Fig. 4.23c.

Example 4.11. Repeat Example 4.7 by the SBM.

Solution: Treating *AC* and *CD* in Fig. 4.24a as separate segments, the *MP/EI* diagram is determined as follows.

First the moment at *A* is calculated as

$$M_A = -50(3) - 60(7)(3.5) = -1620 \text{ kN} \cdot \text{m}$$

M_c is known from Example 4.7. Thus

	AC
SBM	$\dfrac{1}{8}(60)(4)^2 = 120$ kN $\cdot$ m
	$\dfrac{50(3)(1)}{4} = 37.5$ kN $\cdot$ m
M_A	-1620 kN $\cdot$ m
M_C	-270 kN $\cdot$ m

	CD
SBM	$\dfrac{1}{8}(60)(3)^2 = 67.5$
M_C	-270 kN $\cdot$ m
M_D	0

The complete results are drawn in Fig. 4.24b and division by the corresponding *EI* values yields Fig. 4.24c.

Example 4.12. Draw the *MP* diagram for the structure shown in Fig. 4.25 by the SBM.

Solution: Values for the support reactions are given, and verification is left as an exercise for the reader. In this example, end moments are known for all segments with the exception of the moment at

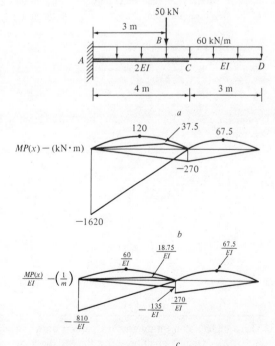

Fig. 4.24.   Results for Example 4.11.

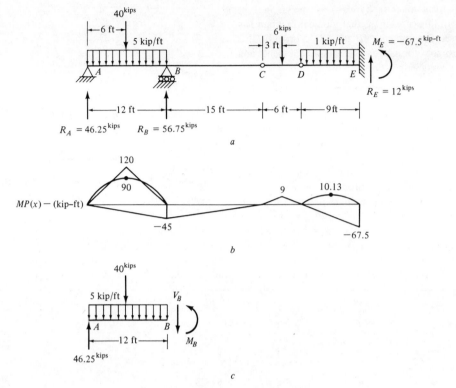

Fig. 4.25.   Beam for Example 4.12.

point B. Moment M_B can be computed with the aid of the free-body diagram drawn in Fig. 4.25c.

$$\Sigma M_B = M_B - 46.25(12)$$

$$+ 5(12)(6) - 40(6) = 0$$

$$\therefore M_B = -45.0 \text{ kip-ft}$$

Values necessary for constructing the MP diagram are:

For any given problem, the MP diagram can be drawn in many different ways. In several examples presented earlier this was actually done. Although each MP diagram had a different appearance, all had a common characteristic—the actual moment diagram is obtained when the individual parts are summed! This statement is a basis for "spot-checking" the results. Any of the diagrams can be employed in displacement computations, and the choice is open to the analyst. It is a good practice to visu-

	AB	BC	CD	DE
SBM	$\frac{1}{8}(5)(12)^2 = 90$ kip-ft	None	$\frac{6(6)}{4} = 9$ kip-ft	$\frac{1}{8}(1)(9)^2 = 10.13$ kip-ft
	$\frac{40(12)}{4} = 120$ kip-ft			
M_L	0	−45.0 kip-ft	0	0
M_R	−45.0 kip-ft	0	0	−67.5 kip-ft

The complete moment diagram by parts is drawn in Fig. 4.25b. One observation is that the MP diagram is less imposing than suggested by the loading and nature of the beam itself.

alize several possible MP diagrams prior to performing any numerical work.

4.6 EXAMPLE PROBLEMS ON DISPLACEMENT COMPUTATIONS

Use of an *MP* diagram in the moment-area method is illustrated in the following example problems.

Example 4.13. Calculate the slope and displacement at point *C* of the beam in Fig. 4.15. Assume *EI* is constant over the entire length of the beam.

Solution: The *MP* diagram in Fig. 4.15e is chosen. The displaced shape shown in Fig. 4.26a is feasible as it satisfies the curvature requirements deduced from the actual moment diagram (Fig. 4.15c.) Other shapes are possible but computations will proceed under the present assumption. Based on this assumption the downward tangential deviation $t_{A/B}$ is calculated. The positive (negative) moment creates upward (downward) curvature and thus a deviation of the member relative to the reference tangent extended from *B*. The deviation is

$$t_{A/B} = |\text{negative moment contribution}|$$
$$- |\text{positive moment contribution}|$$
$$= \frac{1}{3}\left(\frac{80}{EI}\right)(10)\left(\frac{3}{4}\right)(10)$$
$$- \frac{1}{2}\left(\frac{67.2}{EI}\right)(10)\left(\frac{2}{3}\right)(10)$$
$$= \frac{2000}{EI} - 2240 = -\frac{240}{EI}$$

The negative result indicates the deviation must be upward relative to the reference tangent. This is inconsistent with the trial displaced shape. A graphical modification is made in Fig. 4.26b. Thus,

$$10\theta_B = t_{A/B} = \frac{240}{EI}$$

$$\theta_B = \frac{24}{EI}\text{ ccw}$$

Continuing

$$\Delta_C = 4\theta_B - t_{C/B}$$

wherein downward $t_{C/B}$ is assumed positive.

$$t_{C/B} = \frac{1}{2}\left(\frac{12.8}{EI}\right)(4)\left(\frac{3}{4}\right)(4) = \frac{51.2}{EI}\downarrow$$

$$\Delta_C = 4\left(\frac{24}{EI}\right) - \frac{51.2}{EI} = \frac{44.8}{EI}\uparrow$$

Alternate Solution: The *MP* diagram in Fig. 4.20c will be used. Based on Fig. 4.26b,

$$t_{A/B} = |\text{positive moment contribution}|$$
$$- |\text{negative moment contribution}|$$
$$= \frac{2}{3}\left(\frac{20}{EI}\right)(10)(5)$$
$$- \frac{1}{2}\left(\frac{12.8}{EI}\right)(10)\left(\frac{2}{3}\right)(10)$$
$$= \frac{666.67}{EI} - \frac{426.67}{EI} = \frac{240}{EI}\uparrow$$

$$t_{C/B} = |\text{negative moment contribution}|$$
$$- |\text{positive moment contribution}|$$
$$= \frac{1}{2}\left(\frac{12.8}{EI}\right)(4)\left(\frac{2}{3}\right)(4) - \frac{2}{3}\left(\frac{3.2}{EI}\right)(4)(2)$$
$$= \frac{68.27}{EI} - \frac{17.07}{EI} = \frac{51.2}{EI}\uparrow$$

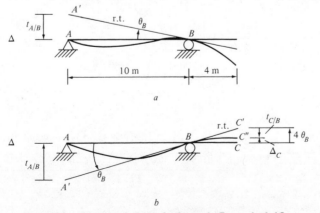

a

b

Fig. 4.26. Displaced shape for beam in Example 4.13.

which confirms the values determined in the preceding solution. θ_B and θ_C follow in the same manner.

Alternate Solution: The MP/EI diagram in Fig. 4.24c will be used.

$$\theta_D = \frac{1}{2}\left(\frac{270}{EI}\right)(3) + \frac{1}{2}\left(\frac{135}{EI}\right)(4) + \frac{1}{2}\left(\frac{810}{EI}\right)(4)$$

$$- \frac{2}{3}\left(\frac{67.5}{EI}\right)(3) - \frac{2}{3}\left(\frac{60}{EI}\right)(4)$$

$$- \frac{1}{2}\left(\frac{18.75}{EI}\right)(1) - \frac{1}{2}\left(\frac{18.75}{EI}\right)(3)$$

$$= \frac{405 + 270 + 1620 - 135 - 160 - 9.375 - 28.12}{EI}$$

$$\theta_D = \frac{1962.5}{EI} \approx \frac{1963}{EI} \text{ cw}$$

$$\Delta_D = \frac{405}{EI}(2) + \frac{270}{EI}\left(\frac{13}{3}\right) + \frac{1620}{EI}\left(\frac{17}{3}\right) - \frac{135}{EI}(3)$$

$$- \frac{160}{EI}(5) - \frac{9.375}{EI}\left(\frac{11}{3}\right) - \frac{28.125}{EI}(5)$$

$$\Delta_D = \frac{9982.5}{EI} \approx \frac{9983}{EI} \downarrow$$

Example 4.14.   Calculate the slope and displacement at the free end of the cantilever beam in Fig. 4.17a.

Solution: The MP diagram in Fig. 4.17c will be used. Areas within segment AC are subdivided in Fig. 4.17d. Area 3 is the entire triangle and area 4 is the chord of a parabola to be subtracted from it, i.e. it is a positive moment effect. The ordinate $a = 60/EI$ and the area of the chord is calculated using Appendix A, Table A.1.

4.7   EFFECT OF CHORD DISPLACEMENT

Evaluation of the integrals embodied in the moment-area theorems does not, in most cases, produce quantities normally considered to be in the category of generalized displacements. It is apparent that the angle t_{AB} in Fig. 4.1a is not a "slope" at any point in the member, nor is $t_{B/A}$ the "deflection" of any point on the beam.

$$\theta_D = t_{AD} = |\text{negative moment effect}| - |\text{positive moment effect}|$$

$$= \frac{1}{2}\left(\frac{75}{EI}\right)(3) + \frac{135}{EI}(4) + \frac{1}{2}\left(\frac{600}{EI}\right)(4)$$

$$+ \frac{1}{3}\left(\frac{270}{EI}\right)(3) - \frac{2}{3}\left(\frac{60}{EI}\right)(4)$$

$$= \frac{112.5 + 540 + 1200 + 270 - 160}{EI}$$

$$\theta_D = \frac{1962.5}{EI} \approx \frac{1963}{EI} \text{ cw}$$

$$\Delta_D = t_{D/A} = |\text{negative moment effect}| - |\text{positive moment effect}|$$

$$= \frac{112.5(6) + 540(5) + 1200(17/3) + 270(9/4) - 160(5)}{EI}$$

$$\Delta_D = \frac{9982.5}{EI} \approx \frac{9983}{EI} \downarrow$$

However, the theorems are a powerful tool, which, with proper application, can be used to determine "true" slope and deflection values. If the perpendicular distances, Δ_A and Δ_B, from the original member axis to points A' and B', respectively, are known, the slope θ_A can be determined in the following steps:

1. Compute $t_{B/A}$ by application of the second moment-area theorem.
2. Obtain ϕ_A as

$$\phi_A = \frac{t_{B/A}}{L} \qquad (4.26)$$

3. Determine R as

$$R = \frac{\Delta_A - \Delta_B}{L} \qquad (4.27)$$

4. Compute θ_A as

$$\theta_A = \phi_A - R + \theta \qquad (4.28)$$

Note that $\phi_A - R$ is the angular rotation of point A' relative to the original axis $A'B'$.

With the position of point A completely established, the location, Δ_x, and slope, θ_x, of any other point on the curve can be easily computed. Quantitatively, this is accomplished by evaluating

$$\theta_x = \theta_A - t_{Ax} \qquad (4.29a)$$

$$\Delta_x = \Delta_A + (\phi_A - R)x - t_{x/A} \qquad (4.29b)$$

where θ_x is measured relative to the horizontal plane and Δ_x is measured perpendicular to the

original member axis. The term t_{Ax} is the change in angle between A and x, and $t_{x/A}$ is the tangential deviation of the point x with respect to a tangent drawn from A. In light of these comments it can be stated that:

If the slope and displacement at any single point on a continuous, elastic curve are known *or* if the displacement of any two points on the elastic curve are known, the moment-area theorems can be employed to determine the position of any other point on the elastic curve relative to its original position.

The preceding statement is a basic constraint inherent in the moment-area theorems. Knowledge of the support conditions in a given problem usually provides sufficient information to overcome this constraint.

It must be noted that Eqs. 4.29 are not intended as equations to be adopted for general use, but as an illustration of one application of the moment-area theorems. In any given problem, it is necessary to carefully examine the geometrical conditions and formulate equations such as Eqs. 4.29 in manner suitable to the problem at hand. Subsequent numerical examples are intended to offer further insight into the moment-area method.

It is significant to recognize that if the angle θ in Fig. 4.1 is either zero or "small," the member is either horizontal or close enough to be considered so in its original position. If so, the displacements are as shown in Fig. 4.27.

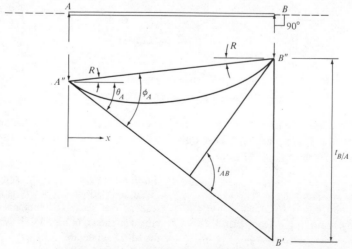

Fig. 4.27.   Geometry of a displaced beam.

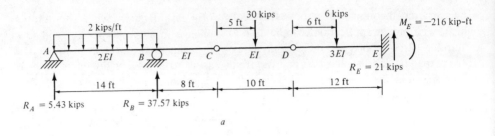

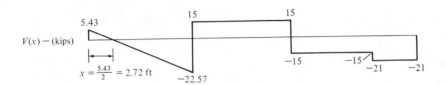

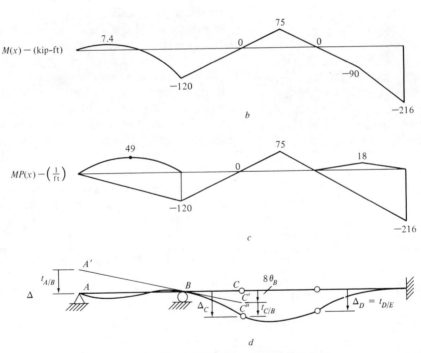

Fig. 4.28.   Beam for Example 4.15.

Specifically, the displacement quantities and tangential deviations are vertical. The only change to Eqs. 4.26 through 4.30 is

$$\theta_A = \phi_A - R \qquad (4.30)$$

which is obtained either by observation from Fig. 4.27 or by removing θ from Eq. 4.28.

Example 4.15.   Calculate the maximum deflection within segment CD of the beam given in Fig. 4.28a.

Solution: Values for the support reactions are given and verification is left as an exercise for the reader. The shear and moment diagrams are obtained as shown in Fig. 4.28b. Values necessary for the MP diagram are

	AB	BC	CD	DE
SBM	$\frac{1}{8}(2)(14)^2 = 49$ kip-ft	0	$\frac{30(10)}{4} = 75$ kip-ft	$\frac{6(12)}{4} = 18$ kip-ft
M_L	0	-120 kip-ft	0	0
M_R	-120 kip-ft	0	0	-216 kip-ft

The complete MP diagram is shown in Fig. 4.28c.

The actual moment diagram suggests the displaced shape shown in Fig. 4.28d. Based on the labels shown, the moment area method is employed to determine the hinge displacements.

$$t_{A/B} = |\text{negative moment effect}|$$
$$- |\text{positive moment effect}|$$
$$= \frac{1}{2}\left(\frac{120}{2EI}\right)(14)\left(\frac{28}{3}\right)$$
$$- \frac{2}{3}\left(\frac{49}{EI}\right)(14)(7) = \frac{2319}{EI} \downarrow$$

$$t_{C/B} = \frac{1}{2}\left(\frac{120}{EI}\right)(8)\left(\frac{16}{3}\right) = \frac{2560}{EI} \downarrow$$

$$\Delta_D = t_{D/E} = |\text{negative moment effect}|$$
$$- |\text{positive moment effect}|$$
$$= \frac{1}{2}\left(\frac{216}{3EI}\right)(12)(8)$$
$$- \frac{1}{2}\left(\frac{18}{3EI}\right)(12)(6) = \frac{3240}{EI}$$

$$\theta_B = \frac{t_{A/B}}{14} = \frac{166}{EI} \text{ cw}$$

$$\Delta_C = 8\theta_B + t_{C/B} = \frac{3888}{EI} \downarrow$$

In addition to a lack of moment resistance, an internal hinge creates a slope discontinuity in the displaced shape of a member. Because integration of the M/EI diagram cannot be performed across a discontinuity, it is necessary to isolate segment CD as shown in Fig. 4.29a. The magnitude of the chord rotation R is needed to apply the moment area theorems. Thus

$$R = \frac{\Delta_C - \Delta_D}{10}$$
$$= \frac{64.8}{EI} \text{ ccw relative to the horizontal}$$

The chord rotation complicates the application of the second moment-area theorem. Continuing

$$t_{D/C} = \frac{1}{2}\left(\frac{75}{EI}\right)(10)(5) = \frac{1875}{EI} \uparrow$$

$$\phi_{CR} = \frac{t_{D/C}}{10} = \frac{187.5}{EI} \text{ cw relative to the displaced chord}$$

By geometry

$$\theta_{CR} = \phi_{CR} - R$$
$$= \frac{122.7}{EI} \text{ cw relative to the horizontal}$$

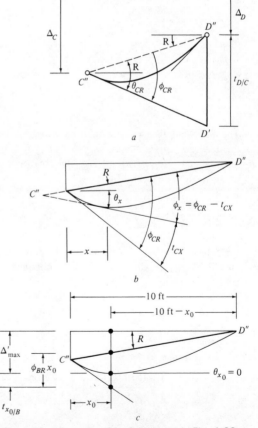

Fig. 4.29.   Segment CD of the beam in Fig. 4.28a.

With θ_{CR} known to be clockwise, from Figs. 4.28d and 4.29a, the maximum displacement in this beam must occur in segment CD.

From Fig. 4.29b it can be observed that the true slope θ_x at any location x can be determined as

$$\theta_x = \phi_x - R$$

The quantity ϕ_x is the angle change between the tangent at x and the displaced chord, $C''D''$ and is given by

$$\phi_x = \phi_{CR} - t_{Cx}$$

where t_{Cx} is the area under the moment diagram between point C and the arbitrary point x. The moment M_x at a distance x from point C is

$$M_x = V_C x = +15x$$

Consequently,

$$t_{Cx} = \frac{1}{2}\left(\frac{15x}{EI}\right)(x) = \frac{7.5}{EI}x^2$$

$$\phi_x = \frac{187.5}{EI} - \frac{7.5}{EI}x^2$$

$$\theta_x = \frac{187.5}{EI} - \frac{7.5x^2}{EI} - \frac{64.8}{EI}$$

$$= \frac{122.7}{EI} - \frac{7.5x^2}{EI}$$

The maximum displacement defines the low point of the displaced curve, and at that point the true slope must be zero. The location, x_0, of the low point is obtained from this condition.

$$\theta_x = \frac{122.7}{EI} - \frac{7.5x_0^2}{EI} = 0 \qquad \therefore x_0 = 4.04 \text{ ft}$$

Consideration of Fig. 4.29a and 4.29c shows that the value of the maximum deflection is obtained according to where $\Delta_{max} = \Delta_D + \Delta'_{max}$

$$\Delta_{max} = \phi_{CR}x_0 - t_{x_0/c} + R(10 - x_0)$$

Numerically,

$$t_{x_0/c} = \frac{1}{2}\left[\frac{15(4.04)}{EI}\right](4.04)\left(\frac{4.04}{3}\right)$$

$$= \frac{164.8}{EI}$$

$$\therefore \Delta'_{max} = \frac{187.5}{EI}(4.04) - \frac{164.8}{EI}$$

$$+ \frac{64.8}{EI}(10 - 4.04)$$

$$= \frac{979}{EI} \qquad \therefore \Delta_{max} = \frac{3240}{EI} + \frac{979}{EI}$$

$$= \frac{4219}{EI} \text{ downward}$$

4.8 CLOSING COMMENTS

Application of the moment-area theorems to the problem of computing beam displacements must be preceded by a careful geometrical analysis of the structure. Formulation of correct physical relationships is an essential step if the theorems are to be successfully applied to a given problem. In this text, no rigorous sign convention has been recommended for use in the moment-area method. Instead, the reader is instructed to rely upon the particular support conditions and shape of the M/EI diagram to gain insight to the characteristics of the deformed shape.

With a feasible elastic curve established, appropriate geometric relationships are written to employ the moment-area theorems. Signs associated with the angle change and tangential deviation quantities that appear in the geometric relationships are determined on the basis of the sense of the curvature the corresponding M/EI areas create. The key to correct formulation of the equations is the realization that a member that is subjected to no moment extends as a line of constant slope, and, when moment does exist, the member curves above or below an imagined straight line depending upon the direction of the induced curvature.

PROBLEMS

Notes: MP/EI diagrams should not be used in Problems 4-1 through 4-8. Use the moment area method in all problems requiring calculation of displacement quantities.

4-1. Calculate the slope and displacement at the free end of each beam.

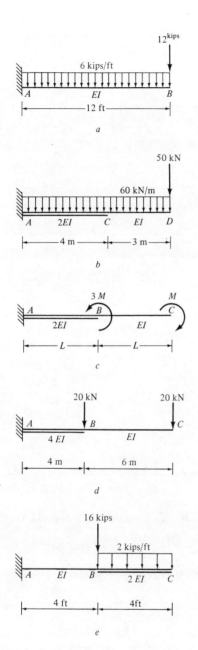

a

b

c

d

16 kips

2 kips/ft

A   EI   B   2 EI   C

4 ft   4ft

e

4-2. Calculate the slope and displacement at point *B*.

15 kN   30 kN

A   B   C   D

Constant *EI*

6 m —— 6 m —— 6 m

4-3. Calculate the maximum displacement for the beam in Problem 4-2.

4-4. Calculate the maximum displacement for each beam.

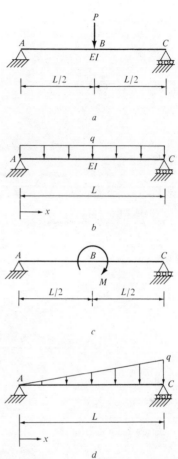

a

b

c

d

4-5. Determine the equation of the elastic curve $y(x)$ for each beam in Problem 4-4, using the moment-area method. Use it to determine the maximum displacement.

4-6. Calculate the midspan displacement for each beam.

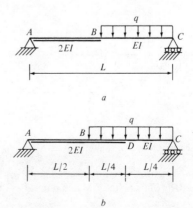

a

b

4-7. Calculate the slope and displacement at the end of each overhang at each support and midway between the supports.

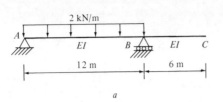

a

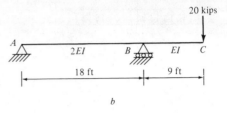

b

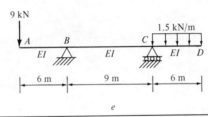

c

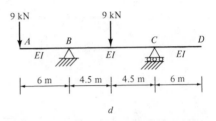

d

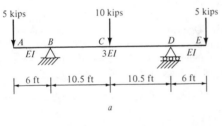

e

4-8. Repeat Problem 4-7 for the given beams.

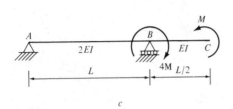

a

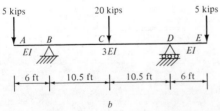

b

4-9. For each beam, draw the following diagrams:

(i)   Actual shear and moment diagrams.
(ii) *MP/EI* using the CM and point *A* as the reference.

(iii) *MP/EI* using the CM and point *B* as the reference.
(iv) *MP/EI* using the CM and point *C* as the reference.
(v)   *MP/EI* using the CM and point *D* as the reference.
(vi) *MP/EI* using the SBM.

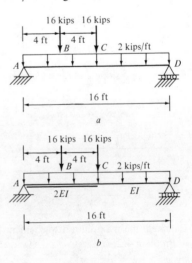

a

b

4-10. Repeat Problem 4-9 for each beam given in Problem 7-20. Note: In some cases point D does not exist. If so, omit diagram (v).

4-11. Draw each of the following diagrams for beams a through j in Problem 3-2:

(i) M/EI using the CM and the left support as the reference.

(ii) MP/EI using the SBM.

4-12. For beams a through j in Problem 3-2, calculate the vertical displacement at the end of the overhang and midway between the supports. Use EI = 200,000 kip-in.^{2} (see Problem 4-11).

4-13. Due to settlement, the support at A displaces downward by 0.25″. Calculate the displacement at B due to the load and the settlement.

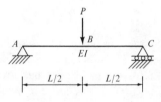

4-14. Determine the location and value of the maximum displacement. The units of the spring constants are kN/m.

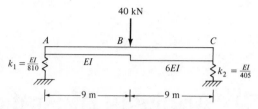

4-15. For the beam in Problem 7-22, calculate the maximum displacement using the moment-area method. EI is constant over the entire length. Hint: Use an MP/EI diagram.

4-16. Do Problem 7-24 by the moment-area method. Hint: Use a MP/EI diagram.

4-17. Do Problem 7-25 by the moment-area method.

4-19. For each beam in Problem 3-3, calculate the displacement and rotation at each free end, support and internal hinge. Unless indicated otherwise, assume EI is constant. Sketch the displaced shape.

5
Analysis of Statically Determinate Planar Trusses

5.1 GENERAL

When man first formed a tripod of branches, he likely invented the simplest (analytically) structural framework—the pinned truss. In attempting to suspend load from the single free joint, he encountered and solved all of the basic structural needs: a sufficient number of members to support the applied load (equilibrium), a suitable connection to maintain connectivity (compatibility), large enough members to maintain rigidity of the structure (material laws), and member orientation and support arrangement suitable to avoid collapse (stability). Until the advent of formed metals, wood members were the primary, if not sole, component in framed structures. The presence of great depth was an inevitable condition in long-span structures during the early ages because of the inherently low modulus of elasticity (a concept that had not been discovered) of wood. Framed structures with pinned connections were probably the only option that existed for incorporating needed depth without resorting to massive domes and stone arches. These factors explain the early predominance of the truss as a structural type.

Classically, the most creative applications of trusses have been in bridges. The development of steel members at the turn of the century spurred engineers to configure a wide array of economic truss arrangements, which usually bore the name of the inventor. Since that time, the use of trusses in bridges has greatly declined, and applications are not as common. Major reasons for the decline are:

1. later development of steel suspension bridges and the contemporary use of cable-stayed bridges for long spans
2. use of structural steel W-shapes, laminated wood, and prestressed concrete for short-to-medium spans
3. general inefficiency and higher construction costs as compared to contemporary bridge systems

However, large planar trusses and elaborate space trusses are popular choices for roof framing systems in many modern buildings. In addition, the use of bracing systems to withstand the transverse effects of wind forces is a common procedure in contemporary high-rise buildings. Usual practice is to idealize the bracing system as a vertically cantilevered framework and employ truss analysis to determine the member forces induced by the applied wind loads.

5.2 TYPES OF TRUSSES

Most trusses are comprised of triangular panels, such as shown in Fig. 5.1a, that are succes-

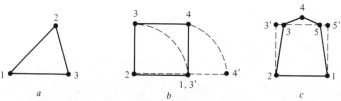

Fig. 5.1.   Stable and unstable panel configurations.

sively connected together to build larger and larger truss systems. This is because the three-sided configuration with pinned joints is inherently stable. Three given members can only be pinned together in one unique shape. In contrast, configurations of four or more sides are internally unstable, they can fold up as exemplified in Fig. 5.1b and c. The configuration in Fig. 5.1b can be flattened, like a cardboard box. The five-sided configuration in Fig. 5.1c can do likewise. This is easily seen if one imagines joint 4 moving downward to form a square to be followed by the motion of Fig. 5.1b.

Trusses constructed entirely of triangles are commonly referred to as "simple trusses." Schematic representations of some examples of simple trusses are shown in Fig. 5.2. The X-shaped truss in Fig. 5.2d illustrates that simple trusses do not necessarily have a simple geometrical layout.

Figure 5.3 illustrates four trusses that would be referred to as "compound trusses." Compound trusses are composed of two or more simple trusses connected together to form a larger truss. The connection is accomplished by linking the trusses together either by three new members or by one member plus a common joint. In more complicated cases the new member(s) might be a (set of) complete truss(es). The presence of nontriangular panels often signals the existence of a compound truss. The truss in Fig. 5.3b exhibits this feature.

In Fig. 5.3a, the simple trusses *ABC* and *CDE* are joined together by the common joint *C* and member *BD*. In Fig. 5.3b, the simple trusses *ABCD* and *EFGH* are joined together by member *AE*, *CI*, and *DF*. A series of arched simple trusses comprise the continuous truss in Fig. 5.3c. In each span a common joint and one member accomplish the interconnection. The

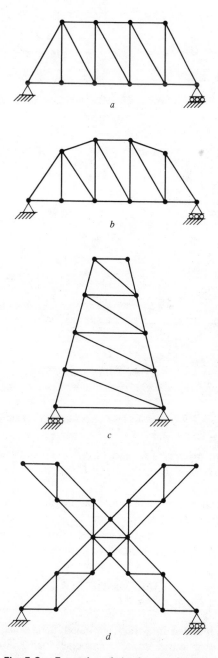

Fig. 5.2.   Examples of simple truss forms.

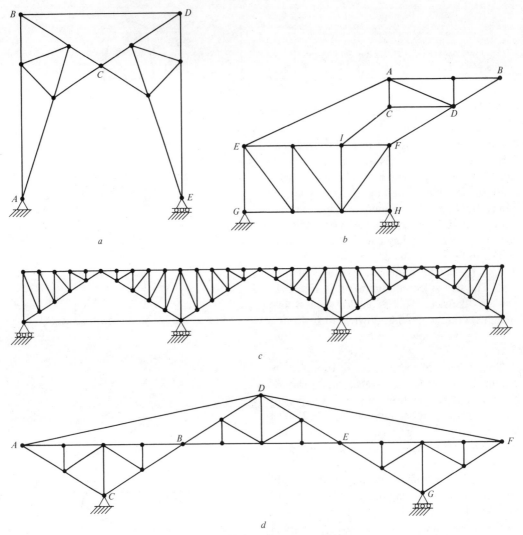

Fig. 5.3. Examples of compound trusses.

two trusses, *ABC* and *EFG*, have been linked by two members, *AD* and *DF*, and a complete truss, *BDE*.

Some trusses do not fit the definitions of either simple or compound. The two trusses in Fig. 5.4 are examples. These types of trusses are termed ''complex'' or ''irregular'' trusses.

5.3 EQUILIBRIUM OF PLANAR TRUSSES

A member in a truss is pinned at its ends and is not subjected to transverse load. Consequently, it carries either an axial tension force

or an axial compressive force. Such an element is shown in Fig. 5.5. By nature of the pinned connections, no moment can exist at either end of the member. Moreover, moment equilibrium about either end indicates that shear forces are nonexistent. Thus, it is a result of statics that a truss member can resist load only by the development of an axial force.

Because trusses are assumed to be loaded only at the joints and the members transmit only axial force, internal equilibrium fundamentally consists of imposing statics at each of the joints in the structure. Fig. 5.6 will be employed to demonstrate the general application of this con-

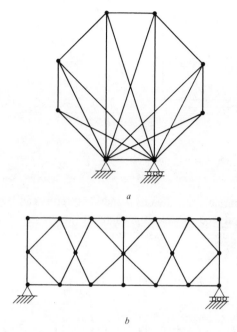

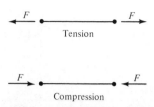

Fig. 5.5.   Free-body diagram of a truss member.

cept. A complete truss is pictured in Fig. 5.6a, and free-body diagrams of each joint are given in Fig. 5.6b. All *possible*, *independent* locations for applied loads have been indicated by the 11 loads, P_1 through P_{11}, which correspond in direction to the 11 degrees of freedom. Because trusses are loaded at the joints, these loads consist of horizontal and vertical components at each free joint and a single force at each roller support. The three support reactions are also shown as forces R_1, R_2, and R_3.

In the free-body diagrams of the joints, the member forces have been arbitrarily numbered

Fig. 5.4.   Examples of complex (irregular) trusses.

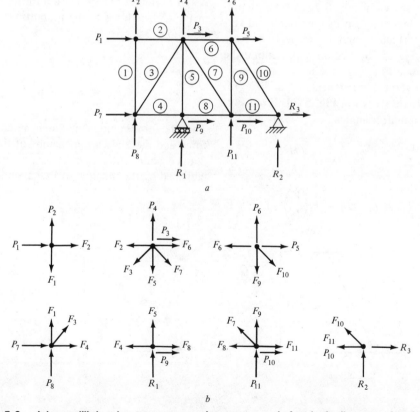

Fig. 5.6.   Joint equilibrium in a truss. *a*, complete structure; *b*, free-body diagrams of each joint.

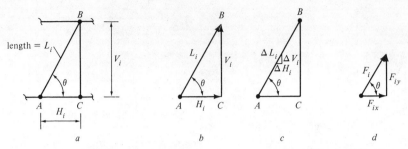

Fig. 5.7.   Proportionality of member force and member length.

F_1 through F_{11}. It is common practice to show each unknown member force as a vector acting in the tensile sense. When its true sense is determined by subsequent computations, its magnitude becomes signed. A positive sign is used for tension and a negative sign for compression.

The direction of the member force vector is coincident with the direction of the member. This observation is useful when components of a member force are needed. A typical truss member is shown in Fig. 5.7a. Vectors representing its length L_i and its horizontal and vertical components H_i and V_i, respectively, are shown in Fig. 5.7b. Vectors representing the member force F_i acting at joint A and its horizontal and vertical components F_{ix} and F_{iy}, respectively, are shown in Fig. 5.7d. From Fig. 5.7d and similar triangles

$$F_{ix} = F_i \cos \theta = F_i \left(\frac{H_i}{L_i} \right)$$

$$F_{iy} = F_i \sin \theta = F_i \left(\frac{V_i}{L_i} \right) \qquad (5.1)$$

Alternatively, the $\cos \theta$ and $\sin \theta$ terms can be expressed in any proportions of the inclination of a member. Thus from Fig. 5.7c

$$F_{ix} = F_i \cos \theta = F_i \left(\frac{\Delta H_i}{\Delta L_i} \right)$$

$$F_{iy} = F_i \sin \theta = F_i \left(\frac{\Delta V_i}{\Delta L_i} \right) \qquad (5.2)$$

As an example, the inclination of the member in Fig. 5.8 reduces to the proportions shown. Equations 5.2 give

$$F_{ix} = .4472 \, F_i$$

$$F_{iy} = .8944 \, F_i$$

Thus the force components can be related to the force and the known dimensions of the member. Of course, the member length can be established by the theorem of Pythagoras as

$$L_i = \sqrt{H_i + V_i} \qquad (5.3)$$

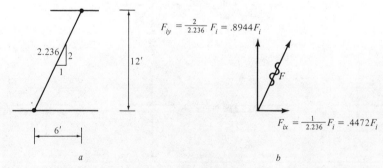

$$F_{iy} = \frac{2}{2.236} F_i = .8944 F_i$$

$$F_{ix} = \frac{1}{2.236} F_i = .4472 F_i$$

Fig. 5.8.   Resolution of a member force into horizontal and vertical components.

Similarly if the force components are known, the member force is given by

$$F_i = \sqrt{F_{ix} + F_{iy}} \qquad (5.4)$$

Because of Eqs. 5.1, each of the unknown forces F_1 through F_{11} in Fig. 5.6b can be resolved into horizontal and vertical components. By applying the statics conditions $\Sigma F_x = 0$ and $\Sigma F_y = 0$ at each of the seven joints, one arrives at a set of 14 simultaneous equations. The 14 equations contain 14 unknown quantities, which consist of the three external reactions and the 11 internal axial forces. In this instance, the complete determination of the unknown quantities is possible by solution of the 14 equilibrium equations. In general, if this is possible the truss is considered to be *statically determinate*. However, it is possible that in any given truss, the number of unknowns could exceed the number of available equations. Such trusses are called *statically indeterminate* and require additional equations from other sources if solution is to be possible. Statically indeterminate trusses arise from two distinct conditions:

1. There is an excessive number of member forces.
2. There is an excessive number of support reactions.

The first condition constitutes trusses that are statically indeterminate *internally*, and the second condition constitutes those that are statically indeterminate *externally*. It is also possible to have both conditions existing at the same time.

5.4 FORCES IN STATICALLY DETERMINATE TRUSSES

The discussions in Section 5.3 indicated that the external reactions and member forces of a statically determinate truss can be determined by solving the set of equilibrium equations obtained by applying statics at each joint of the truss. When the number of equations is large, solution is best done by computer computation. However, for trusses with a moderate number of members, the unknowns can often be computed manually at less expense and expenditure of time than required to produce and submit input needed for a computer solution. Three approaches are possible in a manual solution:

1. Direct solution of equilibrium equations
2. Methods of joints
3. Method of sections

Each of these methods are described in the following sections.

5.4.1 Direction Solution of Equilibrium Equations

Solution of just a moderate number of simultaneous equations is tedious. However, for simple trusses this can often be managed expeditiously.

Example 5.1. Determine the reactions and member forces for the given truss (Fig. 5.9a.)

Solution: A schematic representation of the actual truss is given in Fig. 5.9a. The four joints are labeled by numbers placed within squares. The five members are labeled by numbers placed within a circle.

The loads, internal forces and external reactions are shown in compact form by use of a *general force diagram* (Fig. 5.9b). The reactions are labeled R_1–R_3. As a general sign convention in this textbook, reactions are assumed to be positive if acting either toward the right or upward. The convention adopted for member forces F_1–F_5 is to assume all members are in tension, i.e., they *pull on the joints*.

In the general force diagram each force is shown twice, e.g., F_1 is shown pulling on joints 1 and 2, F_2 is shown pulling on joints 2 and 3, etc. The general force diagrams is a compact presentation of the free-body diagrams drawn in Fig. 5.9c. The isolated joints are used to generate the equations of equilibrium. As the isolated free-body diagrams can be "visualized" by inspecting the general force diagram, Fig. 5.9c is redundant and unnecessary.

Equations of equilibrium are generated by applying statics to each free-body diagram shown in Fig. 5.9c (or visualized from Fig. 5.9b). Because the

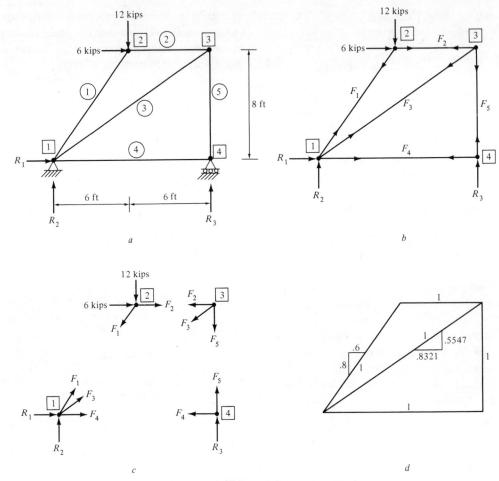

Fig. 5.9.   Truss for Examples 5.1 and 5.3.

member forces and their horizontal and vertical components are in the same proportions as the member lengths and their components, it is useful to draw a *member inclination diagram*, as shown in Fig. 5.9d, to indicate the slope of each member in component form.

Applying statics to joint 1 requires

$$\Sigma F_x = .6F_1 + .8321F_3 + F_4 + R_1 = 0 \qquad (a)$$

$$\Sigma F_y = .8F_1 + .5547F_3 + R_2 = 0 \qquad (b)$$

Applying statics to joint 2 requires

$$\Sigma F_x = -.6F_1 + F_2 + 6 = 0 \qquad (c)$$

$$\Sigma F_y = -.8F_1 - 12 = 0 \qquad (d)$$

Similarly, joints 3 and 4 require

$$\Sigma F_x = -F_2 - .8321F_3 = 0 \qquad (e)$$

$$\Sigma F_y = -.5547F_3 - F_5 \qquad (f)$$

$$\Sigma F_x = -F_4 = 0 \qquad (g)$$

$$\Sigma F_y = F_5 + R_3 = 0 \qquad (h)$$

The eight equilibrium equations contain eight unknowns (five member forces and three reactions) and are solvable simultaneously. A judicious choice of order for treating the equations permits an efficient solution process. Collecting the equations and transferring known constants to the right-hand sides makes it easier to deduce an order of treatment.

$.6F_1$	$+.8321F_3 + 1.0F_4$	$+1.0R_1$	$= 0$ (a)
$.8F_1$	$+.5547F_3$	$+1.0R_2$	$= 0$ (b)
$-.6F_1 + 1.0F_2$			$= -6$ (c)
$-.8F_1$			$= 12$ (d)
$-1.0F_2$	$-.8321F_3$		$= 0$ (e)
	$-.5547F_3$	$-1.0F_5$	$= 0$ (f)
	$-1.0F_4$		$= 0$ (g)
		$1.0F_5$	$+1.0R_3 = 0$ (h)

The sparsity of the above equations is typical of truss equilibrium equations. Also, the coefficients of each member force sums to zero when added vertically (e.g., for column 1, $.6F_1 + .8F_1 - .6F_1 - .8F_1 = 0$.) This is always the case (why?) and offers a check on one's work.

From Eq. d, F_1 is obtained directly:

$$-.8F_1 - 12 = 0$$

$$\therefore F_1 = -15.0 \text{ kips}$$

$$= 15.0 \text{ kips compression}$$

Substitution into Eq. c produces

$$-.6(-15.0) + F_2 + 6 = 0$$

$$\therefore F_2 = -15.0 \text{ kips}$$

$$= 15.0 \text{ kips compression}$$

Note that the negative sign on the computed F_1 value is retained in the subsequent substitution! With F_2 established, Eq. e gives

$$-(-15.0) - .8321F_3 = 0$$

$$\therefore F_3 = 18.03 \text{ kips tension}$$

and substitution into Eq. f gives

$$-.5547(18.03) - F_5 = 0$$

$$\therefore F_5 = -10.00 \text{ kips}$$

$$= 10.00 \text{ kips compression}$$

From Eq. g,

$$F_4 = 0$$

and from Eq. h

$$-10.00 + R_3 = 0$$

$$\therefore R_3 = 10.00 \text{ kips} \uparrow$$

Next, Eqs. a and b produce

$$.6(-15.0) + .8321(18.03) + 0 + R_1 = 0$$

$$\therefore R_1 = -6.00 \text{ kip}$$

$$= 6.00 \text{ kips}$$

$$.8(-15.0) + .5547(18.03) + R_2 = 0$$

$$\therefore R_2 = 2.00 \text{ kips}$$

Note the order of solving the equations is not unique. For example, Eq. g could have been solved at the outset for F_4, which then could be available as a value to substitute in other equations. The order used was chosen to permit a comparison with a solution obtained by the method of joints later in this chapter.

Example 5.2. Determine the reactions and member forces for the given truss (Fig. 5.10a.)

Solution: The truss is similar to the truss in Example 5.1 except the roller support has been moved 6 ft to the right. Proceeding through joints 1 through 4, in order, the statics requirements yield

$.6F_1$	$+.8321F_3 + 1.0F_4$		$+1.0R_1$	$= 0$ (a)
$.8F_1$	$+.5547F_3$		$+1.0R_2$	$= 0$ (b)
$-.6F_1 + 1.0F_2$				$= -6$ (c)
$-.8F_1$				$= 12$ (d)
$-1.0F_2$	$-.8321F_3$	$+.6F_5$		$= 0$ (e)
	$-.5547F_3$	$-.8F_5$		$= 0$ (f)
	$-1.0F_4$	$-.6F_5$		$= 0$ (g)
		$.8F_5$	$+1.0R_3 =$	0 (h)

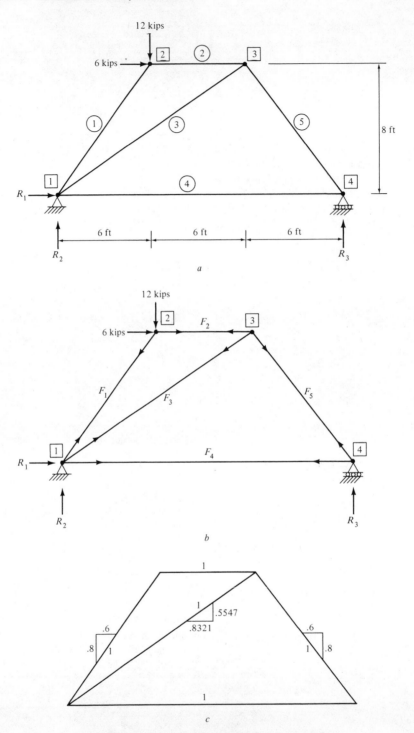

Fig. 5.10.   Truss for Examples 5.2, 5.6, and 5.8.

As in Example 5.1, we begin with Eqs. d and c:

$$-.8F_1 - 12 = 0$$

$$\therefore F_1 = -15.0 \text{ kips}$$

$$= \underline{15.0 \text{ kips compression}}$$

$$-.6(-15.0) + F_2 + 6 = 0$$

$$\therefore F_2 = -15.0 \text{ kips}$$

$$= \underline{15.0 \text{ kips compression}}$$

Proceeding to Eqs. e and f,

$$-(-15.0) - .8321F_3 + .6F_5 = 0$$

$$-.5547F_3 - .8F_5 = 0$$

Unlike in Example 5.1, these conditions cannot be treated separately. Simultaneous solution produces

$$F_3 = \underline{12.01 \text{ kips tension}}$$

$$F_5 = -8.32 \text{ kips}$$

$$= \underline{8.32 \text{ kips compression}}$$

Equations g and h give

$$-1.0F_4 - .6(-8.32) = 0$$

$$\therefore F_4 = -4.99 \text{ kips}$$

$$= \underline{4.99 \text{ kips compression}}$$

$$.8(-8.32) + 1.0R_3 = 0$$

$$\therefore R_3 = \underline{6.66 \text{ kips tension}}$$

Equations a and b give

$$.6(-15.0) + .8321(9.61)$$

$$+ (-4.99) + 1.0R_1 = 0$$

$$\therefore R_1 = 5.99 \text{ kips}$$

$$\approx \underline{6.00 \text{ kips}} \leftarrow$$

$$.8(-15.0) + .5547(12.01) + 1.0R_2 = 0$$

$$\therefore R_2 = -5.34 \text{ kips}$$

$$= \underline{5.34 \text{ kips}} \uparrow$$

5.4.2   Method of Joints

The "method of joints" refers to the process of applying statics, in a selective manner, such as to determine the member forces in some judicious, sequential manner. Each time a joint is treated, any calculated member forces or reactions become known quantities in subsequent calculations. The method of joints is best described by use of example problems.

Example 5.3.   Repeat Example 5.1 by using the method of joints.

Solution: The free-body diagrams of all the joints are shown in Fig. 5.9c. Member inclinations are shown in Fig. 5.9d. These are used as follows. Considering joint 2:

$$\Sigma F_x = -.6F_1 + F_2 + 6 = 0$$

$$\Sigma F_y = -.8F_1 \qquad -12 = 0$$

Solving the second equation for F_1 and substitution into the first equation gives

$$F_1 = -15.0 \text{ kips} = \underline{15.0 \text{ kips compression}}$$

$$F_2 = -15.0 \text{ kips} = \underline{15.0 \text{ kips compression}}$$

Considering joint 3

$$\Sigma F_x = -F_2 - .8321F_3 \qquad = 0$$

$$\Sigma F_y = \qquad -.5547F_3 - F_5 = 0$$

Since F_2 is known,

$$-(-15) - .8321F_3 \qquad = 0$$

$$-.5547F_3 - F_5 = 0$$

and

$$F_3 = \underline{18.03 \text{ kips tension}}$$

$$F_5 = \underline{-10.00 \text{ kips}} = \underline{10.00 \text{ kips compression}}$$

Considering joint 4,

$$\Sigma F_x = -F_4 \qquad = 0$$

$$\Sigma F_y = F_5 + R_3 = 0$$

or

$$-F_4 \qquad\qquad = 0$$

$$(-10.00) + R_3 = 0$$

and

$$F_4 = \underline{0}$$

$$R_3 = \underline{10.00 \text{ kips} \uparrow}$$

Considering joint 1,

$$F_x = .6F_1 + .8321F_3 + 1.0F_4 + 1.0R_1 = 0$$

$$F_y = .8F_1 + .5547F_3 \qquad\quad + 1.0R_3 = 0$$

or

$$.6(-15.00) + .8321(18.03) + 1.0(0)$$

$$+ 1.0R_1 = 0$$

$$.8(-15.00) + .5547(18.03)$$

$$+ 1.0R_3 = 0$$

and

$$R_1 = -6.00 \text{ kips} = \underline{6.00 \text{ kips}} \leftarrow$$

$$R_2 = \underline{2.00 \text{ kips} \uparrow}$$

These results agree with the previous solution.

Note that the sequence of computations executed in Example 5.3 is the same as the sequence used in Example 5.1. Treatment of joint 2 is identical to solving Eqs. c and d in Example 5.1. Treatment of joint 3 is identical to solving Eqs. e and f in Example 5.1. This continues through the solution. The method of joints can be recognized as a visual tool for solving the inherent equilibrium equations of a truss in some orderly fashion. The method of joints involves the following steps:

1. Identify a joint in the truss whose free-body diagram would include no more than two unknowns (either two member forces, two reactions, or one of each).
2. Isolate a free-body diagram of the joint and execute the statics conditions, $\Sigma F_x = 0$ and $\Sigma F_y = 0$, to generate a pair of equi-

librium equations. These will include the unknowns identified in Step 1.
3. Solve the pair of equations simultaneously for the unknowns.
4. Repeat Steps 1 to 3 for the remaining joints until all member forces and reactions have been calculated.

In some cases, it is either advantageous or necessary to precede the method of joints by computation of the reactions separately. This can be done by applying statics to the free-body diagram of the entire truss, provided it is statically determinate. The following example illustrates such a situation.

Example 5.4. Determine the reactions and member forces for the truss given in Fig. 5.11a. Use the method of joints.

Solution: The general force diagram and member inclinations are shown in Fig. 5.11b and c. The free-body diagrams of the joints are drawn in Fig. 5.11d.

A study of the joints indicates each of them has more than two unknown forces (or reactions) acting on it. Thus, the reactions will be calculated by applying statics to Fig. 5.11b.

$$\Sigma F_x = R_3 + 4 = 0$$

$$\therefore R_3 = -4 \text{ kN} = \underline{4 \text{ kN}} \leftarrow$$

$$\Sigma M_{\text{joint 4}} = R_1(32) + 4(12) - 10(16) = 0$$

$$\therefore R_1 = \underline{3.5 \text{ kN} \uparrow}$$

$$\Sigma F_y = R_1 + R_2 - 10 = 0$$

$$3.5 + R_2 - 10 = 0$$

$$\therefore R_2 = \underline{6.5 \text{ kN}}$$

With the reactions known, the method of joints can be executed.

Considering joint 1,

$$\Sigma F_x = F_1 + .8F_2 \qquad\quad = 0$$

$$\Sigma F_y = \qquad - .6F_2 + R_1 = 0$$

or

$$F_1 + .8F_2 = 0$$

$$-.6F_2 + 6.5 = 0$$

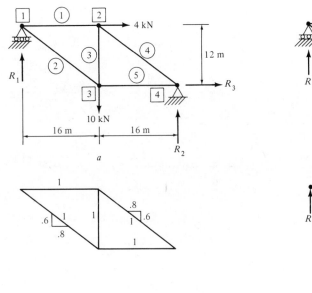

Fig. 5.11.   Truss for Example 5.4.

and

$$\therefore F_1 = -4.66 \text{ kN} = \underline{4.66 \text{ kN compression}}$$

$$\therefore F_2 = 5.83 \text{ kN} \quad = \underline{5.83 \text{ kN tension}}$$

Considering joint 2,

$$\Sigma F_x = -F_1 + .8F_4 + 4 = 0$$

$$\Sigma F_y = -F_3 - .6F_4 \quad\quad = 0$$

or

$$-(-4.66) + .8F_4 + 4 = 0$$

$$-F_3 - .6F_4 = 0$$

and

$$\therefore F_3 = 6.50 \text{ kN} = \underline{6.50 \text{ kN tension}}$$

$$\therefore F_4 = -10.83 \text{ kN} = \underline{10.83 \text{ kN compression}}$$

Considering joint 3,

$$\Sigma F_x = -.8F_2 + F_5 \quad\quad = 0$$

$$\Sigma F_y = \quad .6F_2 + F_3 - 10 = 0$$

or

$$-.8(5.83) + F_5 \quad\quad = 0$$

$$.6(5.83) + 6.50 - 10 = 0$$

The latter equation is unnecessary but gives $0 \approx 0$ as a confirmation of the correctness of earlier calculations. The former equation gives

$$F_5 = \underline{4.66 \text{ kN tension}}$$

Example 5.5.   Repeat Example 5.1 using the method of joints.

Solution: The solution will be executed by a "visual method of joints." In the visual approach the method of joints is performed in place, on a diagram of the truss, with a minimum of written computations. For the present example, this can be done in the manner shown in Fig. 5.12.

Step 1:   Inspecting joint 2 (Fig. 5.12a), it is apparent that for vertical equilibrium a 12-kip vertical component is needed in member 1-2 in order to balance the applied 12-kip force. Also, member 1-2 must push on joint 2 to accomplish this. The 9-kip horizontal component and 15-kip compressive member force are obtained by proportioning; i.e., the member force and its components are proportional

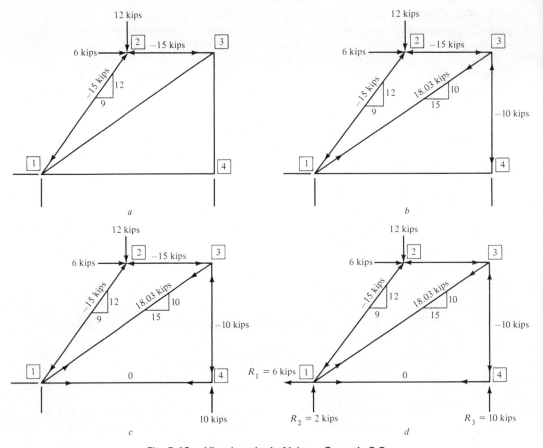

Fig. 5.12. Visual method of joints—Example 5.5.

to the member length and its components as shown in the member inclination diagram (Fig. 5.9d). Also, for horizontal equilibrium at joint 2, member 2-3 must develop a 15-kip compressive force. The results of these observations can be recorded directly as shown in Fig. 5.12a. Compressive forces are labeled with a negative sign to further emphasize that they push on the joints. Note that arrowheads are shown on *both* joints that the member force acts on. Also, until their directions are determined, reactions are indicated by lines without arrowheads.

Step 2: Inspecting joint 3 (Fig. 5.12b), it is apparent that, for horizontal equilibrium to exist, a 15-kip horizontal component is needed in member 1-3. The 10-kip vertical component and the 18.03-kip tensile member force (pulling on the joint) are determined (with a calculator in hand) by proportioning, i.e.,

$$\frac{1}{.8321} = \frac{F_y}{.5547} = \frac{F}{1}$$

$$\therefore F_y = \text{vertical component} = 10 \text{ kips}$$

$$F = \text{member force} = \underline{18.03 \text{ kips tension}}$$

Next, for vertical equilibrium, it is determined that the force in member 3-4 must be 10 kips pushing on the joint. These results are added to the diagram directly as shown in Fig. 5.12b. Note, the arrowheads are added to joints 1 and 4, too.

Step 3: Inspecting joint 4 (Fig. 5.12c), it is determined that member 1-2 has no horizontal force to equilibrate and, thus, is a "zero force member." Vertical equilibrium necessitates an upward R_3 reaction of 10 kips. These results and arrowheads are added in to Fig. 5.12c.

Step 4: Inspecting joint 1, a summation of the known member force components indicates R_1 must be 6 kips acting to the left (for horizontal equilibrium) and R_2 must be 2 kips acting upward for vertical equilibrium. These reactions are added to Fig. 5.12d, with the correct directions shown.

Note that the sequence of figures in Fig. 5.12 are presented to describe the visual method of joints in a step-by-step manner. In actual usage, only one figure of the truss is needed. The entire sequence of computations is done, joint by joint, on one figure. In the present example, Fig. 5.12d represents the single figure needed. Its contents are entered as one proceeds from joint to joint in the sequence and with the rationale described in Steps 1 to 4.

Example 5.6.   Repeat Example 5.2 by the visual method of joints.

Solution: The solution is executed in Fig. 5.13.

Step 1:   In Fig. 5.13a, joint 1 is identical to the preceding example and treated the same way.

Step 2:   Moving to joint 2, progress is stymied because both member forces enter each of the equilibrium conditions. As no other joint has two or fewer unknowns, Fig. 5.13b is used and the usual method of joints executed.

$$F_x = 15 - .8321F_3 + .6F_5 = 0$$

$$F_y = \quad - .5547F_3 - .8F_5 = 0$$

Solving simultaneously

$$F_3 = 12.01 \text{ kips tension}$$

$$F_5 = -8.32 \text{ kips} = 8.32 \text{ kips compression}$$

These forces are entered on Fig. 5.13c. Their components are determined by proportioning. As a check, both horizontal and vertical equilibrium are examined at joint 3 and found to be satisfied.

Step 3:   Proceeding to joint 3, member 1-4 must provide 4.99 kips in tension to satisfy horizontal equilibrium. An upward vertical reaction of 6.66 kips is needed to satisfy vertical equilibrium. Results are added to Fig. 5.13c.

Step 4:   Inspection of joint 1 (Fig. 5.13d), indicates a horizontal reaction of $R_1 = 6$ kips acting to the left and a vertical reaction of $R_2 = 5.34$ kips acting upward are needed. This completes the computations.

In the preceding example problem, simultaneous equations were used at joint 3 (in Step 2) to escape an impasse. This could be avoided by initially applying statics to a free-body diagram of the entire truss to calculate the reactions (actually only R_3 is needed). With the reactions known, the visual method of joints would be executed, without trouble, by examining joints 2, 4, 3, and 1, in that order. However, the visual method of joints is unsuitable for trusses such as shown in Fig. 5.14. At each joint along either an inclined top or bottom chord, simultaneous equations develop, which destroys the simplicity of the solution. In contrast, if the reactions are calculated separately, the visual method of joints is well suited for trusses with both chords horizontal. The truss in Fig. 5.15 is well suited for ready solution by the visual method of joints.

5.4.3 Method of Sections

The method of sections is based on recognizing that in truss analysis, the analyst is not limited to isolating free-body diagrams of the individual joints only. Any portion of a given truss can be isolated as a free-body diagram. When a free-body diagram is isolated, only two equations of equilibrium are useful. This is due to the concurrency of the forces. General free-body diagrams do not necessarily involve concurrent forces. When the forces are nonconcurrent and planar, three equations of equilibrium can be used. Sometimes, this is advantageous in the analysis of planar trusses. The method of sections will be demonstrated by examples.

Example 5.7.   Calculate the forces in the members labeled ①, ②, and ③ in Fig. 5.16a. Use the method of sections.

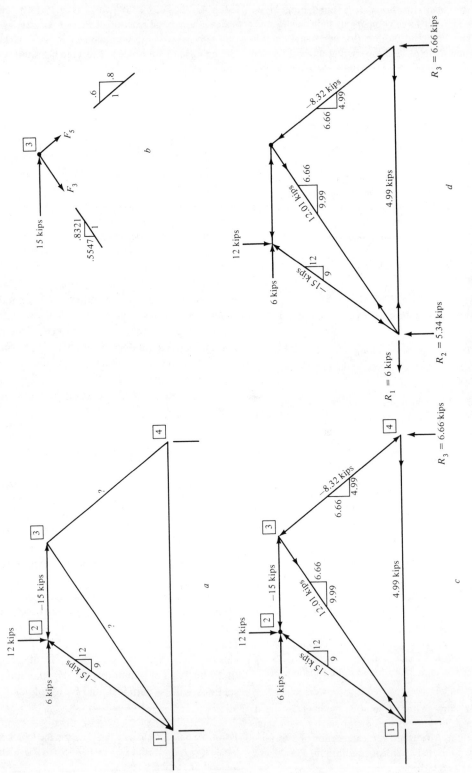

Fig. 5.13.   Truss for Example 5.6 (visual method of joints).

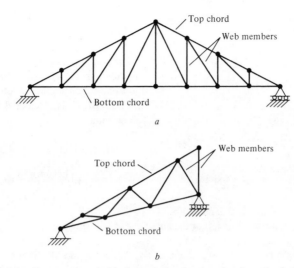

Fig. 5.14.   Trusses unsuitable for solution by the visual method of joints.

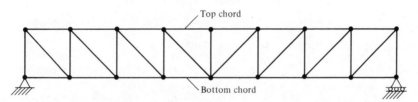

Fig. 5.15.   Truss well suited for solution by the visual method of joints.

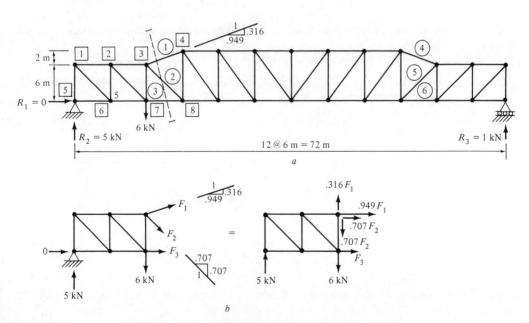

Fig. 5.16.   Truss for Example 5.7 (method of sections).

Solution: The reactions are calculated by applying statics to the complete truss.

$$F_x = R_1 = 0 \qquad \therefore R_1 = \underline{0}$$

$$M_{\text{left support}} = 6(12) - R_3(72) = 0$$

$$\therefore R_3 = \underline{1 \text{ kN} \uparrow}$$

$$F_y = R_2 + R_3 - 6 = 0$$

$$R_2 + R_3 - 6 = 0$$

$$\therefore R_2 = \underline{5 \text{ kN} \uparrow}$$

A section (1-1) is cut through the members of interest. The portion of the truss to the left of the cut is isolated as a free-body diagram (Fig. 5.16b). For convenience, F_1 is replaced by its horizontal and vertical components. The required forces are calculated as follows:

Example 5.8.   Calculate the force F_5 in the truss of Example 5.2.

Solution: A cut is made through members 4 and 5 as shown in Fig. 5.17a. A free-body diagram of the portion of the truss to the left of the cut is shown in Fig. 5.17b. F_5 is calculated by summing moments about joint 1:

$$M_1 = .8F_5(12) + .6F_5(8) + 6(8)$$

$$+ 12(6) = 0$$

$$\therefore F_5 = -8.33 \text{ kips}$$

$$= \underline{8.33 \text{ kips compression}}$$

Within round-off differences, this is the same as the result obtained in Example 5.2. The use of simultaneous equations in Example 5.2 could have

$$M_{\text{joint 3}} = 5(12) - F_3(6) = 0$$

$$\therefore F_3 = \underline{10 \text{ kN tension}}$$

$$M_{\text{joint 8}} = 5(18) - 6(6) + .949F_1(6) + .316F_1(6) + F_2(0) = 0$$

$$\therefore F_1 = -7.11 \text{ kN} = \underline{7.11 \text{ kN compression}}$$

$$F_y = 5 - 6 - .707F_2 + .316F_1 = 0$$

$$-1 - .707F_2 + .316(-7.11) = 0$$

$$\therefore F_2 = -4.59 \text{ kN} = \underline{4.59 \text{ kN compression}}$$

Note that, after similarly calculating the forces in the members labeled 4, 5, and 6, the rest of the truss could be readily analyzed by the visual method of joints.

been avoided by the method of sections in the manner just described.

Example 5.9.   Calculate the member forces in the truss of Fig. 5.18.

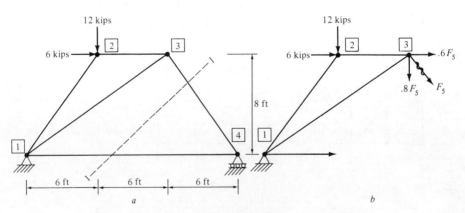

Fig. 5.17.   Truss for Example 5.8 (method of sections).

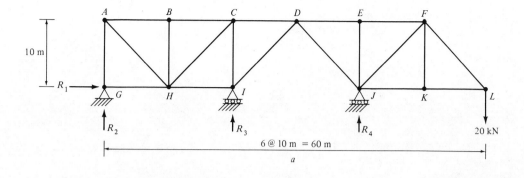

6 @ 10 m = 60 m

a

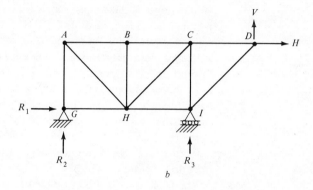

b

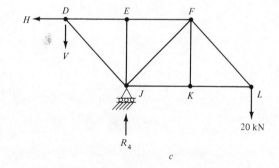

c

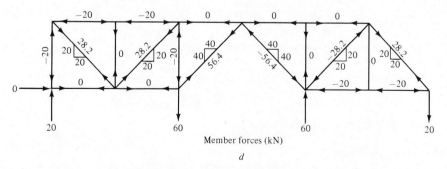

Member forces (kN)

d

Fig. 5.18.   Truss for Example 5.9.

Solution: This compound truss has four external reactions and they cannot all be calculated by applying statics to the complete truss. Although R_1 is obtained by

$$F_x = R_1 = 0 \quad \therefore R_1 = 0$$

the rest of the reactions remain undetermined.

The solution proceeds by separating the component simple trusses into individual free-body diagrams (Figs. 5.18b and c).

From Fig. 5.18b:

$$F_x = R_1 + H = 0$$

$$0 + H = 0 \quad \therefore \underline{H = 0}$$

From Fig. 5.18c:

$$M_{\text{roller}} = -V(10) + 20(20) = 0$$

$$\therefore \underline{V = +40 \text{ kN}}$$

$$F_y = -V + R_4 - 20 = 0$$

$$-(40) + R_4 - 20 = 0$$

$$\therefore \underline{R_4 = 60 \text{ kN} \uparrow}$$

From Fig. 5.18b:

$$M_{\text{roller}} = -R_2(20) - V(10) + H(10) = 0$$

$$-R_2(20) - (40)(10) + 0(10) = 0$$

$$\therefore \underline{R_2 = 20 \text{ kN} \uparrow}$$

$$F_y = R_2 + R_3 + V = 0$$

$$20 + R_3 + 40 = 0$$

$$\therefore R_3 = -60 \text{ kN}$$

$$\underline{= 60 \text{ kN} \downarrow}$$

With the reactions established, member forces are determined by the visual method of joints. The in-place calculations are shown in Fig. 5.18d.

5.5   GEOMETRICAL CONCEPTS

5.5.1   The Basic Truss

Planar trusses have been defined earlier as a configuration of straight-line elements framed together by pinned connections. In a sense, a

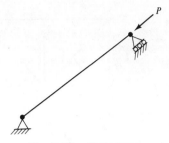

Fig. 5.19.   Basic truss.

single member with adequate support to prevent collapse is the simplest truss form. Such a member is depicted in Fig. 5.19. The pin and roller support condition provides the minimum amount of restraint necessary to maintain equilibrium under all possible loadings. Although it is physically possible to apply transverse loads to such a member, the usual assumption in truss analysis is that loads are applied only at the joints. For the basic truss shown in Fig. 5.19, the load can be applied only in the one direction shown.

The basic truss can be extended to a stable, multimember planar truss in one of two ways. The two possibilities are demonstrated in Fig. 5.20. The addition of a single member, as shown in Fig. 5.20a, must be accompanied by the introduction of an additional roller support at the end of the new member. This is the minimum condition required to maintain stability

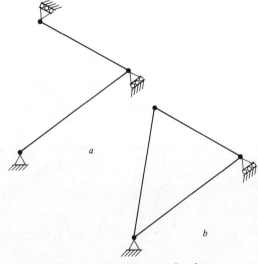

Fig. 5.20.   Extending the size of a planar truss.

in the system. Figure 5.20b illustrates the alternative step that might be taken to extend the size of the basic truss. By adding two members that interconnect at a common joint, stability is already maintained, and an additional support becomes unnecessary. In either Fig. 5.20a or b, either an additional support or member may be added if desired, as shown in Fig. 5.21, but they are actually unnecessary for geometric stability. The various truss forms shown in Fig. 5.20 and 5.21 will, of course, respond to load in different ways. Selection of a particular configuration is a part of design and depends upon the structural purpose of the system and the magnitude and direction of the loads that are to be resisted by the chosen truss.

5.5.2   Stability and Statical Determinacy

5.5.2.1   General Criteria

Recognition of the two concepts for extending trusses permits one to establish a criterion for assessing the stability of simple trusses. The basic truss has one member, two joints, and three reaction forces, and satisfies the relationship:

$$NM = 2 \times NJ - NRF \qquad (5.5)$$

in which

$$NM = \text{total number of truss members}$$
$$NJ = \text{total number of joints (including those that are supported)}$$
$$NRF = \text{total number of reaction forces}$$

By extending the basic truss as shown earlier in Fig. 5.20a, one member, one joint, and one support reaction have been added to the original structure. By extending the basic truss as shown in Fig. 5.20b, two members, one joint, and no reaction forces have been added to the original structure. Inspection of Eq. 5.5 indicates that the equality of the relationship is maintained despite altering the basic truss in either of these ways.

For case a,

$$NM + 1 = 2 \times (NJ + 1) - (NRF + 1)$$
$$\therefore NM = 2 \times NJ - NRF$$

For case b,

$$NM + 2 = 2 \times (NJ + 1) - (NRF + 0)$$
$$\therefore NM = 2 \times NJ - NRF$$

Consequently, satisfaction of Eq. 5.5 indicates the stability of truss forms created by any such extensions of the basic truss form.

The trusses in Fig. 5.21 are examples of structures that have additional members or supports in "excess" of the minimal number needed to extend the basic stable truss. In both cases, Eq. 5.5 becomes an inequality such that the left-hand side exceeds the right-hand side. This occurs because either NM is increased or NRF is decreased with no change in NJ. The observation derived from this study is that satisfaction of the inequality

$$NM \geq 2 \times NJ - NRF \qquad (5.6)$$

is an indicator of the stability of simple trusses.

If any member is removed from a stable truss it becomes statically unstable *internally*. In Eq. 5.6, the left side is reduced (NM decreases)

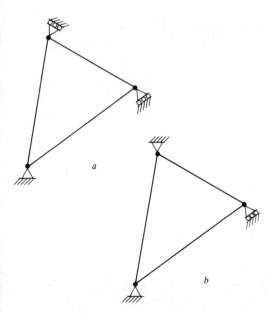

Fig. 5.21.   Multimember trusses.

while the right side remains the same. Similarly, if any reaction is removed from a stable truss it becomes statically unstable *externally*. In Eq. 5.6, the right side is increased (*NRF* decreases) while the left side remains the same. The inference is that the condition

$$NM < 2 \times NJ - NRF \qquad (5.7)$$

is an indicator of statical instability.

Equation 5.7 can be formulated by alternative reasoning. For statically determinate trusses, it has been demonstrated that the number of equations obtained by applying statics independently to each joint of the truss are sufficient in number to permit determination of all unknown force quantities. In the general case of *NF* unknown member forces and *NRF* unknown reactions, this sufficiently can be stated in equation form as

$$NRF + NF = 2 \times NF \qquad (5.8)$$

Thus,

$$NF = 2 \times NF - NRF \qquad (5.9)$$

Because $NF = NM$,

$$NM = 2 \times NJ - NRF$$

Thus, Eq. 5.7 is, indeed, a basis for establishing the statical determinacy of a planar truss.

The truss in Fig. 5.22a is statically determinate. By adding a member between joints A and C, the truss in Fig. 5.22b is obtained. One more unknown member force exists, but the number of joint equilibrium conditions remains the same. Consequently, the truss is statically indeterminate to the first degree.

If a pin support is added at joint C of the truss in Fig. 5.22b, the truss in Fig. 5.22c is obtained. One more reaction exists, but the number of joint equilibrium conditions is unchanged. Consequently, the truss is statically indeterminate to the second degree.

Observe, for case c in Fig. 5.22, Eq. 5.7 gives

$$3 > 2(3) - 4 = 2$$

and for case b it gives

$$3 > 2(3) - 5 = 1$$

The inference is that the degree of statical indeterminacy is the amount by which the left side of Eq. 5.7 exceeds the right side.

In summary, Eq. 5.7 can be used as an indicator of the classification of a simple truss according to

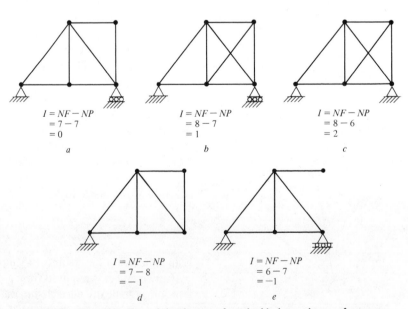

$$
\begin{array}{ccc}
I = NF - NP & I = NF - NP & I = NF - NP \\
= 7 - 7 & = 8 - 7 & = 8 - 6 \\
= 0 & = 1 & = 2 \\
a & b & c
\end{array}
$$

$$
\begin{array}{cc}
I = NF - NP & I = NF - NP \\
= 7 - 8 & = 6 - 7 \\
= -1 & = -1 \\
d & e
\end{array}
$$

Fig. 5.22. Determination of the degree of statical indeterminacy of a truss.

$NM = 2 \times NJ - NRF$;

truss is stable and statically

determinate

$NM > 2 \times NJ - NRF$;

truss is stable and statically

indeterminate to

$(NM - 2 \times NJ - NRF)$

degrees

$NM < 2 \times NJ - NRF$;

truss is unstable (5.10)

The above conditions can be expressed as

$$I = NM - (2 \times NJ - NRF) \quad (5.11)$$

in which I is the total degree of statical inde-terminacy. For unstable trusses, I is negative.

An alternate equation for I can be developed. For all trusses it is apparent that

$$2 \times NJ = NP + NRF \quad (5.12)$$

in which NP, the number of independent load directions, was defined earlier. In words, Eq. 5.12 states that each joint must have either two load directions (free joint), or one load direc-tion plus one reaction (roller support) or two reactions (pin support). Thus when $NP + NRF$ are added, the result must equal twice the num-ber of joints. Substitution of Eq. 5.8 into Eq. 5.12 establishes that

$$NP = NF \quad (5.13)$$

for statically determinate planar trusses.

For statically indeterminate trusses the de-gree of statical indeterminacy is given by

$$I = NF - NP \quad (5.14)$$

and the conditions

$I = 0$;   truss is stable and statically

determinate

$I > 0$;   truss is stable and statically

indeterminate

$I < 0$;   truss is unstable

are applicable in lieu of Eqs. 5.6 and 5.7. The author prefers Eq. 5.14 because it applies to all framed structures, if NF and NP are properly defined. Conditions 5.10 apply only to trusses. The reader is advised to use both conditions as a means of cross-checking the outcome. Re-gardless of which criterion is chosen, excep-tions to the rules do exist and caution is urged.

The five trusses shown in Fig. 5.22 demon-strate the validity and rationale of Eq. 5.14. The truss in Fig. 5.22a is clearly determinate and I is correctly zero. By adding a single member to the truss, as done in Fig. 5.22b, the truss becomes statically indeterminate internally to the first degree. The effect on Eq. 5.14 is to increase NF by one without changing NP, yielding a value of $I = 1$. If the roller support is then converted to a pin support (Fig. 5.22c), the truss becomes statically indeterminate to the second degree. This additional degree of stati-cal indeterminacy is one of external indeter-minacy because the number of unknown reac-tions has been raised by one. In Eq. 5.14, this change reduces NP by one (one degree of free-dom has been removed), and because NF is un-affected, I becomes 2.

The truss in Fig. 5.22e is obviously stati-cally unstable externally. The truss in Fig. 5.22e is obviously statically unstable inter-nally. Equation 5.14 gives the correct indica-tion in both cases.

5.5.2.2 Stability of Simple Trusses

Consider the trusses in Fig. 5.23. Extending the concept of building upon the basic truss by the addition of triangular panels, truss config-urations such as the one shown in case a can be constructed. The numbering of the joints is in the order in which the framing is accom-plished. For case a, panel 1-2-3 is erected, then two members are connected to joint 4, two members are connected to joint 5, etc. Because a triangular panel is a stable assembly, the re-sulting simple trusses are inherently stable in-

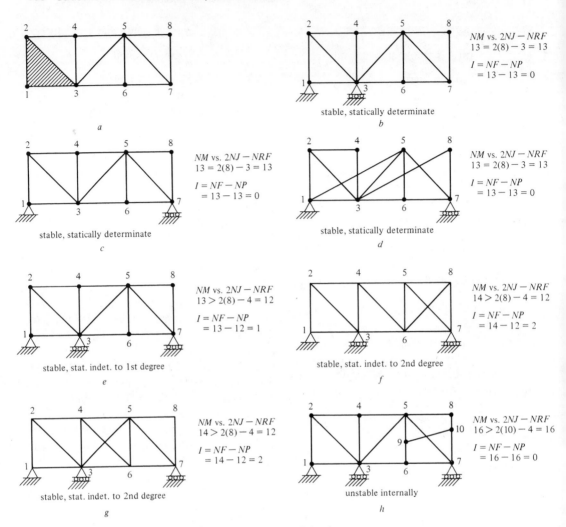

Fig. 5.23.   Stability of simple trusses.

ternally. To insure the external stability, it is only necessary to support the framework in such a way that at least three properly oriented support reactions can be developed. For example, the assembly in case a can be safely supported as shown in either case b or case c. In each case, Eqs. 5.6 and 5.14 indicate the existing stability and statical determinacy. Case d demonstrates the framing pattern could be irregular, but a stable, statically determinate truss still results. Again, the joint numbering indicates the order of assembly.

Following the explanation given earlier, the statically determinate truss in case b becomes statically indeterminate if either additional support reactions, case e, or new members framing between existing joints, cases f and g, are cre-

ated. Case h illustrates that the addition of a member (9-10) which creates new joints (9 and 10) can lead to instability. The truss in Fig. 5.23h may appear to be stable, but closer examination shows it is not. A horizontal load applied to joint 9 would require a force to develop in member 9-10, because the joint condition $\Sigma F_x = 0$ requires it. However, at joint 10, the $\Sigma F_x = 0$ condition indicates that member 9-10 must be a zero force member. This is impossible.

5.5.3   Stability of Compound Trusses

Compound trusses present additional stability considerations. Consider the trusses in Fig. 5.24. Each case is a compound truss config-

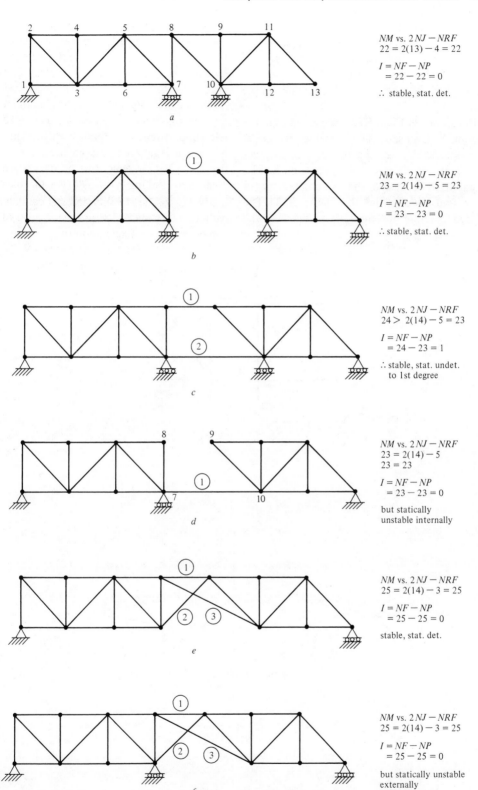

Fig. 5.24.   Stability of compound trusses.

ured, in different ways, from the same basic pair of simple trusses. For ease of discussion, the truss with joints numbered 1 through 8 is called simple truss 1 (ST1) and the truss with joints numbered 8 through 13 is called simple truss 2 (ST2).

In case a, ST1 and ST2 are pinned together at joint 8. By itself, ST1 is stable. By itself, ST2 is unstable, needing two additional reaction forces. By interconnection to joint 8, ST2 is stabilized because of the two components of force transmitted from the stable ST2. The needed external reactions exist because the force components are ultimately transmitted to the supports of ST1. Overall, the compound truss has four properly arranged support reactions and is statically stable externally and internally (being comprised of triangles). The two criteria established (Eqs. 5.6 and 5.14) each indicate the stability and statical indeterminacy.

Case b is identical to case a except ST2 has two support reactions, both vertical. Thus, an horizontal reaction would be needed to stabilize it if standing alone. Instead, this is accomplished by linking it to ST1 via the member marked ①. The stability and statical determinacy (evident internally and externally) of the compound truss is again confirmed by the two

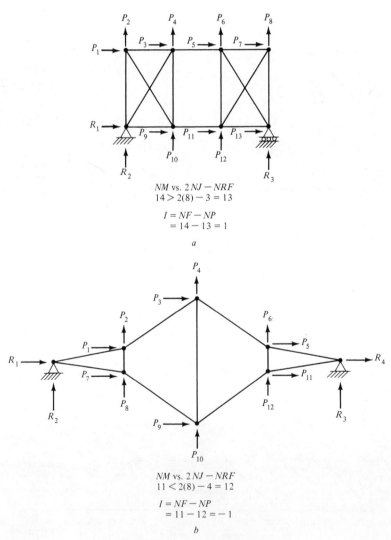

NM vs. $2NJ - NRF$
$14 > 2(8) - 3 = 13$

$I = NF - NP$
$= 14 - 13 = 1$

a

NM vs. $2NJ - NRF$
$11 < 2(8) - 4 = 12$

$I = NF - NP$
$= 11 - 12 = -1$

b

Fig. 5.25.   Trusses for Example 5.10.

criteria. Case c is the same as case b, except member ② has been added. As expected, the criteria confirm it is stable and statically indeterminate to the first degree.

In case d, ST2 by itself has a single pinned support and thus requires an additional reaction force. Instead, member 1 is added to link ST2 to the stable ST1. Both criteria suggest this stabilizes the system. However, cutting through member 1 by the method of sections reveals the instability of the truss. Using the resulting free-body diagram of ST2 and summing moments about the pin support, shows the force in member ① and the reactions at the pin are concurrent. There is no way to equilibrate any load applied to ST2, except those passing through the support. The problem can be overcome either by replacing member ① by a member connected to joints 7 and 9 or, as in case b, between joints 8 and 9.

In case e, ST1 and ST2 are both unstable if standing alone. ST1 requires an additional reaction and ST2 requires two additional reactions. Adding members ①, ②, and ③, stabilizes the trusses. The resulting compound truss is statically determinate. By contrast, the

truss in case f is not stabilized by this action. Despite ST1 and ST2 requiring one additional reaction and two additional reactions, respectively, as in case e, there is no possibility to link them into a stable compound truss. No matter how they might be linked, the compound truss is geometrically unstable because it is supported by three horizontal rollers (i.e., by parallel reactions) and cannot resist horizontal loads. Cases d and f demonstrate that each of the two criteria for stability and statical determinacy is a necessary but not a sufficient condition for such to exist.

5.5.4 Example Problems

Example 5.10.   Assess the stability of the trusses given in Fig. 5.25. If unstable, show a physical reason.

Solution: In each figure, the possible load directions *NP* have been shown. The two criteria for assessing stability by equations are also shown and utilized.

Case a: Both criteria indicate the truss is stable. However, Fig. 5.26a demonstrates the truss is un-

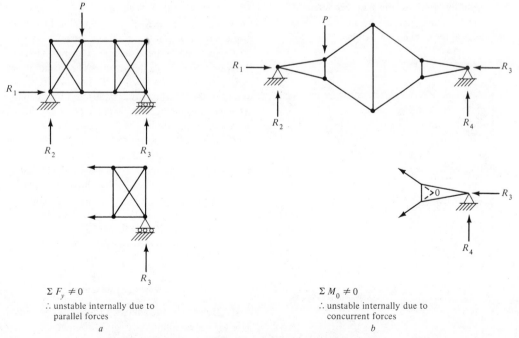

$\Sigma F_y \neq 0$
∴ unstable internally due to
  parallel forces
a

$\Sigma M_0 \neq 0$
∴ unstable internally due to
  concurrent forces
b

Fig. 5.26.   Sources of instability—Example 5.10.

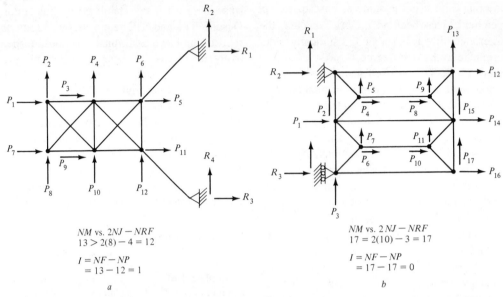

$$NM \text{ vs. } 2NJ - NRF$$
$$13 > 2(8) - 4 = 12$$

$$I = NF - NP$$
$$= 13 - 12 = 1$$

a

$$NM \text{ vs. } 2NJ - NRF$$
$$17 = 2(10) - 3 = 17$$

$$I = NF - NP$$
$$= 17 - 17 = 0$$

b

Fig. 5.27.   Trusses for Example 5.11.

stable. If loaded as shown, R_3 has a nonzero value. Cutting a section through the middle panel reveals the instability. The parallel member forces cannot resist R_3.

Case b: Both criteria indicate an unstable truss. Figure 5.26b shows the reason. If loaded as shown, R_4 is nonzero. The section shown removed reveals the instability. Because the member forces and R_3 are concurrent at point 0, the moment about that point caused by R_4 cannot be resisted.

Both trusses are statically stable externally but statically unstable internally.

Example 5.11.   Assess the stability of the trusses given in Fig. 5.27. If unstable, show a physical reason.

Solution: In each figure, the possible load directions and the stability criteria are shown.

Case a: Both criteria indicate the truss is stable. However, Fig. 5.28a demonstrates an instability exists due to concurrency of the internal forces exposed in the free-body diagram in the lower figure. The moment created about point *O* by the load *P* cannot be resisted.

Case b: Both criteria indicate the truss is stable. However, Fig. 5.28b demonstrates instability exists. The section removed in the lower figure cannot be in equilibrium when *P* is applied. The parallel member forces provide no vertical component.

Both trusses are statically stable externally but statically unstable internally.

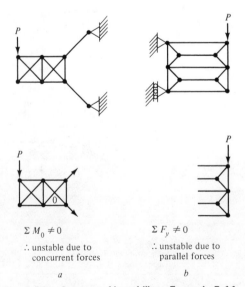

$$\Sigma M_0 \neq 0$$
∴ unstable due to concurrent forces

a

$$\Sigma F_y \neq 0$$
∴ unstable due to parallel forces

b

Fig. 5.28.   Sources of instability — Example 5.11.

PROBLEMS

5-1.   Determine the member forces in each truss. Use the direct solution of equilibrium equations approach to truss analysis (see Section 5.4.1).

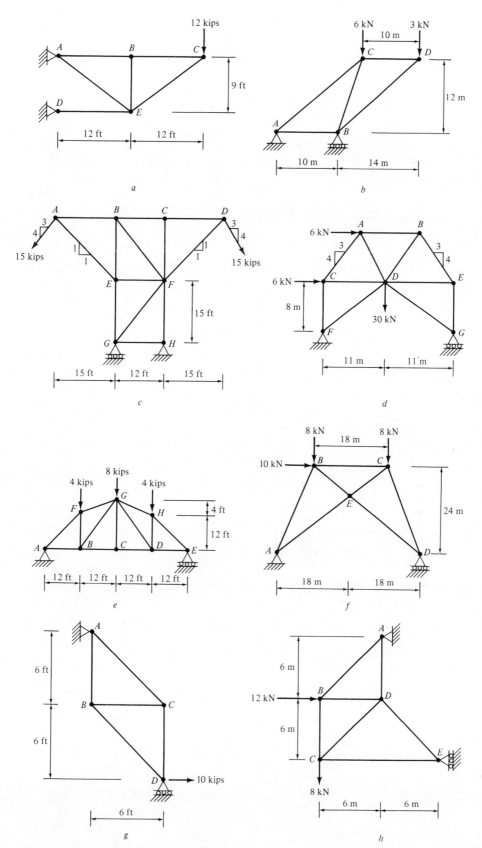

a

b

c

d

e

f

g

h

163

5-2. Repeat Problem 5-1, but use the analytical method of joints (e.g., see Example 5.3).

5-3. Repeat Problem 5-1, but use the visual method of joints (e.g., see Example 5.5).

5-4. Determine the member forces in each truss by the method of joints.

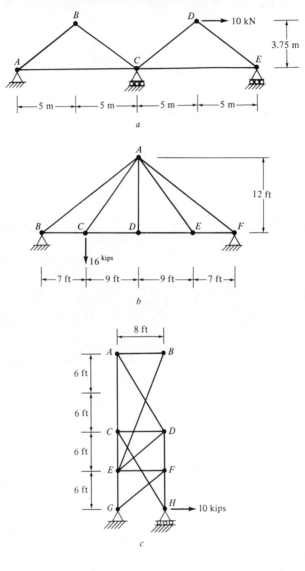

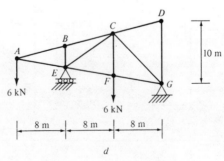

5-5. Determine the member forces in each truss by the method of joints. Where helpful, use the method of sections.

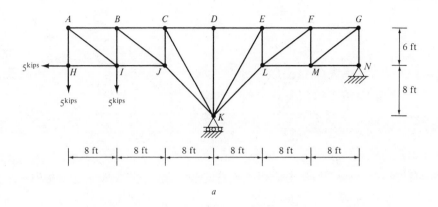

a

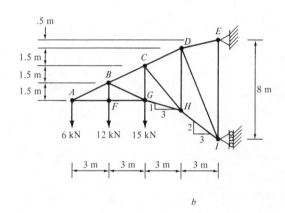

b

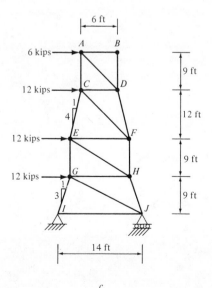

c

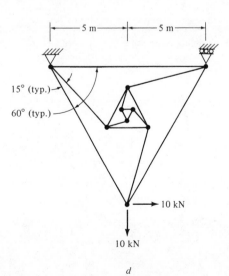

d

5-6. A unit load acting downward is placed at joint B_1, then moved to joint B_2, then moved to joint B_3, and so on. Determine the force in member $T_2 T_3$, $T_3 B_4$, and $T_3 B_3$ for each loading, and plot the results. Show the magnitude of the force on the y-axis, treating tension as positive. Connect the ordinates with straight lines. (Note: The result is commonly referred to as an *influence diagram*. Comment on the usefulness of such a diagram.) (Hint: Use the method of sections.)

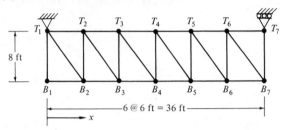

5-7. Repeat Problem 5-6 for the given truss, except determine the forces in members $B_3 B_4$ and $M_1 B_4$.

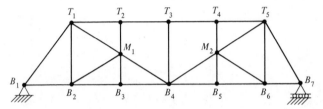

5-8. Repeat Problem 5-6 for the given truss.

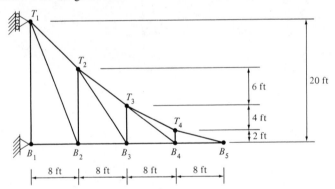

5-9. Calculate the force in members *CD*, *CP*, *JP*, and *JK* of the given truss.

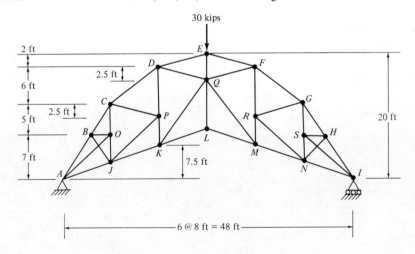

5-10. Repeat Problem 5-9 for members *EQ* and *LQ*.

5-11. Determine if each truss is statically determinate and stable. If so, calculate the member forces.

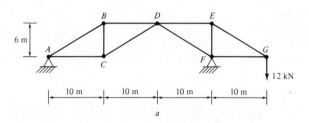

a

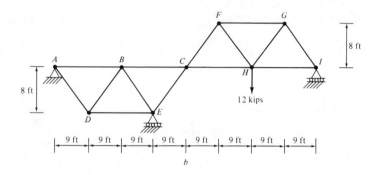

b

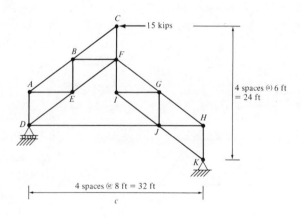

c

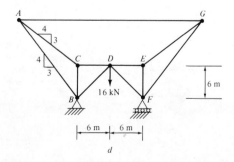

d

5-12. Assess the stability and statical determinacy of each truss. If unstable, show a physical reason using sketches.

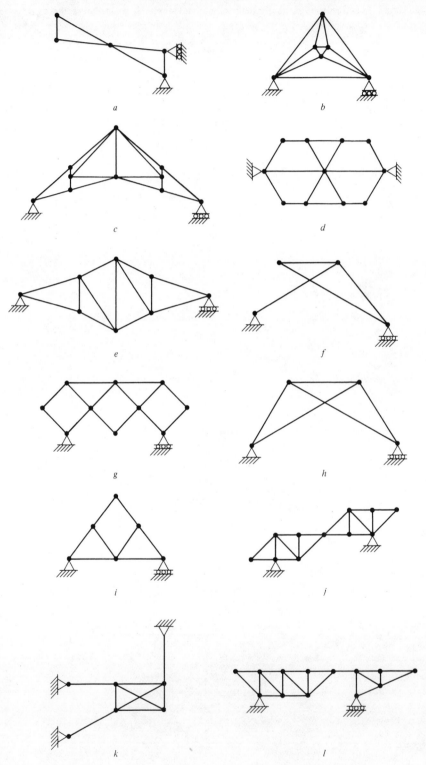

6

Analysis of Determinate Planar Frames

6.1 GENERAL COMMENTS

Planar frames are comprised of an array of interconnected, flexural members that lie in and deform in a common plane. Their use as a basic component in the skeletal form of tiered buildings was described in Chapter 2. Unlike trusses, members in a frame are depended upon to provide flexural as well as axial strength and stiffness. In essence, the beam element is the building block for all planar frames. Mathematically, the elastic curve of a beam can be established if the loading and support conditions are known. Quadruple integration, as represented by Eq. 3.27, is the mode for accomplishing the task. If the moment variation is known in advance, double integration or its numerical equivalent, the moment-area method, is a suitable mode. In this chapter, the mathematical and geometrical concepts that underlie beam behavior will be extended to have applicability to planar frames.

Planar frames must be provided sufficient support to insure statical geometrical stability under all possible loadings. As a minimum, the presence of a single fixed support or an appropriately placed combination of a pin support and a roller support is required. Several examples of stable frames are depicted in Fig. 6.1. Each of these structures has at least three reaction forces, and a support arrangement that insures geometric stability. Herein, all frames will be configured to be stable, and the reader need not be concerned with this determination.

In Chapter 2, it was stated that members in a planar frame are capable of developing three types of internal forces: axial and shear forces, and flexural moment. Determination of the variation in these internal forces and the corresponding deformations for each member in any framework are both part of the objectives of frame analysis. As a further aim, it is necessary to study how the various members interact to produce the displaced shape taken by the loaded structure. Calculation of the displaced position of each joint will be shown to be fundamental to that aim.

In this chapter, only the analysis of statically determinate frames is described. Structural analysis of determinate frames is a direct extension of the concepts developed for the analysis of beams. Indeterminate frames of either type require special methods of analysis treated in several subsequent chapters.

6.2 STATICALLY DETERMINATE FRAMES

6.2.1 Introduction

In earlier chapters, statical determinacy has been defined, in general, and explained for beams and planar trusses. A planar frame is statically determinate if all the force quantities

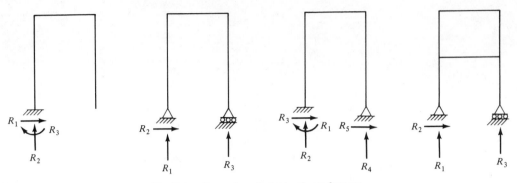

Fig. 6.1.   Examples of stable planar frames.

acting at any point on or within a frame can be determined by the sole application of the conditions of statics. Force quantities that occur in planar frames consist of the reaction forces that maintain external equilibrium, and the three internal forces that create internal equilibrium at any given location. If an individual member in a frame is extracted and its free-body diagram drawn, it will appear as shown in Fig. 6.2. Generally, an axial force, shear force, and flexural moment will exist at each end of the member. Theoretically, with the end forces and loading known for all members, all other response quantities for each member can be determined by sequential use of the quadruple integration method. In this way the complete displaced shape of the framework can be established. Furthermore, it will be demonstrated that for any statically determinate frame, the moment-area theorems also provide a means for determining the generalized displacements at any point in the system. Hence, one aim in the analysis is to compute the end forces for all members. In a statically determinate frame, this can be accomplished by applying the principles of statics.

Quadruple integration and the moment-area

method permit the determination of flexural displacements. The axial force in each member causes axial deformation, which contributes to the overall displaced shape of the complete framework. In classical methods of structural analysis of ordinary frameworks, axial deformation is neglected. Usually the first-order effects of axial deformation are incorporated in computer-based matrix methods of structural analysis. Second-order effects, such as interaction of axial force with the transverse displacement (the P-Δ effect), the coupled member stiffness, etc. can also be included in computer-based methods. Because nonlinear analysis is required, these refinements are made on a selective basis. The present chapter is limited to first-order treatment of planar frames with axial deformation neglected. The conjugate frame method, slope-deflection method, and moment-distribution method covered in subsequent chapters include the same limitation. Computer methods for either including or neglecting axial deformation are also presented subsequently.

6.2.2   Independent Unknown Forces

A statically determinate frame is shown in Fig. 6.3a. A simple formula for establishing determinacy in a planar frame will be developed in Section 6.6, but for this frame it will be demonstrated by investigation. Free-body diagrams of each member and the connecting joints are drawn in Fig. 6.3b. At each section where the frame has been cut, three internal forces exist unless support conditions or internal hinges

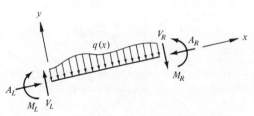

Fig. 6.2.   Internal forces in a framed member.

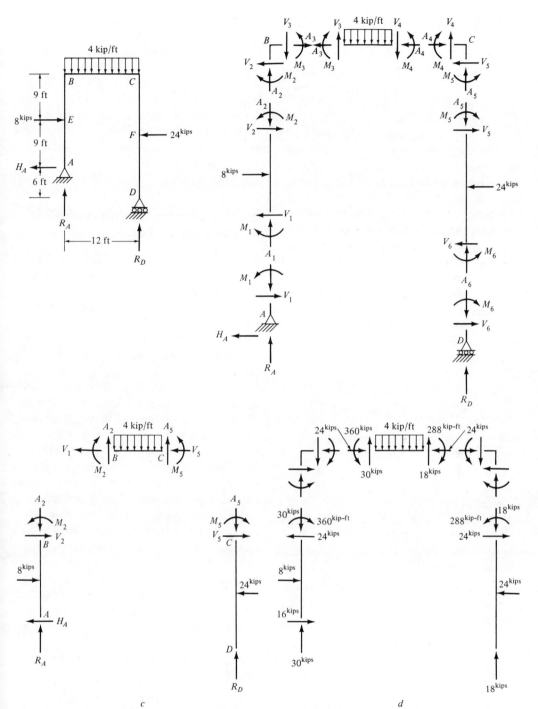

Fig. 6.3.   Internal forces in the planar frame of Example 6.1.

dictate the absence of some of them. For example, member AB has no internal moment at end A by virtue of the pin support at that point.

Similarly, member CD has neither an internal moment nor a shear force because of the roller support at point D.

Twenty-one individual forces are indicated in Fig. 6.3b: the six moments M_1–M_6, the six shear forces V_1–V_6, the six axial forces A_1–A_6, and the forces due to reactions, H_A, R_A, and R_D. Knowledge of the magnitudes of these 21 force quantities enables the determination of any other force response quantity. The statical determinacy of the frame is established by counting the number of statics equations available for use. Seven separate free-body diagrams are drawn in Fig. 6.3b. Consequently, 21 equilibrium equations can be formed by applying statics to each of the different free-body diagrams. A simultaneous solution of these equations would yield the desired force values, and it is evident that the frame is, indeed, statically determinate.

The 21 forces identified in Fig. 6.3b are not independent. It is possible to reduce the number of unknowns by initially applying statics to each free joint in the frame as follows:

At joint A,

$$\Sigma F_x = V_1 - H_A = 0 \qquad \therefore V_1 = H_A$$

$$\Sigma F_y = R_A - A_1 = 0 \qquad \therefore A_1 = R_A$$

$$\Sigma M_A = M_1 = 0 \qquad \therefore M_1 = 0$$

At joint B,

$$\Sigma F_x = V_2 - A_3 = 0 \qquad \therefore A_3 = V_2$$

$$\Sigma F_y = A_2 - A_3 = 0 \qquad \therefore V_3 = A_2$$

$$\Sigma M_B = M_2 - M_3 = 0 \qquad \therefore M_3 = M_2$$

At joint C,

$$\Sigma F_x = A_4 - V_5 = 0 \qquad \therefore A_4 = V_5$$

$$\Sigma F_y = A_5 + V_4 = 0 \qquad \therefore V_4 = -A_5$$

$$\Sigma M_C = M_4 - M_5 = 0 \qquad \therefore M_4 = M_5$$

At joint D,

$$\Sigma F_x = V_6 = 0 \qquad \therefore V_6 = 0$$

$$\Sigma F_y = R_D - A_6 = 0 \qquad \therefore A_6 = 0$$

$$\Sigma M_D = M_6 = 0 \qquad \therefore M_6 = 0$$

As a consequence the member free-body diagrams are redrawn as shown in Fig. 6.3c. With the free-body diagrams of Fig. 6.3c available, the number of unknown quantities has been reduced to nine independent forces. The unknown forces are moments M_2 and M_5, shear forces V_2 and V_5, axial forces A_2 and A_5, and the three reaction forces. The number of usable statics conditions also have been reduced to nine so the system remains statically determinate, as expected.

6.2.3   Calculation of the Independent Unknown Forces

To identify the independent unknown forces in a statically determinate frame, all conditions of statics pertinent to the joints are exhausted. Therefore, calculation of the magnitude of the independent unknown forces must involve the application of statics to the individual members and the frame as a whole. Three approaches are possible.

The first approach is fundamental. Statics is applied to each member and the set of equations collected. In the case of Fig. 6.3c, three members exist and nine equations result. The nine equations contain the nine independent unknown forces which are determined by simultaneous solution. This is a cumbersome technique and alternatives are warranted.

As a second approach, all unknown forces can be determined by applying statics to each member, one at a time, in a judicious, orderly fashion. Each time a member is chosen for isolation as a free body diagram, it must contain no more than three unknown forces. After each force is calculated, it becomes known in subsequent statics computations. Once all members are treated, no unknowns remain.

In the third approach, a two-step procedure is used:

1. Apply statics to the frame as a whole to determine external reactions.
2. Apply statics to each member, one at a time, in a judicious orderly fashion as described for the second approach. Continue until all remaining unknowns have been determined.

The preliminary use of external statics to calculate the reactions has the subtle implication that there is an extra member free body diagram remaining after completion of step 2. The need for this free-body diagram to be in equilibrium is a check on the solution. Conversely, in the second approach to determining the independent unknown forces, the free-body diagram of the frame as a whole is available at the conclusion. It can be used in checking the results obtained for the external reactions.

As a reminder, the application of statics to the joints to reduce the unknowns to the independent quantities is prerequisite to each of the three approaches just described. Application of statics at each joint can be done visually and Fig. 6.3c then drawn immediately. If so, the analyst would likely use the labeling A_1, V_1, M_1, A_2, V_2, and M_2 in place of A_2, V_2, M_2, A_5, V_5, and M_5, respectively.

Example 6.1. Calculate the reactions and internal forces for the planar frame depicted in Fig. 6.3a.

Solution: Applying statics to the members in Fig. 6.3c yields the following:
Member AB:

$$\Sigma F_x = -H_A - V_2 + 8 = 0 \qquad (a)$$

$$\Sigma F_y = R_A - A_2 = 0 \qquad (b)$$

$$\Sigma M_A = M_2 - V_2(18) - 8(9) = 0 \qquad (c)$$

Member BC:

$$\Sigma F_x = V_2 + V_5 = 0 \qquad (d)$$

$$\Sigma F_y = A_2 + A_5 - 4(12)(6) = 0 \qquad (e)$$

$$\Sigma M_B = M_2 - M_5 + 4(12)(6) = 0 \qquad (f)$$

Member CD:

$$\Sigma F_x = V_5 - 24 = 0 \qquad (g)$$

$$\Sigma F_y = R_D - A_5 = 0 \qquad (h)$$

$$\Sigma M_C = M_5 + 24(12) = 0 \qquad (i)$$

Equations a through i have been developed for illustration of the procedure. They can be solved simultaneously for the unknowns, but this is bypassed in favor of other solution methods.

Alternate Solution: The member forces can be established by a selective application of statics to the member free-body diagrams given in Fig. 6.3c.
Member CD:

$$\Sigma F_x = 24 - V_5 = 0$$
+ left $\qquad \therefore V_5 = 24$ kips

$$\Sigma M_C = M_5 + 24(12) = 0$$
+ cw $\qquad \therefore M_5 = -288$ kip-ft

Member BC:

$$\Sigma F_x = V_5 + V_2 = (24) + V_2 = 0$$
+ left $\qquad \therefore V_2 = -24$ kips

Member AB:

$$\Sigma F_x = H_A - V_2 - 8 = H_A + (-24) - 8 = 0$$
left $\qquad \therefore H_A = -16$ kips $= 16$ kips $\rightarrow$

$$\Sigma M_A = M_2 - V_2(18) - 8(9)$$
+ ccw
$$= M_2 - (-24)(18) - 8(9) = 0$$
$$\therefore M_2 = -360 \text{ kip-ft}$$

Member BC:

$$\Sigma M_B = M_2 - M_5 + 4(12)(6) - A_5(12) = 0$$
+ cw
$$(-360) - (-288) + (288) - A_5(12) = 0$$
$$\therefore A_5 = 18 \text{ kips}$$

$$\Sigma F_y = A_2 + A_5 - 4(12) = 0$$
+ up
$$= A_2 + (18) - 48 = 0 \qquad \therefore A_2 = 30 \text{ kips}$$

Member AB:

$$\Sigma F_y = R_A - A_2 = R_A - (30) = 0$$
+ up
$$\therefore R_A = 30 \text{ kips} \uparrow$$

Member CD:

$$\Sigma F_y = R_D - A_5 = R_D - (18) = 0$$
+ up
$$\therefore R_D = 18 \text{ kips} \uparrow$$

In effect, this alternate solution corresponds to the solution of Eqs. a through i in the particular order g, i, d, a, c, f, e, b, and h. Physical examination of the members was employed to deduce this order. Thus, explicit ''simultaneous'' solution of the equations (e.g., by matrix inversion) is unnecessary. Instead, the equations can be solved in selective fashion, such as done in Example 5.3 for implementing the method of joints for planar trusses. Usually, more than one order of treating the equations is apparent and any one of them is satisfactory.

Alternate Solution: Statics is applied first to the frame as a free body itself. (Fig. 6.3a)

$$\Sigma M_A = 12R_D - 8(9) - 4(12)(6) + 24(6) = 0$$
+ ccw $$\therefore R_D = 18 \text{ kips } \uparrow$$

$$\Sigma F_y = R_A + R_D - 4(12) = 0$$
+ up $$\therefore R_A = 30 \text{ kips } \uparrow$$

$$\Sigma F_x = H_A - 8 + 24 = 0$$
left $$\therefore H_A = -16 \text{ kips } = 16 \text{ kips } \rightarrow$$

The calculated reactions are shown in Fig. 6.3d.
   Next, the equilibrium of each member is investigated. See Fig. 6.3c.
   Member AB:

$$\Sigma F_x = -V_2 + H_A - 8 = V_1 + (-16) - 8 = 0$$
+ left $$\therefore V_2 = -24 \text{ kips}$$

$$\Sigma F_y = R_A - A_2 = 30 - A_2 = 0$$
+ up $$\therefore A_2 = 30 \text{ kips compression}$$

$$\Sigma M_B = M_2 - H_A(18) + 8(9) = 0$$
+ ccw

$$= M_2 - (-16)(18) + 8(9) = 0$$
$$\therefore M_2 = -360 \text{ kip-ft}$$

   Member CD:

$$\Sigma F_x = 24 + V_5 = 0$$
+ left $$\therefore V_5 = 24 \text{ kips}$$

$$\Sigma F_y = R_D - A_5 = 18 - A_5 = 0$$
+ up $$\therefore A_5 = 18 \text{ kips compression}$$

$$\Sigma M_C = M_5 + 24(12) = 0$$
+ cw $$\therefore M_5 = -288 \text{ kip-ft}$$

As a check on the solution, member BC (not used in the preceding computations) will be investigated.
   Member BC:

$$\Sigma F_x = V_2 + V_5 = 24 + (-24) = 0$$
+ right

$$\Sigma F_y = A_2 + A_5 - 4(12)$$
+ up

$$= 30 + 18 - 48$$

$$= 0$$

$$\Sigma M_B = M_2 - M_5 + 4(12)(6) - A_2(12)$$
+ cw

$$= -360 - (-288) + (288)$$
$$- (18)(12)$$

$$= 0$$

Both alternate solutions, as expected, produce the same results, and the analyst must judge the relative expediency of the two methods for any given problem. The calculated values are presented pictorially in Fig. 6.3d. All forces are shown to be acting in their proper directions.

Example 6.2.   Determine the reactions and member end forces for the planar frame given in Fig. 6.4a.

Solution: Applying statics to the frame taken as a whole,

$$\Sigma F_x = H_A = 0$$
+ right $$\therefore H_A = 0$$

$$\Sigma F_y = R_A - 12 = 0$$
+ up $$\therefore R_A = 12 \text{ kN } \uparrow$$

$$\Sigma M_A = M_A + 12(20) = 0$$
+ cw

$$\therefore M_A = -240 \text{ kN } \cdot \text{ m } = 240 \text{ kN } \cdot \text{ m ccw}$$

Free-body diagrams of each joint and member are drawn in Fig. 6.4b. The reactions are shown in their correct directions. By inspection of equilibrium at joints A and C, the member end forces at A and C are determined. The unknown internal forces at end B of member AB are labeled A_1, V_1, and M_1. Inspection of equilibrium at joint B allows these forces to be transferred to end B of member B and in the directions shown. Thus, only three unknown independent internal forces exist. Their magnitudes are calculated as follows:

Member AB:

$\Sigma F_x = -A_1 = 0$
+ right $\therefore A_1 = 0$
$\Sigma F_y = 12 - V_1 = 0$
+ up

$\qquad\qquad \therefore V_1 = 12$ kN ↑

$\Sigma M_B = -240 - 12(14) + M_1 = 0$
+ ccw

$\qquad \therefore M_1 = -72$ kN · m $= 72$ kN · m ccw

These end forces can also be calculated using member BC and the reader should do so to verify the above results. Fig. 6.4c shows these forces, in their appropriate directions, acting on the member. Finally, when shown acting on member BC, the forces A_1 and V_1 are not acting as axial and shear forces, respectively. An additional sketch of member BC is shown in Fig. 6.4c with these forces replaced by components acting parallel to (axial component) and perpendicular to (shear component) the member.

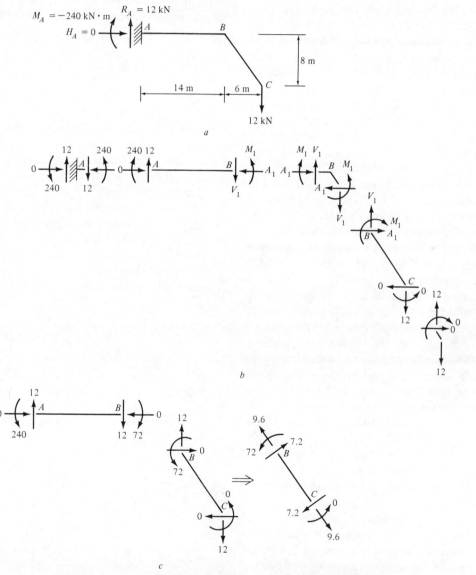

Fig. 6.4. Frame for Example 6.2. (Forces in kN; moments in kN · m.)

6.3   MATHEMATICAL ANALYSIS OF FLEXURAL BEHAVIOR

Solution of any of the differential equations that govern the flexural behavior of an individual framed member is a boundary value problem. The exact form of the resulting equation depends upon the specific loading and boundary conditions associated with the member under consideration. Known values of shear, moment, slope (or rotation), and displacement constitute boundary conditions. These values are used to evaluate the constants of integration. In classical methods of frame analysis, torsional loading is not considered and axial and shear deformations are ignored. Consequently, establishment of the equation of the elastic curve, $y(x)$, for a member constitutes the major objective. Other important quantities are then obtained from appropriate derivatives of $y(x)$.

Governing mathematical relationships that permit the determination of the equation of the elastic curve were formulated in Section 3.3. Knowledge of a member's loading and its end shears, moments, and flexural deformations is sufficient information to solve the applicable equations and apply the necessary boundary conditions. Generalized forces, created by subjecting a member to load, are computed by application of Eqs. 3.5 and 3.7. Consecutive integration of these equations yields the shear and moment equations for the member. Evaluation of the constants of integration requires known shear and moment values. In a determinate frame, both values are easily determined at the ends of every member and, if known, these are more than sufficient to complete the integration.

Subsequent to the determination of shear and moment equations, the equation of the elastic curve of the member can be obtained by one of two approaches. The first procedure involves use of the moment equation and Eq. 3.22. The double-integration procedure for execution of this step was described in Section 3.4. Alternatively, the quadruple integration method, Eq. 3.27, can be employed. Regardless of the integration method selected, member end defor-

mations must be known if the constants of integration are to be evaluated. The usual output of frame analysis includes these values.

Under certain conditions, a complete determination of the equation of the elastic curve for a planar frame member is possible without resorting to special frame-analysis methods. This occurs if the frame is statically determinate and the needed member end deformations are known by nature of the external support conditions (i.e., can be established without the use of frame analysis). For many statically determinate frames, this condition exists for at least one member. More significantly, for many statically determinate frames, the displaced shape of all members in the frame can be determined if at least one member has two known boundary conditions. Clearly, once the first member is analyzed, its end deformations become boundary conditions for the attached members. Sequential analysis, in some judicious order, of the remaining members is made possible as additional end deformations become known.

Examples 3.6 through 3.8 and 3.12 through 3.19 illustrated numerical application of the various governing differential equations to simple beams. Sections 6.3.1 and 6.3.2, which follow, discuss a similar mathematical analysis of the flexural effects in an individual member extracted from a planar frame. Fig. 6.5a de-

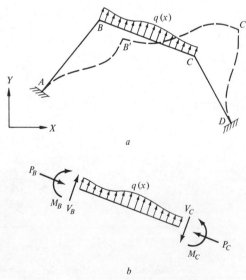

Fig. 6.5.   Individual member in a planar frame.

picts a planar frame, of arbitrary configuration and loading, in both its original and displaced positions. In particular, member BC shifts to location $B'C'$ and develops the internal forces indicated in Fig. 6.5b. It will be shown that knowledge of the loading and force quantities and end deformations in Fig. 6.5b is sufficient for a complete determination of the response of member BC.

6.3.1 Shear and Moment Equations

The free-body diagram of a member contains all quantities needed for a mathematical determination of its shear and moment equations. Member BC from Fig. 6.5a is shown in Fig. 6.6 in a reoriented position. Coordinate axes shown are local in nature; i.e., the x-axis aligns with the undeformed member axis. For the given loading, $q(x)$, the internal forces acting throughout the length of the member can be described by shear and moment equations written in terms of the local x coordinate. For the general case, Eq. 3.5 states

$$dV(x) = q(x)\,dx \qquad (3.5)$$

which implies

$$V(x) = \int q(x)\,dx + C_1 \qquad (6.1)$$

Furthermore, Eq. 3.7 requires

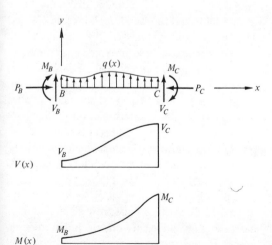

Fig. 6.6. Shear and moment diagrams.

$$dM(x) = V(x)\,dx \qquad (3.7)$$

or, by integration,

$$M(x) = \int V(x)\,dx + C_2 \qquad (6.2)$$

For discussion, it is preferable to express Eq. 6.2 in the form

$$M(x) = \int \int q(x)\,dx + C_1 x + C_2 \qquad (6.3)$$

Constants C_1 and C_2 are to be determined from known conditions regarding shear and moment in the given member. These conditions usually consist of known shear and moment values at the ends of the member. Examination of Eqs. 6.1 and 6.2 indicate that knowledge, either of both end moments or of one end moment and one end shear, is sufficient for a solution. However, if both shears and neither end moment have been determined, insufficient information exists for the evaluation of the constants of integration. Functions $V(x)$ and $M(x)$ are the desired shear and moment equations and can be plotted as shown in Fig. 6.6.

6.3.2 Equation of the Elastic Curve

With $M(x)$ for member BC in Fig. 6.6a determined according to Eq. 6.3, it would seem to be a straightforward matter to establish the member's curvilinear shape by the double-integration method

$$EIy'(x) = \int M(x)\,dx + C_3 \qquad (6.4)$$

and

$$EIy(x) = \int \int M(x)\,dx + C_3 x + C_4 \qquad (6.5)$$

However, care must be taken in assessing the subtle meaning of Eqs. 6.4 and 6.5. Specifically, because deformations predicted by these equations are in terms of local member coordinates, i.e., the x direction corresponds to the longitudinal axis of the member and $y(x)$ is measured transverse to the member. Conditions used for computation of C_3 and C_4 must be chosen properly. Two possibilities are evident:

1. Substitution of the member end displacements (measured in the local y direction) into Eq. 6.5.
2. Substitution of one end slope and one end displacement (measured in local coordinates) into each of Eqs. 6.4 and 6.5.

Unfortunately, neither option is possible until the member end deformations are known. Computation of these values has not yet been discussed in the text. Methods for computing the displacements at the joints of a planar frame are treated in subsequent sections. With knowledge of these methods, determination of the constants in Eqs. 6.4 and 6.5 becomes possible. It should be observed that the sequential execution of Eqs. 6.1, 6.2, 6.4, and 6.5 is, in fact, a particular form of the quadruple-integration procedure described in Section 3.3. The quadruple-integration procedure consists of the steps

$$Ely^{IV}(x) = q(x) \qquad (6.6)$$

$$Ely'''(x) = \int q(x)\, dx + C_1 \qquad (6.7)$$

$$Ely''(x) = \int \int q(x)\, dx + C_1 x + C_2 \quad (6.8)$$

$$Ely'(x) = \int \int \int q(x)\, dx$$
$$+ C_1 \frac{x^2}{2} C_2 x + C_3 \qquad (6.9)$$

$$Ely(x) = \int \int \int \int q(x)\, dx + C_1 \frac{x^3}{6}$$
$$+ C_2 \frac{x^2}{2} + C_3 x + C_4 \qquad (6.10)$$

The procedure inferred in Eqs. 6.1 through 6.5 is specialized in the sense that determination of the four constants of integration is based upon a set of conditions comprised of a com-

bination of two known internal forces and two known member deformations. The general quadruple-integration procedure is not limited in this way.

The particular pattern of quadruple integration described in the preceding paragraphs has been included as background for Section 6.5. Of immediate importance is the fact that sign conventions established in Chapter 3 for loads, forces, and deformations remain the same in the context of the present chapter. For clarity, these conventions are briefly restated here:

1. Positive shear acts upward on the left end of a member and downward on the right end.
2. Positive moment causes compression in the upper fibers and tension in the lower fibers; i.e., it causes positive curvature in the mathematical sense.
3. Positive slope is in agreement with the mathematical definition of slope (local coordinates).
4. Positive displacement is in the direction of the positive y-axis (local coordinates).

6.3.3   Calculation of Frame Displacements

A complete structural analysis of a planar frame includes a determination of its displaced position. It is possible to accomplish this task by either the quadruple- or double-integration method if shear and moment diagrams have been established for each member of the frame. The integration methods can be applied to all members one at a time provided that they are chosen in a judicious sequence. Each time a member is selected, it is necessary to insure that a sufficient number of known boundary conditions exist to allow determination of the two constants of integration. Support conditions and previously determined displacement quantities comprise the source of the boundary conditions for the member at hand. For determinate frames, the double-integration method is always directly applicable because shear and moment diagrams are easily drawn. Indeterminate frames require preliminary investigations using one of many *frame analysis methods* before such diagrams can be drawn. The usual outcome of frame analysis methods is either the member end forces or the member end deformations. With the former results, shear and moment diagrams can be drawn directly and the double-integration method is viable. With the latter results, shear and moment diagrams *cannot* be drawn directly, and double integration cannot be applied expeditiously. However, knowledge of each member's end deformations and loading instead enables direct use of the quadruple-integration method.

Example 6.3.   Analysis of the frame shown in Fig. 6.7a indicates that end forces for member BC are as shown in Fig. 6.7b. For member BC, determine (a) the shear and moment equations by mathematical integration, and (b) assuming EI is constant, the equation of the elastic curve. Neglect axial deformation in all members. Also describe how the remaining members can be analyzed.

Solution:
a. Inspection of Fig. 6.7b indicates that the end shears have not been specified. This is usually the case in frame-analysis output. However, these can be determined by applying statics to the free-body diagram of the member. Indeed for this statically determinate frame the given moments could have been calculated by statics, too:

$$\Sigma M_B = 8V_c + 2(8)(4) + 210 = 0$$
$$+ \text{ cw}$$
$$\therefore V_c = -34.25 \text{ kips}$$
$$\Sigma F_y = V_B - V_c - 2(8) = 0$$
$$+ \text{ up}$$
$$\therefore V_B = -18.25 \text{ kips}$$

The shear and moment equations are determined by application of Eqs. 6.1 and 6.2.

$$V(x) = \int (-2) \, dx + C_1$$
$$= -2x + C_1$$

$$M(x) = \int (-2x + C_1) \, dx + C_2$$
$$= -x^2 + C_1 x + C_2$$

Known end forces permit determination of C_1 and C_2.

$$M(0) = +210 \qquad\qquad \therefore C_2 = +210$$
$$V(0) = -18.25 \qquad\qquad \therefore C_1 = -18.25$$
$$\therefore V(x) = -2x - 18.25$$
$$\therefore M(x) = -x^2 - 18.25x + 210$$

These equations are plotted in Fig. 6.7c.
b. The moment equation is employed to determine the elastic shape as follows:

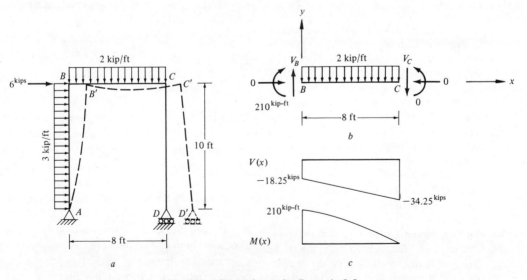

Fig. 6.7.   Planar frame for Example 6.3.

$$Ely'(x) = \int (-x^2 - 18.25x + 210) \, dx + C_3$$

$$Ely'(x) = -\tfrac{1}{3}x^3 - 9.125x^2 + 210x + C_3$$

$$Ely(x) = \int Ely'(x) \, dx + C_4$$

$$Ely(x) = -\tfrac{1}{12}x^4 - 3.04x^3 + 105x^2 + C_3x + C_4$$

Because axial deformations are ignored and first-order analysis is used, the columns in the frame do not change in length. Points B and C do not displace vertically, and the local boundary conditions for member BC are that the end displacements are zero. Consequently,

$$y(0) = 0 \qquad\qquad \therefore C_4 = 0$$

$$y(8) = 0$$

$$\therefore -\tfrac{1}{12}(8)^4 - 3.04(8)^3 + 105(8)^2 + C_3(8) = 0$$

$$\therefore C_3 = -603$$

$$\therefore Ely(x) = -\tfrac{1}{12}x^4 - 3.04x^3 + 105x^2 - 603x$$

The function $y(x)$ gives the displaced shape of member BC except that an unknown rigid-body sway to the right has also occurred.

c. Shear and moment equations are easily obtained for members AB and CD because the frame is statically determinate. For each member, the moment equation can be integrated twice to obtain the equation of the elastic curve. Boundary conditions are needed to evaluate the constants of integration obtained as follows:

1. Differentiate the equation of the elastic curve for member BC. Evaluate the end rotations by substitution of the local coordinates ($x = 0$ and $x = 8$ ft).
2. Because joint B is rigid, members AB and BC have the same slope (in local coordinates) at point B. This known rotation and the known zero transverse displacement at A (in local coordinates) permits the determination of the elastic curve for member AB.
3. The result of step 2 permits determination of distance BB'. With axial deformation neglected point C translates horizontally by an amount equal to the distance BB'. Furthermore, the rotation (in local coordinates) of end C of member CD is known by virtue of the rigid joint at that location. With the local rotation and displacement of point C known, suf-

ficient boundary condition exist to complete the analysis of the member.

Despite the theoretical possibility, mathematical integration is clearly impractical for calculating the displacements of a planar frame. Example 6.3 offers ample evidence of the validity of this statement. Historically, the use of closed-form integration has been uncommon. A variety of "classical" alternatives exist and have had wide usage in the past.

For determinate frame analysis, numerical techniques (e.g., the moment-area method) and work and energy methods (e.g., the unit-load method) are common classical approaches. Application of the moment-area method will be described in Section 6.5, and a conversion to the use of the more expeditious "angle change principle" and "conjugate frame method" is addressed subsequently. Work and energy methods are treated in Chapters 8 and 9.

A variety of classical frame analysis procedures, characterized as either flexibility methods or stiffness methods, exists for the investigation of indeterminate frames. Because of the recent decline in their usage, presentation of classical approaches is reserved for Chapters 11, 12, and 13. In the contemporary state of the art, indeterminate frame analysis is performed more often by matrix computer methods. Introductory and intermediate treatments of matrix methods of structural analysis are presented in later chapters.

6.4 SHEAR AND MOMENT DIAGRAMS

Variation in shear and moment within a beam is conveniently expressed by plotting shear and moment diagrams. No new concepts are needed to extend this mode of expression to planar frames. In the context of framed structures, a shear diagram shows the variation in internal shear that exists for each member of the frame. A moment diagram shows the corresponding variation in internal moment. Basically, the diagrams are drawn by considering the members as individual flexural members in space and plotting the variation in shear and moment for each of them according to a designated sign convention. Additionally, a standard orientation for the complete diagrams for the whole framework should be established for general use.

Several conventions exist for plotting shear and moment diagrams for frames, and the reader is cautioned that other textbooks might employ rules that differ from those established herein. All of the various methods are equally valid because the appearance of the diagram is not the relevant issue. The important aspect is the ability to interpret the physical counterpart to the diagrams, i.e., to correctly associate the numerical values shown in the diagram with the directions of action for the corresponding forces. The steps to be followed in drawing shear and moment diagrams for planar frames follow:

1. Draw free-body diagrams for each member, and numerically determine the magnitude of the end forces.
2. Schematically orient each member in the horizontal position. For vertical columns, the members can be folded up (or down) to the horizontal position.
3. Construct shear and moment diagrams for each member following the rules outlined for beams.
4. Assemble the complete shear and moment diagrams by depicting the diagrams obtained in step 3 on a schematic diagram of the entire frame. Loosely speaking, for frames with vertical columns, this implies "folding the 'legs' of the shear and moment diagrams" back to their original positions.

Example 6.4. Plot the shear and moment diagrams for the frame (see Fig. 6.3a) studied in Example 6.1.

Solution: Internal forces were computed in Example 6.1 and drawn in Fig. 6.3d. By "folding the columns up" into a horizontal position, the free-body diagrams shown at the top of Fig. 6.8a are obtained. Plotting shear and moment diagrams begins with the known values at the left end of the member. Then,

for each member, the loading diagram can be numerically integrated to produce a shear diagram. The shear diagram, in turn, is numerically integrated to produce the moment diagram. Signs assigned to the numerical values are consistent with rules established for beams. For example, the end moments for each member cause compression in the bottom fibers. Thus they plot as negative moments. Also, a shear force that acts upward (downward) on the left (right) end of a member plots as a positive (negative) value. In this example, the 16 kip force at A acts downward and plots as a negative value. In this analysis, the members oriented in the horizontal position are viewed as "beams," and the shear and moment diagrams drawn accordingly. In Fig. 6.8b, the assembled shear and moment diagrams, obtained by "folding the columns down," are drawn adjacent to the loading diagram. With experience, it is possible to omit the preliminary steps for simple frames and directly construct the three diagrams shown in Fig. 6.8b if the loading and reactions are known. The ability to do this is valuable, but the task should not be undertaken until proficiency in the detailed procedure is attained.

Example 6.5. Plot shear and moment diagrams for the frame (see Fig. 6.4a) analyzed in Example 6.2.

Solution: Internal forces were determined in Example 6.2 and drawn in Fig. 6.4c. The members are shown, reoriented as beams, in Fig. 6.9. Shear and moment diagrams are then plotted below each member, starting with the left end in each case. The reassembled results are shown in Fig. 6.10.

6.5 DISPLACEMENTS BY THE MOMENT-AREA METHOD

Moment-area theorems express two fundamental relationships between the M/EI diagram of an individual flexural member and its displaced shape. Although a frame is comprised of an array of members, the moment-area theorems are still useful. The deformed position of a frame, although complex as a whole, can be visualized as a series of deformed members interconnected at their ends. By applying the moment-area theorems to the members one at a time, it is possible to assemble the displaced shape of the frame in a piecewise fashion. In pursuing

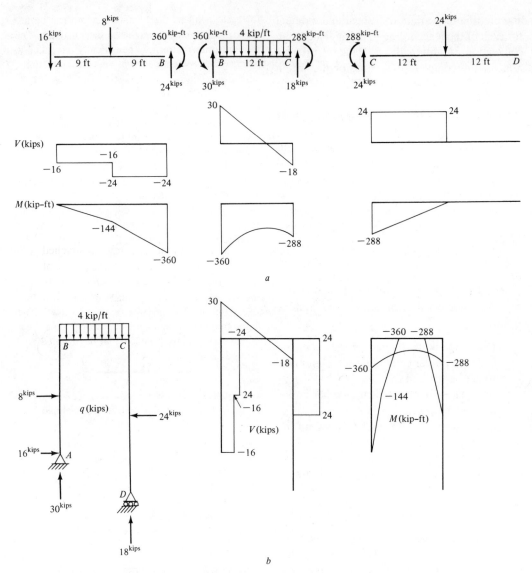

a

b

Fig. 6.8.   Shear and moment diagrams for the frame of Fig. 6.3a.

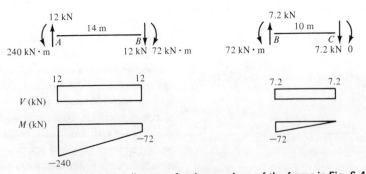

Fig. 6.9.   Shear and moment diagrams for the members of the frame in Fig. 6.4a.

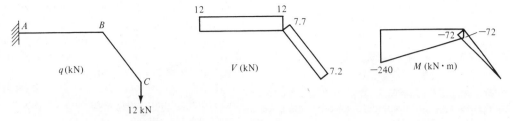

Fig. 6.10.   Assembled shear and moment diagrams for the frame of Fig. 6.4a.

this course, the stepwise treatment of the members cannot be done in an arbitrary order. Knowledge of either the rotation at two points or the rotation and displacement at a single point on a member is a fundamental requirement of the moment-area theorems. Indeed, this prerequisite condition dictates the order of member selection. The analysis of frames has the further complication of requiring the analyst to visualize the general appearance of the deformed shape in advance of the numerical computation. This must be done by using the M/EI diagram and support conditions as the only evidence. Normally, sketching a feasible displaced shape is the most difficult task for the beginning student.

To introduce the features of applying the moment-area theorems to planar frames, it is useful to study the general characteristics of a particular type of structure, namely, the *cantilevered frame.*

6.5.1   Cantilevered Frames

A cantilevered frame of an arbitrarily chosen configuration is shown in Fig. 6.11a. The aim of the following discussion is to trace the influence of the individual member M/EI diagrams on the displacements that occur at the free joints of the frame. The moment diagram for the given loading is obtained by the procedure outlined in Section 6.3. After division by the indicated EI values, the result is the M/EI diagram constructed in Fig. 6.11b. An approximate form of the displaced shape (A–B'–C'–D') is drawn in Fig. 6.11c. The general characteristics are determined by assessing the curvature created in each member and the sup-

port condition at A. In this case, each member is subjected to moments that induce "inward" curvature with respect to the original position of the frame. Because of the physical requirement that the frame must remain attached to the fixed support at A, the frame must displace as shown in Fig. 6.11c. The immediate objective is to examine, in detail, the manner by which each member's curvature contributes to the deformed shape of the framework.

A more definitive display of the displaced frame geometry is given in Fig. 6.11d. The deformations are shown to an exaggerated scale for ease of labeling. In reality, the displaced position A–B'–C'–D' differs only slightly from the original position A–B–C–D, and small deflection theory applies. In addition, the effect of axial deformations was not included in the derivation of the moment-area theorems and must be neglected in the analysis. Normally, such deformations are small and can be neglected safely. Figure 6.11d can be discussed in light of these realizations.

Joint B experiences a rotation, θ_B, and a vertical displacement, Δ_{BV}. No horizontal motion occurs ($\Delta_{BH} = 0$) because axial deformation is ignored in member AB, and, in small-deflection theory, the transverse motion of the member does not alter the longitudinal position of point B. Joint C undergoes a rotation, θ_C, and translations Δ_{CH} and Δ_{CV}. However, member BC does not change in length, and Δ_{CV} must equal Δ_{BV}. Transverse motion has no effect on the longitudinal position of point C. Finally, joint D experiences the three indicated displacements, θ_D, Δ_{DV}, and Δ_{DH}. Again, it can be reasoned that Δ_{DH} must equal Δ_{CH}. The three compatibility conditions

$$\Delta_{BH} = 0$$

$$\Delta_{CV} = \Delta_{BV}$$

$$\Delta_{DH} = \Delta_{CH}$$

infer that θ_B, θ_C, θ_D, Δ_{BV}, Δ_{CH}, and Δ_{DV} are the independent joint displacement quantities for this framework. Knowledge of these quantities establishes the other joint displacements. In other words, there are three rotational and three translational degrees of freedom in the framework. Determination of the magnitude of the displacements in these six directions for any given loading establishes the position of all joints and provides a basis for determining the location of any other point. The effect of each member on these displacements is discussed in the next paragraphs.

The influence of member AB is shown in Fig. 6.11e. This diagram depicts the consequence of allowing the curvature of member AB to take place while restraining the curvature of the other members. In other words, the effects of the M/EI diagram for member AB are isolated. Under these conditions, members BC and CD change orientation but remain straight. Because point A is fixed in space,

$$\theta'_A = 0 \tag{6.11}$$

$$\Delta'_{AH} = \Delta'_{AV} = 0 \tag{6.12}$$

and it is a convenient starting point for the moment-area theorems. The three displacements at joints B are given by

$$\theta'_B = t_{AB} \tag{6.13}$$

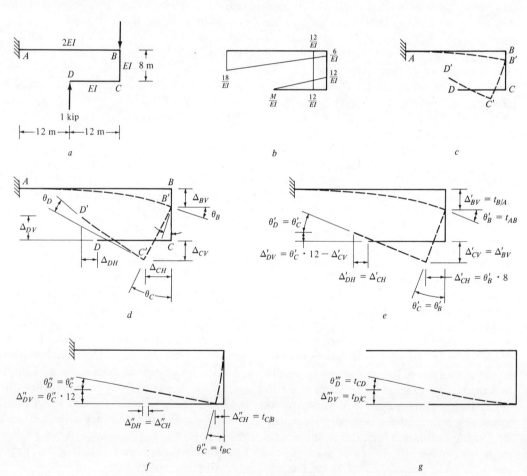

Fig. 6.11.   Frame displacements by the moment-area method.

$$\Delta'_{BH} = 0 \qquad (6.14)$$

$$\Delta'_{BV} = t_{B/A} \qquad (6.15)$$

With the position of joint B established, joint C can be located by applying the moment-area theorems to member BC using point B as a reference point. In this stage of the analysis, no M/EI diagram exists for this member, and

$$\theta'_C = \theta'_B + t_{BC} = \theta'_B \qquad (6.16)$$

$$\Delta'_{CH} = \theta'_B \cdot 8 + t_{C/B} = \theta'_B \cdot 8 \qquad (6.17)$$

Neglecting axial deformation requires

$$\Delta'_{CV} = \Delta'_{BV} \qquad (6.18)$$

With the position of joint C established, moment-area theorems can be applied to member CD to yield

$$\theta'_D = \theta'_C + t_{CD} = \theta'_C \qquad (6.19)$$

$$\Delta'_{DV} = \theta'_C \cdot 12 + t_{D/C} - \Delta'_{CV}$$

$$= \theta'_C \cdot 12 - \Delta'_{CV} \qquad (6.20)$$

and

$$\Delta'_{DH} = \Delta'_{CH} \qquad (6.21)$$

by neglect of axial deformation in the member.

As a second stage of analysis, consider the influence of member BC. The isolated effects of its curvature are illustrated in Fig. 6.11f. Because no M/EI diagrams exists for member AB, it experiences no deformation, and joint B remains in its initial position. As a result,

$$\theta''_A = \Delta''_{AH} = \Delta''_{AV} = 0 \qquad (6.22)$$

$$\theta''_B = \Delta''_{BH} = \Delta''_{BV} = 0 \qquad (6.23)$$

Motions at joint C are determined by application of the moment-area theorems to member BC and recognition of the effects of neglecting axial deformation.

$$\theta''_C = \theta''_B + t_{BC} = t_{BC} \qquad (6.24)$$

$$\Delta''_{CH} = \theta''_B \cdot 8 + t_{C/B} = t_{C/B} \qquad (6.25)$$

$$\Delta''_{CV} = 0 \qquad (6.26)$$

Joint D is located by further application of the moment-area theorems to member CD:

$$\theta''_D = \theta''_C + t_{CD} = \theta''_C \qquad (6.27)$$

$$\Delta''_{DV} = \theta''_C \cdot 12 + t_{D/C} = \theta''_C \cdot 12 \qquad (6.28)$$

$$\Delta''_{DH} = \Delta''_{CH} \qquad (6.29)$$

Curvature in member CD and its effect on the frame are isolated as the final stage of analysis. By considering the displaced shape shown in Fig. 6.11g,

$$\theta'''_A = \Delta'''_{AH} = \Delta'''_{AV} = 0 \qquad (6.30)$$

$$\theta'''_B = \Delta'''_{BH} = \Delta'''_{BV} = 0 \qquad (6.31)$$

$$\theta'''_C = \Delta'''_{CH} = \Delta'''_{CV} = 0 \qquad (6.32)$$

$$\theta'''_D = t_{CD} \qquad (6.33)$$

$$\Delta'''_D = t_{D/C} \qquad (6.34)$$

$$\Delta'''_{DH} = 0 \qquad (6.35)$$

The complete framework experiences the combined effects of the entire M/EI diagram. Quantitatively, a generalized displacement quantity, δ, at some point in the frame is given by

$$\delta = \delta' + \delta'' + \delta'''$$

where δ', δ'', and δ''' are values of the displacement quantity extracted from Figs. 6.11e through g, respectively. In particular, the total joint rotations are evaluated as

$$\theta_A = \theta'_A + \theta''_A + \theta'''_A = 0$$

$$\theta_B = \theta'_B + \theta''_B + \theta'''_C$$

$$= \theta'_B$$

$$\theta_B = t_{AB} \qquad (6.36)$$

$$\theta_C = \theta'_C + \theta''_C + \theta'''_C$$

$$= \theta'_B + t_{BC}$$

$$\theta_C = t_{AB} + t_{BC} \qquad (6.37)$$

$$\theta_D = \theta'_D + \theta''_D + \theta'''_D$$

$$= \theta'_C + \theta''_C + t_{CD}$$

$$= \theta'_B + t_{BC} + t_{CD}$$

$$\theta_D = t_{AB} + t_{BC} + t_{CD} \qquad (6.38)$$

Equations 6.36 through 6.38 suggest that the slope at any point in the given frame is equal to the summation of all angle changes (area under the M/EI diagrams) that occur between point A and the point under consideration. Indeed, this is true but *only because point A is fixed in space.*

Horizontal joint translations are obtained as

$$\Delta_{AH} = \Delta'_{AH} + \Delta''_{AH} + \Delta'''_{AH} = 0 \qquad (6.39)$$

$$\Delta_{BH} = \Delta_{AH} = 0 \qquad (6.40)$$

$$\Delta_{CH} = \Delta'_{CH} + \Delta''_{CH} + \Delta'''_{CH}$$

$$= \theta'_B \cdot 8 + t_{C/B} \qquad (6.41)$$

$$\Delta_{DH} = \Delta'_{DH} + \Delta''_{DH} + \Delta'''_{DH}$$

$$= \Delta'_{CH} + \Delta''_{CH} + 0$$

$$= \theta'_B \cdot 8 + t_{C/B}$$

$$\Delta_{DH} = \Delta_{CH} \qquad (6.42)$$

and the vertical translations as

$$\Delta_{AV} = \Delta'_{AV} + \Delta''_{AV} + \Delta'''_{AV} = 0 \qquad (6.43)$$

$$\Delta_{BV} = \Delta'_{BV} + \Delta''_{BV} + \Delta'''_{BV}$$

$$= t_{B/A} \qquad (6.44)$$

$$\Delta_{CV} = \Delta'_{CV} + \Delta''_{CV} + \Delta'''_{CV}$$

$$= \Delta'_{CV}$$

$$= \Delta'_{BV} \qquad (6.45)$$

$$\Delta_{DV} = \Delta'_{DV} + \Delta''_{DV} + \Delta'''_{DV}$$

$$= (\theta'_C \cdot 12 - \Delta'_{CV}) + (\theta''_C \cdot 12) + (t_{D/C})$$

$$= (\theta'_C + \theta''_C)(12) + t_{D/C} - \Delta'_{CV}$$

$$\Delta_{DV} = \theta_C \cdot 12 + t_{D/C} - \Delta_{CV} \qquad (6.46)$$

Numerical evaluation of Eqs. 6.36 through 6.46 constitutes a solution by the moment-area method for the joint displacements in the frame of Fig. 6.11a. Some observations regarding the development of these relationships are relevant. As the analysis progressed, the complexity of the deformations and associated equations decreased for each successive stage. The reason for this, which is of fundamental importance, is expressed by the following general statement: An angle change at a single point in a member causes a relative rigid body motion between the two portions of the structure that are attached to that point. This concept is depicted in Fig. 6.12.

Curvature is the accumulation of angle changes, which explains the increasing ease of computations at various stages. In simple terms, the curvature in member AB affects all three members in the frame; the curvature in member BC affects members BC and CD; and the curvature in member CD affects only the member CD itself. Viewing the response in this sense, it is perhaps easier to study Figs. 6.11e through g in reverse order. Understanding the influence of an isolated angle change in a frame is the basis for deriving the *conjugate-grid method* of frame analysis. This very powerful classical method for computing frame displacements is presented in Chapter 7. With a conceptual discussion of the application of the moment-area theorems to cantilevered frames completed, a numerical example is valuable as the next step.

Example 6.6. Determine the joint displacements for the frame given in Fig. 6.11a. Use the moment-area method.

Solution: Governing relationships are Eqs. 6.36 through 6.46. The nontrivial equations (those with nonzero values) are repeated here.

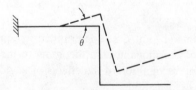

Fig. 6.12. Effect of an angle change at a point in a frame.

$$\theta_B = t_{AB}$$

$$\theta_C = t_{AB} + t_{BC} = \theta_B + t_{BC}$$

$$\theta_D = t_{AB} + t_{BC} + t_{CD} = \theta_C + t_{BC}$$

$$\Delta_{BV} = t_{B/A}$$

$$\Delta_{CH} = \theta_B \cdot 8 + t_{C/B}$$

$$\Delta_{CV} = \Delta_{BV}$$

$$\Delta_{DH} = \Delta_{CH}$$

$$\Delta_{DV} = \theta_C \cdot 12 + t_{D/C} - \Delta_{CV}$$

A stage-by-stage development of these relationships is not a necessity in the solution process. The usual procedure is to draw the complete deformed shape with labeling, as shown exaggerated in Fig. 6.13, by mentally performing the stepwise development. With this figure available, the needed moment-area relationships can be obtained directly. It is useful for the reader to recreate the relationships with the aid of Fig. 6.13 alone.

Quantities required in these relationships are evaluated using the M/EI diagram drawn in Fig. 6.11b.

$$t_{AB} = \frac{6}{EI}(24) + \frac{1}{2}\left(\frac{12}{EI}\right)(24)$$

$$= \frac{144}{EI} + \frac{144}{EI} = \frac{288}{EI}$$

$$t_{BC} = \frac{12}{EI}(8) = \frac{96}{EI}$$

$$t_{CD} = \frac{1}{2}\left(\frac{12}{EI}\right)(12) = \frac{72}{EI}$$

$$t_{B/A} = \frac{144}{EI}(12) + \frac{144}{EI}(16) = \frac{4032}{EI}$$

$$t_{C/B} = \frac{96}{EI}(4) = \frac{384}{EI}$$

$$t_{D/C} = \frac{72}{EI}(8) = \frac{576}{EI}$$

Joint displacements are determined by direct substitution into the governing relationships given earlier. The results are

$$\theta_B = \frac{288}{EI} \text{ cw}$$

$$\theta_C = \frac{384}{EI} \text{ cw}$$

$$\theta_D = \frac{456}{EI} \text{ cw}$$

$$\Delta_{BV} = \frac{4032}{EI} \downarrow$$

$$\Delta_{CH} = \frac{2688}{EI} \leftarrow$$

$$\Delta_{CV} = \frac{4032}{EI} \downarrow$$

$$\Delta_{DH} = \frac{2688}{EI} \leftarrow$$

$$\Delta_{DV} = \frac{1152}{EI} \uparrow$$

6.5.2 General Frames

A cantilever frame has a natural feature that simplifies the implementation of the moment-area theorems. The fixed support is a point of known slope and displacement, and presents an obvious starting point for the analysis. The observation that the angle changes are additive as one proceeds from member to member is one indication of the advantage of this support condition. Unfortunately, most frames are not cantilevered (although many do have fixed supports). In the absence of a fixed support, determination of a starting point requires more careful study of the structure. In making this

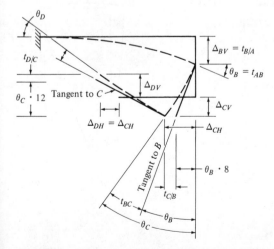

Fig. 6.13. Displaced shape for the frame in Fig. 6.8a.

determination, it is necessary to remember the prerequisite of two known boundary conditions for any member. Example 6.7 is more illustrative of the general aspects of frame analysis by the moment-area method.

Example 6.7.   Employ the moment-area theorems to determine the joint displacements in the frame of Fig. 6.3a. Sketch and label the deformed shape. Assume EI is constant and the same for all members.

Solution: Initially, the moment diagram must be constructed for the given loading. This was accomplished in Example 6.2, and the diagram is shown in Fig. 6.8b. Because the frame members constitute individual beam elements, it is possible to convert the moment diagram for any one of them to a moment diagram by parts. In this example, it is advantageous to make this conversion for members AB and BC.

Member AB:

$$SBM = \frac{8(18)}{4} = 36 \text{ kip-ft}$$

Member BC:

$$SBM = \tfrac{1}{8}(4)(12)^2 = 72 \text{ kip-ft}$$

After extracting the end moments from Fig. 6.8b, the MP diagram can be plotted easily. Division by EI yields the diagram shown in Fig. 6.14a. The actual moment diagram for member CD is a simple triangle, and no advantage is gained by a conversion to an MP diagram.

As the next step, it is necessary to determine a starting point for an analysis by the moment-area method. Observation of the actual moment diagram suggests a displaced shape similar to that shown in Fig. 6.14b. If axial deformation is neglected, joints B and C do not displace vertically. This observation is the basis for rejecting the displaced shape in Fig. 6.15 as infeasible. For this shape to exist, member CD would have to experience an axial deformation, which is not permitted. With the displacements at ends known, member BC is a suitable starting point. Tangents to the displaced joints, B' and C', have been drawn in Fig. 6.14b. It is evident that the end slopes can be computed as

$$\theta_B = t_{B/C} \div 12$$

$$\theta_C = t_{C/B} \div 12$$

As an alternative, θ_C can be determined from

$$\theta_C = \theta_B - t_{BC}$$

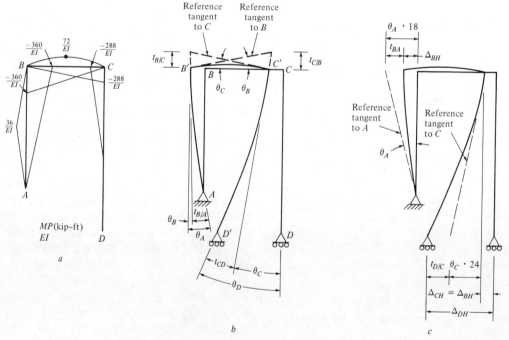

Fig. 6.14.   Displaced shape for the frame in Fig. 6.3a.

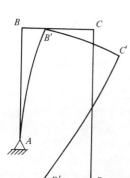

Fig. 6.15. Inadmissible displaced shape.

From Fig. 6.14a, (for the assumed displaced shape, positive M/EI is taken as negative when computing the tangential deviations).

$$t_{BC} = +\frac{1}{2}\left(\frac{360}{EI}\right)(12) + \frac{1}{2}\left(\frac{288}{EI}\right)(12)$$

$$-\frac{2}{3}\left(\frac{72}{EI}\right)(12)$$

$$= +\frac{2160}{EI} + \frac{1728}{EI} - \frac{576}{EI} = \frac{3312}{EI}$$

$$t_{C/B} = \left(\frac{2160}{EI}\right)(8) + \left(\frac{1728}{EI}\right)(4) - \left(\frac{576}{EI}\right)(6)$$

$$= \frac{20736}{EI}$$

$$\therefore \theta_B = \frac{20736}{EI} \div 12 = \frac{1728}{EI}$$

$$\therefore \theta_B = \frac{1728}{EI} \text{ cw}$$

$$\therefore \theta_C = \frac{1728}{EI} - \frac{3312}{EI} = -\frac{1584}{EI}$$

$$\therefore \theta_C = \frac{1584}{EI} \text{ ccw}$$

The rotation at joint A is determined as

$$\theta_A = \theta_B + t_{BA}$$

where negative M/EI is considered to create positive angle changes. Consequently,

$$t_{BA} = \frac{1}{2}\left(\frac{360}{EI}\right)(18) - \frac{1}{2}\left(\frac{36}{EI}\right)(18)$$

$$= \frac{3240}{EI} - \frac{324}{EI} = \frac{2916}{EI}$$

$$\theta_A = \frac{1728}{EI} + \frac{2916}{EI} = \frac{4644}{EI}$$

$$\therefore \theta_A = \frac{4644}{EI} \text{ cw}$$

Similarly, the rotation θ_D is given by

$$\theta_D = \theta_C + t_{CD}$$

$$= \frac{1584}{EI} + \frac{1}{2}\left(\frac{288}{EI}\right)(12)$$

$$= \frac{1584}{EI} + \frac{1728}{EI} = \frac{3312}{EI}$$

$$\therefore \theta_D = \frac{3312}{EI} \text{ ccw}$$

Remaining displacements are determined after consideration of Fig. 6.14c.

$$\Delta_{BH} = \theta_A \cdot 18 - t_{B/A}$$

$$\Delta_{DH} = \Delta_{BH} + \theta_C \cdot 24 + t_{D/C}$$

and

$$t_{B/A} = \left(\frac{3240}{EI}\right)(6) - \left(\frac{324}{EI}\right)(9) = \frac{16,524}{EI}$$

$$\Delta_{BH} = \left(\frac{4644}{EI}\right)(18) - \frac{16,524}{EI} = \frac{67,068}{EI}$$

$$\therefore \Delta_{BH} = \frac{67,068}{EI} \leftarrow$$

$$t_{D/C} = \left(\frac{1728}{EI}\right)(20) = \frac{34,560}{EI}$$

$$\Delta_{DH} = \frac{67,068}{EI} + \left(\frac{1584}{EI}\right)(24) + \frac{34,560}{EI}$$

$$= \frac{139,644}{EI}$$

$$\therefore \Delta_{BH} = \frac{139,644}{EI} \leftarrow$$

It should be emphasized that in making these computations, the signs given the M/EI areas are determined by the sense of the curvatures. For example, by examining Fig. 6.14c, if no M/EI existed for member AB, its deformed position would be that of the tangent to A. For the tangential deviation $t_{B/A}$ shown, the negative M/EI area of member BC causes curvature away from the tangent. Thus, the negative M/EI area must be taken as positive. The positive M/EI area causes the opposite curvature and must be considered negative in computing $t_{B/A}$.

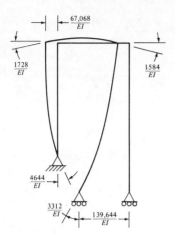

Fig. 6.16. Computed displacements—Example 6.7.

The computed displaced position of the frame is depicted in Fig. 6.16. As an indication of the small magnitude of the displacements, the chord rotation angle, R, for member CD will be computed for a W24 × 76 standard steel member ($I = 2160$ in.^{4}).

$$R = \frac{t_{D/C} + \theta_C (24)}{18}$$

$$= \frac{[34{,}560 + (1584)(24)]}{18 \, EI} = \frac{4032}{EI}$$

$$= \frac{(4032 \text{ ft}^2\text{-kips})(144 \text{ in.}^2/\text{ft}^2)}{(30{,}000 \text{ kips/in.}^2)(2160 \text{ in.}^4)}$$

$$= 0.0090 \text{ rad}$$

$$\tan R = \tan (0.0090 \text{ rad}) = 0.0090$$

6.6 STATICAL INDETERMINACY

Section 6.2 defined statically determinate planar frames and demonstrated that the available equilibrium equations are sufficient in number to calculate all external reactions and internal member end forces (axial force, shear force, and bending moment) for such structures. If this cannot be done, the structure is termed *statically indeterminate*. The indeterminacy is evident as either an inability to determine all external reactions (*external indeterminacy*) or an inability to determine all end forces (*internal indeterminacy*). As a prerequisite to later chapters, criteria for assessing the degree of indeterminacy of frames are developed in this section.

6.6.1 Expressions for Degree of Statical Indeterminacy

A number of statically indeterminate planar frames are depicted in Fig. 6.17. In Fig. 6.17a, the indeterminacy is immediately evident from the existence of six unknown reaction forces. The number of unknown reactions exceeds the external statics conditions by three and indicates that the frame is statically indeterminate (externally) to the third degree. Internally, no additional degrees of indeterminacy exist because, if all the reactions are known, shear and moment diagrams are easily drawn.

The frame in Fig. 6.17b is an internally indeterminate planar frame. The external statics

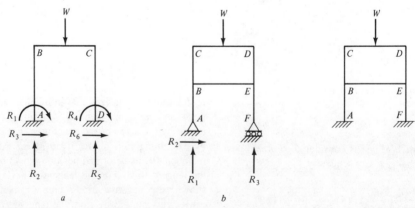

Fig. 6.17. Statically indeterminate planar frames. *a*, externally indeterminate; *b*, internally indeterminate; *c*, combination.

conditions are sufficient in number to determine the three reaction quantities. However, even with the reactions known, complete shear and moment diagrams cannot be drawn. This dilemma is evident by the following consideration. If member EF is isolated as a free-body diagram, the internal forces at end E can be determined if R_3 is known. These forces are shown transferred to the free-body diagram of joint E drawn in Fig. 6.18. Despite knowledge of the end forces for member EF, six unknown force quantities exist at the joint and cannot be determined by mere application of statics. This condition, which is also true at joint B, demonstrates the internal indeterminacy of the structure.

Many planar frames are statically indeterminate both externally and internally. It is clear then that the structure in Fig. 6.17c fits that category. Distinguishing the degree of internal indeterminacy by itself is not easily done. Instead, it is usual practice to establish the combined degree of indeterminacy and subtract the external degree of indeterminacy.

The external degree of statical indeterminacy, I_e, is simply determined as

$$I_e = NRF - NIH - 3 \qquad (6.47)$$

where NRF is the number of reaction forces and NIH is the number of internal hinges in the structure. The explanation for this formula is straightforward. Each internal hinge represents a point of known moment which can be used to develop internal equilibrium conditions to

supplement the three normally available by external application of statics. In an externally determinate structure, these equations ($NIH + 3$) must be equal in number to the number of unknown reactions (NRF), i.e., I_e must be zero.

Note that Eq. 6.47 is identical to Eq. 3.1 developed for beams. Because all beams are statically determinate internally, $I = I_e$ and Eq. 3.2 is valid, i.e.,

$$I = NRF - NIH - 3 \qquad (3.2)$$

Unlike beams, planar frames can also be statically indeterminate internally. Thus Eq. 3.2 does not apply to planar frames. Criteria for establishing the total *degree of statical indeterminacy*, is addressed in the following paragraphs. A commonly used expression for the total degree of statical indeterminacy, I, of a planar frame is

$$I = NRF + 3(NM - NJ) - NIH \qquad (6.48)$$

in which NJ is the number of joints (including supports and internal hinges). The basis for this equation can be described with the aid of Fig. 6.19. Each member of a planar frame (e.g., Fig. 6.19a) and the joints that connect them can be isolated as separate free-body diagrams (e.g., Fig. 6.19b). If statics is applied to each free-body diagram, a set of equations is obtained for use in determining the unknown forces that appear in the diagrams. The total number of equations obtained is given by

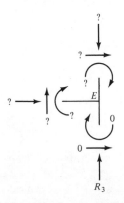

Fig. 6.18.   Joint free-body diagram.

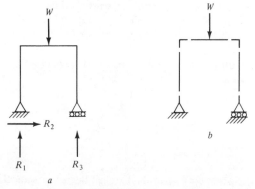

Fig. 6.19.   Determinate planar frame.

$$NE = 3(NJ + NM) \qquad (6.49)$$

The total number of unknown forces, NU, is given by

$$NU = NRF + 6NM \qquad (6.50)$$

For a statically determinate structure,

$$NE = NU$$

and by substitution of Eqs. 6.49 and 6.50

$$3(NJ + NM) = NRF + 6(NM)$$

$$NRF + 3(NM - NJ) = 0 \qquad (6.51)$$

In the particular case of Fig. 6.19a, Eq. 6.51 gives

$$3 + 3(3 - 4) = 0$$

$$0 = 0$$

as expected.

If a planar frame is statically indeterminate, Eq. 6.51 becomes an inequality, and the left side gives the total degree of statical indeterminacy

$$I = NRF + 3(NM - NJ) \qquad (6.52)$$

of a frame that has *no internal hinges*. Equation 6.52 is converted to the general expression, Eq. 6.48, by subtracting NIH. It is left to the reader to justify this subtraction.

Example 6.8. Determine the nature of the indeterminacy of the frames shown in Fig. 6.17a and in Fig. 6.17b.

Solution (Fig. 6.17a): The total degree of statical indeterminacy is given by Eq. 6.48:

$$I = NRF + 3(NM - NJ) - NIH$$

$$= 6 + 3(3 - 4) - 0$$

$$I = 3$$

Equation 6.47 gives the external degree of statical indeterminacy:

$$I_e = NRF - NIH - 3$$

$$= 6 - 0 - 3$$

$$I_e = 3$$

Thus, all degrees of statical indeterminacy are *external*.

Solution (Fig. 6.17b): The total degree of statical indeterminacy is

$$I = NRF + 3(NM - NJ) - NIH$$

$$= 3 + 3(6 - 6) - 0$$

$$I = 3$$

In this case, Eq. 6.48 gives

$$I_e = NRF - NIH - 3$$

$$= 3 - 0 - 3$$

$$I_e = 0$$

and indicates that all degrees of indeterminacy are *internal*.

Example 6.9. Determine the degree of statical indeterminacy for the planar frame in Fig. 6.20. Indicate the nature of the indeterminacy (external vs. internal).

Solution:

Figure 6.20a:

$$I = NRF + 3(NM - NJ) - NIH$$

$$= 4 + 3(9 - 8) - 0$$

$$I = 7$$

$$I_e = NRF - NIH - 3$$

$$= 4 - 0 - 3$$

$$I_e = 1$$

statically indeterminate to the seventh degree: six degrees internally and one degree externally.

Figure 6.20b:

$$I = 5 + 3(4 - 5) - 2 = 0$$

$$\therefore \text{ statically determinate.}$$

Figure 6.20c:

$$I = 12 + 3(5 - 6) - 0 = 9$$

$$I_e = 12 - 0 - 3 = 9$$

∴ statically indeterminate externally to the ninth degree.

Figure 6.20d:

$$I = 6 + 3(3 - 4) - 0 = 3$$

$$I_e = 6 - 0 - 3 = 3$$

∴ statically indeterminate externally to the third degree.

___

6.6.2 Reduction to a Statically Determinate Frame

A visual determination of the degree of statical indeterminacy of a planar frame is possible. This is done by assessing the number of external reactions and internal forces one must remove to reduce the given frame to a statically determinate, stable frame. Consider the structure shown in Fig. 6.21a.

By removing the external reactions at supports A and D and the internal force at any point

in member AE, the statically determinate cantilevered frame in Fig. 6.21b is obtained. As a total of six forces (three external reactions and three internal forces) were removed, the original frame is statically indeterminate to the sixth degree: three degrees externally and three degrees internally. This is confirmed by use of Eq. 6.48

$$I = NRF + 3(NM - NJ) - NIH$$

$$= 6 + 3(8 - 8) = 6$$

$$I_e = NRF - NIH - 3 = 6 - 0 - 3 = 3$$

Alternatively, the original frame could have been made statically determinate by either of the modifications shown in Figs. 6.21c and d. In each case, three external reactions and three internal forces have been removed to reduce the frame to a statically determinate, stable structure. Because the member configuration in each of the resulting structures in Figs. 6.21b, c, and d is such that the members "branch out" from points of support, this method is sometimes referred to as the "method of trees."

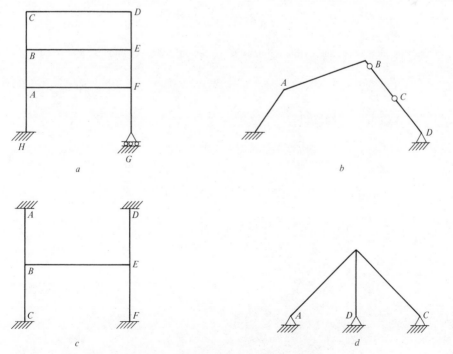

Fig. 6.20. Planar frames for Example 6.8.

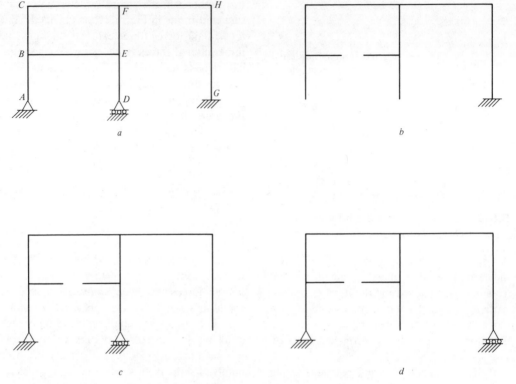

Fig. 6.21.   Reduction to a statically determinate frame.

PROBLEMS

6-1.   Draw complete shear and moment diagrams for each of the given frames. Sketch a feasible displaced shape.

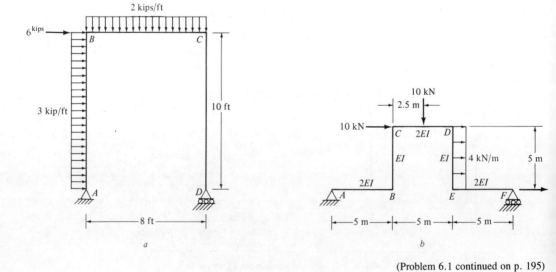

(Problem 6.1 continued on p. 195)

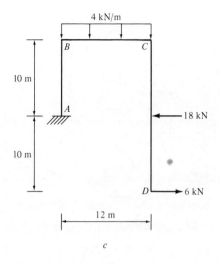

c

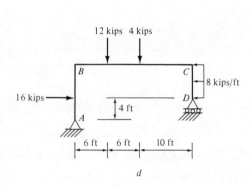

d

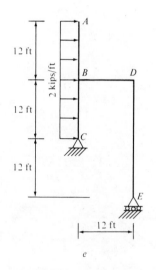

e

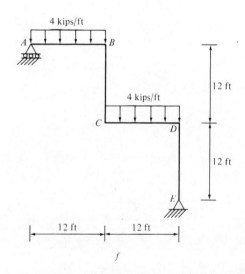

f

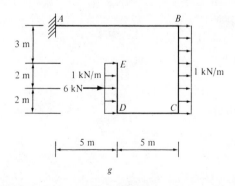

g

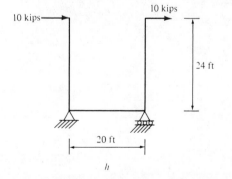

h

(Problem 6.1 continued on p. 196)

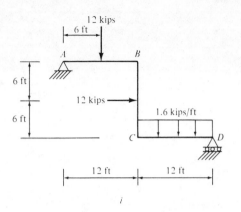

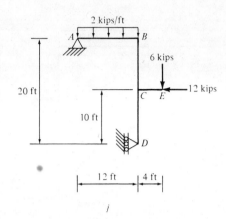

i

j

6-2. Draw complete shear and moment diagrams for each of the given frames. Sketch a feasible displaced shape.

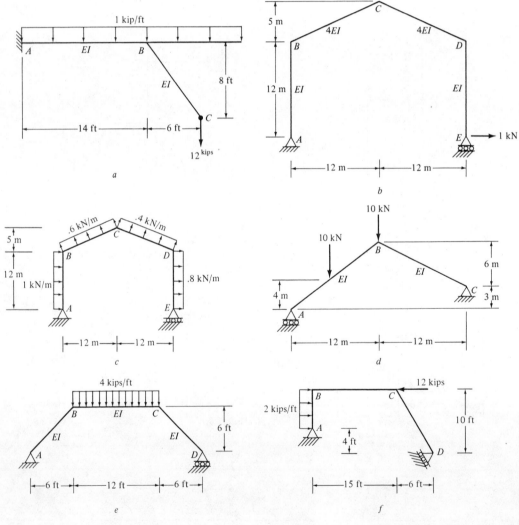

a

b

c

d

e

f

(Problem 6.2 continued on p. 197)

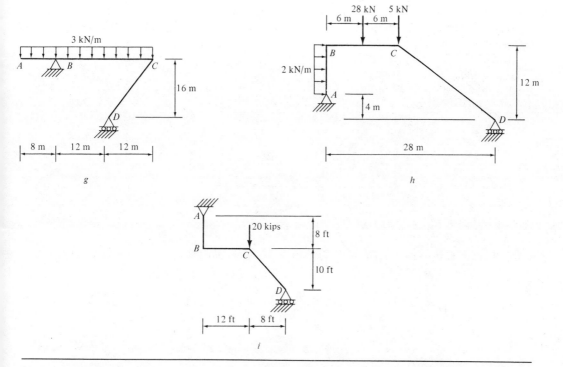

g

h

i

6-3. Structural analysis of the given frame indicates the following results.

$$M_B = -71.80 \text{ kip-ft} \qquad M_C = -78.52 \text{ kip-ft}$$

$$M_D = 36.31 \text{ kip-ft}$$

Plot the moment diagram the frame. (Note: Positive moment implies compression on the outside fibers.)

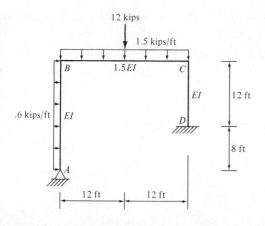

6-4. Draw the moment diagram by parts for the frames given in Problem 6-1.

6-5. Draw the moment diagram by parts for the frames in Problem 6-2.

6-6. Use the moment-area method to compute the rotation and displacement at each joint of the frames in Problem 6-1. Sketch and label the displaced shape in each case.

6-7. Analyze the frame given in Problem 6-2a by the moment-area method. Sketch the deformed shape, and label the movements and rotation at each joint.

6-8. Analyze the frame in Problem 6-2b by the moment-area method. Sketch and label the displaced shape.

6-9. Determine the displaced shape of the given frame. Label the joint movements and rotations. Use the moment-area method.

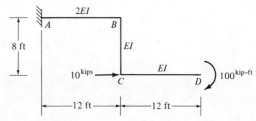

6-10.   A trough of height H is partially filled with water. As an approximation, a 12-in. width of its length is isolated as a frame having the configuration shown in the accompanying figure. Draw the moment diagram for each of the following cases:

(a) $a = 60$ ft, $b = 6$ ft, $H = 6$ ft
(b) $a = 30$ ft, $b = 6$ ft, $H = 6$ ft
(c) $a = 15$ ft, $b = 6$ ft, $H = 6$ ft.

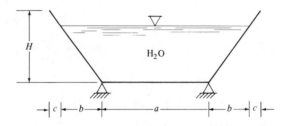

6-11.   Determine the relative horizontal displacement between the tops of the two sidewalls of the trough in Problem 6-10. Assume EI is constant. Use the moment-area method.

7
Angle Changes and Conjugate Structures

7.1 INTRODUCTION

With the advent of computer-based structural analysis methodology, manual computation of displacements of beams and frames has less importance in everyday engineering practice. Occasionally, spot checks of computer output or direct analysis of small frameworks may be performed. Chapter 4 developed the moment-area method as a basic tool of mechanics and illustrated its use for calculating beam displacements. In Chapter 6, the application to planar frames was illustrated. It was shown that no theoretical extension of the basic theorems is necessary for planar frame problems. However, the interconnection of members creates geometrical conditions that complicate the use of these basic theorems. Analysis for the displaced position necessitates a sequential tracing of the effects of curvature in each of the frame members and recognition of support constraints and joint compatibility. Furthermore, visualization of the general shape of the displaced frames is a critical prerequisite to the numerical work. Unless the frame is simple in form and loading, the resulting computational effort is tedious and time-consuming. For general frames, it is desirable to employ procedures that eliminate the inefficiencies of the moment-area method. This is possible by introduction of the

angle-change principle. Subsequent to the discussion of the angle-change principle, the conjugate-grid method will be developed. This powerful technique virtually eliminates the geometry features of displacement computation.

7.2 THE ANGLE-CHANGE PRINCIPLE

7.2.1 Fundamentals

As an introduction, consider the fundamental meaning of the first moment-area theorem. This theorem essentially states that the total angle change between two points is equal to the area under the M/EI diagram between the two points. More specifically, the total angle change is the summation of all the discrete angle changes, $(M(s)/EI)ds$, that occur between the two points. Several useful relationships, referred to as angle-change principles, can be developed by consideration of the effect of a single, discrete angle change on the deformation of a framework. Angle-change principles will prove to be valuable in bypassing the cumbersome geometric aspects of the moment-area method.

An arbitrary planar frame is depicted in Fig. 7.1a. The effect of inducing a discrete angle change ϕ at a point P in member BC will be studied. If the portion of the frame between

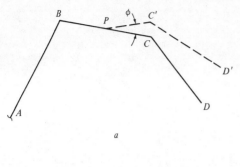

a

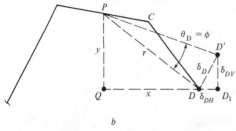

b

Fig. 7.1.   Effects of a discrete angle change.

points A and P is held rigidly in place, the portion between points P and D will experience a rigid body rotation of magnitude ϕ. As indicated in Fig. 7.1a, joints C and D will move to new locations C' and D', respectively. Angle-change principles can be developed by considering the motion of any point on the frame that lies *at or between points P and D*, i.e., any point beyond the angle-change location. For convenience, the motion of point D will be examined.

A detailed illustration of the motion of point D is drawn in Fig. 7.1b. As a direct consequence of the discrete angle change, point D moves from D to D'. The direction of this movement is perpendicular to the radial line drawn from point P to point D, and its magnitude, δ_D, is given by

$$\delta_D = r\phi \tag{7.1}$$

where r is the length of line PD. It is desirable to determine the horizontal component δ_{DH} and the vertical component δ_{DV} of the movement of joint D. Further examination of Fig. 7.1b permits this determination.

If x and y are the horizontal and vertical distances, respectively, from point P to point D,

it is evident that triangles PQD and DD_1D' are geometrically similar. From this observation, δ_{DH} and δ_{DV} can be expressed as

$$\delta_{DV} = \left(\frac{x}{r}\right)\delta \tag{7.2}$$

$$\delta_{DH} = \left(\frac{y}{r}\right)\delta \tag{7.3}$$

Substitution of Eq. 7.1 yields

$$\delta_{DV} = x\phi \tag{7.4}$$

$$\delta_{DH} = y\phi \tag{7.5}$$

The rotation experienced at point D, θ_D, is equal to the discrete angle change and is expressed as

$$\theta_D = \phi \tag{7.6}$$

Equations 7.1 through 7.6 constitute basic geometric relationships in small-displacement theory and permit three important statements:

1. Vertical displacement at a point, as caused by the rigid-body rotation associated with a discrete angle change, is equal to the product of the angle change magnitude and the horizontal distance measured from the location of the angle change to the point of interest.
2. Horizontal displacement at a point, as caused by the rigid-body rotation associated with a discrete angle change, is equal to the product of the angle change magnitude and the vertical distance measured from the location of the angle change to the point of interest.
3. Rotation at a point, as caused by the rigid-body rotation associated with a discrete angle change, is equal to the discrete angle change.

When combined, these three statements represent the "angle-change principle." It is significant to note again that this principle relates a discrete angle change to the relative motion of certain points on the frame; i.e., it describes the effect of a rigid-body motion of a portion of the frame, assuming that the remaining portion is fixed in place.

Equations 7.1 through 7.6 are not directly applicable to the analysis of frames subjected to general loads. It is necessary to expand their meaning to account for continuous angle changes such as produced by a complete M/EI diagram. Consider the cantilever frame depicted in Fig. 7.2. By extension of Eqs. 7.1 through 7.6, it is possible to determine the movement of point D (or any other point on the structure). The curvilinear shape $AB'C'D'$ is attributed to the existence of flexural moments throughout the structure or, more specifically, to the M/EI diagram shown in Fig. 7.2b. Because point A is fixed in space, the displacement of point D is caused by the additive effects of all the discrete angle changes that occur between points A and D. By investigating the effect of the incremental angle change, $d\theta$, shown in Fig. 7.2b and c, the total motion can be determined by integration.

From Fig. 7.2b and the first moment-area theorem, the incremental angle change associated with a small segment of the M/EI diagram from member BC has a magnitude given by

$$d\theta = \frac{M(s)}{EI} ds = dA \qquad (7.7)$$

In other words, $d\theta$ is the area under the infinitesimal portion of the M/EI diagram of length ds. The contributions made to the displacements Δ_{DV} and Δ_{DH} at point D by this discrete angle change are

$$d\Delta_{DV_{BC}} = x(s)\, d\theta = \frac{M(s)}{EI} x(s)\, ds$$

$$= x(s)\, dA$$

$$d\Delta_{DH_{BC}} = y(s)\, d\theta = \frac{M(s)}{EI} y(s)\, ds$$

$$= y(s)\, dA$$

In which $x(s)$ and $y(s)$ are measured from the location of $d\theta$ to point D.

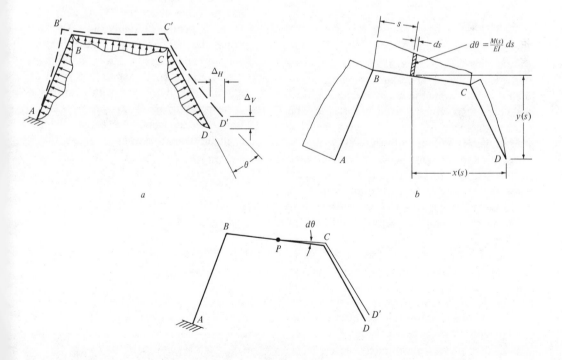

Fig. 7.2. Effects of continuous angle changes. *a*, loaded frame; *b*, M/EI diagram.

By integrating over the length of the member, the total influence of the M/EI diagram for member BC is obtained from

$$\Delta_{DV_{BC}} = \int_0^l x(s)\, dA \qquad (7.8)$$

and

$$\Delta_{DH_{BC}} = \int_0^l y(s)\, dA \qquad (7.9)$$

The right-hand side of Eq. 7.8 is recognized as the first moment of the M/EI area for member BC, taken about a vertical axis through point D. Equation 7.9 represents the corresponding first moment about the horizontal axis through point D. It is vital to reiterate that in Fig. 7.2b, $x(s)$ and $y(s)$ are measured from the *base* of the differential M/EI area, i.e., from the location of the angle change itself, to point D. Equivalent forms of Eqs. 7.8 and 7.9 are

$$\Delta_{DV_{BC}} = t_{BC}\bar{x}_{BC} = \phi_{BC}\bar{x}_{BC} \qquad (7.10)$$

and

$$\Delta_{DH_{BC}} = t_{BC}\bar{y}_{BC} = \phi_{BC}\bar{y}_{BC} \qquad (7.11)$$

where t_{BC} is the area under the M/EI diagram for member BC. This symbol has the same meaning as in the moment-area method. Quantities $\bar{x}_{BC}$ and $\bar{y}_{BC}$ are the horizontal and vertical distances, respectively, from point D to the point on member BC that lies below the centroid of the M/EI area. Substitution of ϕ_{BC} for t_{BC} is a matter of nomenclature to emphasize the fact that the area under an M/EI diagram is an angle change. In this discussion, ϕ_{BC}, which will be referred to as a *concentrated angle change*, constitutes the total difference in rotation that exists between the end points of member BC. The meaning of ϕ_{BC} is shown in Fig. 7.3. For computational purposes the actual curvature is replaced by chords BP and PC'. Since the actual curvature is upward relative to BC the kink introduced at P is counterclockwise.

The effect of the concentrated angle change on point D is the creation of a rotation θ_{BC} expressed as

$$\theta_{BC} = \phi_{BC} \qquad (7.12)$$

Expressions similar to Eqs. 7.10 through 7.12 can be developed for members AB and CD. The rotation and the components of displacement at joint D are then obtained by summing the separate effects of each member.

$$\Delta_{DV} = \Delta_{DV_{AB}} + \Delta_{DV_{BC}} + \Delta_{DV_{CD}}$$

$$= \phi_{AB}\bar{x}_{AB} + \phi_{BC}\bar{x}_{BC} + \phi_{CD}\bar{x}_{CD} \qquad (7.13)$$

$$\Delta_{DH} = \Delta_{DH_{AB}} + \Delta_{DH_{BC}} + \Delta_{DH_{CD}}$$

$$= \phi_{AB}\bar{y}_{AB} + \phi_{BC}\bar{y}_{BC} + \phi_{CD}\bar{y}_{BC} \qquad (7.14)$$

$$\theta_D = t_{AB} + t_{BC} + t_{CD}$$

$$= \phi_{AB} + \phi_{BC} + \phi_{CD} \qquad (7.15)$$

With the preceding development as background, a general procedure for calculating displacements for cantilevered frames can be enumerated. Assuming displacement and rotation values are desired at some point of interest:

1. Calculate ϕ, the concentrated angle change (area under the M/EI diagram), for each member. Usually, it is expedient to use the MP/EI diagram to compute a concentrated angle, ϕ_i, for each part of the diagram. ϕ_i is located at a point *on the member* that aligns with centroid of corresponding M/EI area.

2. For each concentrated angle change, evaluate the expressions

$$\Delta_{V_i} = \phi_i\bar{x}_i$$

$$\Delta_{H_i} = \phi_i\bar{y}_i$$

$$\theta_i = \phi_i$$

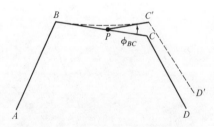

Fig. 7.3.   Discrete angle change ϕ_{BC}.

where $\bar{x}_i$ and $\bar{y}_i$ are the horizontal and vertical distances, respectively, from the location of ϕ_i to the point of interest. Quantities $\bar{x}_i$ and $\bar{y}_i$ are zero for portions of the MP/EI diagram that do not lie between the fixed support and the point of interest.

3. Determine the vertical displacement, Δ_V, horizontal displacement, Δ_H, and the rotation, θ, at the point of interest from

$$\Delta_V = \sum_{i=1}^{NMP} \Delta_{Vi}$$

$$\Delta_H = \sum_{i=1}^{NMP} \Delta_{Hi}$$

$$\theta = \sum_{i=1}^{NMP} \theta_i$$

where NMP is the number of individual M/EI areas in the MP/EI diagram. Directions must be assigned to each Δ_{Vi}, Δ_{Hi}, and θ_i quantity. It is useful to imagine that a rigid bar extends directly from the concentrated angle change to the point of interest. For example, in Fig. 7.4, the

counterclockwise angle change at P causes end D of the rigid bar to move upward and to the right. Recall, from Fig. 7.3, that the direction of ϕ was deduced from the actual curvature implied by the MP/EI diagram.

Example 7.1.   Determine displacements for joint D for the frame given in Fig. 7.5a. Use Angle-Change Principle.

Solution: Initially, the moment diagram must be constructed and converted to an M/EI diagram. This is easily done, and the result is shown in Fig. 7.5b. The diagram is labeled as an MP/EI because the moment diagram for member BC is drawn by parts. Diagrams shown for members AB and CD are actual moment diagrams. Each component area of the MP/EI diagram represents a concentrated angle change. Figure 7.5c offers a mode for visually depicting the concentrated angle changes. Any particular angle change is indicated by a dot placed on the frame and is located directly below the centroid of the corresponding M/EI area.

A tabular form can be employed for the numerical computations. The displacements at point D are determined first:

Angle Change	$\phi_i = M/EI$ area	x	y	$\phi_i x$	$\phi_i y$
1	$\left(\dfrac{10}{EI}\right)(12) = \dfrac{120}{EI}$ cw	18	8	$\dfrac{2160}{EI}$ ↓	$\dfrac{960}{EI}$ ←
2	$\dfrac{1}{2}\left(\dfrac{20}{EI}\right)(8) = \dfrac{80}{EI}$ cw	12	16/3	$\dfrac{960}{EI}$ ↓	$\dfrac{427}{EI}$ ←
3	$\dfrac{1}{2}\left(\dfrac{100}{EI}\right)(8) = \dfrac{400}{EI}$ cw	12	8/3	$\dfrac{4800}{EI}$ ↓	$\dfrac{1067}{EI}$ ←
4	$\left(\dfrac{100}{EI}\right)(12) = \dfrac{1200}{EI}$ cw	6	0	$\dfrac{7200}{EI}$ ↓	0
	$\Sigma\ \dfrac{1800}{EI}$ cw			$\Sigma\ \dfrac{15120}{EI}$ ↓	$\Sigma\ \dfrac{2454}{EI}$ ←

As indicated at the base of the table, quantities given in the second, fifth, and sixth columns of the preceding table are summed to obtain

$$\theta_D = \frac{1800}{EI}\ \text{cw}$$

$$\Delta_{DV} = \frac{15{,}120}{EI}\ \downarrow$$

$$\Delta_{DH} = \frac{2454}{EI}\ \downarrow$$

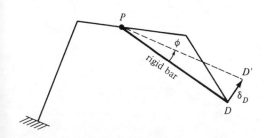

Fig. 7.4. Direction of motion caused by a concentrated angle change.

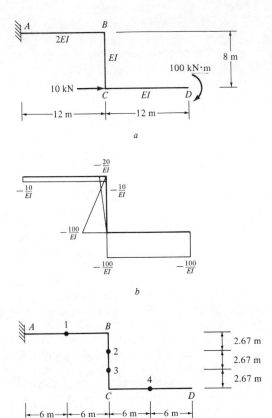

a

b

c

Fig. 7.5. Frame for Example 7.1. *a*, loading; *b*, *MP/EI* diagram.

The direction of the rotation and two displacements caused by the individual angle changes are established by qualitative inspection of Fig. 7.6 and Fig. 7.7. As an example, Fig. 7.6a depicts the consequences of angle change 1. The *MP/EI* diagram that corresponds to this angle change is entirely negative in sense and indicates downward curvature. The angle-change principle infers that the entire angle change occurs at a distinct location and pro-

duces a "kink" in the structure. In this case, the kink causes point *D* to move downward and to the left and to experience a clockwise rotation. Figure 7.7a shows the use of a rigid bar to visualize the displacement of point *D*. In this given problem, the same observation can be made regarding the remaining concentrated angle changes; i.e., all angle changes cause point *D* to move downward and to the left. This is evident in Fig. 7.6 and 7.7, parts b through d. The inference is that the frame deforms into the dashed shape shown in Fig. 7.8 instead of the actual curved shape. Although this is not the true shape of the frame, the correct displacement and rotation values are determined for point *D*.

Example 7.2. Determine the rotation and components of displacement at joint *B* of the frame given in Fig. 7.5a. Use the angle-change principle.

Solution: Joint *B* is affected only by the M/EI diagram between points *A* and *B*. The portion of the M/EI diagram beyond *B* does not enter into the computations. (It is erroneous to conclude that the loads on members *BC* and *CD* have no effect on the motion of joint *B*. Without their presence, the M/EI diagram for member *AB* would be altered, and different displacement values would result.) Physically, this implies that member *AB* behaves as if it were an independent cantilever beam. The desired quantities are calculated using the angle change ϕ_1 computed in Example 7.1.

$$\theta_B = \phi_1 = \frac{120}{EI} \text{ cw}$$

$$\Delta_{BV} = \phi_1 x = \frac{120}{EI} (6) = \frac{720}{EI} \downarrow$$

$$\Delta_{BH} = \phi_1 y = \frac{120}{EI} (0) = 0$$

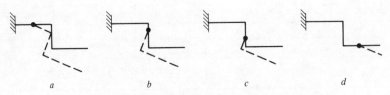

a   b   c   d

Fig. 7.6. Concentrated angle changes in Example 6.6.

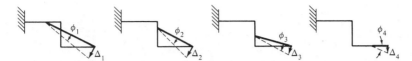

Fig. 7.7.   Concentrated angle changes in Example 7.1.

As expected, it may be observed that the above computations are identical in form to a solution by the moment-area method.

7.2.2 Corrective Rigid-Body Rotation

A three-member plane frame is shown in Fig. 7.7a. The structure is supported by a pin at A and a roller at D. It is desired to compute the rotation θ_A, the horizontal displacement, δ_H and the angular rotation, θ_D, at joint D due to the introduction of a concentrated angle change ϕ at any point P. Because the type of supports do not meet the conditions of the angle-change principle, this principle cannot be directly applied to the problem. An indirect method, based upon the principle of superposition (thus limited to first-order elastic analysis), shall be developed.

Consider Fig. 7.9b. Here the frame of Fig. 7.9a has been temporarily altered by completely freeing joint D and completely fixing joint A, but the angle change ϕ is still imposed at point P. This structure meets the boundary conditions used in developing the angle-change principle, and the displacements at joint D can be determined directly as

$$\Delta'_{DV} = \phi x \qquad (7.16)$$

$$\Delta'_{DH} = \phi y \qquad (7.17)$$

$$\theta'_D = \phi \qquad (7.18)$$

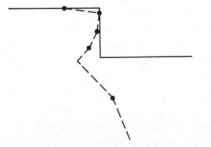

Fig. 7.8.   Displaced shape caused by a series of concentrated angle changes.

Figure 7.9c illustrates how the structure in Fig. 7.9b can be converted to that in Fig. 7.9a by imposing the rigid-body rotation, θ_A. The magnitude of the rotation θ_A is unknown but must be sufficient to restore point D to a position of zero vertical movement. By the angle-change principle, the displacements Δ''_{DV} and Δ''_{DH} are expressed as

$$\Delta''_{DV} = \theta_A b \qquad (7.19)$$

$$\Delta''_{DH} = \theta_A a \qquad (7.20)$$

$$\theta''_D = \theta_A \qquad (7.21)$$

The condition of no vertical translation at point D in Fig. 7.9a requires

$$\Delta'_{DV} - \Delta''_{DV} = 0 \qquad (7.22)$$

or by substitution of Eq. 7.19

$$\Delta'_{DV} - \theta_A b = 0 \qquad (7.23)$$

Thus, the magnitude of the rigid-body rotation θ_A in Fig. 7.9a and c is

$$\theta_A = \frac{\Delta'_{DV}}{b} \qquad (7.24)$$

The corrected horizontal displacement in Fig. 7.9a is obtained by the superposition

$$\Delta_{DH} = \Delta'_{DH} - \Delta''_{DH} \qquad (7.25)$$

By substitution of Eq. 7.20 into Eq. 7.25,

$$\Delta_{DH} = \Delta'_{DH} - \theta_A a \qquad (7.26)$$

Further substitution of Eq. 7.24 gives

$$\Delta_{DH} = \Delta'_{DH} - \Delta'_{DV} \frac{a}{b} \qquad (7.27)$$

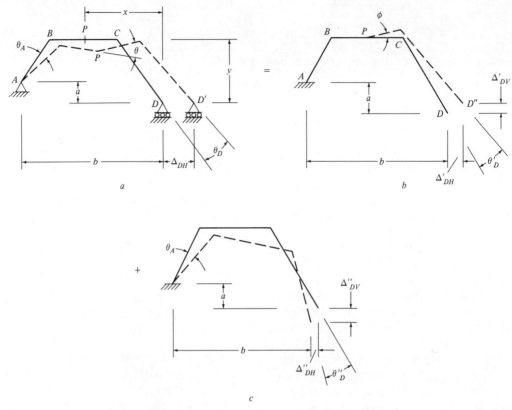

Fig. 7.9. Application of corrective rigid-body rotation. *a*, actual displacements; *b*, cantilevered displacements; *c*, rigid-body rotation.

The two terms on the right side of Eq. 7.25 can be viewed separately as the horizontal displacement predicted by the angle-change principle and a displacement correction to account for the actual support conditions. When the supports at A and D in Fig. 7.9a are at the same level ($a = 0$), Eq. 7.27 reduces to $\Delta_{DH} = \Delta'_{DH}$. That is, no correction to the horizontal displacement is necessary.

The true rotation θ_D can be obtained as

$$\theta_D = \theta'_D - \theta''_D = \theta'_D - \theta_A \quad (7.28)$$

Equations 7.24, 7.27, and 7.28 represent the solution to the problem posed at the beginning of this section—i.e., determining the rotation at point A and the horizontal displacement and rotation at point D of the frame in Fig. 7.9a. The equations were derived for the case of a single concentrated angle change. Frame loadings are such that many concentrated angle changes exist for a given frame. Thus, the equations are to be employed for each concentrated angle change, and the individual effects summed. This is best done in tabular format.

Although the solution procedure is general, the equations developed for this example are limited to certain conditions of the problem. Specifically, the roller support lies to the right of and below the pin support. As a consequence, the corrective horizontal motion Δ''_{DH}, in Eq. 7.26 is negative; i.e., it reduces Δ'_{DH}. For other conditions, the corrective rotation can be additive, and the analyst must be careful to recognize the appropriate sign. With the introduction of an appropriate sign convention, one could develop formulas having general applicability. The author prefers to omit such a development in favor of a solution procedure that is based upon physical understanding of the concept of a corrective rigid-body rotation.

At this point, it is appropriate to summarize the preceding concepts:

If the angle-change principle is applied to

frames of arbitrary support conditions, the computed displacements may violate the condition of compatibility. True displacement values are obtained by adjusting the results to include the effects of a corrective rigid-body rotation. The magnitude and direction of the rigid-body rotation are chosen in a manner that restores compatibility to the structure. A frame that is cantilevered from a fixed support does not require the application of a corrective rigid-body rotation. Also, if the supports in the given frame are at the same vertical level ($a = 0$), no correction of the horizontal displacement of the roller is necessary.

An analytical procedure for frames similar to the one given in Fig. 7.9a can be enumerated as follows.

1. Construct the M/EI diagram (or MP/EI diagram) for the actual frame.
2. Compute the concentrated angle changes and their respective locations.
3. Compute the rotation, vertical displacement, and horizontal displacement at the roller support by using the angle-change principle. In effect, the pin and roller supports are mentally converted to fixed and free joints, respectively.
4. Calculate the magnitude and direction of

joint to restore the fictitious free end to a zero vertical-displacement value.
5. By use of the angle change principle, determine the horizontal motion and rotation induced at the fictitious free end by the corrective rigid-body rotation.
6. Combine the results of the initial angle-change principle solution and the corrective rigid-body rotation.

For frames that differ from Fig. 7.9a in support arrangement, the procedure is altered only by the nature of the corrective rigid-body rotation. It is the analyst's task to determine the appropriate corrective rotation as dictated by the location of the pinned and roller supports and the nature of the incompatibility that results from the initial application of the angle-change principle.

Example 7.3. Using the angle-change principle, determine the displacement of the roller support for the frame given in Fig. 7.10. Assume EI is constant and the same value for all members.

Solution: The MP/EI diagram (Fig. 7.10b) is established using Example 6.7. With the various parts of the diagram numbered as shown, the angle-change principle is applied. The computations are:

Angle Change	ϕ	x	y	ϕx	ϕy
①	$0.5(36)(18)/EI = \dfrac{324}{EI}$ ccw	12	15	$\dfrac{3888}{EI}$ ↑	$\dfrac{4860}{EI}$ →
②	$0.5(360)(18)/EI = \dfrac{324}{EI}$ cw	12	18	$\dfrac{38,880}{EI}$ ↓	$\dfrac{58,320}{EI}$ ←
③	$0.5(360)(12)/EI = \dfrac{2160}{EI}$ cw	8	24	$\dfrac{17,280}{EI}$ ↓	$\dfrac{51,840}{EI}$ ←
④	$\dfrac{2}{3}(72)(12)/EI = \dfrac{576}{EI}$ ccw	6	24	$\dfrac{3456}{EI}$ ↑	$\dfrac{13,824}{EI}$ →
⑤	$0.5(288)(12)/EI = \dfrac{1728}{EI}$ cw	4	24	$\dfrac{6912}{EI}$ ↓	$\dfrac{41,472}{EI}$ ←
⑥	$0.5(288)(24)/EI = \dfrac{3456}{EI}$ cw	0	16	0	$\dfrac{55,296}{EI}$ ←
⑦	$0.5(144)(24)/EI = \dfrac{1728}{EI}$ ccw	0	12	0	$\dfrac{20,736}{EI}$ →
	$\theta'_D = \dfrac{7956}{EI}$ cw			$\delta'_{Dv} = \dfrac{55,728}{EI}$ ↓	$\delta'_{DH} = \dfrac{167,508}{EI}$ ←

the corrective rigid-body rotation which must be imposed at the fictitious fixed

Because the roller support is incapable of a vertical displacement, the computed downward movement is

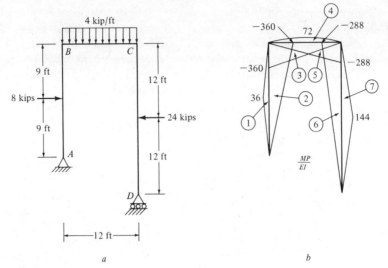

Fig. 7.10.   Frame for Example 7.3.

an incompatible outcome. A rigid-body rotation is required to correct this inconsistency. The concentrated angle change, θ_A, necessary at joint A is determined by application of the angle-change principle.

$$\theta_A x = \Delta'_{DV}$$

$$\theta_A(12) = \frac{55,728}{EI}$$

$$\therefore \theta_A = \frac{4644}{EI} \text{ ccw}$$

This counterclockwise rotation restores the roller to its proper place and induces a secondary rotation θ''_D and a secondary horizontal motion Δ''_{DH} at joint D, which are computed as

$$\theta''_D = \theta_A = \frac{4644}{EI} \text{ ccw}$$

$$\Delta''_{DH} = \theta_A y = \frac{4644}{EI}(6) = \frac{27,864}{EI} \rightarrow$$

The final solution is determined by combining the displacements

$$\theta_D = \theta'_D + \theta''_D$$

$$= \frac{7956}{EI} \text{ cw} + \frac{4644}{EI} \text{ ccw}$$

$$\therefore \theta_D = \frac{3312}{EI} \text{ cw}$$

$$\Delta_{DH} = \Delta'_{DH} + \Delta''_{DH}$$

$$= \frac{167,508}{EI} \leftarrow + \frac{27,864}{EI} \rightarrow$$

$$\therefore \delta_{DH} = \frac{139,644}{EI} \leftarrow$$

As a means of verification, the reader can compare these answers with the moment-area solution obtained earlier in Example 6.7. The reader should realize that the computed θ_A value is indeed the total rotation at joint A. (Why?)

7.2.3   Remarks

The use of physical interpretation to establish directions for the various displacement quantities tabulated in Examples 7.1 through 7.3 must be emphasized. This is a necessity because of the manner in which the angle-change relationships were derived. For example, the terms x and y in Eqs. 7.2 and 7.3 and other subsequent relationships are not mathematical coordinates. This is evident from Fig. 7.1 in which x and y are shown to be measured from point P to point D. Since P is any point on the frame, no fixed origin is used for these distances. Also, no sign convention has been stated for the quantities in the developed equations. A general set of angle-change relationships can be derived but the author has a strong preference for the physical approach. The angle-change principle was derived for frames having particular support conditions. However, the principle is sufficient for the analysis of any planar frame. General application often requires ingenuity on the part of the analyst. In other words: use of the angle-change principle is similar to the moment-area theorems. The principle has general validity but application is not automatic. A careful study of the frame geometry and support conditions is a prerequisite.

7.3   THE CONJUGATE-GRID METHOD

The moment-area method gives the best "physical feeling" for the effect of flexure on framed structures; the inherent use of geometry requires careful thought and sketching on the part of the analyst. Partial relief from the geometric complexities of the moment-area method is gained by employing the angle-change principle, which offers simplicity in the form of a tabulated procedure. However, it is desirable to develop an approach that *completely removes the geometry aspects*. The conjugate-grid method, which essentially accomplishes this aim, is the subject of this section. (In most classical texts, this method was referred to as the "conjugate-frame method." Because the terminology "conjugate-grid" is a more definitive descriptor, it is adopted in this text.) In this method, generalized displacements are computed on the basis of replacing the actual frame with an appropriate *conjugate grid*. The loading on the conjugate grid consists of the M/EI diagram of the actual beam. Desired displacements are computed by creating an analogy between internal shears and moments of the conjugate grid and external rotations and translations, respectively, in the actual frame.

7.3.1   Analogy to the Angle-Change Principle

The method of constructing the appropriate conjugate grid is drawn from a consideration of the angle-change principle. Figure 7.11a illustrates a cantilever frame. To simplify the discussion, a piece of the M/EI diagram (a single concentrated angle change) is shown rather than the complete diagram. The conjugate grid for this structure is shown in Fig. 7.11b. Its construction is based upon the following procedure:

1. The supports are removed from the actual frame and it is placed in a flat position. The dimensions of the real frame are maintained.
2. The loading is the angle-change diagram (areas under the M/EI diagram) of the real frame, and it is applied perpendicular to the plane of the conjugate grid (i.e., in

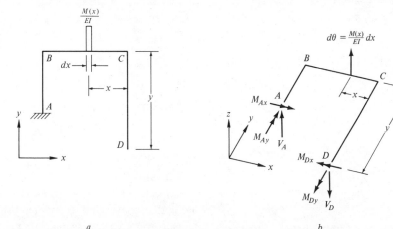

a   *b*

Fig. 7.11.   Conversion of a real frame to a conjugate grid. *a*, the real frame; *b*, the conjugate grid.

the z-direction). In this case, the angle change diagram is the single angle change $d\theta$. An M/EI area that causes counterclockwise curvature relative to the near end in the real frame is considered a positive load and is applied in the positive z-direction. Thus, $d\theta$ is shown acting upward.

3. The member ends of the conjugate grid are considered to be capable of developing a resisting moment about its longitudinal axis (torsional moment), a moment about transverse axis that lies in the x-y plane (flexural moment), and a shear force in the z-direction. (This is the usual condition for a grid.)

4. It will be shown that moments about the x and y axes and shear in the z-direction at any point in the conjugate grid are analogous to displacement in the x and y directions and rotation about the z-axis, respectively, at the corresponding point in the real frame. *Consequently, known displacement conditions in the real frame convert to known internal force conditions in the conjugate grid.* In this instance, the fixed condition at joint A of the real frame is modeled by declaring M_{Ay}, M_{Ax}, and V_A of Fig. 7.11b to have zero value. This is done in Fig. 7.12.

5. *Unknown displacement quantities* at the supports of the real frame are converted to *unknown internal force quantities* at the corresponding points in the conjugate

frame. In the present case, only joint D is affected (as shown in Fig. 7.12).

Displacements Δ_H, Δ_V, and θ at any point in the actual structure are obtained by computing the M_x, M_y, and V values at the same point in the conjugate grid. On a "front face" (or "far end") in the conjugate grid such as joint D, positive moment and shear vectors point in the directions of the negative coordinate axes. On a "back face" (or "rear end") such as joint A, positive directions for these vectors are the reverse, i.e., they point in the directions of the positive coordinate axes.

A sign convention also must be established for determining the direction of the computed displacements. In the real system, positive rotations are counterclockwise, and positive x and y translations are in the direction of the positive axes (i.e., to the right and upward). Positive values for the corresponding force vectors in the conjugate grid indicate the displacements in the real structure are also positive.

Based upon the preceding discussion, the moment and shear values at point D of the conjugate grid of Fig. 7.12 can be calculated. In the following computations, the nomenclature ΣM_{Dx} implies summing moments about the x-axis at point D. Similarly, ΣM_{Dy} implies summing moments about the y-axis at point D.

$$\Sigma F_z = \quad \theta_D - d\theta = 0$$
$$+ \; DN$$

$$\theta_D = d\theta \qquad (7.29)$$

$$\Sigma M_{Dx} = \Delta_{DH} - y\,d\theta = 0$$
$$+ \; \leftarrow\!\leftarrow$$

$$\Delta_{DH} = y\,d\theta \qquad (7.30)$$

$$\Sigma M_{Dy} = \Delta_{DV} + x\,d\theta = 0$$
$$+ \; \uparrow$$

$$\Delta_{DV} = x\,d\theta \qquad (7.31)$$

If Eqs. 7.29 through 7.31 are compared with Eqs. 7.4 through 7.6, the two sets of formulas are seen to be identical. The observation to be made is that conversion to and analysis of the conjugate grid give displacement values that are in exact agreement with the angle-change prin-

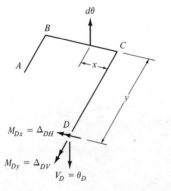

Fig. 7.12. Conjugate grid with boundary conditions.

ciple. This comparison demonstrates that conversion of a real *cantilever* frame to its corresponding conjugate grid, and using the latter to compute desired displacements in the real frame is an analytical procedure that has general validity. In the following paragraphs, the conjugate-grid concept is extended to another type of frame. The conjugate-grid analogy has been developed for a single angle change. In application, the loading on the conjugate frame is a set of concentrated angle changes. The principle of superposition permits treatment of these angle changes as a combined loading.

Figure 7.13a illustrates a portal type frame subjected to a concentrated angle change, ϕ. Following the rules established earlier, the appropriate conjugate frame is constructed as shown in Fig. 7.13b. Boundary conditions for the conjugate frame are established by recognizing that in the real frame, no translations are permitted at joint A and no vertical translation is permitted at joint D. By conversion, moments M_{Ax}, M_{Ay}, and M_{Dy} are specified to have zero value and need not be shown in Fig. 7.13b. Remaining unknown reactions in the conjugate frame are the shear forces V_A and V_D and the flexural moment M_{Dx}, which correspond to the rotations at joints A and D and the horizontal displacement at joint D, respectively, in the real frame. A general solution for the displacement values is obtained by applying statics to the conjugate frame.

$$\Sigma M_{Dy} = \theta_A(b) + \phi(x) = 0$$
$$+ \uparrow$$

$$\theta_A = \frac{\phi x}{b} \tag{7.32}$$

$$\Sigma F_z = \theta_A + \theta_D + \phi = 0$$
$$+ \text{ UP}$$

$$\theta_D = \phi + \theta_A \tag{7.33}$$

$$\Sigma M_{Dx} = \Delta_{DH} + \theta_A(a) + \phi(y) = 0$$
$$+ \longrightarrow$$

$$\Delta_{DH} = \phi y + \theta_A a \tag{7.34}$$

Equations 7.32 through 7.34 are identical to the solution obtained in Section 7.2. Inspection of Eqs. 7.16 through 7.18 and 7.24, 7.26, and 7.28 is a verification of this fact. In making this assessment, it is necessary to realize that a negative value of θ_A obtained from Eq. 7.32 implies a clockwise rotation at A. Substitution of a negative value for θ_A in Eqs. 7.33 and 7.34 produces positive values of θ_D and Δ_{DV}, which imply a counterclockwise rotation and a horizontal movement to the right, respectively. Thus, the magnitudes and directions are in agreement with the angle-change principle. These observations demonstrate the generality of the conjugate frame analogy for the two types of frames for which this principle has been demonstrated. As stated earlier, with proper care, the angle-change principles can be

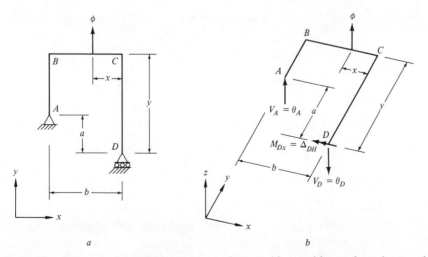

Fig. 7.13.   Conversion of a real frame to a conjugate grid. *a*, real frame; *b*, conjugate grid.

applied to any frame. Thus, the conjugate frame analogy also has general applicability by virtue of the proven equivalence to the angle-change principle.

7.3.2 Example Problems

Example 7.4. Draw conjugate grids for the frames given in Fig. 7.14a. Show the conjugate grid loadings. Determine the displacements of the supports.

Solution: The loaded frame and associated M/EI diagram are shown in Figs. 7.14a and b, respectively. The conjugate grid is drawn in Fig. 7.14c. Joint A is the near end and joint D is the far end. Resisting forces shown for the conjugate frame are in recognition of the rotational degrees of freedom at joints A and D, and the horizontal degree of freedom in translation at joint D. An upward loading is indicated for member BC because the M/EI diagram produces counterclockwise curvature in the real frame.

The areas under the triangular loads are concentrated angle changes whose magnitudes are

$$\phi_1 = \frac{1}{2}\left(\frac{24}{EI}\right)(6) = \frac{72}{EI}$$

$$\phi_2 = \frac{1}{2}\left(\frac{24}{EI}\right)(4) = \frac{48}{EI}$$

These are shown in the redrawn conjugate grid shown in Fig. 7.14d. Each acts below the centroid of the corresponding triangle. (It should be realized that the concentrated angle changes are *resultants* of distributed loads and *do not replace them* entirely. Caution is advised to be aware that, in any given free-body diagram, the "ϕ loads" used are those that replace those distributed loads, or portions thereof, that actually act on the body).

Using the conjugate grid constructed in Fig. 7.14d, the desired displacements are computed by applications of statics.

Step 1: By summing moments about the y-axis at point D.

$$\Sigma M_{Dy} = \theta_A(10) + \phi_1 + \phi_2 = 0$$
$$+\uparrow$$

$$= \theta_A(10) + \frac{72}{EI}(6)$$

$$+ \frac{48}{EI}(4)\left(\frac{8}{3}\right) = 0$$

$$\theta_A = -\frac{56}{EI}$$

$$\therefore \theta_A = \frac{56}{EI}\ \text{cw}$$

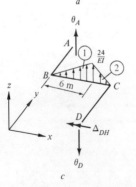

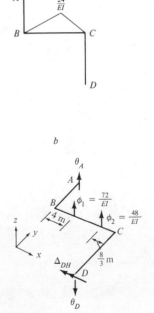

Fig. 7.14. Frame for Example 7.4.

Step 2: By summing forces in the direction of the z-axis.

$$\Sigma F_z = -\theta_D + \theta_A + (\phi_1 + \phi_2) = 0$$
$$+\,UP$$

$$-\theta_D + \left(-\frac{56}{EI}\right) + \frac{120}{EI} = 0$$

$$\theta_D = \frac{64}{EI}$$

$$\therefore \theta_D = \frac{64}{EI}\ \text{ccw}$$

Step 3: By summing moments about the x-axis at point D.

$$\Sigma M_{Dx} = -\Delta_{DH} + \theta_A(16)$$
$$+\ \longrightarrow$$

$$+\frac{120}{EI}(10) = 0$$

$$\Delta_{DH} - \left(-\frac{56}{EI}\right)(16) - \frac{1200}{EI} = 0$$

$$\Delta_{DH} = \frac{304}{EI}$$

$$\therefore \Delta_{DH} = \frac{304}{EI}\ \longrightarrow$$

Step 4: By combining these results with a study of the curvature dictated by M/EI diagram, the general displaced shape depicted in Fig. 7.15 is obtained.

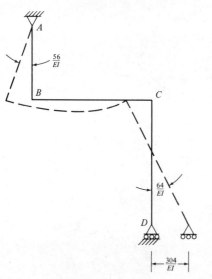

Fig. 7.15. Displaced shape of the frame given in Fig. 7.14.

Example 7.5 Calculate the displacements at the free joint of the frame shown in Fig. 7.16a. Use the conjugate-grid method.

Solution: The cantilever frame shown in Fig. 7.16a is subjected to the indicated pair of loads. This loading produces the MP/EI diagram given in Fig. 7.16b. Displacements in the real frame are computed by employing the conjugate grid depicted in Fig. 7.16c. No resisting forces exist at joint A of the conjugate grid because this joint is fixed in the actual frame. Three internal forces are shown at joint C in Fig. 7.16c to represent the three degrees of freedom, θ_C, Δ_{CH}, and Δ_{CV}, which exist at joint C of the real frame. Load directions in Fig. 7.16c are consistent with a clockwise and counterclockwise curvature produced by the negative and positive M/EI areas, respectively.

Next, each of the triangular loads is replaced by a concentrated angle change as shown in Fig. 7.16d. The magnitudes are

$$\phi_1 = \frac{1}{2}\left(\frac{2}{EI}\right)(10) = \frac{10}{EI}$$

$$\phi_2 = \frac{1}{2}\left(\frac{6}{EI}\right)(10) = \frac{30}{EI}$$

$$\phi_3 = \frac{1}{2}\left(\frac{6}{EI}\right)(7.21) = \frac{21.6}{EI}$$

Application of statics to this conjugate grid permits termination of the desired displacements.

$$\Sigma F_z = -\theta_c + \phi_1 - \phi_2 - \phi_3 = 0$$
$$+\,UP$$

$$-\theta_C + \frac{10}{EI} - \frac{30}{EI} - \frac{21.6}{EI} = 0$$

$$\theta_C = -\frac{41.6}{EI}$$

$$\therefore \theta_C = \frac{41.6}{EI}\ \text{cw}$$

$$\Sigma M_{CX} = -\Delta_{CH} + \phi_1(0) - \phi_2(2) - \phi_3\left(\frac{8}{3}\right)$$
$$+\ \longrightarrow$$

$$= -\Delta_{CH} + \frac{10}{EI}(0) - \frac{30}{EI}(2)$$

$$-\frac{21.6}{EI}\left(\frac{8}{3}\right) = 0$$

$$\Delta_{CH} = -\frac{118}{EI}$$

$$\therefore \Delta_{CH} = \frac{118}{EI}\ \leftarrow$$

$$\Sigma M_{CY}$$
$$+ \uparrow$$

$$= -\Delta_{CV} + \phi_1\left(6 + \frac{16}{3}\right)$$

$$- \phi_2\left(6 + \frac{8}{3}\right) - \phi_3(4) = 0$$

$$= -\Delta_{CV} + \frac{10}{EI}(11.33) - \frac{30}{EI}(8.67)$$

$$- \frac{21.6}{EI}(4) = 0$$

$$\Delta_{CV} = -\frac{233}{EI}$$

$$\therefore \Delta_{CV} = \frac{233}{EI} \downarrow$$

Computed displacements and the general displaced shape are shown in Fig. 7.17. By combining the MP/EI diagram parts to obtain the actual moment diagram, member AB is seen to have the double curvature shown.

Example 7.6 Repeat Example 7.3 by the conjugate-grid method.

Solution: The conjugate grid is drawn in Fig. 7.18. The loading was obtained from the MP/EI diagram in Fig. 7.10b. The magnitudes and directions of the concentrated angle changes are obtained from the seven areas indicated in Fig. 7.10k:

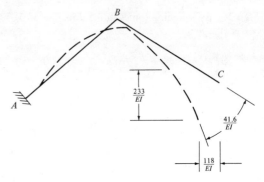

Fig. 7.17. Displaced shape of the frame given in Fig. 7.16a.

$$\phi_1 = .5\left(\frac{36}{EI}\right)(18) = \frac{324}{EI} \uparrow$$

$$\phi_2 = .5\left(\frac{360}{EI}\right)(18) = \frac{3240}{EI} \downarrow$$

$$\phi_3 = .5\left(\frac{360}{EI}\right)(12) = \frac{2160}{EI} \downarrow$$

$$\phi_4 = \frac{2}{3}\left(\frac{72}{EI}\right)(12) = \frac{576}{EI} \uparrow$$

$$\phi_5 = .5\left(\frac{288}{EI}\right)(12) = \frac{1728}{EI} \downarrow$$

$$\phi_6 = .5\left(\frac{288}{EI}\right)(24) = \frac{3456}{EI} \downarrow$$

$$\phi_7 = .5\left(\frac{144}{EI}\right)(24) = \frac{1728}{EI} \uparrow$$

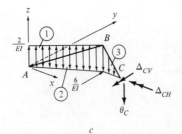

a

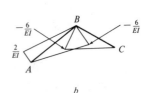

b

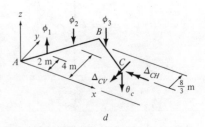

c

d

Fig. 7.16. Frames for Example 7.5.

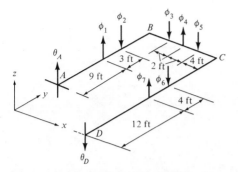

Fig. 7.18.   Conjugate grid for Example 7.6.

statics problem. Analogies drawn between resisting forces in the conjugate grid and the displacements of the real frame virtually eliminate the physical analysis of the structure, which is a fundamental prerequisite to procedures that rely upon pure geometry. Indeed, with a careful enumeration of the rules governing the transformation to a conjugate grid, it is conceivable that one could determine frame displacements with no background in the geometric methods. Because physical understanding of structural behavior is essential to a structural engineer, such an approach is undesirable. In addition, as demonstrated herein, knowledge of other methods has been used to derive—and

The displacement quantities are calculated as follows:

$$\Sigma M_{D_y} = \theta_A(12) + \phi_1(12) - \phi_2(12) - \phi_3(8) + \phi_4(6) - \phi_5(4) = 0$$

$$\theta_A(12) + \frac{324}{EI}(12) - \frac{3240}{EI}(12) - \frac{2160}{EI}(8) + \frac{576}{EI}(6) - \frac{1728}{EI}(4) = 0$$

$$\therefore \theta_A = \frac{4644}{EI} = \frac{4644}{EI} \text{ ccw}$$

$$\Sigma F_z = \theta_A - \theta_D + \phi_1 - \phi_2 - \phi_3 + \phi_4 - \phi_5 - \phi_6 + \phi_7 = 0$$

$$\frac{132}{EI} - \theta_D + \frac{324 - 3240 - 2160 + 576 - 1728 - 3456 + 1728}{EI} = 0$$

$$\therefore \theta_D = -\frac{3312}{EI} = \frac{3312}{EI} \text{ cw}$$

$$\Sigma M_{D_x} = \Delta_{AH} + \theta_A(6) + \phi_1(15) - \phi_2(18) - \phi_3(24) + \phi_4(24) - \phi_5(24) - \phi_6(16) + \phi_7(12) = 0$$

$$\Delta_{AH} + \left(\frac{4644}{EI}\right)(6) + \frac{324(15) - 3240(18) - 216(24) + 576(24) - 1728(24)}{EI}$$

$$-\frac{3456(16) + 1728(12)}{EI} = 0$$

$$\therefore \Delta_{AH} = -\frac{139644}{EI} = \frac{139644}{EI} \leftarrow$$

These results confirm the earlier solution based on the angle-change principle.

7.4   THE CONJUGATE BEAM METHOD

7.4.1   Relation to the Conjugate Grid Method

Structural analysis of planar frames by the conjugate-grid method is a technique that reduces the determination of frame displacements to a

therefore justify—the general validity of the conjugate grid method. However, subsequent to developing a clear understanding of the basic approaches to structural analysis, methods that make the computations routine are desirable. The conjugate-grid method is one method that has this characteristic. A further simplification, which permits routine computation of beam deflections, will be described in the succeeding paragraphs. Generally termed the *conjugate-beam method*, this method of analysis is a degenerate form of the conjugate-grid method and simply introduces the concept of *transformed support conditions*.

In rudimentary form, the concepts involved in the conjugate-grid method can be stated as

1. Converting the real frame to a conjugate grid.
2. Modeling known displacement conditions at the supports of the real frame by an appropriate set of resisting forces in the conjugate grid.
3. Recognizing a mathematical equivalence between the statics of the conjugate grid and the geometry of the real frame.

Preliminary to a discussion of the support transformation concept, it is useful to consider the conjugate-grid method applied directly to beam problems. Theoretically, three resisting forces are needed at any given point in the conjugate frame: a shear and two orthogonal moments. If the framed structure is a beam, the conjugate grid is also a beam and can be referred to as the *conjugate beam*. Furthermore, when axial loads are excluded from the real beam, no horizontal displacements occur, and the corresponding moment vectors can be excluded from the conjugate beam.

As an introduction, consider constructing the conjugate grid for a simply supported beam subjected to a single concentrated load applied at midspan. This real beam and its loading are illustrated in Fig. 7.19a. Rules developed in earlier sections permit a conversion to the conjugate structure shown in Fig. 7.19b. The beam itself is replaced by a line element, which lies in the x-y plane of the three-dimensional space and along the x-axis. Loading for the conjugate structure consists of the real M/EI diagram. In this example, a single, upward, triangular load exists with a maximum value of $PL/4EI$ at midspan. At the supports of the real beam, vertical displacement and horizontal displacement do not occur. Resisting moments, which replace these displacements in the conjugate system, must have zero value and need not be shown. Rotation does occur at either support of the real beam, and the two quantities are modeled in the conjugate system by shear forces, θ_A and θ_B, at its ends.

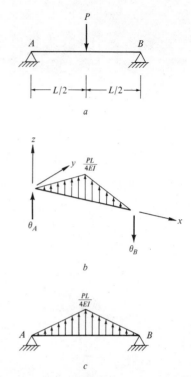

Fig. 7.19. Conjugate-beam analysis of a simple beam.

It is apparent that the conjugate structure drawn in Fig. 7.19b is, in reality, a beam element with particular support conditions. Internal force conditions that exist at the ends of this particular conjugate structure dictate the absence of moment resistance but not the absence of shear resistance. In effect, the ends of the beam in Fig. 7.19b can be viewed as being simply supported. Seen in this way, it is proper to depict the conjugate structure in the manner shown in Fig. 7.19c and refer to it as a conjugate beam. Ordinary beam analysis can be employed to determine the shear and moment at any point in the conjugate beam and, by analogy, the slope and transverse displacement at the same point in the real beam. Indeed, the complete displaced shape of the real beam can be established by constructing a moment diagram for the conjugate beam. In addition, the maximum displacement occurs at a point of zero slope in the real beam and can be located by calculating the maximum moment in the conjugate beam.

Further insight into the conjugate-beam procedure is gained by considering the beam drawn in Fig. 7.20a. This structure is similar to the beam in Fig. 7.19a except for the added feature of overhanging elements. Employing the rules for the conjugate-grid method, the real beam converts to the structure shown in Fig. 7.20b. The reader should note several key features concerning this structure. As before, the real M/EI diagram becomes the loading for conjugate structure. In this instance, it is a single triangular loading centered between points B and C with a maximum intensity of $3PL/4EI$. At the ends of the conjugate member, shear forces θ_A and θ_D and bending moments Δ_{A_y} and Δ_{D_y} are needed in recognition of the free ends of the real beam. Additional conditions must be specified to describe the interior pin supports at points B and C. At both locations, a rotation occurs, but no vertical translation is permitted

because of the pin supports. In the conjugate structure, these constraints require the specification of a zero value of moment capacity at points B and C. However, a unique shear value must exist at these points if the existence of rotations at points B and C in the real beam is to be properly incorporated. These requirements are satisfied if the *internal* moment at points B and C are specified to be zero. In Fig. 7.20b, these restraints at intermediate locations along the span are indicated by showing a dashed vector of zero magnitude for Δ_{BV} and Δ_{CV}. Nonzero rotation values are specified at points B and C by use of dashed vectors for the associated shear forces, θ_B and θ_C.

Statics can be directly applied to the conjugate structure in Fig. 7.20b to compute the various displacement quantities if the internal conditions are properly noted. A more suitable procedure is available by recognizing the conjugate structure in Fig. 7.20b to be equivalent to the conjugate beam shown in Fig. 7.20c. Careful study of the latter figure indicates that the shear and moment constraints of Fig. 7.20b have been represented by a transformation of support conditions. The free ends and interior pin supports of Fig. 7.20a have been converted to fixed ends and internal hinges, respectively. Free ends are capable of developing both shear and moment resistance as necessary at points A and D of the conjugate structure in Fig. 7.20b. Internal hinges transmit an internal shear force of unique value but offer no moment resistance. In this way, the internal force conditions at points B and C in Fig. 7.20b are properly reflected in the conjugate beam. Clearly, the conjugate beam in Fig. 7.20c maintains static equilibrium in a manner identical to the conjugate structure of Fig. 7.20b and constitutes a correct analogy for determining the displacement quantities for the real beam shown in Fig. 7.20a.

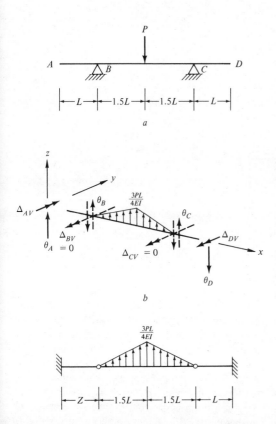

Fig. 7.20. Conjugate-beam analysis of a simple beam with overhanging ends.

7.4.2 Rules for Transforming to a Conjugate Beam

For completeness, the transformed support conditions commonly required in the conjugate-beam method, are assembled in the table

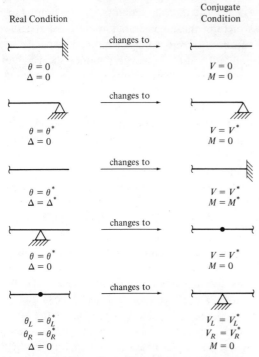

Fig. 7.21. Transformation of real-beam support conditions to conjugate-beam support conditions.

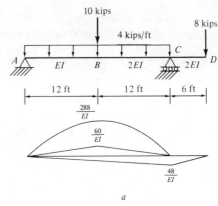

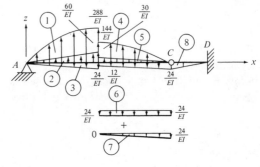

a

b

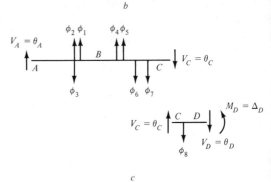

c

Fig. 7.22. Beam for Example 7.7.

given in Fig. 7.21. The left-hand column of schematic diagrams indicates the possible real-beam conditions and their inherent displacement constraints. Corresponding support conditions, which must be incorporated in the conjugate beam, are listed in the right-hand column. The reader is urged to verify the shear and moment conditions associated with each conjugate support and insure that the correct correspondence with the displacement constraints of the real condition is achieved. Moreover, it is not the writer's intent to have the reader employ Fig. 7.21 as a reference table. Instead, the analyst should understand the analogy behind these conversions and deduce the support transformations appropriate to a given problem without referring to Fig. 7.21. Steps involved in the conjugate-beam method are best demonstrated by example problems.

7.4.3 Example Problems

Example 7.7. Calculate the slope and deflection at point D of the beam shown in Fig. 7.22a. Use the conjugate beam method.

Solution: The M/EI diagram is given below the beam. Its determination by the SBM is left as an exercise for the reader. The conjugate beam is drawn in Fig. 7.22b. Points C and D of the real beam are converted to an internal hinge and fixed end, respectively. The exterior pin support at A remains a pinned support in the conjugate beam. The loading on the conjugate beam is the MP/EI diagram obtained by scaling the M/EI diagram to reflect the variable EI of the beam.

In Fig. 7.22c, the beam has been divided (at the hinge) into two segments. Distributed loads have been replaced with concentrated angle changes, given by

$$\phi_1 = \frac{2}{3}\left(\frac{288}{EI}\right)(12) = \frac{2304}{EI} \uparrow$$

$$\phi_2 = .5\left(\frac{60}{EI}\right)(12) = \frac{360}{EI} \uparrow$$

$$\phi_3 = .5\left(\frac{24}{EI}\right)(12) = \frac{144}{EI} \downarrow$$

$$\phi_4 = \frac{2}{3}\left(\frac{144}{EI}\right)(12) = \frac{1152}{EI} \uparrow$$

$$\phi_5 = .5\left(\frac{30}{EI}\right)(12) = \frac{180}{EI} \uparrow$$

$$\phi_6 = \frac{24}{EI}(12) = \frac{288}{EI} \downarrow$$

$$\phi_7 = .5\left(\frac{24}{EI}\right)(12) = \frac{144}{EI} \downarrow$$

$$\phi_8 = .5\left(\frac{24}{EI}\right)(6) = \frac{72}{EI} \downarrow$$

The locations for ϕ_1 and ϕ_2 are not indicated in Fig. 7.22c, but each is below the centroid of its corresponding load area.

From segment AC:

$$\Sigma M_A = \theta_C(24) + \phi_3(8) + \phi_6(18) + \phi_7(20)$$
$$+cw$$
$$\qquad - \phi_1(7.5) - \phi_2(8) - \phi_4(16.5)$$
$$\qquad - \phi_6(6) = 0$$

$$\therefore \theta_C = \frac{1536}{EI} = \frac{1536}{EI} \; ccw$$

From segment CD:

$$\Sigma F_z = \theta_C - \theta_D - \phi_8 = 0$$

$$\frac{1536}{EI} - \theta_D - \frac{72}{EI} = 0$$

$$\theta_D = \frac{1464}{EI} = \frac{1464}{EI} \; ccw$$

$$\Sigma M_D = \theta_c(6) - \phi_8(4) - \Delta_D = 0$$
$$+cw$$

$$\frac{1536}{EI}(6) - \left(\frac{72}{EI}\right)(4) - \Delta_D = 0$$

$$\Delta_D = \frac{8928}{EI} = \frac{8928}{EI} \uparrow$$

Note that the sign convention is unchanged from that established for the conjugate-grid method. Also, caution is needed in replacing distributed loads by concentrated angle changes. Concentrated angle changes are resultants of distributed loads. Whenever any segment of a conjugate beam is isolated as a free body diagram, the distributed loads acting on that segment should be shown first and then replaced by their resultants. For example, if a segment 4 feet long, beginning at point A, is isolated, Fig. 7.22c suggests no angle change would act on the segment. This is erroneous (why?). This same caution applies to the use of the angle-change principle itself and the conjugate-grid method.

―――――――――

Example 7.8. Calculate the maximum displacement in the beam shown in Fig. 7.23a. Use the conjugate beam method.

Solution: By symmetry of the statically determinate real beam

$$R_A = R_D = 2(15) = 30 \text{ kips} \uparrow$$

Isolating segment AB as a free-body diagram (Fig. 7.23b)

$$\Sigma M_B = M_A + 30(9) - 2(9)(4.5)$$
$$+cw$$

$$\qquad = -189 \text{ kip-ft}$$

Based on these values the MP diagram is established (Fig. 7.23c.) The MP/EI is applied to the appropriate conjugate beam as depicted in Fig. 7.23d and converted to the concentrated angle changes (Fig. 7.23e):

$$\phi_1 = .5\left(\frac{756}{EI}\right)(9) = \frac{3402}{EI}$$

$$\phi_2 = \frac{2}{3}\left(\frac{81}{EI}\right)(9) = \frac{486}{EI}$$

$$\phi_3 = \frac{2}{3}\left(\frac{36}{EI}\right)(12) = \frac{288}{EI}$$

$$\phi_4 = \frac{2}{3}\left(\frac{243}{EI}\right)(9) = \frac{1458}{EI}$$

$$\phi_5 = .5\left(\frac{2268}{EI}\right)(9) = \frac{10206}{EI}$$

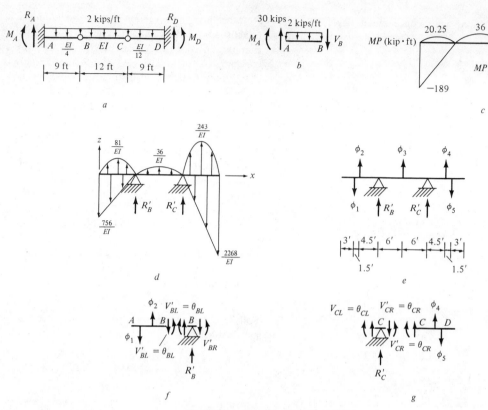

Fig. 7.23. Beam for Example 7.8.

Thus

$$\Sigma M_C = R'_B(12) - \phi_1(18) + \phi_2(4.5) + \phi_3(6)$$
$$+cw$$

$$- \phi_4(4.5) + \phi_5(6) = 0$$

and

$$\Sigma F_z + R'_B + R'_C + \phi_2 + \phi_3$$

$$+ \phi_4 - \phi_1 - \phi_5 = 0$$

Substituting the concentrated angle changes yields

$$R'_B = -\frac{226.5}{EI} \qquad R'_C = \frac{11641.5}{EI}$$

The shear to the left of B, V_{BL}, and to the right of B, V_{BR}, are calculated from Fig. 7.23f:

$$V'_{BL} = \theta_{BL} = \phi_2 - \phi_1 = -\frac{2916}{EI} = \frac{2916}{EI} \text{ cw}$$

$$V'_{BR} = V'_{BL} + R'_B = -\frac{3142.5}{EI} = \frac{3142.5}{EI} \text{ cw}$$

Similarly, from Fig. 7.23g,

$$V_{CR} = \theta_{CR} = \phi_5 - \phi_4 = \frac{8748}{EI} = \frac{8748}{EI} \text{ ccw}$$

$$V'_{CL} = \theta_{CL} = V'_{CR} - R'_C = -\frac{2893.5}{EI}$$

$$= \frac{2893.5}{EI} \text{ cw}$$

Because θ_{BR} and θ_{CL} are both clockwise, it is deduced that the maximum displacement is not within segment BC. Instead, it occurs at point C. The unlabeled moment vector in Fig. 7.21g is M_C and thus

$$M'_C = \Delta_C = \phi_4(4.5) - \phi_5(6)$$

$$= -\frac{54675}{EI} = \frac{54675}{EI}$$

The displaced shape of the real beam is drawn in Fig. 7.24. The value $\Delta_B = 18225/EI$ is determined by deduction.

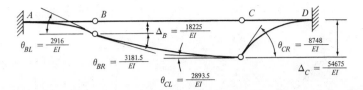

Fig. 7.24.   Results for Example 7.8.

PROBLEMS

7-1.   Determine the rotation and displacements at the free end of each frame. Use the angle-change principle. Assume constant EI, unless shown otherwise.

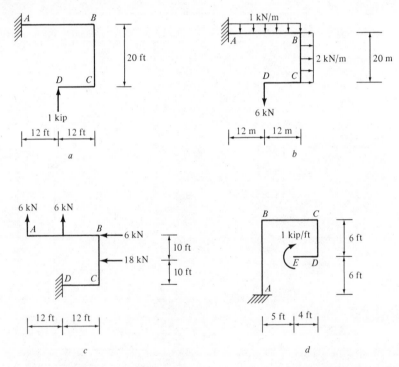

7-2.   Determine the rotation and displacements at the free end of each frame. Use the angle-change principle. Assume constant EI.

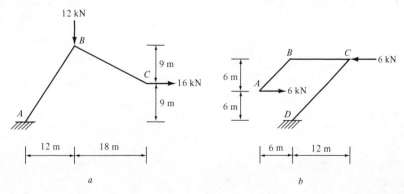

(Problem 7-2 continued on p. 222)

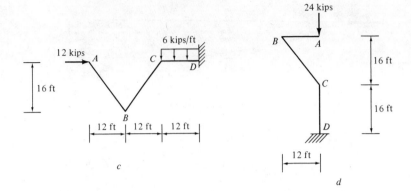

c

d

7-3. Determine the rotations and displacements at the supports of each frame. Use the angle-change principle. Assume constant *EI*.

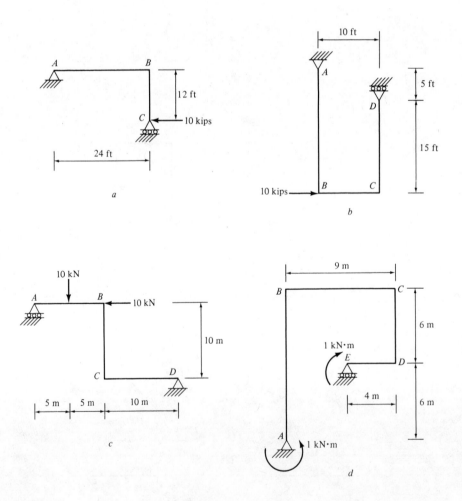

a

b

c

d

(Problem 7-3 continued on p. 223)

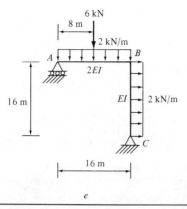

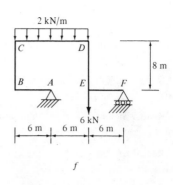

e

f

7-4. Determine the rotation and displacement at the supports of each frame. Use the angle-change principle.

7-5. Determine the rotation and displacements at joint *D* for each frame. Use the angle-change principle.

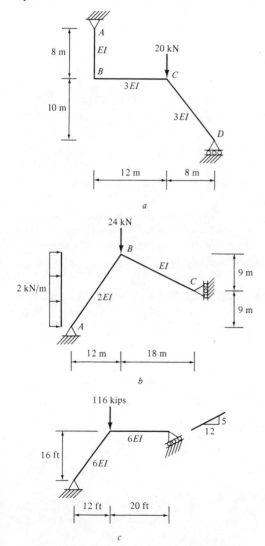

a

b

c

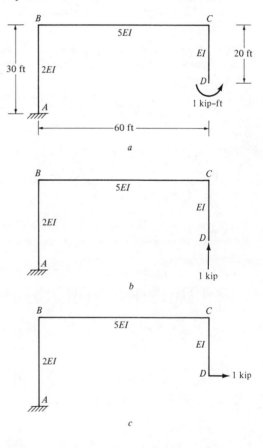

a

b

c

7-6. For each frame in Problem 6-1, determine the rotations and displacements at the supports and the free ends. Use the angle-change principle.

7-7. For each frame in Problem 6-2, determine the rotations and displacements at the supports. Use the angle-change principle.

7-8. Redo Problem 7-1 using the conjugate-grid method.

7-9. Redo Problem 7-2 using the conjugate-grid method.

7-10. Redo Problem 7-3 using the conjugate-grid method.

7-11. Redo Problem 7-4 using the conjugate-grid method.

7-12. Redo Problem 7-5 using the conjugate-grid method.

7-14. For each frame in Problem 6-1, determine the rotation and displacements at the supports. Use the conjugate-grid method.

7-15. For each frame in Problem 6-2, determine the rotations and displacements at the supports. Use the conjugate-grid method.

7-16. Calculate the displacements and rotation at joint B by using the conjugate-grid method. The arch is circular and has a constant EI value.

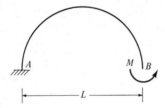

7-17. Calculate the rotations and displacements at the supports of the given frame. Use the conjugate-grid method.

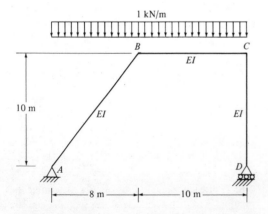

7-18. Calculate the rotations and displacements at the supports of the given frame. Use the conjugate-grid method.

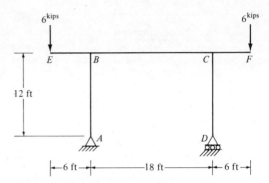

7-19. Calculate the rotations at A, B, and C and the displacements at B and C for the given frame. Use the conjugate-grid method.

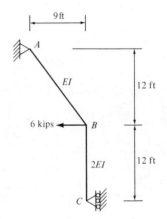

7-20. Calculate the displacement at the free end(s) of each beam. Use the conjugate-beam method. (Suggestion: Use the MP/EI diagram requested in Problem 4-10(vi).)

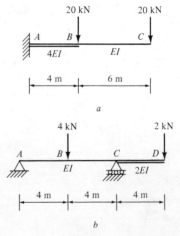

(Problem 7-20 continued on p. 225)

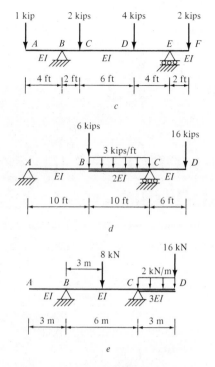

c

d

e

7-21. Repeat Problem 7-20b through e, except calculate the displacement midway between the pin supports.

7-22. For each beam, calculate the slope and displacement at each support, each internal hinge and the load point. Use the conjugate-beam method.

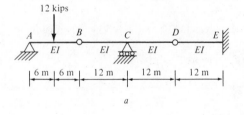

a

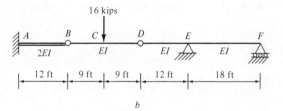

b

7-23. Calculate the maximum displacement for the beam in Fig. 7.22b of Prob. 7-22. Use the conjugate-beam method. EI is constant over the entire length.

7-24. Calculate the displacements midway between points A and B, at point C and midway between points C and D. Use the conjugate-beam method. EI is constant over the entire length.

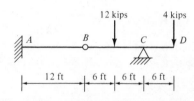

7-25. Calculate Δ_D by the conjugate-beam method. EI is constant over the entire length.

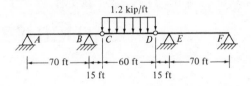

8

Work and Energy Principles in Structural Analysis

8.1 INTRODUCTION

In the context of a closed, conservative structural system, the Law of Conservation of Work and Energy can be stated as:

> For elastic structures, the external work performed by the loads applied to the structure is balanced by an equal amount of energy stored within the components of the structure.

The internal energy is termed *strain energy* because it relates directly to the internal strains created by application of the loads. Work and internal strain energy were shown earlier to be of two forms: real and virtual. The real and virtual work were expressed by Eqs. 1.12 and 1.8, respectively. Many fundamental theorems in structure mechanics are derived by use of work and energy concepts. Some of the more important theorems are described in this chapter. In many cases, solutions that would otherwise be extremely tedious or difficult to obtain follow very easily when work and energy concepts are used. As a final note, work and energy approaches to structural analysis are frequently termed simply *energy methods*. The author prefers to avoid such inexact terminology in order to maintain a consistent distinction between *work* and *energy*.

8.2 REAL WORK AND STRAIN ENERGY

If a load is applied to a structure, the structure will deform into a new shape. As the load displaces, it performs external work as defined by Eq. 1.7. The mathematical interpretation of Eq. 1.7 is that external work is the area under the curve depicting the load-displacement behavior (Fig. 8.1a). If the component material behaves in a linear, elastic manner, the load-displacement behavior is triangular (Fig. 8.1b) and the work performed is given by

$$W = \tfrac{1}{2} P\Delta \qquad (1.12)$$

In Eq. 1.12, Δ is the displacement in the direction of the load P *caused by the application of the load*.

When more than one load is applied simultaneously, the external work is

$$W = \tfrac{1}{2} \sum_{i=1}^{NP} P_i\Delta_i \qquad (8.1)$$

in which NP is the total number of loads, and Δ_i is a displacement in the direction of P_i. To apply Eq. 8.1, the load system must be ''pro-

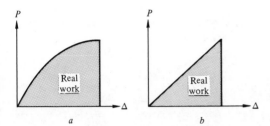

Fig. 8.1.   Concept of real work. *a*, general; *b*, linear structure.

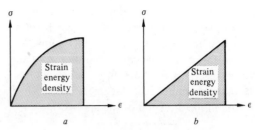

Fig. 8.2.   Concept of strain energy density. *a*, general; *b*, linear material.

portional''—i.e., the relative proportions of the loads are maintained as the general load level is slowly increased. Equation 8.1 is also dependent upon the validity of the principle of superposition (Section 3.7).

Internal strain is the direct consequence of loading a structure. The stresses that are created by resistance to strain are also a source of work. This work is internal in nature. If uniaxial stress results, the work done at any point by the stress σ is given by

$$dU = \int_0^\epsilon (\sigma \, dA) \, d\epsilon \, dL \qquad (8.2)$$

in which dA is a unit area, $d\epsilon$ is a differential increment of strain, and dL is a unit length. Because a unit volume d vol is equal to $dA \, dL$,

$$dU = \left[\int_0^\epsilon \sigma \, d\epsilon \right] d \text{ vol} = U_0 \, d \text{ vol} \qquad (8.3)$$

in which U_0 is called the *strain energy density* and is represented by the area under the curve shown in Fig. 8.2a. For a linear, elastic material, the behavior is as shown in Fig. 8.2b, and the strain energy density is

$$U_0 = \tfrac{1}{2} \sigma\epsilon \qquad (8.4)$$

Integration of the strain energy density over the volume of the structure gives

$$U = \int_{\text{vol}} U_0 \, d \text{ vol} \qquad (8.5)$$

which is the *total strain energy* or simply the "strain energy" absorbed by the structure. Because the strain energy is internal work, it is often termed the "internal strain energy."

In the most general case, normal and shear components of stress and strain exist in all three principal directions. Under these conditions, the strain energy density is given by

$$U_0 = \tfrac{1}{2} \sigma_x \epsilon_x + \tfrac{1}{2} \sigma_y \epsilon_y + \tfrac{1}{2} \sigma_z \epsilon_z$$
$$+ \tfrac{1}{2} \tau_{xy} \gamma_{xy} + \tfrac{1}{2} \tau_{yz} \gamma_{yz} + \tfrac{1}{2} \tau_{zx} \gamma_{zx} \qquad (8.6)$$

Substituting Eq. 2.11 into Eq. 8.6 and neglecting Poisson's ratio yields

$$U = \frac{1}{2E} \int_{\text{vol}} (\sigma_x^2 + \sigma_y^2 + \sigma_z^2) \, dV$$
$$+ \frac{1}{2G} \int_{\text{vol}} (\tau_{xy}^2 + \tau_{yz}^2 + \tau_{zx}^2) \, d \text{ vol}$$
$$\qquad (8.7)$$

Setting Eqs. 8.1 and 8.7 equal to each other produces Eq. 8.8, which is the mathematical statement of the Law of Conservation of Work and Energy for a linear, elastic, isotropic material, that is

$$\frac{1}{2} \sum_{i=1}^{NP} P_i \Delta_i$$
$$= \frac{1}{2E} \int_{\text{vol}} (\sigma_x^2 + \sigma_y^2 + \sigma_z^2) \, d \text{ vol}$$
$$+ \frac{1}{2G} \int_{\text{vol}} (\tau_{xy}^2 + \tau_{yz}^2 + \tau_{zx}^2) \, d \text{ vol}$$
$$\qquad (8.8)$$

Normally, Eq. 8.8 is of limited direct value in computing displacement quantities. If more than one load exists, the corresponding displacements cannot be calculated by use of Eq. 8.8 by itself. If only a single load is applied, the only determinate displacement is that which occurs in the direction of the applied load.

Example 8.1.   Determine expressions for the total strain energy due each of the following effects: (a) constant uniaxial force F; (b) torsional moment T; (c) flexural moment M.

Solution: For constant cross section, member properties, and internal forces, the results are:

(a) Axial force:

$$U = \frac{1}{2E} \int_{vol} \sigma_x^2 \, dV = \frac{1}{2E} \int_0^L \left(\frac{F}{A}\right)^2 A \, dx$$

$$U = \frac{1}{2} \frac{F^2 L}{EA} \tag{8.9}$$

(b) Torsional moment:

$$U = \frac{1}{2G} \int_{vol} \tau^2 \, d \, vol$$

$$= \frac{1}{2G} \int_0^r \int_0^L \left(\frac{T\rho}{J}\right)^2 (2\pi\rho \, d\rho) \, dx$$

$$= \frac{T^2 L}{GJ^2} \int_0^r \pi\rho^3 \, d\rho$$

$$U = \frac{1}{2} \frac{T^2 L}{GJ} \tag{8.10}$$

(c) Flexural moment:

$$U = \frac{1}{2E} \int_{vol} \sigma_x^2 \, d \, vol$$

$$= \frac{1}{2E} \int_0^A \int_0^L \left(\frac{My}{I}\right)^2 dA \, dl$$

$$= \frac{M^2 L}{2EI^2} \int_0^A y^2 \, dA$$

$$U = \frac{1}{2} \frac{M^2 L}{EI} \tag{8.11}$$

In each of the preceding cases, a constant cross section (including material properties) was presumed.

If the cross section, material properties, and internal forces vary with distance x along the member span, the general results are

Axial force:

$$U = \frac{1}{2} \int_0^L \frac{F(x)F(x)}{E(x)A(x)} \, dx \tag{8.12}$$

Torsional moment:

$$U = \frac{1}{2} \int_0^L \frac{T(x)T(x)}{G(x)J(x)} \, dx \tag{8.13}$$

Flexural moment:

$$U = \frac{1}{2} \int_0^L \frac{M(x)M(x)}{E(x)I(x)} \, dx \tag{8.14}$$

Work and energy theorems generally prove to be particularly useful in the analysis of curved members. An example of this type of application is presented in Ex. 8.2.

Example 8.2.   Determine the displacement Δ in the direction of the applied load for the structure illustrated in Fig. 8.3a. Assume $EI = 2 \, GJ$ and both are constant.

Solution: The internal forces for the circular member are depicted in Fig. 8.3b (shear is neglected).

$$M(\theta) = \frac{PD}{2} \sin \theta$$

$$T(\theta) = \frac{PD}{2} (1 - \cos \theta)$$

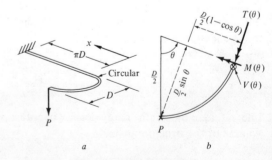

Fig. 8.3.   Structure for Example 8.2. a, structure; b, internal forces, curved segment.

The corresponding strain energy is

$$U_1 = \frac{1}{2} \int_0^L \frac{M(\theta)M(\theta)}{EI} \, dl$$

$$+ \frac{1}{2} \int_0^L \frac{T(\theta)T(\theta)}{GJ} \, dl$$

$$= \frac{1}{2EI} \int_0^\pi \left(\frac{PD}{2} \sin \theta\right)^2 \left(\frac{D}{2} \, d\theta\right)$$

$$+ \frac{1}{EI} \int_0^\pi \left(\frac{PD}{2}\right)^2 (1 - \cos \theta)^2 \left(\frac{D}{2} \, d\theta\right)$$

$$U = \frac{\pi P^2 D^3}{32EI} + \frac{3\pi P^2 D^3}{16EI} = \frac{7\pi P^2 D^3}{32EI}$$

Within the straight piece,

$$M(x) = Px$$

$$T(x) = -PD$$

and the corresponding strain energy is

$$U_2 = \frac{1}{2} \int_0^L \frac{M(x)M(x)}{EI} \, dx$$

$$+ \frac{1}{2} \int_0^L \frac{T(x)T(x)}{GJ} \, dx$$

$$= \frac{1}{2EI} \int_0^{\pi D} (Px)^2 \, dx$$

$$+ \frac{1}{EI} \int_0^{\pi D} (-PD)^2 \, dx$$

$$U_2 = \frac{\pi^3 P^2 D^3}{6EI} + \frac{\pi P^2 D^3}{EI}$$

By the Law of Conservation of Work and Energy

$$\tfrac{1}{2}P\Delta = U_1 + U_2$$

$$\Delta = \frac{\pi^3 PD^3}{3EI} + \frac{39\pi PD^3}{16EI}$$

8.3 COMPLEMENTARY WORK AND STRAIN ENERGY

The load-displacement behavior of a structure can be examined from an alternative viewpoint. The curve of Fig. 8.1a has been repeated in Fig. 8.4a, where a region above the curve

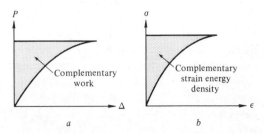

Fig. 8.4. Complementary work and complementary strain energy density.

has been shaded and labeled *complementary work*. This region is bounded by the load axis on the left and by the curve on the right. Mathematically, the shaded area is obtained by the integration

$$CW = \int_0^P \Delta \, dP \qquad (8.15)$$

in which CW is the nomenclature for *complementary work*. The work is complementary in the sense that it has the same units as real work, and when complementary and real work are combined, the *sum* is the rectangular area $P \cdot \Delta$.

A similar definition can be formulated with regard to strain-energy density. The curve of Fig. 8.2a is repeated in Fig. 8.4b, except that a different area is shaded. Adopting a terminology similar to that in Fig. 8.4a, the shaded area (bounded on the left by the stress axis and on the right by the stress-strain curve) is termed *complementary strain energy density*, CU_0. The total *complementary strain energy* is obtained when the complementary strain energy density is integrated over the entire volume of the structure. Thus, it is expressed as

$$CU = \int_{vol} CU_0 \, d \, vol \qquad (8.16)$$

in which

$$CU_0 = \int_0^\sigma \epsilon \, d\sigma \qquad (8.17)$$

Complementary work and complementary strain energy are strictly mathematical quantities. Unlike real work and energy, no physical counterparts exist for describing the shaded areas shown in Fig. 8.4. However, these quantities play a fundamental role in the historical development of many well established and powerful energy theorems in the study of structural analysis.

8.4   PSEUDO-VIRTUAL WORK AND PSEUDO-VIRTUAL STRAIN ENERGY

Consider a state in which a linear, elastic structure has been subjected to a load P_1 and, as a consequence, has displaced an amount, Δ_{11}, in the direction of the load. This behavior is depicted by the initial portion of the graph drawn in Fig. 8.5a. The triangular area, W_1, represents the real work performed by the load and is given by

$$W_1 = \tfrac{1}{2}P_1\Delta_{11} \qquad (8.18)$$

In Δ_{11}, the first subscript indicates the displacement is in the direction of P_1. The second subscript indicates the displacement is due to the application of P_1. The corresponding real strain

energy density, U_{0_1}, at a point in the structure is represented by the triangular area in Fig. 8.5b. In the figure, σ_{P_1} and ϵ_{P_1} are the stress and strain, respectively, created by application of the load P_1. Integration of the strain energy density over the volume produces the total strain energy density, U_1.

Next, without removing P_1, consider that the structure is subsequently subjected to a second load, P_2, applied at some other point. One consequence of this action is the superposition of additional displacements atop those created by the original load P_1. These displacements will satisfy all geometric constraints of the structure and thus are compatible (the load causing the displacements and the additional support reactions will also satisfy equilibrium conditions, but this is unimportant here).

The shaded areas in Fig. 8.5 depict part of the effects caused by imposing the second compatible displaced shape on the previously existing state. Load P_1 does not change in magnitude, but is *caused* to move along its line of action by an amount Δ_{12}. The shaded, rectangular area in Fig. 8.5a constitutes the work VW performed by the load during this motion. Because the imposed displacement state exists only by virtue of the imposition of the additional displacement, VW is called *virtual work* and is expressed as

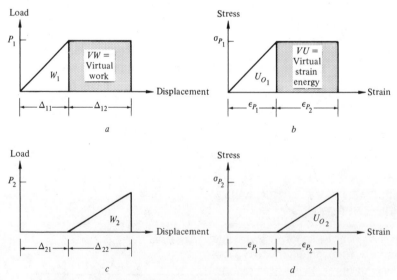

Fig. 8.5.   Concepts of virtual work and virtual strain energy density.

$$VW = P_1 \Delta_{12} \qquad (8.19)$$

Virtual work is distinguished from real work by the absence of a "$\frac{1}{2}$" on the right side of Eq. 8.19.

Coincident with the virtual work, incremental strains inherent in the imposed displaced state cause the previously existing stresses to perform internal work (create strain energy). This energy is called *virtual strain energy*. The *virtual strain energy density*, VU_0, at a point is represented by the shaded area in Fig. 8.5b. Thus,

$$VU_0 = \sigma_{P_1} \epsilon_{P_2} \qquad (8.20)$$

in which σ_{P_1} is the stress at a point due to the application of P_1 and ϵ_{P_2} is the strain at that point caused by the application of P_2. As with virtual work, the absence of a "$\frac{1}{2}$" in Eq. 8.20 is conspicuous. Integration of the virtual strain energy density over the volume of the structure yields the total virtual strain energy

$$VU = \int_{vol} VU_0 \, d \, vol \qquad (8.21)$$

In addition to the effects just described, the external work performed by P_2 itself and the corresponding strain energy must be considered. The work, W_2, performed by P_2 is depicted in Fig. 8.5c. Note that the point at which P_2 is applied had previously been displaced by an amount Δ_{21} by the application of P_1. W_2 is given by

$$W_2 = \frac{1}{2} P_2 \Delta_{22} \qquad (8.22)$$

The stresses and strains created by P_2 result in strain energy density, U_{0_2}, as depicted in Fig. 8.5d. In the figure, σ_{P_2} is the stress at a point due to the application of P_2. The triangular area gives the strain energy density U_{0_2} and, when integrated over the volume of the structure, it gives the total strain energy, U_2. Both W_2 and U_2 are "real" and recognized as having the same magnitude as the external work and internal strain energy that would have been created had P_2 been applied *by itself*.

Under the conditions of sequential loading just described, the work energy balance is expressed as

$$W_1 + VW + W_2 = U_1 + VU + U_2 \qquad (8.23)$$

By the Law of Conservation of Work and Energy, $W_1 = U_1$ and $W_2 = U_2$. Thus, Eq. 8.23 reduces to

$$VW = VU \qquad (8.24)$$

The validity of Eq. 8.24 is an outcome of the Principle of Virtual Work, which can be stated as:

When an elastic structure, in equilibrium under a given set of loads, is subjected to a virtual displacement state, the virtual work of the existing loads going through the imposed displacements is equal to the virtual strain energy of the existing internal stress resultants moving through the imposed internal deformations.

Equation 8.24 was developed for a *linear*, elastic structure, but the Principle of Virtual Work only requires an elastic material.

Consider two different states of the *same* structure. Let the first state be one in which the loads and internal forces conform to

$$\{P_1\} = [E_P] \{F_1\} \qquad (8.25)$$

Equation 8.25 expresses the equilibrium between the loads $\{P_1\}$ and internal forces $\{F_1\}$ in matrix form. $[E_P]$ is termed an "equilibrium matrix." These matrices are described in chapter 14.

Let the second state be one in which the member deformations and independent external displacements conform to

$$\{\Delta_2\} = [C] \{X_2\} \qquad (8.26)$$

Equation 8.26 expresses the compatibility between the member deformations $\{\Delta_2\}$ and the joint displacements $\{X_2\}$ in matrix form. $[C]$ is termed a "compatibility matrix." These matrices are also described in chapter 14.

If the first state is subjected to the displacements of the second state, the *Principal of Virtual Work* (Eq. 8.24) states that

$$\{P_1\}^T \{X_2\} = \{F_1\}^T \{\Delta_2\} \quad (8.27)$$

By substitution of Eqs. 8.25 and 8.26,

$$\{F_1\}^T [E_P]^T \{X_2\} = \{F_1\}^T [C] \{X_2\}$$

$$[E_P]^T = [C] \quad (8.28)$$

which proves that, in general, the equilibrium matrix and the compatibility matrix for any structure are the transpose of each other. This will be observed to be true in later example problems solved by the basic matrix analysis method (Chapter 14).

In truth, the virtual displacements described here are incorrectly named. Virtual displacements are more properly defined as

disturbances of an equilibrium state created by the imposition of an *imaginary* compatible displacement state.

The virtual displacements described in the preceding paragraphs, however, were created by the *actual* application of a supplementary load, and, in that sense, the imposed displacements are not imagined. In recognition of this distinction, it is appropriate in this section to place "pseudo-" in front of the terms *virtual displacement*, *virtual work*, and *virtual strain energy*. The reader should keep this subtle, but important aspect in mind.

8.5 MAXWELL'S RECIPROCAL THEOREM

Examine the general structure, depicted in Fig. 8.6, which is subjected to a pair of concentrated loads at arbitrary locations. Furthermore, assume the structure is composed of a linear, elastic material. Let δ_{ii} and δ_{ji} be the displacements in the directions of loads P_i and P_j, respectively, *if load P_i is applied by itself*. Conversely, let δ_{ij} and δ_{jj} be the displacements in the directions of loads P_i and P_j, respectively, *if load P_j is applied by itself.*

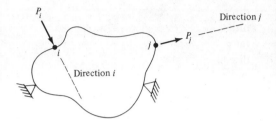

Fig. 8.6. General structure subjected to a pair of loads.

If load P_i is applied to the structure and is followed by the application of load P_j, a certain amount of work, W_{ij}, will be performed. The work done by P_i will be of the nature depicted in Fig. 8.5a in which $\Delta_{11} = \delta_{ii}$ and $\Delta_{12} = \delta_{ij}$. P_j does work of the nature shown in Fig. 8.1b in which $\Delta_{22} = \delta_{jj}$. The total work performed is

$$W_{ij} = \tfrac{1}{2} P_i \delta_{ii} + P_i \delta_{ij} + \tfrac{1}{2} P_j \delta_{jj} \quad (8.29)$$

Conversely, if the loads were applied in reverse order—i.e., P_j applied first and followed by P_i—in the total work, W_{ji}, would be

$$W_{ji} = \tfrac{1}{2} P_j \delta_{jj} + P_j \delta_{ji} + \tfrac{1}{2} P_i \delta_{ii} \quad (8.30)$$

Regardless of the order of application, the total work performed should be the same. On this basis, Eqs. 8.29 and 8.30 can be equated to give

$$P_i \delta_{ij} = P_j \delta_{ji} \quad (8.31)$$

The magnitude of the loads is arbitrary, and if each is assigned a value of unity, Eq. 8.31 reduces to

$$\delta_{ij} = \delta_{ji} \quad (8.32)$$

which is *Maxwell's reciprocal theorem*. This theorem can be stated as:

For a structure behaving in a linear, elastic manner, the displacement in a direction i (at some point i) due to a unit load applied in a direction j (at some point j) is equal in magnitude to the displacement in the direction j (at point j) due to a unit load applied in the direction i (at point i)." The quantities δ_{ij} and δ_{ji} are commonly called *flexibility coefficients*.

Two facts regarding Eqs. 8.31 and 8.32 deserve emphasis. First, the relationships hold regardless of the type of structure (i.e., for all planar structures, space structures, and continua) provided that the component material has linear, elastic properties. Second, the nature of the forces and displacements is irrelevant. Either force could be a moment provided that the corresponding displacement was the appropriate rotation.

8.6   BETTI'S RECIPROCAL ENERGY THEOREM

Equation 8.31, which is the basis of Maxwell's reciprocal theorem, has a particular interpretation. This equation shows that for a given structure, the virtual work performed by a load P_i moving through a displacement caused by a load P_j is equal to the virtual work of the load P_j moving through a displacements caused by the load P_i. It is possible to extend this observation to a more general case involving two *sets* of loads.

Consider a general linear, elastic structure subjected to two separate loadings as depicted in Fig. 8.7. In Fig. 8.7a, loads are applied at points 1, 2, and 3 causing the corresponding displacements X_{11}, X_{21}, and X_{31}. Let the displacements at the additional (unloaded) points 4 and 5 be X_{41} and X_{51}, respectively. Conversely, in Fig. 8.7b loads are applied at points 4 and 5 and cause displacements X_{42} and X_{52} at the load points. Let X_{12}, X_{22}, and X_{32} be the displacements caused at points 1, 2, and 3, respectively. The loads in Fig. 8.7a and b will be denoted as *loading 1* and *loading 2*, respectively.

If loading 1 is applied first and then followed by loading 2, the external work is given by

$$W_{12} = \tfrac{1}{2}\sum_{i=1}^{3} P_i X_{i1} + \sum_{i=1}^{3} P_i X_{i2} + \tfrac{1}{2}\sum_{j=4}^{5} P_j X_{j2}$$

$$(8.33)$$

Reversing the order of loading changes the work expression to

$$W_{21} = \tfrac{1}{2}\sum_{j=4}^{5} P_j X_{j2} + \sum_{j=4}^{5} P_j X_{j1} + \tfrac{1}{2}\sum_{i=1}^{3} P_i X_{i1}$$

$$(8.34)$$

Equations 8.33 and 8.34 can be equated because the total work should be the same in both cases. Doing this yields

$$\sum_{i=1}^{3} P_i X_{i2} = \sum_{j=4}^{5} P_j X_{j1} \qquad (8.35)$$

Equation 8.35 is generalized by expressing it in matrix form

$$\{P_1\}^T \{X_2\} = \{P_2\}^T \{X_1\} \qquad (8.36)$$

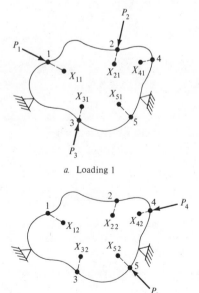

a. Loading 1

b. Loading 2

Fig. 8.7.   General structure subjected to separate loadings.

$\{P_1\}$ and $\{X_1\}$ are loads and displacements, respectively, in the first loading and $\{P_2\}$ and $\{X_2\}$ are loads and displacements, respectively, in the second loading.

Equation 8.36 represents Betti's Reciprocal Energy Theorem. Loosely stated, the theorem indicates that:

For any given structure behaving under linear, elastic conditions, the virtual work of a load system 1 moving through the displacements created by a load system 2 is equal to the virtual work of load system 2 moving through the displacements created by load system 1.

Later in this textbook, the relationship between the loads and displacements of a structure is defined in terms of a structure stiffness matrix $[K]$.

Betti's theorem provides a means to prove that the structure stiffness matrix is symmetric for all types of structures. Equation 8.36 can be rearranged to give

$$\{P_1\}^T \{X_2\} = \{X_1\}^T \{P_2\} \quad (8.37)$$

Let $\{P_1\}$ and $\{X_1\}$ correspond to the state

$$\{P_1\} = [K] \{X_1\} \quad (8.38)$$

and $\{P_2\}$ and $\{X_2\}$ correspond to the state

$$\{P_2\} = [K] \{X_2\} \quad (8.39)$$

Substitution into Eq. 8.37 yields

$$\{X_1\}^T [K]^T \{X_2\} = \{X_1\}^T [K] \{X_2\}$$
$$[K]^T = [K] \quad (8.40)$$

Betti's theorem is one of the most useful energy theorems with a wide range of applications in both elementary and advanced structural analysis. A simple example of its validity will be given.

Example 8.3. Figure 8.8 depicts two springs in series. The linear spring constants are k_1 and k_2 for the top and bottom springs, respectively. In Fig.

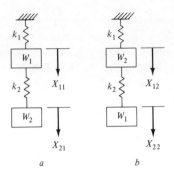

Fig. 8.8.   Linear springs in series.

8.8a, two weights, W_1 and W_2, are placed as shown. In Fig. 8.8b, these same weights have been applied but reversed in location. Demonstrate that Betti's reciprocal energy theorem is satisfied.

Solution: For the system in Fig. 8.8a, the indicated displacements are easily calculated as

$$X_{11} = \frac{W_1 + W_2}{k_1}$$

$$X_{21} = \frac{W_1 + W_2}{k_1} + \frac{W_2}{k_2}$$

For the system in Fig. 8.8b,

$$X_{12} = \frac{W_1 + W_2}{k_1}$$

$$X_{22} = \frac{W_1 + W_2}{k_1} + \frac{W_1}{k_2}$$

Betti's theorem (Eq. 8.36) requires that

$$W_1 X_{12} + W_2 X_{22} = W_2 X_{11} + W_1 X_{21}$$

or

$$W_1 \left(\frac{W_1}{k_1} + \frac{W_2}{k_1} \right) + W_2 \left(\frac{W_1}{k_1} + \frac{W_2}{k_1} + \frac{W_1}{k_2} \right)$$

$$= W_2 \left(\frac{W_1}{k_1} + \frac{W_2}{k_1} \right) + W_1 \left(\frac{W_1}{k_1} + \frac{W_2}{k_1} + \frac{W_2}{k_2} \right)$$

Multiplication and combining similar terms yield

$$\frac{W_1^2}{k_1} + 2\frac{W_1 W_2}{k_1} + \frac{W_2^2}{k_1} + \frac{W_1 W_2}{k_2}$$

$$= \frac{W_1^2}{k_1} + 2\frac{W_1 W_2}{k_1} + \frac{W_1 W_2}{k_2} + \frac{W_2^2}{k_1}$$

The equality is evident and demonstrates the validity of Betti's theorem for this system.

PROBLEMS

8-1. Determine the expression for the total strain energy due to a constant shear force, V, in a beam of span L.

8-2. Determine the displacement, Δ, in the direction of the applied load. Neglect axial and shear deformations. Use the Law of Conservation of Work and Energy.

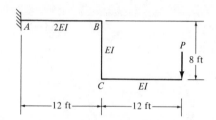

8-3. Repeat Problem 8-2 for the given planar truss. Assumed EA is the same for all members.

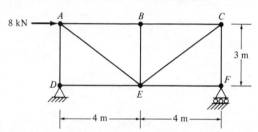

8-4. Repeat Problem 8-2 for the given space frame. Assume $EI_z = EI_y = 1.25GJ$ and all members are the same. Neglect axial and shear deformations.

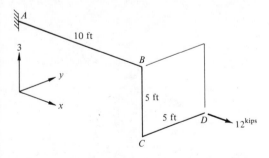

8-5. Derive a general expression for the vertical displacement at joint C of the given plane frame. Segment BC is three-quarters of a circle. Use the Law of Conservation of Work and Energy, and neglect axial and shear deformations.

8-6. Repeat Example 8.2 for the given grid structure. Use $EI = GJ$.

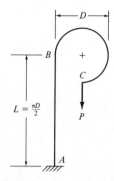

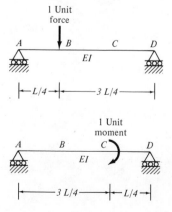

8-7. Prove that the external work done by the loads is the same for the following two conditions:

Condition 1: Both loads are applied simultaneously.

Condition 2: Load P_1 is applied first, and then load P_2 is added to the beam.

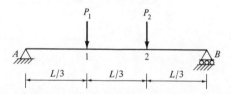

8-8. Demonstrate that Maxwell's reciprocal theorem is valid for the two given load systems.

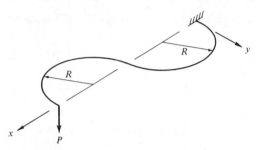

8-9. Repeat Problem 8-8 for the two given load systems.

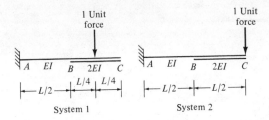

8-10. Demonstrate that Betti's theorem is satisfied by the two given load systems. Assume $k_1 > k_2$.

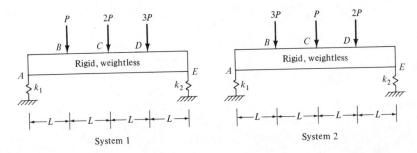

9
Unit-Load Method

9.1 INTRODUCTORY REMARKS

Structural analysis is commonly divided into two groups: classical methods and computer methods. As stated previously, this division is primarily chronological. Most classical analysis methods were developed prior to the availability of electronic computation and can be considered manual methods. Analytical methods intended for and formulated for solution by use of computer algorithms comprise the "computer methods."

So far, the purpose of this book has been to establish a fundamental background in both classical and computer methods of analysis. Treatment of classical methods has been purposely limited in scope. An understanding of the governing differential equations underlying first-order structural analysis of static systems and the physical behavior that is their counterpart was a major goal. Material on the moment-area method and the angle-change principles was presented in order to emphasize the physical behavior of structural members. In subsequent material, the conjugate-grid method was shown to replace direct use of differential equations and tedious geometry and considerations. The basic work and energy theorems in structural analysis were also introduced. Work and energy theorems serve two purposes. First, they directly provide several supplementary methods of classical structure analysis. Second, they

form a basis of derivation for the established computer methods of analysis, namely, matrix methods and the finite-element method.

The subject matter presented in the earlier chapters constitutes a fundamental base of knowledge for a contemporary beginning student in structural analysis. The reader who has attained a sound understanding of the material and the underlying concepts has the tools necessary to proceed a practical application and advanced studies. However, several well-established classical methods that were widely used before the current age of computer technology have not yet been presented in this textbook. Several of these methods will be presented in Chapters 9 through 12. All of the methods described in these chapters are intended for manual computations and are not conducive to structures that either have a large number of degrees of freedom or are highly statically indeterminate. Consequently, in recent years everyday usage of such classical methods has declined greatly. Nevertheless, classical methods continue to have an important role, both in the analysis of small structures and in the preliminary analysis of large structures. This chapter develops the "unit-load method" for calculating the displacement at any point in a framed structure. Generalized displacements in trusses can be determined by the matrix procedures described subsequently in Chapters 13 through 16. If the number of displacement values needed is limited to a few, manual computation is sometimes practical. The predominant classical approach employed in the past has been the unit-load method. A primary use of this method is to generate displacement values needed as part of the flexibility method presented in Chapter 11.

9.2   DEVELOPMENT FOR PLANAR TRUSSES

9.2.1   General Equations

Fundamentally, the unit-load method is based upon the concept of pseudo-virtual work discussed in Chapter 8. Consider the problem of computing any particular joint displacement, X_j, of the truss in Fig. 9.1a due to a general set of P-loads. In order to arrive at a formulation for determining this value, an auxiliary unit load will be employed in the manner shown in Fig. 9.1b.

The truss shown is identical with the structure shown in Fig. 9.1a except that the P-loads are removed and replaced by a single unit force applied in the particular direction X_j. The internal member forces caused by the auxiliary unit load can be determined by structural analysis. In particular, let the force that is created in the ith member be f_i and its resulting deformation (elongation or contraction) be δ_i. In order to accommodate the member deformation, the truss as a whole must undergo displacements (i.e., the joints and roller supports must move). In particular, joint 5 moves an amount d_j in the direction of the unit load. This joint displacement takes place as the applied unit load slowly increases from zero value to its full value, unity. Figure 9.2 illustrates this behavior for a linearly elastic structure. The area under the curves shown in Figs. 9.2a and 9.2b represent external work and internal strain energy, respectively. By the Law of Conservation of Work and Energy,

$$\tfrac{1}{2} \cdot 1 \cdot d_j = \tfrac{1}{2} \sum_{i=1}^{NM} f_i \cdot \delta_i \qquad (9.1)$$

where NM = the number of truss members.

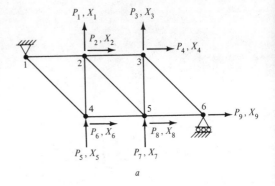

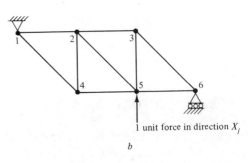

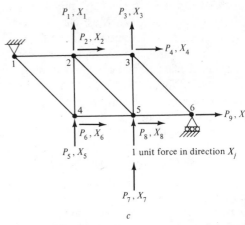

Fig. 9.1.   Truss analysis by unit-load method. a, truss subjected to a set of actual loads; b, truss subjected to the auxiliary unit load; c, truss with actual loads applied subsequent to auxiliary unit load.

After removing the unit load from the truss, apply the general set of P-loads shown in Fig. 9.1a to the truss. The P-loads create member forces, member deformations, and joint displacements. In particular, the ith member develops a force F_i and undergoes a deformation Δ_i. In the direction of the particular applied load P_k, the joint at which it is applied dis-

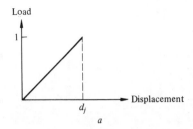

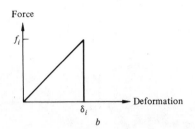

Fig. 9.2.   Effects of auxiliary unit load. *a*, behavior in direction X_j; *b*, behavior of member *i*.

places an amount X_k. Each joint displacement takes place as the applied actual forces slowly, and simultaneously, increase from zero value to their full intensity. Figures 9.3a and 9.3b illustrate the behavior for load P_k and internal force F_i, respectively. By equating total external work to total internal strain energy

$$\tfrac{1}{2} \sum_{k=1}^{NP} P_k \cdot X_k = \tfrac{1}{2} \sum_{i=1}^{NM} F_i \cdot \Delta_i \quad (9.2)$$

where *NP* is equal to the number of loads acting on the truss.

Finally, consider the unloaded truss subject to the following loading pattern:

1. Apply the auxiliary unit load.
2. Without removing the unit load, superimpose the actual loads (*P*-loads).

This load sequence is pictorially shown in Fig. 9.1c. The combined load-displacement behavior is plotted in Fig. 9.4. Figure 9.4a illustrates the behavior in the direction of the auxiliary unit load. Figure 9.4b shows the behavior in the direction of any other degree of freedom. Note that the effect of P_j is excluded from Fig. 9.4b. This is because its effect is shown in Fig. 9.4a. Similarly, the induced member force de-

formation behavior is shown in Fig. 9.5. By equating the external work to the internal strain energy,

$$\tfrac{1}{2} \cdot 1 \cdot d_j + 1 \cdot X_j + \tfrac{1}{2} \sum_{k=1}^{NP} P_k \cdot X_k$$

$$= \tfrac{1}{2} \sum_{i=1}^{NM} f_i \cdot \delta_i + \tfrac{1}{2} \sum_{i=1}^{NM} F_i \cdot \Delta_i$$

$$+ \sum_{i=1}^{NM} f_i \cdot \Delta_i \quad (9.3)$$

By virtue of Eqs. 9.1 and 9.2, Eq. 9.3 reduces to:

$$1 \cdot X_j = \sum_{i=1}^{NM} f_i \cdot \Delta_i \quad (9.4)$$

Equation 9.4 offers a procedure for computing joint displacements in a loaded truss. The procedure is termed the *unit-load method*. Equation 9.4 essentially states that the displacement of a joint in a particular direction, X_j, as caused by the actual loads, is numerically equal to the summation over all members, of the product $f_i \cdot \Delta_i$. The f_i values are the member forces created in the truss when it is

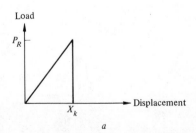

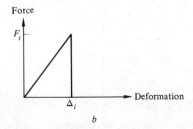

Fig. 9.3.   Effects of actual loads. *a*, behavior in direction X_k; *b*, behavior of member *i*.

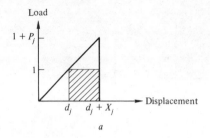

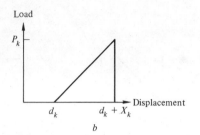

Fig. 9.4. External effects of combined loading pattern. *a*, behavior in direction X_j; *b*, behavior in direction X_k (direction X_j excluded).

subjected to a unit load *applied alone* and *in the direction of the desired displacement.* The Δ_i values are the member deformations created by the actual loads *acting alone*; that is,

$$\Delta_i = \frac{F_i L_i}{E_i A_i}$$

The most efficient procedure is to compute the Δ_i values separately and then a series of joint displacements can be computed by evaluating Eq. 9.4 for several different sets of f_i forces, i.e., several unit-load cases.

9.2.2 Example Problems

Example 9.1. A statically determinate truss is shown in Fig. 9.6. By employing the unit-load method, calculate the vertical and horizontal translations of joint U_0.

Solution: Figure 9.6b indicates the member forces caused by applying the actual loads to the truss. These were obtained by the visual method of joints and are needed in both of the required translation

computations. Prior to applying Eq. 9.4, it is necessary to compute the individual member deformations caused by the actual loads:

Member	F_i (kips)	L_i (in.)	A_i (in.^{2})	$\Delta_i = \dfrac{F_i L_i}{E_i A_i}$ (in.)
$U_0 U_1$	+16	96	0.5	3072/E
$U_1 U_2$	+8	96	0.5	1536/E
$U_0 L_0$	−20	120	1.0	−2400/E
$U_1 L_0$	−6	72	0.5	−864/E
$U_1 L_1$	+10	120	1.0	1200/E
$U_2 L_1$	0	—	—	—
$L_0 L_1$	−16	96	0.5	−3072/E

Auxiliary unit loadings are needed for each of the desired joint displacement computations. These loadings are shown in Fig. 9.6c and d. For determination of the vertical displacement at joint U_0, an auxiliary unit load is applied in the vertical direction at U_0, and the resulting set of f_i member forces are determined by the visual method of joints. The vertical unit load may be placed in the upward or downward position. A sign convention for interpreting the results is discussed subsequently. Results are recorded directly on the truss shown in Fig. 9.6c. Application of Eq. 9.4 permits the computation of the vertical deflection, X_1

$$1 \cdot X_1 = \sum_{i=1}^{NM=7} f_i \cdot \Delta_i$$

$$1 \cdot X_1 = \left(-\frac{4}{3}\right)\left(-\frac{3072}{E}\right) + (0)\left(\frac{1536}{E}\right)$$

$$+ \left(+\frac{5}{3}\right)\left(-\frac{2400}{E}\right) + (+1)\left(-\frac{864}{E}\right)$$

$$+ \left(-\frac{5}{3}\right)\left(\frac{1200}{E}\right) + (0)(0)$$

Force

$f_i + F_i$

f_i

Deformation

δ_i $\delta_i + \Delta_i$

Fig. 9.5. Internal effects of combined loading pattern.

$$+ \left(+\frac{4}{3}\right)\left(\frac{-3072}{E}\right)$$

$$= \frac{15,060}{E} = \frac{15,060}{30,000} = -0.502$$

$X_1 = 0.502$ in. downward

The direction of the movement is established by noting whether the virtual strain energy is positive or negative. A negative result means that the corresponding virtual work is negative and the unit load moves backwards. In the case at hand, a negative answer results, and therefore joint U_0 must move downward.

Horizontal movement at joint U_0 is computed by employing the auxiliary unit-load system depicted in Fig. 9.6d. Internal member forces are again computed by the visual method of joints. The horizontal displacement, X_2, is determined by evaluating Eq. 9.4 for the new set of f_i values. Elongations, Δ_i values, to be substituted in the equation are unchanged and

$$1 \cdot X_2 = \overset{NM=7}{\underset{i=1}{\Sigma}} f_i \cdot \Delta_i$$

$$X_2 = (-1)\left(\frac{3072}{E}\right) + (0)\left(\frac{1536}{E}\right)$$

$$+ (0)\left(\frac{2400}{E}\right) + \left(+\frac{3}{4}\right)\left(-\frac{864}{E}\right)$$

$$+ \left(-\frac{5}{4}\right)\left(\frac{1200}{E}\right) + (0)(0) + (0)(0)$$

$$= -\frac{5220}{E} = -\frac{5220}{30,000} = -0.174 \text{ in.}$$

$$= 0.174 \text{ in. to the left}$$

Negative virtual strain energy results indicating the unit load moves opposite to its direction of application.

Example 9.2. Calculate the relative motion D_1 between joints L_0 and U_2 in the truss of Example 9.1. Use the unit-load method.

Solution: The auxiliary unit loading is a pair of 1 kip loads applied as shown in Fig. 9.7. Member forces, determined by the visual method of joints, are also shown. Eq. 9.4 gives

$$1 \times D_1 = \overset{NM}{\underset{i=1}{\Sigma}} f_i \cdot \Delta_i$$

where the set of f_i member forces are those shown in Fig. 9.7 and the set of Δ_i displacements are unchanged from Example 9.1. Substitution of known quantities gives

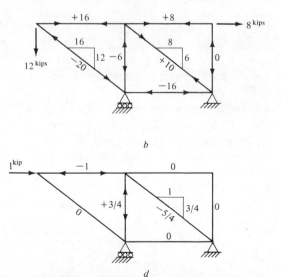

Fig. 9.6. Computation of truss displacements by the unit-load method. *a*, truss with actual loads; *b*, member forces due to actual loads; *c*, member forces due to first auxiliary unit load; *d*, member forces due to second auxiliary unit load.

$$1 \cdot D_1 = \sum_{i=1}^{7} f_i \cdot \Delta_i$$

$$D_1 = (-0.8)\left(\frac{1536}{E}\right) + (-0.6)\left(-\frac{864}{E}\right)$$

$$+ (-0.6)(0) + (-0.8)\left(-\frac{3072}{E}\right)$$

$$+ (+1.0)\left(\frac{1200}{E}\right)$$

$$= +\frac{2947}{E} = \frac{2947}{30000}$$

$$= 0.0982 \text{ in. together}$$

The positive value for the vertical strain energy indicates that in Fig. 9.7, the joints L_0 and U_2 move together.

9.3 DEVELOPMENT FOR BEAMS

Displacement due to flexure for framed structures can be computed by numerous methods of analysis. Equations 3.22 and 3.27 constitute two elementary differential equations that permit the determination of the elastic curve of a framed member. The double- and quadruple-integration procedures are the mathematical approaches employed to produce closed-form solutions from these fundamental equations. Geometric concepts embodied in the moment-area theorems provide a more versatile alternative to the direct-integration methods when specific displacement quantities are desired instead of the entire elastic curve. In effect, the double-integration method is replaced by its physical meaning. The conjugate-grid and -beam methods offers additional ease by removing the geometry considerations. Matrix

methods provide automated or semiautomated computational procedures that bypass the tedious computations of both of these classical methods. An extension of the subject matter of Section 9.2, the unit-load method will be described in the context of the analysis of planar frames.

9.3.1 General Equations

Consider the problem of determining the displacement Δ at any point, C, along the span of the beam given in Fig. 9.8a. The beam is statically loaded in some arbitrary manner, and the corresponding moment diagram is given in Fig. 9.8b. The unit-load method for computing Δ can be formulated in a manner similar to the derivation for trusses given in Section 9.2. *After removing the actual loading, $q(x)$, apply an auxiliary unit load as shown in Fig. 9.8d.* By choice, the unit load is applied at point C and in the direction of the desired displacement Δ. The consequence of applying the auxiliary unit load is the development of shear and moment (see Fig. 9.8e), throughout the span and a deflected shape $\delta(x)$. In particular, a differential element at location x develops a shear $v(x)$ and a moment $m(x)$, and at point C a displacement, δ, occurs. In a first-order analysis, the deformations caused by the shear force can be safely neglected. The moment $m(x)$ produces an angular rotation $d\phi$, as shown in Fig. 9.8f, and by the first moment-area theorem,

$$d\phi = \frac{m(x)\,dx}{EI}$$

The flexural strain energy, du_1, at this point in the member is given by

$$du_1 = \tfrac{1}{2}m(x)\,d\phi$$

and the total strain energy in the beam is

$$u_1 = \tfrac{1}{2}\int_0^L m(x)\,d\phi \qquad (9.5)$$

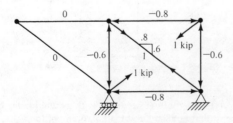

Fig. 9.7. Auxiliary unit load for Example 9.2.

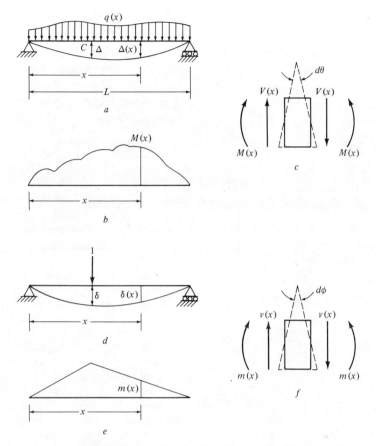

Fig. 9.8. Beam analysis by the unit-load method.

Because the displacement δ is *produced by the slow application of the auxiliary unit load*, the work, w_1, performed is

$$w_1 = \tfrac{1}{2}(1)(\delta) \qquad (9.6)$$

By the Law of Conservation of Work and Energy, Eqs. 9.5 and 9.6 can be equated giving

$$\frac{\delta}{2} = \frac{1}{2} \int_0^L m(x)\, d\phi \qquad (9.7)$$

After removing the auxiliary unit load from the beam, reapply the general loading $q(x)$. This loading produces the moment variation $M(x)$ depicted in Fig. 9.8b, causing a differential element to deform as shown in Fig. 9.8c. Again, shear deformations are considered neg-

ligible. Angular rotation, $d\theta$, at a point in the span is given by

$$d\theta = \frac{M(x)\, dx}{EI} \qquad (9.8)$$

and the corresponding flexural strain energy produced in the differential element is

$$du = \tfrac{1}{2} M(x)\, d\theta$$

When integrated over the entire span, the total strain energy becomes

$$u = \tfrac{1}{2} \int_0^L M(x)\, d\theta \qquad (9.9)$$

The static loading performs an equivalent amount of external work as given by

$$w = \frac{1}{2} \int_0^L q(x)\Delta(x)\, dx \qquad (9.10)$$

where $\Delta(x)$ = the deflected shape caused by $q(x)$. Consequently, due to the general loading applied by itself,

$$\frac{1}{2} \int_0^L q(x)\Delta(x)\, dx = \frac{1}{2} \int_0^L M(x)\, d\theta \qquad (9.11)$$

As a final loading sequence, consider the beam loaded according to the following pattern:

1. Apply the auxiliary unit load.
2. *Without removing the auxiliary unit load,* add the general loading $q(x)$.

This loading sequence produces a total strain energy:

$$U = \frac{1}{2} \int_0^L m(x)\, d\phi + \int_0^L m(x)\, d\theta$$

$$+ \frac{1}{2} \int_0^L M(x)\, d\theta \qquad (9.12)$$

as caused by the external work

$$W = \frac{1}{2}(1)(\delta) + 1(\Delta) + \frac{1}{2} \int_0^L q(x)\Delta(x)\, dx$$

$$(9.13)$$

U and W can be equated, and after observing Eqs. 9.7 and 9.11,

$$1 \cdot \Delta = \int_0^L m(x)\, d\theta \qquad (9.14)$$

The left-hand side is the virtual work created by the unit auxiliary load moving through the displacement Δ *caused by the actual load,* $q(x)$. Similarly, the right-hand side is the virtual strain energy created by the auxiliary unit-load moments being subjected to the internal deformations *caused by* $q(x)$. By substitution of Eq. 9.8 into Eq. 9.14,

$$1 \cdot \Delta = \int_0^L \frac{M(x)m(x)}{EI}\, dx \qquad (9.15)$$

which is the desired unit-load relationship. By similar derivation, the rotation, θ, at any point in the beam as caused by the actual loading can be determined as

$$1 \cdot \theta = \int_0^L \frac{M(x)m'(x)}{EI}\, dx \qquad (9.16)$$

where $m'(x)$ is the moment variation created by a unit couple applied, by itself, at the point where the desired rotation occurs.

Example 9.3.   Use the unit-load method to determine the vertical displacement and slope at midspan of the beam given in Fig. 9.9a.

Solution for midspan displacement: It is a straightforward matter to determine the moment variation, $M(x)$, caused by the actual loading. Because the moment diagram has a sudden change at the load point, each half must be treated separately, as shown in Fig. 9.9b. With a reference taken at the left support,

$$0 \le x \le L/2 \qquad M(x) = \frac{P}{2}x_1$$

$$L/2 \le x \le L \qquad M(x) = \frac{P}{2}(L - x_1)$$

Computation of the vertical displacement, $\Delta_{\textrm{\$}}$, requires the auxiliary unit loading and moment variation, $m_1(x)$, depicted in Fig. 9.9d and e:

$$0 \le x \le L/2 \qquad m(x) = \frac{1}{2}x_1$$

$$L/2 \le x \le L \qquad m(x) = \frac{1}{2}(L - x_1)$$

Consequently,

$$1 \cdot \Delta_{\textrm{\$}} = \int_0^L \frac{M(x)m(x)}{EI}\, dx$$

$$= \int_0^{L/2} \frac{\left(\dfrac{P}{2}x_1\right)\left(\dfrac{1}{2}x_1\right)}{EI}\, dx_1$$

$$+ \int_{L/2}^{L} \frac{\left(\frac{P}{2}\right)(L - x_1)\left(\frac{1}{2}\right)(L - x_1)}{EI} \, dx_1$$

$$= \frac{P}{4EI} \int_0^{L/2} x_1^2 \, dx$$

$$+ \frac{P}{4EI} \int_{L/2}^{L} (L^2 - 2x_1 L + x_1^2) \, dx_1$$

$$= \frac{PL^3}{48EI}$$

$$\therefore \Delta_{\mathfrak{c}} = \frac{PL^3}{48EI} \text{ down}$$

Because positive work results, the displacement is in the direction of the auxiliary unit load. Also, a simpler integration often results from a redefinition of the origin. In this case a redefinition of the origin, as shown in Figs. 9.9c and f, gives

$$1 \cdot \Delta_{\mathfrak{c}} = \int_0^{L/2} \frac{\left(\frac{P}{2} x_1\right)\left(\frac{1}{2} x_1\right)}{EI} \, dx_1$$

$$+ \int_0^{L/2} \frac{\left(\frac{P}{2} x_2\right)\left(\frac{1}{2} x_2\right)}{EI} \, dx_2$$

which integrates to the same result.

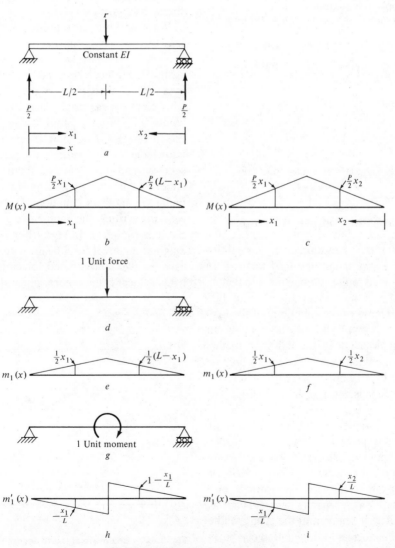

Fig. 9.9.   Beam for Example 9.4.

Solution for θ_{ξ}: The auxiliary unit load and moment variation, $m'(x)$, needed for computation of the midspan slope are given in Fig. 9.9g and h. Using a reference at the left support,

$$1 \cdot \theta_{\xi} = \int_0^{L/2} \frac{M(x)m'(x)}{EI} \, dx = 0$$

$$= \int_0^{L/2} \frac{\left(\dfrac{P}{2}x_1\right)\left(-\dfrac{x_1}{L}\right)}{EI} \, dx_1$$

$$+ \int_{L/2}^{L} \frac{\left(\dfrac{P}{2}\right)(L - x_1)\left(1 - \dfrac{x_1}{L}\right)}{EI} \, dx_1$$

$$= 0$$

Using the redefined references shown in Fig. 9.8i,

$$1 \cdot \theta_c = \int_0^{L/2} \frac{\left(\dfrac{P}{2}x_1\right)\left(-\dfrac{x_1}{L}\right)}{EI} \, dx_1$$

$$+ \int_0^{L/2} \frac{\left(\dfrac{P}{2}x_2\right)\left(\dfrac{x_2}{L}\right)}{EI} \, dx_2$$

which clearly integrates to a zero slope value.

9.3.2 Simplified Moment Equations

Example 9.3 illustrates a relatively straightforward application of the unit-load method. Integrals involved in the computations in that example were set up and evaluated with little difficulty because of the simplicity of the loading. As the complexity of the loading increases, employment of the unit-load method becomes a more extensive and tedious process. In the case of a beam subjected to a series of concentrated loads, such as shown in Fig. 9.10, the moment variation, $M(x)$, cannot be described by a single function. Each time the beam is interrupted by a reaction force or a load, the moment diagram changes shape. Consequently, the integration required in the unit-load method must be performed in a piecewise fashion. A separate integral is needed for each different segment of the span. Mathematically, this format for integration can be expressed by

$$\int_0^L \frac{M(x)m(x)}{EI} \, dx = \sum_{i=1}^{NS} \int_{x_L}^{x_R} \frac{M_i(x)m_i(x)}{EI} \, dx$$

$$(9.17)$$

where NS is the number of different segments. In Eq. 9.17, $M_i(x)$ and $m_i(x)$ are the moment variations in the ith segment of the beam as caused by actual loading and the auxiliary unit loading, respectively. If a rotation is being calculated, $m_i'(x)$ would be employed in place of $m(x)$. The coordinates of the left end, x_L, and right end, x_R, of the segment constitute the limits of integration and, of course, are different for each segment.

If a single reference point is used as the origin of the x-axis, the nature of the integrals gets more complex as one moves along the span. This difficulty can be partially alleviated by a periodic redefinition of the reference point. This concept was briefly demonstrated in Example 9.3. Further simplification of the analytical process if possible by recognizing that knowledge of the loading on a particular beam segment and the shear and moment values at either end of the segment is sufficient to formulate its moment equation. The general process for writing the moment equation is depicted in Fig. 9.11. In Fig. 9.11a, the free-body diagram is drawn for a beam segment that lies between two concentrated forces (reactions or loads). Shear and moment values at the left end

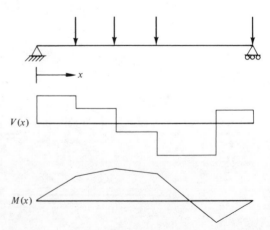

Fig. 9.10. Beam subjected to several concentrated loads.

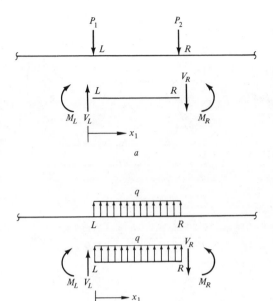

Fig. 9.11.   Loaded beam segments.

where v_L and m_L are the shear and moment at the left end of the segment caused by the unit load. Similarly, moment variation in a segment, as caused by a unit-couple, can be expressed as

$$m'(x_1) = m'_L + v'_L x_1 \qquad (9.21)$$

in which v'_L and m'_L are shear and moment at the left end of the segment.

Equations 9.18 through 9.21 permit a modification of the mathematical statement of the unit-load method. If moment equations are written in the manner just described, Eq. 9.17 becomes

$$\int_0^L \frac{M(x)m(x)}{EI}\,dx$$

$$= \sum_{i=1}^{NS} \int_0^{l_i} \frac{M_i(x_1)m_i(x_1)}{E_i I_i}\,dx_1 \qquad (9.22)$$

Again, if the quantity being calculated is a rotation, $m'_i(x_1)$ replaces $m_i(x_1)$. Thus

$$\int_0^L \frac{M(x)\,m'(x)}{EI}\,dx$$

$$= \sum_{i=1}^{NS} \int_0^{l_i} \frac{M_i(x_i)\,m'_i(x_i)}{E_i I_i}\,dx_i \qquad (9.23)$$

It is vital to note that the use of a local reference axis affects the limits of integration. Specifically, the integration for any particular segment is performed between the limits $x_1 = 0$ and $x_1 = l_i$ where l_i is the length of the segment. This general procedure will be demonstrated by an example.

(subscript L) and at the right end (subscript R) are the only forces that appear in the free body. Consequently, using a local reference, x_1, at the left end of the beam, the moment variation in the segment is expressed as

$$M(x_1) = M_L + V_L x_1 \qquad (9.18)$$

A beam segment subject to a uniform loading is shown in Fig. 9.11b. For this situation, the moment variation is given by

$$M(x_1) = M_L + V_L x_1 + qx_1^2/2 \qquad (9.19)$$

if q is taken as positive when acting upward. Because the values, V_L and M_L, can be extracted from the shear and moment diagrams, respectively, formation of integrals for the unit-load method is expedited by employment of Eqs. 9.18 and 9.19. Similar expressions can be written for all necessary auxiliary unit-loadings. For the case of a unit load, no transverse load, $q(x)$, exists other than the unit-load or couple itself. Thus, the moment variation for a beam segment is

$$m(x_1) = m_L + v_L x_1 \qquad (9.20)$$

Example 9.4.   Develop the integrals necessary to determine the rotation at point A in the beam of Fig. 9.12a by the unit-load method. Assume EI is constant over the entire span.

Solution: Shear and moment diagrams corresponding to the loading of Fig. 9.12a are drawn in Fig. 9.12b. Partial free-body diagrams of the different segments of the beam are drawn in Fig. 9.12c. These diagrams are incomplete because shear and

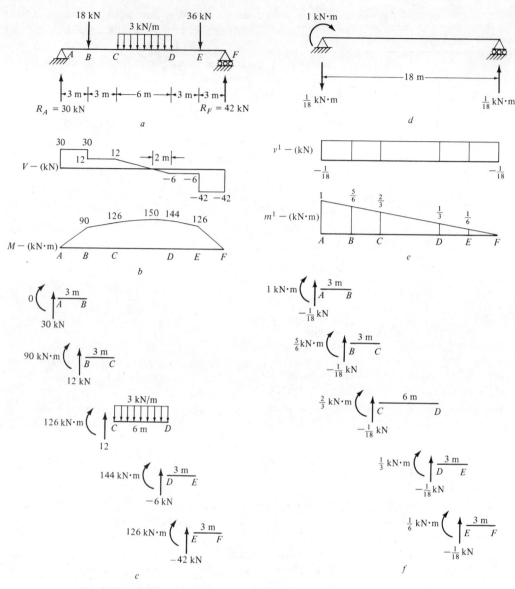

Fig. 9.12.   Beam for Example 9.4. *a*, actual loading; *b*, auxiliary unit load.

moment values at the right end of each segment have been omitted from the drawings. In practice, these diagrams can be visualized and need not actually be drawn. They are presented in this context for ease of comprehension by the reader. Moment equations for the five segments are obtained by substitution into Eqs. 9.18 or 9.19, as dictated by the loadings. For the problem at hand, $M(x)$ is replaced by

$$0 \le x \le 3 \quad M_1(x_1) = 30x_1$$

$$3 \le x \le 6 \quad M_2(x_1) = 90 + 12x_1$$

$$6 \le x \le 12 \quad M_3(x_1) = 126 + 12x_1 - 3x_1^2/2$$

$$12 \le x \le 15 \quad M_4(x_1) = 144 - 6x_1$$

$$15 \le x \le 18 \quad M_5(x_1) = 126 - 42x_1$$

where x_1 is a local axis with an origin at the left end of the beam segment.

Determination of the rotation at point A requires the auxiliary unit loading shown in Fig. 9.12d. Shear and moment diagrams corresponding to this unit couple are drawn in Fig. 9.12e. Moment values at the ends of each segment defined earlier have been labeled on the moment diagram. By drawing (or visualizing) the free-body diagrams shown in Fig. 9.12f, moment equations associated with the auxiliary unit load can be written in the form given in Eq. 9.21:

$$0 \leq x \leq 3 \qquad m_1'(x_1) = 1 - \tfrac{1}{18} x_1$$

$$3 \leq x \leq 6 \qquad m_2'(x_1) = \tfrac{5}{6} - \tfrac{1}{18} x_1$$

$$6 \leq x \leq 12 \qquad m_3'(x_1) = \tfrac{2}{3} - \tfrac{1}{18} x_1$$

$$12 \leq x \leq 15 \qquad m_4'(x_1) = \tfrac{1}{3} - \tfrac{1}{18} x_1$$

$$15 \leq x \leq 18 \qquad m_5'(x_1) = \tfrac{1}{6} - \tfrac{1}{18} x_1$$

Rotation, θ_A, at point A is computed by evaluating

ergy that must exist in the structural system. To extend the relationships to the analysis of planar frames it is only necessary to sum the total virtual work done by the loads acting on the framework and equate it to the sum of the virtual strain energy stored in all its members. If desired, the effect of axial deformation can be either neglected or included in the total virtual strain energy.

9.4.1. Planar Frames—Axial Deformation Neglected

It has already been shown that a beam can be treated as a series of *segments* that are inter-

$$1 \times \theta_A = \int_0^L \frac{M(x)m'(x)}{EI} \, dx$$

$$= \sum_{i=1}^{5} \int_0^{l_i} \frac{M_i(x_1)m_i'(x_1)}{E_i I_i} \, dx_1$$

$$1 \times \theta_A = \int_0^3 \frac{(30x_1)\,(1 - x_1/18)}{EI} \, dx_1$$

$$+ \int_0^3 \frac{(90 + 12x_1)\,(5/6 - x_1/18)}{EI} \, dx_1$$

$$+ \int_0^6 \frac{(126 + 12x_1 - 3x_1^2/2)\,(2/3 - x_1/18)}{EI} \, dx_1$$

$$+ \int_0^3 \frac{(144 - 6x_1)\,(1/3 - x_1/18)}{EI} \, dx_1$$

$$+ \int_0^3 (126 - 42x_1)\,(1/6 - x_1/18) \, dx_1$$

As noted earlier, the limits used in the integration correspond to the local reference point for each segment. Because the lower limit is *always zero* when the local reference is used, the integration is greatly speeded. Evaluation of the integrals is left as an exercise for the (energetic!) reader.

9.4 EXTENSION TO PLANAR FRAMES

Extension of the unit-load method to the treatment of planar frames is a direct development. Theoretical concepts presented during the derivation of Eqs. 9.15 and 9.16 are unchanged by a change in the nature of the framed structure. These relationships simply account for a balance of virtual work and virtual strain en-

connected to form the continuous beam. The powerful implication of this observation is that the virtual strain energy integrals can be evaluated segmentally. Physically, a planar frame is an assemblage of interconnected beam elements. Mathematically, application of the unit-load method to planar frames requires summing the virtual strain energy for all beam elements that comprise the framework. If virtual strain energy due to flexure only is considered, a modification to Eqs. 9.15 and 9.16 is sufficient to apply them to planar frames. In essence, the basic integration limits, 0 to L, in Eqs. 9.15 and 9.16, imply "over the length of the frame." In actual computation, this means that each beam element of the framework can be visualized as a separate piece of the overall system. Complexity of loading or variation in EI normally necessitate a further segmentation of each member. Either equation 9.22 or 9.23 can be readily employed to sum the contributions that the individual segments make to the total virtual strain energy.

Example 9.5. Using the unit-load method, compute the rotation, θ_D, at point D of the frame shown in Fig. 9.13a. Assume EI is constant and the same all members.

Solution: Shear and moment diagrams for the given loading are shown in Fig. 9.13b and c, respectively. Moment equations for the individual members are formulated using Eq. 9.18.

Member AB:

$$0 \le x \le 18 \qquad M_1(x_1) = -6x_1$$

Member BC:

$$0 \le x \le 6 \qquad M_2(x_1) = -108 + 8x_1$$

$$6 \le x \le 12 \qquad M_3(x_1) = -60 - 2x_1$$

Member CD:

$$0 \le x \le 12 \qquad M_4(x_1) = -72 + 6x_1$$

$$12 \le x \le 24 \qquad M_5(x_1) = 0$$

For determination of the slope, θ_D, moment equations are needed for the auxiliary unit-loading shown in Fig. 9.13d. This loading consists of a counter-

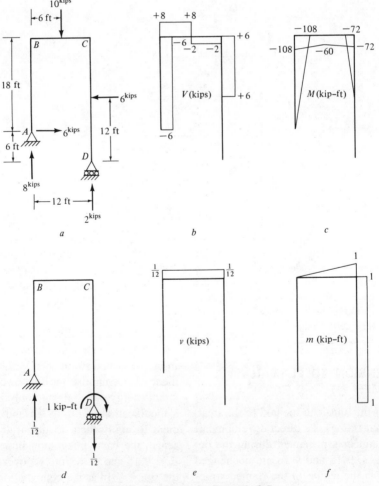

Fig. 9.13. Frame for Example 9.5.

clockwise unit couple applied at point D. Corresponding shear and moment diagrams are presented in Figs. 9.13e and f, respectively. By application of Eq. 9.18,

Member AB:

$$0 \le x \le 18 \qquad m_1'(x_1) = 0$$

Member BC:

$$0 \le x \le 6 \qquad m_2'(x_1) = +\tfrac{1}{12} x_1$$

$$6 \le x \le 12 \qquad m_3'(x_1) = \tfrac{1}{2} + \tfrac{1}{12} x_1$$

Member CD:

$$0 \le x \le 12 \qquad m_4'(x_1) = 1$$

$$12 \le x \le 24 \qquad m_5'(x_1) = 1$$

The desired slope value is obtained by substitution into Eq. 9.22, except $m_i'(x_1)$ replaces $m_i(x_1)$:

$$1(\theta_D) = \int_0^6 \frac{(-108 + 8x_1)\left(\tfrac{1}{12} x_1\right)}{EI} \, dx_1$$

$$+ \int_0^{12} \frac{(-60 - 2x_1)\left(\tfrac{1}{2} + 12x_1\right)}{EI} \, dx_1$$

$$+ \int_0^{12} \frac{(-72 + 6x_1)(1)}{EI} \, dx_1$$

$$= -\frac{1242}{EI}$$

$$\therefore \theta_D = \frac{1242}{EI} \text{ cw}$$

9.4.2   Planar Frames—Axial Deformation Included

In first-order structural analysis, axial deformation in a planar frame is easily accounted for in the unit-load method. It is simply necessary to include the virtual strain energy due to axial forces when establishing the virtual work/virtual strain energy balance. This amounts to combining Eqs. 9.4 with either Eq. 9.15 or Eq. 9.16, as necessary. Thus, if axial deformation is included

$$1 \cdot \Delta = \int_0^L \frac{M(x)m(x)}{EI} \, dx + \sum_{i=1}^{NF} f_i \, \Delta_i$$

$$(9.24)$$

$$1 \cdot \theta = \int_0^L \frac{M(x)m'(x)}{EI} \, dx + \sum_{i=1}^{NF} f_i \, \Delta_i$$

$$(9.25)$$

in which Δ and θ are the displacement and rotation, respectively, at some point in the frame.

If the frame is divided into segments, Eqs. 9.22 and 9.23 can be used. Thus

$$1 \cdot \Delta = \sum_{i=1}^{NS} \int_0^{l_i} \frac{M_i(x_i)m_i(x_i)}{E_i I_i} \, dx_i + \sum_{i=1}^{NS} f_i \Delta_i$$

$$(9.26)$$

$$1 \cdot \theta = \sum_{i=1}^{NS} \int_0^{l_i} \frac{M_i(x_i)m_i'(x_i)}{E_i I_i} \, dx_i + \sum_{i=1}^{NS} f_i' \, \Delta_i$$

$$(9.27)$$

Example 9.6. Determine the vertical displacement at point C of the frame shown in Fig. 9.14a.

Solution: The internal distribution of moment due to the actual loading is established in Fig. 9.14b. Thus,

Member AB:

$$M_i(x) = M_1(x) = \frac{wL^2}{2}$$

$$\Delta_i = \Delta_1 = \frac{-wL(L)}{EA} = \frac{-wL^2}{EA}$$

Member BC:

$$M_i(x) = M_2(x) = \frac{-wx^2}{2}$$

$$\Delta_i = \Delta_2 = 0$$

The unit auxiliary loading and the corresponding internal moment variation are shown in Figs. 9.14c and d, respectively. Thus

Member AB:

$$m_i(x) = m_1(x) = -L$$

$$f_i = f_1 = -1$$

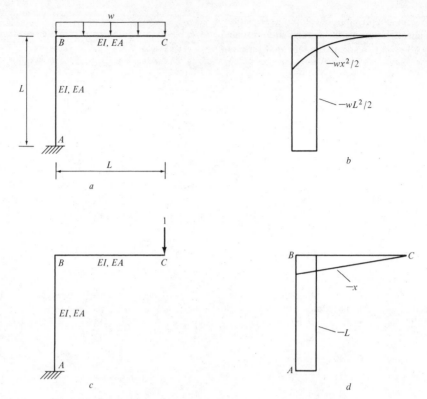

Fig. 9.14. Frame for Example 9.6. *a*, actual loading; *b*, $M(x)$ variation; *c*, auxiliary unit load; *d*, $m(x)$ variation.

Member BC:

$$m_i(x) = m_2(x) = -x$$

$$f_i = f_2 = 0$$

From Eq. 9.26 (recognizing that for member BC the reference for x is at the right end of the member):

$$1 \cdot \Delta = \int_0^L \frac{\left(-\dfrac{wL^2}{2}\right)(-L)}{EI} \, dx$$

$$+ \int_0^L \frac{\left(-\dfrac{wx^2}{2}\right)(-x)}{EI} \, dx$$

$$+ (-1)\left(-\frac{wL^2}{EA}\right)$$

$$= \frac{5wL^4}{8EI} + \frac{wL^2}{EA}$$

It is evident from Example 9.6 that the determination of displacements in a planar frame by the unit-load method can be tedious. The number of separate integrals required in a solution increases rapidly as the loading pattern becomes more complicated. Therefore, this analytical method is an impractical approach for most frame problems. However, in the flexibility method to be described in Chapter 11, many simple loadings are encountered for which the unit-load method is well suited and its usage has merit in that method.

9.5 GENERAL STRUCTURES

The unit-load method has been described within the context of planar trusses (Section 9.2), beams (Section 9.3) and planar frames (Section 9.4). Fundamentally, the method recognizes the existence of a balance between (pseudo) virtual work and (pseudo) virtual strain energy.

In Eq. 9.4, which applies to trusses and planar frames, the virtual strain energy is due to the axial deformation of the members. In Eqs. 9.15 and 9.16, which apply to a member of a beam or planar frame, the virtual strain energy is due to the flexural deformation of the member. Conceptually, extension of the applicability of the unit-load method to general structures is a simple matter. Basically, one must account for all the particular forms of virtual strain energy, which exist in the given structure under the action of the sequential application of a unit load and the actual loads.

Regardless of the nature of a given structure, the virtual work term in the unit-load method is always the product of an auxiliary unit load and a displacement (caused by the actual loads) in its direction. Conversely, the strain energy depends upon the structural type and can exist in several forms in the same structure. In the most general type of structural framework, the space frame, virtual strain energy is the consequence of four effects. These effects are:

1. Flexural deformation
2. Shear deformation
3. Torsional deformation
4. Axial deformation

Treatment of flexural deformation and axial deformation as sources of virtual strain energy has been described already. The latter was discussed in the context of a planar truss. Axial deformation in general structures and the remaining sources of virtual strain energy are discussed in the following paragraphs.

9.5.1   Shear Deformation

Deformation due to internal shear strains was discussed in Section 3.5. The consequence of shear strain in a differential element (see Fig. 3.27) is a relative displacement of the end faces of the element. The absolute magnitude of the displacement dy is given by

$$dy = \frac{n}{GA} V(x)dx \qquad (9.28)$$

in which $V(x)$ is the shear force in the differential element, n is a shape factor, G is the

shearing modulus of elasticity, and A is the cross-sectional area.

Let $V(x)$ be the shear force due to the actual loading and $v(x)$ be the shear force due to the preapplied unit load. Under these circumstances, the shear force $v(x)$, when subjected to the deformation dy caused by $V(x)$, would perform an increment of virtual strain energy dVU given by

$$dVU = v(x)dy = \frac{nV(x)v(x)}{GA}dx \qquad (9.29)$$

The total virtual strain energy in the member is obtained by integrating Eq. 9.29 over the length of the member to give

$$VU = \int_0^L \frac{nV(x)v(x)}{GA} dx \qquad (9.30)$$

Normally, $V(x)$ and/or $v(x)$ cannot be described by functions applicable to the entire member length. In this circumstance, a separate integral is needed for each different segment of the span. When this occurs, the right side of Eq. 9.30 becomes

$$\int_0^L \frac{nV(x)v(x)}{GA} dx = \sum_{i=1}^{NS} \int_{xL}^{xR} \frac{nV_i(x)v_i(x)}{GA} dx$$

$$(9.31)$$

which is similar to Eq. 9.17. The subscript i infers the shear functions are valid within the ith segment.

9.5.2   Torsional Deformation

Deformation due to internal torsional shear strains was described in Section 2.11.4. Torsional shear strain in a differential element (see Fig. 2.21) produces a relative rotation of the end faces of the element. The absolute magnitude of the relative rotation $d\phi$ is given by

$$d\phi = \frac{T(x)}{GK} dx \qquad (9.32)$$

in which $T(x)$ is the torsional moment in the differential element and K is the torsional constant for the given cross section.

If the torsional moment $t(x)$ due to the unit load is subjected to the deformation caused by $T(x)$, the torsional moment induced by the actual loads virtual strain energy occurs. The increment of strain energy dVU is given by

$$dVU = t(x)\,d\phi = \frac{T(x)t(x)}{GK}\,dx \quad (9.33)$$

Integration over the member length produces the total virtual strain energy

$$VU = \int_0^L \frac{T(x)t(x)}{GK}\,dx \quad (9.34)$$

When $T(x)$ and/or $t(x)$ vary segmentally,

$$\int_0^L \frac{T(x)t(x)}{GK}\,dx = \sum_{i=1}^{NS}\int_{x_L}^{x_R} \frac{T_i(x)t_i(x)}{GK}\,dx \quad (9.35)$$

9.5.3 Axial Deformation

In a truss member, the axial force is constant over the length of the member, and Eq. 9.4 is applicable. General structures may have members in which the axial force is a function $P(x)$ of the location along the member length. In this circumstance, virtual strain energy in the unit load method is given by

$$VU = \int_0^L \frac{P(x)p(x)}{EA}\,dx \quad (9.36)$$

in which $p(x)$ is the variation in axial force due to the unit load. If $P(x)$ and/or $p(x)$ vary segmentally

$$\int_0^L \frac{P(x)p(x)}{EA}\,dx = \sum_{i=1}^{NS}\int_{x_L}^{x_R} \frac{P_i(x)p(x)}{EA}\,dx \quad (9.37)$$

9.5.4 General Equation

Equations 9.15, 9.30, 9.34, and 9.36 represent the four sources of virtual strain energy in the unit-load method. Combining these effects and equating the result to the virtual work yield the general equation

$$1\cdot\Delta = \int_0^L \frac{M(x)m(x)}{EI}\,dx$$
$$+ \int_0^L \frac{nV(x)v(x)}{GA}\,dx$$
$$+ \int_0^L \frac{T(x)t(x)}{GJ}\,dx$$
$$+ \int_0^L \frac{P(x)p(x)}{EA}\,dx \quad (9.38)$$

If necessary, each of the terms on the right side of Eq. 9.38 can be replaced by their counterparts, Eqs. 9.17, 9.31, 9.35, and 9.37, respectively.

In classical structural analysis, axial deformation is neglected (except in trusses, of course), and the last term in Eq. 9.38 is omitted. However, by including this term, exact comparisons can be made with the matrix stiffness method (Chapters 15 and 16). Also, if the member has a variable cross section, I, A, and J can be replaced by the functions $I(x)$, $A(x)$, and $J(x)$, respectively. Typically, this leads to complex integrals and is an undesirable approach to the analysis of such members.

Finally, if the desired displacement quantity is a rotation θ, Eq. 9.38 becomes

$$1\cdot\theta = \int_0^L \frac{M(x)m'(x)}{EI}\,dx$$
$$+ \int_0^L \frac{nV(x)v'(x)}{GA}\,dx$$
$$+ \int_0^L \frac{T(x)t'(x)}{GJ}\,dx$$
$$+ \int_0^L \frac{P(x)p'(x)}{EA}\,dx \quad (9.39)$$

in which "primed" quantities are due to the appropriate auxiliary-unit moment.

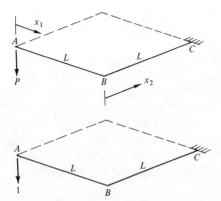

Fig. 9.15.   Structure for Example 9.7.

9.5.5   Example Problems

Example 9.7. Determine the displacement in the direction of the load P for the structure illustrated in Fig. 9.15a. Use the unit-load method. Both members are identical. EI is the same about each axis of the cross section.

Solution:

Step 1: Using the independent reference points indicated in Fig. 9.15a:

Member AB:

$$M(x_1) = Px_1$$
$$T(x_1) = 0$$

Member BC:

$$M(x_2) = Px_2$$
$$T(x_2) = PL$$

Step 2: Using the auxiliary unit load shown in Fig. 9.15b:

Member AB:

$$m(x_1) = x_1$$
$$t(x_1) = 0$$

Member BC:

$$m(x_2) = x_2$$
$$t(x_2) = L$$

Step 3: Substitution into Eq. 9.38 yields

$$1 \times \Delta = \int_0^L \frac{(Px_1)\,(x_1)}{EI}\,dx_1$$

$$+ \int_0^L \frac{(Px_2)\,(x_2)}{EI}\,dx_2 + \int_0^L \frac{PL(L)}{GJ}\,dx_2$$

$$\Delta = \frac{2PL^3}{3EI} + \frac{PL^3}{GJ}$$

Example 9.8. Determine the displacement in the direction of the load for the space frame in Fig. 9.16. Use the unit-load method. EI, GJ, GA, and EA are the same for all members. EI is the same about each cross-sectional principal axis and $GJ = EI$, $EA = .1EI$ and $GA = .05EI$. Assume $n = 1$.

Solution: For the actual loading,
   Member AB:

$$M_y(x_1) = -60$$

$$M_x(x_1) = -60$$

$$V_y(x_1) = V_z(x_1) = T_x(x_1) = 0$$

$$P_x(x_1) = 12$$

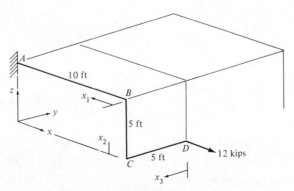

Fig. 9.16.   Frame for Example 9.8.

Member BC:

$$M_x(x_2) = 0$$

$$M_y(x_2) = -12x_2$$

$$V_x(x_2) = 12$$

$$V_y(x_2) = 0$$

$$T_z(x_2) = -60$$

$$P_z(x_2) = 0$$

Member CD:

$$M_x(x_3) = 0$$

$$M_z(x_3) = -12x_3$$

$$V_x(x_3) = T_y(x_3) = P_y(x_3) = 0$$

Since the auxiliary unit load is 1 kip at joint D and coincides with the direction of the actual load, the response to the unit load is one-twelfth of the response of the actual load. Thus, from Eq. 9.38,

$$1 \cdot \Delta = \int_0^{10} \frac{(-60)(-5)}{EI} dx_3$$

$$+ \int_0^{10} \frac{(-60)(-5)}{EI} dx_3$$

$$+ \int_0^{10} \frac{(12)(1)}{EA} dx_3$$

$$+ \int_0^5 \frac{(-12x)(-x)}{EI} dx_2$$

$$+ \int_0^5 \frac{(1)(12)(1)}{EA} dx_2$$

$$+ \int_0^5 \frac{(-12x)(-x)}{EI} dx_1$$

$$+ \int_0^5 \frac{(1)(12)(1)}{GA} dx_1$$

$$= \frac{7000}{EI} + \frac{120}{EI} + \frac{120}{EI}$$

$$\Delta = \frac{10600}{EI}$$

The positive result indicates the calculated displacement is in the direction of the 12 kip load.

9.6   UNIT-LOAD METHOD FOR CONTINUA

When a general continuum is subjected to a loading applied, both normal and shear stresses (and corresponding strains) develop throughout the body. The displacement Δ at any point on the surface of the continuum can be obtained by application of the unit-load method. The equivalence between virtual work and corresponding virtual strain energy requires

$$1 \cdot \Delta = \int_{vol} VU_0 \, d \text{ vol}$$

where VU_0 is the virtual strain energy density (defined in Chapter 8). VU_0 is given by

$$VU_0 = \sigma_x' \epsilon_x + \sigma_y' \epsilon_y + \sigma_z' \epsilon_z$$

$$+ \tau_{xy}' \gamma_{xy} + \tau_{yz}' \gamma_{yz} + \tau_{zx}' \gamma_{zx} \quad (9.40)$$

where the primed quantities are stresses created by an auxiliary unit load acting (in the direction of the desired Δ) alone. Unprimed quantities are strains created by the actual loads acting alone. For an isotropic material, these strains are related to the stresses according to Eqs. 2.11. Computation of stresses for general continua is accomplished by methods developed in the theory of elasticity and its special forms (e.g., the theory of plate bending, the theory of shells, etc.)

For isotropic members of framed structures (Fig. 9.17), Eq. 9.40 can be used to verify Eq. 9.38. For example, for the flexural term, assuming bending about the z-axis and $\nu = 0$, Eqs. 2.11 and 9.40 yield

$$VU_0 = \sigma_x' \epsilon_x = \left(\frac{m_z(x) \cdot y}{I_z} \right) \left(\frac{M_z(x) \cdot y}{EI_z} \right)$$

Thus

$$1 \cdot \Delta = \int_{vol} \frac{M_z(x)}{I_z} \frac{m_z(x)}{EI_3} y^2 \, d \text{ vol}$$

$$= \int_0^L \int_0^A \frac{M_z(x) m_z(x)}{EI_z^2} y^2 \, dx \, dA$$

$$1 \cdot \Delta = \int_0^L \frac{M_z(x) m_z(x)}{EI_z} dx \quad (9.41)$$

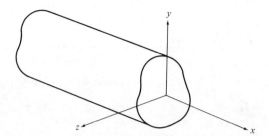

Fig. 9.17.   Axes of a framed member.

For shear V_y the same conditions produce

$$VU_0 = \tau'_{yz}(x)\, \gamma_{yz}(x)$$

$$= \frac{nv_y(x)\, nV_y(x)}{GA^2}$$

and

$$1 \cdot \Delta = \int_{vol} \frac{nV_y(x)\, v_y(x)}{GA^2}\, d\,vol$$

$$= \int_0^L \int_0^A \frac{nV_y(x)\, v_y(x)}{GA}\, dx\, dA$$

$$1 \cdot \Delta = \int_0^L \frac{nV_y(x)\, v_y(x)}{GA}\, dx \qquad (9.42)$$

Similarly, for torsion,

$$VU_0 = \tau'_{yz}(x) \cdot \gamma_{yz}(x) = \frac{t_x(x) \cdot r}{J}\, \frac{T_x(x) \cdot r}{GJ}$$

and

$$1 \cdot \Delta = \int_{vol} \frac{T_x(x)\, t_x(x)}{GJ^2}\, r^2\, d\,vol$$

$$= \int_0^L \int_0^A \frac{T_x(x)\, t_x(x)}{GJ^2}\, r^2\, dx\, dA$$

$$1 \cdot \Delta = \int_0^L \frac{T_x(x)\, t_x(x)}{GJ}\, dx \qquad (9.43)$$

Finally, for axial force,

$$VU_0 = \sigma'_x \cdot \epsilon_x = \frac{p(x)\, P(x)}{EA^2}$$

and

$$1 \cdot \Delta = \int_{vol} \frac{P_x(x) p_x(x)}{EA^2}\, d\,vol$$

$$= \int_0^L \int_0^A \frac{P_x(x) p_x(x)}{EA^2}\, dx\, dA$$

$$1 \cdot \Delta = \int_0^L \frac{P_x(x) p_x(x)}{EA}\, dx \qquad (9.44)$$

Removal of the subscripts in Eqs. 9.41 through 9.44 confirms the generic unit-load (virtual work) expression given as Eq. 9.38. Also, Eqs. 9.41 and 9.42 assume simple bending. For more complex situations, e.g., biaxial bending, unsymmetrical bending, etc. (see Chapter 2), Eq. 9.40 still applies but the stress and strains are given by more complex expressions.

PROBLEMS

Note: Unless otherwise stated, all members have the same EI value.

9-1.   Calculate the displacement of the roller support by the unit-load method. Bar areas (in.^{2}) are given in parentheses.

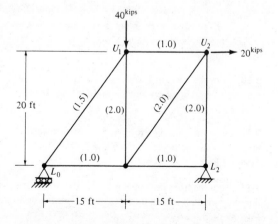

9-2.   Determine the horizontal and vertical displacements at joint C by the unit-load method. Bar areas (cm^2) are given in parentheses.

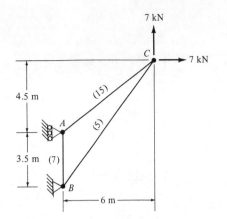

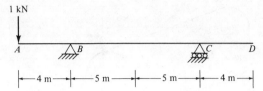

9-3. For the truss in Problem 5-1a, determine the horizontal and vertical displacements at joint C. Use the unit-load method.

9-4. For the truss in Problem 5-1d, determine the horizontal and vertical displacements at joint D. Use the unit-load method.

9-5. For the truss in Problem 5-1e, determine the horizontal and vertical displacements at joint C. Use the unit-load method.

9-6. For the truss in Problem 5-1g, determine the horizontal and vertical displacements at joint C. Use the unit-load method.

9-7. For the truss in Problem 5-11a, determine the horizontal and vertical displacements at joint C. Use the unit-load method.

9-8. For the truss in Problem 5-1a, determine the relative motion between joints B and D. Use the unit-load method.

9-9. For the truss in Problem 5-1e, determine the relative motion between joints C and F. Use the unit-load method.

9-10. For the truss in Problem 5-1g, determine the relative motion between joints A and D. Use the unit-load method.

9-11. For the truss in Problem 5-4a, determine the relative motion between joints B and D. Use the unit-load method.

9-12. For the truss in Problem 5-11c, determine the relative motion between joints C and F. Use the unit-load method.

9-13. Determine the rotation and displacement at point D. Use the unit-load method and assume EI is constant over the entire member length.

9-14. For each cantilever beam in Problem 4-1, calculate the slope and displacement at the free end. Use the unit-load method.

9-15. Calculate the slope and displacement at point B of the beam given in Problem 4-2. Use the unit-load method.

9-16. Calculate the midspan displacement for each beam in Problem 4-6 by the unit-load method.

9-17. Calculate the displacement at the free ends for each beam in Problem 4-7 by the unit-load method.

9-18. Calculate the midspan displacement for each beam in Problem 4-9. Use the unit-load method.

9-19. For each beam in Problem 7-20, calculate the displacement at the free ends. Use the unit-load method.

9-20. Analyze the given tapered beam (of constant width) by the unit-load method. Express the answer in terms of I_0, the moment of inertia at the small end.

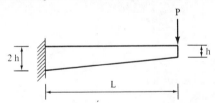

9-21. Calculate the displacement of the internal hinge in the beams of Problem 3-3 by the unit-load method. In cases where more than one hinge exists, use the leftmost hinge.

9-22. Calculate the slope, horizontal displacement, and vertical displacement at the free end of each planar frame in Problem 7-1. Use the unit-load method.

9-23. Calculate the slope, horizontal displacement and vertical displacement at the free end of each planar frame in Problem 7-2. Use the unit-load method.

9-24. Calculate the displacement of the roller support for each planar frame in Problem 7-3. Use the unit-load method.

9-25. Calculate the displacement of the roller support for each planar frame in Problem 7-4. Use the unit-load method.

9-26. Do Problem 7-5 by the unit-load method.

9-27. Compare the vertical displacement of the columns in the given frame using the unit load method:

(a) $w = 4$ kips/ft, $L_1 = 20$ ft, $L_2 = 20$ ft, neglect axial effects

(b) $w = 4$ kips/ft, $L_1 = 20$ ft, $L_2 = 20$ ft, include axial effects

(c) $w = 4$ kips/ft, $L_1 = 20$ ft, $L_2 = 40$ ft, neglect axial effects

(d) $w = 4$ kips/ft, $L_1 = 20$ ft, $L_2 = 40$ ft, include axial effects

Use $E = 30,000$ kips/in.^{2}, $I = 1000$ in.^{4}, and $A = 20$ in.^{2}.

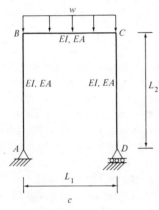

9-28. Do Problem 7-16 by the unit-load method.

9-29. Do Problem 8-5 by the unit-load method.

9-30. Do Problem 8-6 by the unit-load method.

9-31. Repeat Example 9.8 for the given space frame. Use the properties specified in the example.

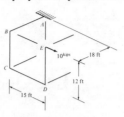

9-32. Repeat Problem 9-31, but calculate the vertical displacement of joint E.

9-33. Determine the horizontal and vertical displacements of joint B by the unit-load method. Use $EA = 200,000$ kips and $EI = 15 \times 10^6$ kip-in.^{2} for the beam. Use $EA = 20,000$ kips for the cable.

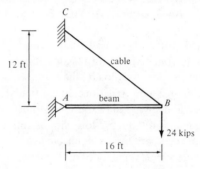

10

Degrees of Freedom

10.1 GENERALIZED DISPLACEMENTS AND DEFORMATIONS

Framed structures consist of individual members interconnected to form a structural configuration. Under static loads, the framework will move into a new position. In particular, the joints will displace and experience translational motion and rotation. More importantly, the ends of members rigidly connected to the joint will undergo identical motions. A member connected to a joint at both of its ends will experience the effect of both joint motions. If, by pure chance, both joints of an unloaded member translated in the same direction by an equal amount and experienced no rotation, the attached member would experience no deformation. Such movement is termed *rigid body motion* and usually occurs only if a structure is unrestrained against motion in a particular direction. Normally, this is not the case, and all the joints have different motions. It is these differential joint movements that cause strain in the members. An example of rigid body motion of a planar frame is depicted in Fig. 10.1.

As an illustration of the effect of differential joint motion on the deformation of a member, consider the planar frame in Fig. 10.2. Under the action of the loading shown, the entire frame assumes a new position defined by points A', B', C, and D. The exact position taken by the frame is dependent upon the dimensions and

material properties of the members and the nature and intensity of the loading. The approximate displaced shape in Fig. 10.2, shown to an exaggerated scale, is sufficient for discussion. Because of the presence of applied loads and the boundary conditions imposed by the supports at C and D, rigid body motion cannot occur in the frame of Fig. 10.2. In this instance, joints A and B translate and rotate by differential amounts, as indicated in Fig. 10.3. Member AB is attached to these joints and, as a result, moves to position $A'B'$. The new position of the joints is completely established if the rotations, θ_A and θ_B, and displacements, x_A, y_A, x_B, and y_B, are known. It is common practice to refer to these joint motions as *generalized displacements*. Generalized displacements are the independent translations and rotations that define the external change in position of any differential element in the structural framework. For ease of expression, a differential element is usually referred to as a *point in the structure*.

All structural types are characterized by the number of independent generalized displacements that occur at any given point in the structure. In the case of a planar frame, there are three independent generalized displacements at each point in the frame—the two orthogonal translations and a single rotation, all of which occur in the plane of the frame itself. A point in a space frame undergoes six independent

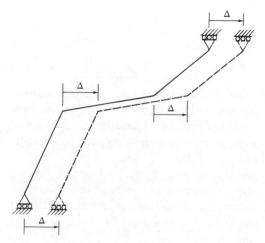

Fig. 10.1.   Rigid-body motion of a planar frame.

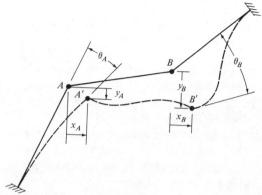

Fig. 10.3.   Displacement of joints *A* and *B*.

generalized displacements—translation in the direction of, and rotation about, each of the three orthogonal coordinate axes.

Member *AB* in the frame of Fig. 10.2 is shown in more detail in Fig. 10.4. Movement of the member can be separated into three distinct effects:

1. orthogonal translations, x'_A and y'_B, which are parallel to, and y'_A and y'_B, which are perpendicular to, the original member axis *AB*
2. rotations of the member ends, ϕ_A and ϕ_B, relative to the displaced member chord
3. flexing action caused by the applied loads

The translations, by themselves, simply move the member from position *AB* to *A'B'*, causing

a change of member length (*axial deformation*) without inducing any curvature. Although the member remains straight, its axis rotates relative to its original position as indicated by the angle *R* in Fig. 10.4. Angular rotations, ϕ_A and ϕ_B, and the transverse load cause the member to curve away from the straight line position *A'B'* into its final curvilinear position. For the planar frame, there are three independent deformations at each end of a member—the orthogonal translations, *x'* and *y'*, and the rotation, ϕ, relative to the displaced chord. In a space frame, six independent deformations occur at each member end—namely, three orthogonal translations and three orthogonal rotations.

The independent internal deformations at a point in the member are commonly termed *generalized deformations*. A clear distinction should be made between generalized displacements and generalized deformations. The generalized displacements occur at each point in

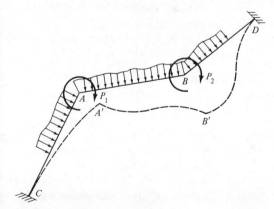

Fig. 10.2.   Planar frame subjected to an arbitrary loading.

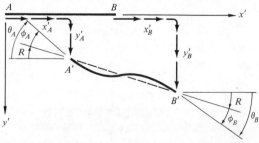

Fig. 10.4.   Deformation of member *AB*.

the structure when the structure is viewed in its entirety. Their magnitude and direction are given in terms of a coordinate system established for the structure as a whole. Generalized deformations, on the other hand, occur at each point in a member and are generally specified in terms of a local coordinate system established for the isolated member.

For admissible displaced shapes of a structure, a relationship must exist between the generalized displacements at two adjacent joints of the structure and the generalized deformations of the member ends that frame into the joints. For the planar frame at hand, examination of Figs. 10.3 and 10.4 indicates

$$x_A^2 + y_A^2 = x_A'^2 + y_A'^2 \quad (10.1)$$

$$x_B^2 + y_B^2 = x_B'^2 + y_B'^2 \quad (10.2)$$

$$\theta_A = \phi_A + R \quad (10.3)$$

$$\theta_B = \phi_B + R \quad (10.4)$$

10.2 DEGREES OF FREEDOM

10.2.1 Definition

The importance of determining the movements of the joints in a structure has been described in Section 6.3.3. Determination of the joint displacements has the inherent feature of producing the information needed to compute deformations at the ends of the members that frame into them. Although analytical treatment of the subject matter is presented later in the text, the calculation of deformations at the ends of any given member is of central importance to structural analysis. Indeed, specification of their numerical values and the loading on and material properties of the member establishes a unique structural response for the entire member. No additional information is needed to determine the complete response of each member and, consequently, the response of the structure as a whole. Therefore, it is vital that the structural analyst be capable of establishing the number of independent generalized joint displacements that occur in a given structure. This paramount before any discussion of the computation of their values is undertaken.

A more common term for independent generalized joint displacements is *degrees of freedom*. This phrase is, perhaps, preferable because it more clearly connotes the physical implication of these displacements. In real-world structures, each joint of a structure has the potential to translate in each of the three Cartesian coordinate directions and to rotate about each of the three orthogonal axes that define this coordinate system. Unless restraint is provided against some of the motions, all six displacements occur freely. Each of the six movements is independent in the sense that no single one affects the others. For example, if support conditions completely restrain five of the movements, the sixth is still free to occur, and, within the limits of material strength, its magnitude is also unrestricted. When completely unrestricted by supports, a joint in a three-dimensional structure is termed a *free joint* and is said to have 6 degrees of freedom. The presence of supports at any particular joint would reduce that joint's degrees of freedom to a number less than 6 (the exact number depends on the nature of the support). Consequently, it is a simple matter to count the total number of degrees of freedom in any real structure.

As part of the analytical process, real framed structures are replaced by idealized models. In the schematic diagram that represents a structure, six degrees of freedom do not always exist at each free joint. Their number is normally reduced for one of two reasons:

1. The idealized model is a structural type other than a space frame.
2. Some member deformations are considered negligible.

An understanding of these two effects is necessary before and analyst can create an adequate structural model.

10.2.2 Determination of Degrees of Freedom

Each of the types of framed structures was depicted earlier in Fig. 1.4. After reviewing this figure, the number and nature of the degrees of freedom that exist at a free joint can be discussed for each type. The six types are:

1. *Space frames* (see Fig. 1.4f). Each free joint of a space frame has 6 degrees of freedom. As stated earlier these consist of a translation in each coordinate direction and a rotation about each coordinate axis. At restrained joints, the number is less than six, and four conditions are common: (a) completely fixed—no degrees of freedom, (b) completely pinned—three rotational degrees of freedom, (c) pinned in two planes and a roller in one plane—one translational and three rotational degrees of freedom, and (d) pinned in one plane and a roller in two planes—three rotational and two translational degrees of freedom. Conditions (c) and (d) will be referred to as a "type 1 roller" (R_1) and a "type 2 roller" (R_2), respectively. Schematic representations for each of the four common support conditions are given in Fig. 10.5. A general formula is easily written for calculating the number of degrees of freedom in a space frame. Letting

NX = degrees of freedom
NJ = number of joints of all types
NFJ = number of fixed joints
NPJ = number of pinned joints
NR_1 = number of type 1 rollers
NR_2 = number of type 2 rollers

the applicable equation is

$$NX = 6NJ - 6NFJ - 3NPJ$$
$$- 2NR_1 - NR_2 \qquad (10.5)$$

However, it is normally more expedient to simply count the degrees of freedom directly from inspection of the schematic diagram.

2. *Space trusses* (see Fig. 1.4b). As explained in Chapter 1, a space truss is a degenerate form of a space frame. All members are assumed to be pin-connected at the joints. With the absence of rigid connections, the rotations of the member ends are all independent quantities. Theoretically, three rotational degrees of freedom exist at any free joint *for each member framing into the joint*. However, in the analytical model these rotations have no effect on the internal forces and need not be computed. The only degrees of freedom of interest are the joint translations. Unlike the rotations, all member ends that frame into the same joint must undergo the same translational motion as the joint itself. In a space truss, three orthogonal translational degrees of freedom exist at a joint unless it is restrained in some way. Except for the absence of fixed supports, the common support conditions are the same as for a space frame. The general expression for the number of degrees of freedom in a space truss is easily deduced as

$$NX = 3NJ - 3NPJ - 2NR_1 - NR_2 \qquad (10.6)$$

Alternatively, the degrees of freedom can be counted by inspection of schematic diagram of any given space truss.

3. *Planar frames* (see Fig. 1.4e). In a planar frame, all movements are assumed to occur within the plane that contains the frame. Fundamentally, this assumption implies that all out-of-plane motion is negligible. A brief discussion of the reduction of space frames to planar frames is presented in Section 2.12, and comments are made regarding the validity of this assumption. Nonetheless, given that the assumption is valid, only two orthogonal translations and a single rotation can occur at a free joint. If supports of the type listed in Table 1.1 exist, some of the three degrees of freedom at the joint are removed. Fixed supports have no degrees of freedom, pinned supports can only have a single rotational motion, and a roller support has both a rotational and a translational degree of freedom. Mathematically, the total number of degrees of freedom for a planar frame that has common support conditions can be determined from

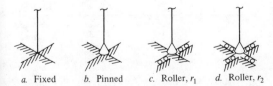

a. Fixed   *b.* Pinned   *c.* Roller, r_1 *d.* Roller, r_2

Fig. 10.5.   Support conditions for space structures.

$$NX = 3NJ - 3NFJ - 2NPJ - NR \qquad (10.7)$$

which can also be determined by visual count. No subscript is shown on NR because only one type of roller support is possible.

4. *Planar trusses* (see Fig. 1.4a). A planar truss is a degenerate form of a planar frame. Similar to space trusses, the members are assumed to be pinned to the joints, but the rotations of their ends have no effect on the internal forces. Translational motion of the joints in the two orthogonal coordinate directions constitute the only significant degrees of freedom. Accounting for pinned and roller supports, the total number of degrees of freedom of a planar truss are given by

$$NX = 2NJ - 2NPJ - NR \quad (10.8)$$

or can be simply counted by inspecting each joint condition.

5. *Grids* (see Fig. 1.4d). Three displacements are considered significant in a grid structure. In contrast to a planar frame, all translations that take place in the plane containing a grid are treated as negligible. Consequently, at a free joint, the translation transverse to the plane of the grid, and rotations about the two axes that define this plane are the only degrees of freedom. Except for their orientation, the same support conditions that are common to planar frames are usually encountered in grid structures. Joints may be fixed or pinned. Recognizing these possibilities, the degrees of freedom of a grid are computed using

$$NX = 3NJ - 3NFJ - NPJ \quad (10.9)$$

in lieu of a visual counting.

6. *Beams* (see Fig. 1.4c). Continuous beams are the simplest structural form, but care must be taken in determining the degrees of freedom for such systems. The beam structure is best viewed as a degenerate form of a grid structure. In essence, a beam is a grid whose members all lie along a common axis, i.e., the "plane of the grid" degenerates to a longitudinal axis. As with any grid, in-plane motion (axial deformation) is normally neglected, and the presence of roller supports does not introduce any translational (i.e., longitudinal) degrees of

freedom. Normally, torsional rotation of the member is also neglected and need not be considered when counting degrees of freedom. A free joint is limited to just two degrees of freedom: a single rotation and a single transverse displacement. The usual form of a free joint is a free end, such as depicted at joint D in Fig. 10.6. A fixed joint has no degrees of freedom, and a pin support (or roller) is considered to have a single rotational degree of freedom. An alternative type of free joint in a beam is the fictitious joint indicated at point E in Fig. 10.6b. This structure is identical with that of Fig. 10.6a, except point E has been "declared a joint." The intent is to consider the single member AB as two separate members that meet at point E. Viewed in this way, two additional degrees of freedom must be included to account for the "declared" free joint. A physical situation that corresponds to this condition is when two members are actually connected in this manner. Such a connection is termed a *splice*. A splice can be designed to produce complete continuity or result in an internal hinge. Internal hinges require special treatment and are discussed in the next section. Equation 10.10 permits computation of the degrees of freedom of a beam *that has no internal hinges*

$$NX = 2NJ - 2NFJ - NPJ - NR \quad (10.10)$$

Equations 10.5 through 10.10 are tabulated in Table 10.1. The equations for beams, grids, and planar and space frames neglect axial deformation.

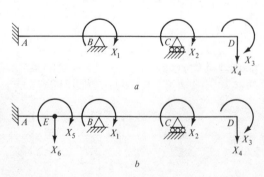

Fig. 10.6. Degrees of freedom for continuous beams.

Table 10.1   Equations for Determining Degrees of Freedom.*

Space Frame	$NX = 6NJ - 6NFJ - 3NPJ - 2NR_1 - NR_2$
Space Truss	$NX = 3NJ - 3NPJ - 2NR_1 - NR_2$
Planar Frames	$NX = 3NJ - 3NFJ - 2NPJ - NR$
Planar Truss	$NX = 2NJ - 2NPJ - NR$
Grid	$NX = 3NJ - 3NFJ - NPJ$
Beams	$NX = 2NJ - 2NFJ - NPJ - NR$

*Axial deformation included, no internal hinges

Example 10.1.   Determine the number of degrees of freedom for each structure shown in Fig. 10.7 and the beam shown in Fig. 10.6a. In each case, indicate them in a figure of the structure.

Solution:

Space frame (Fig. 10.7a)
$$NX = 6NJ - 6NFJ - 3NPJ - 2NR_1 - NR_2 \qquad (10.5)$$
$$= 6(7) - 6(1) - 3(1) - 2(0) - (1)$$
= 32 degrees of freedom (as shown in Fig. 10.7a). Note that the letter X will be used throughout the text to label the degrees of freedom.

Space truss (Fig. 10.7b)
$$NX = 3NJ - 3NPJ - 2NR_1 - NR_2 \qquad (10.6)$$
$$= 3(8) - 3(2) - 2(1) - (1)$$
= 15 degrees of freedom (as shown in Fig. 10.7b)

Planar frame (Fig. 10.7c)
$$NX = 3NJ - 3NFJ - 2NPJ - NR \qquad (10.7)$$
$$= 3(7) - 3(1) - 2(1) - (1)$$
= 15 degrees of freedom (as shown in Fig. 10.7b)

Planar truss (Fig. 10.7d)
$$NX = 2NJ - 2NPJ - NR \qquad (10.8)$$
$$= 2(6) - 2(1) - (1)$$
= 9 degrees of freedom (as shown in Fig. 10.7d)

Grid (Fig. 10.7e)
$$NX = 3NJ - 3NFJ - NPJ \qquad (10.9)$$
$$= 3(5) - 3(0) - (4)$$
= 11 degrees of freedom (as shown in Fig. 10.7c)

Beam (Fig. 10.6a)
$$NX = 2NJ - 2NFJ - NPJ - NR \qquad (10.10)$$
$$= 2(4) - 2(1) - (1) - (1)$$
= 4 degrees of freedom (as shown in Fig. 10.6a)

10.3   EFFECT OF INTERNAL HINGES

In a pin-connected joint, no resistance is offered to the rotation of member ends that frame into the joint. In all types of framed structures, such connections often occur at the support locations. In trusses, the presence of a pin connection at every joint is an implicit feature. For structures composed of flexural members (planar frames, space frames, beams, and grids), a pin connection sometimes exists between adjacent members. In Chapter 1, this particular form of pinned connection was termed an internal hinge (see Table 1.1).

In determining the degrees of freedom, it is necessary to take into account joints where two or more members are interconnected by a pin (including internal hinges). All pinned connections introduce a common feature to structures; namely, the continuity of rotation between adjacent members is interrupted (released) by their presence. Members that frame into a pinned joint are free to undergo relative rotations at that point. Rotation at the end of any particular member is unaffected by the presence of other members that frame into the pin connection. In this sense, the end rotations of the individual members are "independent" and are not related to each other. Each is a separate quantity and must be included in counting the structure's degrees of freedom. This aspect of

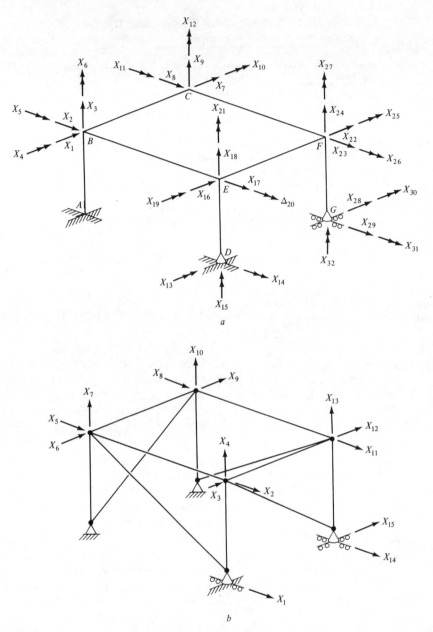

Fig. 10.7. Degrees of freedom in various structures. *a*, space frame; *b*, space truss; *c*, planar frame; *d*, planar truss; *e*, grid.

joint behavior for planar structures can be stated in general terms: *One rotational degree of freedom exists for each member end that independently frames into a pinned connection (including internal hinges).* In space structures, it is necessary to consider whether each pinned connection permits rotation of each attached mem-

ber in either one plane or in two orthogonal planes.

In planar trusses, where all members are pinned at both ends, the number of rotational degrees of freedom is twice the number of members. However, as noted earlier, it is unnecessary to include the rotation of the member

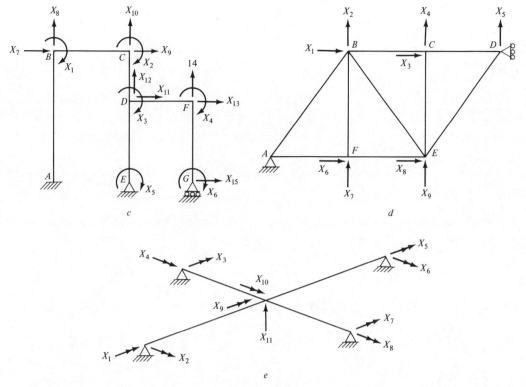

Fig. 10.7.   (*Continued*)

ends in performing a structural analysis on this type of structure. This is a basic assumption in the analysis of trusses.

Members in a truss are considered to remain straight and only undergo rigid body rotation. Knowledge of the joint translations at each end of a member is sufficient to establish its new position in space. This points out the significant difference between external displacements and internal deformations. In a truss, each member as a whole does undergo a rotation in space, but no rotational *deformations* occur at its ends. The inference is that it is unnecessary to compute the rigid body rotation of the member because it induces no strain (or stress) in the member. All strain is due purely to the change in the length of the member. A more subtle observation is inferred from this discussion: *In counting a structure's degrees of freedom, it is only necessary to include those independent external displacements that are considered to induce internal deformations in the members.* One implication of this statement

is that at pinned joints, it is the rotational *deformation* of the member ends that constitute the rotational degrees of freedom at that location. Indeed, at a pinned support, the support does not rotate. Figure 10.8 demonstrates important concepts regarding pinned joints.

By definition, when a rigid joint undergoes a rotation, the ends of the attached members must experience no relative angular deformation. As an example, when the joint in Fig. 10.8a rotates an amount ϕ, the three members that meet at the joint must rotate by the same amount. In Fig. 10.8b, this means the angular rotation θ of each member must equal the joint rotation ϕ. Knowledge of the magnitude of ϕ is sufficient to establish the rotation of the three member ends. Consequently, the joint has a single rotational degree of freedom.

In Fig. 10.8c, the joint is changed to a pinned connection. Because rotational resistance is releasd at each of the three member ends, a joint rotation causes the deformations illustrated in Fig. 10.8d. Unlike a rigid connection, the

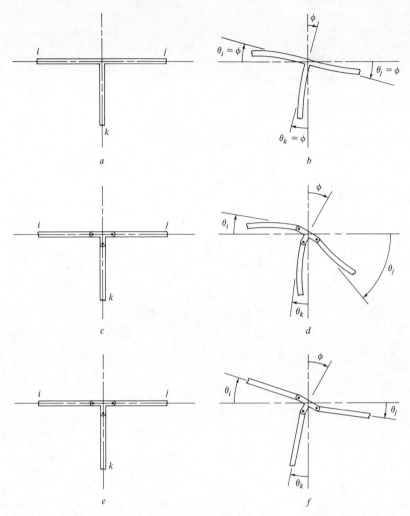

Fig. 10.8. Effects of internal hinges. *a*, undeformed rigid joint; *b*, deformed rigid joint; *c*, undeformed pinned joint; *d*, deformed pinned joint; *e*, undeformed truss joint; *f*, deformed truss joint.

member deformations θ_i, θ_j, and θ_k bear no relationship to one another or to the joint rotation ϕ. The curvilinear shape of any particular member is related, in part, to its end rotations.

In the case shown in Fig. 10.8d, it would be necessary to determine θ_i, θ_j, and θ_k as part of the structural analysis process. Knowledge of the magnitude of the joint rotation ϕ would be no aid. Each member rotation must be established separately. As a result, the joint has three rotational degrees of freedom, which consist of the three member end rotations.

In the case of a truss, it was stated earlier that the number of rotational degrees of free-

dom at each pinned joint would degenerate to zero. A physical explanation is seen by inspecting Figs. 10.8e and f. If the joint is rotated by an amount ϕ, each member would experience a separate and independent rotation. By condition of the assumptions in truss analysis, these rotations induce no member deformations, and their magnitudes are of no importance. The underlying concept is that the joints in a truss must translate in such a way that each member undergoes rigid-body translation and remains straight. This set of physical constraints (*compatibility conditions*) permits the use of a mathematical model that does not require computation of the member rotations. While such motions actually occur, their magnitude is of no importance to the analysis. Consequently, the rotations shown in Fig. 10.8f need not be calculated and *do not represent degrees of freedom*. This is unlike the situation depicted in Fig. 10.8d, which is representative of an internal hinge in a planar frame.

Example 10.2. Determine the number of degrees in each of the following structures:

  Case a. Beam (Fig. 10.9a)
  Case b. Planar frame (Fig. 10.9c)

Solution:
Case a. The beam has internal hinges at points B and C. Declaring these as joints permits designating the 6 degrees of freedom shown in Fig. 10.9b. At each internal hinge, each member end connected to the hinge has an independent rotational degree of freedom. In addition, each hinge can undergo a vertical translation, and all member ends attached to it must experience this same movement.

Case b. The internal hinge shown at joint D is intended to indicate that member DF is pin-connected to the rigid joint defined by the continuity of members BD and DC shown in the blowup of the joint. Connected in this manner, end D of member DF undergoes rotation independently of the rotation of the joint itself. Consequently, two rotational degrees of freedom must be declared at point D. (The inter-

nal hinge is considered to be very close to joint D, and its vertical translation *relative to the joint* is considered negligible. No additional translational degree of freedom exists at the internal hinge.) The complete set of degrees of freedom is indicated in Fig. 10.9d.

10.4 NEGLECT OF AXIAL DEFORMATION

10.4.1 Effect on Degrees of Freedom

In Section 1.9.3, a distinction was made between classical and matrix computer methods of analysis. The difference was stated to be one of chronology, with the beginning of the age of computer methods marked by the advent of the use of computer technology in the field of structural analysis. In addition, computer technology has removed certain technical barriers in structural analysis. Prior to this time, analytical methods were constrained to procedures that were suited to manual computation, with the aid of mechanical calculators at best. Classical methods were by nature restricted to relatively small frameworks with very few degrees of freedom. Large structural systems had to be reduced in size by physical approximations if expedient solutions were to be possible by such methods. The nature of these approx-

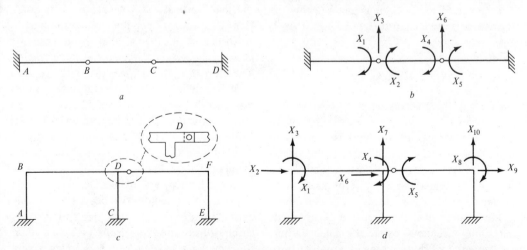

Fig. 10.9. Structures for Example 10.2. *a*, beam; *b*, degrees of freedom; *c*, planar frame; *d*, degrees of freedom.

imations was discussed briefly in Section 2.12. In addition to the limitation created by the lack of digital computation, classical methods are marked by a fundamental theoretical simplification that is unnecessary in modern computer methods. When classical methods are applied to planar frames, it is common practice to *neglect the axial deformation* of the members.

Consider the member shown earlier in Fig. 10.4. In small-deflection theory, transverse displacements do not contribute to the axial deformation of the member (recall Section 3.6.3). Thus, the axial deformation Δ_H is given by

$$\Delta_H = x_B' - x_A' \qquad (10.11)$$

The chord rotation R is given by

$$R = \frac{y_B' - y_A'}{L} \qquad (10.12)$$

where L is the length of the member. Knowing ϕ_A and ϕ_B would establish θ_A and θ_B from Eqs. 10.3 and 10.4. In this circumstance, the translations x_A', y_A', x_B', and y_A' and the rotations ϕ_A and ϕ_B are six independent degrees of freedom. Knowing their magnitudes establishes both the axial deformation and the boundary conditions needed for the flexural analysis of the member (recall Section 6.3.3).

Under normal loadings, the axial deformation is much smaller than the transverse displacements. The effect of neglecting this deformation is evident from Eq. 10.11. If Δ_H is zero,

$$x_B' = x_A' \qquad (10.13)$$

As a result, the independence of these two displacements is removed, and the number of independent degrees of freedom is reduced to five. There is one apparent value of this observation in the analysis of planar and space frames. By neglecting axial deformation, one degree of freedom is removed for each member in the frame. Often this is a true observation, but some exceptions exist. As stated earlier, neglecting axial deformation is a normal procedure in classical structral analysis because

of a need to minimize the size of the problem. In contemporary computer methods, there is no need to neglect axial deformations, regardless of their magnitude. In fact, neglecting axial deformations actually creates great difficulty in the development of so-called *automated computer methods*. However, unless a computer solution is based on the neglect of axial deformation, the calculated results will differ from classical solutions. Computer methods exist for either including or neglecting axial deformation. For the latter, it is necessary to determine degrees of freedom when axial deformation is neglected.

10.4.2   Identifying Degrees of Freedom

As a preliminary to a complete discussion of all types of structures, examine the planar frame depicted in Fig. 10.10. If axial deformation is included, 11 degrees of freedom must be prescribed as shown in Fig. 10.10a. Determination of these 11 quantities for any given loading is necessary and sufficient to establish the deformed position of the entire structure.

In any classical method of analysis, the axial deformation of the members would be ignored and the degrees of freedom reduced to the 7 shown in Fig. 10.10b. Physically, these can be explained as follows. Joint A is fixed and has no degrees of freedom. Member A does not change in length and therefore joint B cannot translate vertically. (Note again: This condition is dependent upon the basic assumption of small-deflection theory that *transverse displacement does not change the member length!*) Therefore only two degrees of freedom exist at joint B, rotation X_1, and the translation X_2. To establish the location of joint C, two orthogonal components of its motion must be known. If the position of joint B is known, the movement of joint C in the longitudinal direction of member BC is known because the member length is known, and axial deformation is neglected. Consequently, two degrees of freedom exist at joint C—the joint rotation and the translation—*in the direction transverse to member BC*, which are labeled X_3 and X_4, respectively.

Next, consider joint D. With joint C estab-

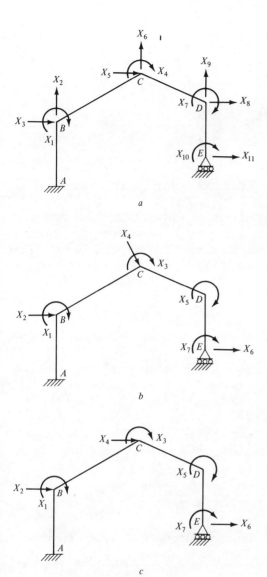

Fig. 10.10.   Degrees of freedom in a planar frame.
a, axial deformation included; *b*, axial deformation
neglected; *c*, axial deformation neglected.

degree of freedom in rotation, X_5, exists at joint
D. Finally, at joint E rotation of the joint, X_7,
and the motion of the roller (transverse to
member DE), X_6, are the only degrees of freedom.

This completes the physical determination of
the degrees of freedom for the given frame. As
a final comment, consider the single translational degree of freedom, X_4, shown at joint C
in Fig. 10.10b. In truth, knowledge of any
component of the motion represented by this
vector establishes the vector motion itself.
Consequently, the degree of freedom could be
depicted as the horizontal component shown in
Fig. 10.10c. For ease of future discussion, it is
preferable to replace any inclined degree of
freedom with either a horizontal or a vertical
component. Doing this permits a tabulation of
the degrees of freedom as shown in Table 10.2.

Table 10.2 is arranged to coincide, as best
as possible, with the pattern one would follow
in visually determining the degrees of freedom
for the frame of Fig. 10.10c. Each member is
inspected one end at a time, and a consecutive
number entered as each degree of freedom is
established. An asterisk indicates either a support condition or that a degree of freedom has
already been established. In the case of Table
10.2, member AB is fixed at end A so three asterisks are entered. At end B, two degrees of
freedom are entered to coincide with the rotation and horizontal motion that can occur at the
joint. For member BC, three asterisks are en-

**Table 10.2.   Degrees of Freedom for
Fig. 10.10c.**

Member	Member End	θ_z	Δ_x	Δ_y
AB	A	*	*	*
	B	1	2	*
BC	B	*	*	*
	C	3	4	*
CD	C	*	*	*
	D	5	*	*
DE	D	*	*	*
	E	6	7	*

lished in position, sufficient information would
exist to locate joint D. Joint E is a roller support and cannot move vertically. Consequently, joint D cannot move vertically because the axial deformation in member DE is
neglected. Knowing the position of joint C, it
is a simple matter to intersect the known length
of member CD with a horizontal line through
joint D. This would locate the horizontal position of joint D. In this light, only the single

tered for end B because joint B has already been inspected. Two degrees of freedom exist at end C and are entered in the next line of the table. Completion of the remainder of the table requires no further discussion. It is not the author's intent that the reader complete such a table as part of the solution process. This type of table is simply used herein as to convey the author's procedure for sequentially establishing the degrees of freedom. Normally, the lines of the table cannot be filled in consecutive order. In studying such tables, the reader should understand

1. Asterisks are first entered for each direction of restraint as dictated by the support conditions.
2. When a numerical entry or an asterisk is entered for a particular member end, it implies a condition at a particular joint. An asterisk is placed in the same column for all other member ends that frame into that joint.
3. Numbers are entered in the order that the degrees of freedom are discovered.

Discussion of the computation of degrees of freedom when axial deformation is neglected is necessary only for space frames and planar frames. In truss-type structures, axial deformation is the only deformation and cannot be ignored. In this text, beams are considered as a degenerate form of a grid. Axial deformation has no meaning in grids because no in-plane loads or in-plane motions are considered. Therefore, no further discussion of grids or beams is necessary, as Eqs. 10.9 and 10.10 already presume no axial deformation. Earlier, it was shown that neglect of axial deformation in frames reduces the number of degrees of freedom by one for each member in the structure. Based upon this, general formulas can be obtained by modification of Eqs. 10.5 and 10.7. In the following development, the term NM represents the number of members in the structures. Its presence reflects the reduction in the number of degrees of freedom by one for each member as just discussed.

For space frames:

$$NX = 6NJ - 6NFJ - 3NPJ$$
$$- 2NR_1 - NR_2 - NM$$
$$= 3NJ - 3NFJ + (3NJ - 3NFJ$$
$$- 3NPJ - 2NR_1 - NR_2 - NM)$$

$$(10.14)$$

$$NX = 3NJ - 3NFJ + NT \qquad (10.15)$$

Equations 10.14 and 10.15 are equivalent, but the latter has the advantage of separating out the number of translational degrees of freedom, NT, which are also called *sidesway motions*. These are given by

$$NT = 3NJ - (3NFJ + 3NPJ$$
$$+ 2NR_1 + NR_2 + NM) \qquad (10.16)$$

The reader should *physically* justify the conversion of Eqs. 10.14 to 10.16.

For planar frames:

$$NX = 3NJ - 3NFJ - 2NPJ - NR - NM$$
$$= NJ - NFJ + (2NJ - 2NFJ - 2NPJ$$
$$- NR - NM) \qquad (10.17)$$

$$NX = NJ - NFJ + NT \qquad (10.18)$$

Again the sidesway motions have been distinguished separately as

$$NT = 2NJ - (2NFJ + 2NPJ + NR + NM) \qquad (10.19)$$

The conversion of Eq. 10.17 to Eq. 10.18 also should be physically justified by the reader.

Equations 10.15 and 10.18 are primarily intended for use as a check on the visual process of determining the number of degrees of freedom. It is important to know the direction of each degree of freedom, and sole use of equations does not provide this information. Again, the equations are not foolproof. Under certain circumstances, erroneous results can be obtained. Finally, none of the equations is generally applicable when internal hinges are present in the structure.

Example 10.3. Determine the number of degrees of freedom and their directions for the space frame in Fig. 10.7a if axial deformation is neglected.

Solution: The number of independent *translational* degrees of freedom are given by Eq. 10.16.

$$NT = 3NJ - (3NFJ + 3NPJ + 2NR_1$$

$$+ NR_2 + NM)$$

$$= 3(7) - [3(1) + 3(1) + 2(0) + (1)$$

$$+ (7)]$$

$$= 7$$

Equation 10.15 gives the total number of degrees of freedom:

$$NX = 3NJ - 3NFJ + NT$$

$$= 3(7) - 3(1) + 7$$

$$= 25$$

Table 10.3.   Degrees of Freedom for Fig. 10.11.

Member	Member End	θ_x	θ_y	θ_z	Δ_x	Δ_y	Δ_z
AB	A	*	*	*	*	*	*
	B	1	2	3	4	*	5
BC	B	*	*	*	*	*	*
	C	6	7	8	*	9	10
BE	B	*	*	*	*	*	*
	E	14	15	16	17	*	*
CF	C	*	*	*	*	*	*
	F	18	19	20	*	*	*
EF	E	*	*	*	*	*	*
	F	*	*	*	*	*	*
DE	D	11	12	13	*	*	*
	E	*	*	*	*	*	*
FG	F	*	*	*	*	*	*
	G	21	22	23	24	*	25

The 25 degrees of freedom are shown in Fig. 10.11 as determined by visual analysis. Table 10.3 indicates the sequence in which the degrees of freedom are determined. The vectors in Fig. 10.11 are numbered in the same order as entered in Table 10.3. Several helpful comments follow:

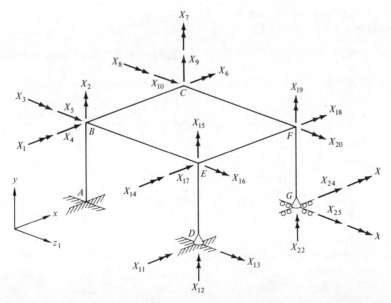

Fig. 10.11.   Degrees of freedom in a space frame–axial deformation neglected.

Line 2: At joint B translation in the y-direction is impossible because member AB has no axial deformation and joint A is fixed.

Line 4: If the location of joint B is known, the translation of joint C in the x-direction is established because member BC has no axial deformation.

Line 6: If the position of joint B is known, the translation of joint E in the z-direction is established by knowing the length of member BE. With joint D pinned and no axial deformation permitted in member DE, joint E cannot translate in the y-direction.

Line 8: With the position of joint C and E known and joint G restrained against motion in the y-direction, the length of members CF, EF, and FG can be intersected to locate joint F. Consequently, no translational degrees of freedom exist at joint F. (*Three lines must be intersected because the joint is in three-dimensional space!*).

Line 14: Joint G is a type 2 roller with two translational degrees of freedom.

It should be noted that the 25 degrees of freedom shown in Fig. 10.11 are not a unique answer. Had the process been started at another joint, an entirely different set of vectors would result but the total would still be 25. The reader should ponder why this is so.

Example 10.4. Determine the number and directions of the degrees of freedom for the planar frame of Fig. 10.7c. Neglect axial deformation.

Solution: The number of independent translational degrees of freedom are given by Eq. 10.19.

$$NT = 2NJ - (2NFJ + 2NPJ + NR + NM)$$

$$= 2(7) - [2(1) + 2(1) + 1 + 6]$$

$$= 3$$

Substitution into Eq. 10.18 yields the total number of degrees of freedom

$$NX = NJ - NFJ + NT$$

$$= 7 - (1) + 3$$

$$= 9$$

Visual inspection serves to cumulatively assign numbers to the degrees of freedom. The results are shown in Fig. 10.12 with the vectors numbered according to the sequence by which they were deduced. The reader should verify these and the corresponding entries in Table 10.4.

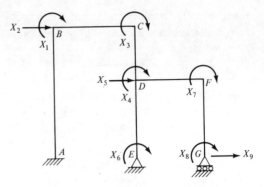

Fig. 10.12.   Degrees of freedom in a planar frame—axial deformation neglected.

Example 10.5. Determine the number and directions of the degrees of freedom for the planar frames in Figs. 10.13a and b. Neglect axial deformation.

Fig. 10.13a Solution: The degrees of freedom are shown in Fig. 10.13c. One translational degree of freedom exists at the roller support to accommodate transverse motion of member BD. Joint B has no vertical translational degree of freedom because member BD cannot change in length. Joint B has no horizontal motion because neither member AB nor BC can change in length.

Note that Eq. 10.19 gives

$$NT = 2NJ - (2NFJ + 2NPJ + NR + NM)$$

$$= 2(4) - [2(2) + 2(0) + (1) + 3]$$

$$= 0$$

Table 10.4.   Degrees of Freedom for Fig. 10.12.

Member	Member End	θ_z	Δ_x	Δ_y
AB	A	*	*	*
	B	1	2	*
BC	B	*	*	*
	C	3	*	*
CD	C	*	*	*
	D	4	5	*
DE	D	*	*	*
	E	6	*	*
DF	D	*	*	*
	F	7	*	*
FG	F	*	*	*
	G	8	9	*

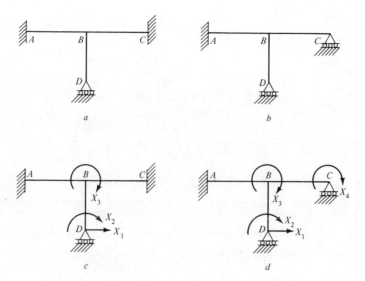

Fig. 10.13.   Planar frames for Example 10.5.

which is one less than the correct result. The source of the error is the entry of $NM = 3$. Physically, no translations occur at the fixed joints and the roller at D removes the vertical degree of freedom at joint D. Lack of axial deformation in member BD removes the *vertical* degree of freedom at joint B. Lack of axial deformation in member AB removes the *horizontal* degree of freedom at joint B. The presence of member BC (i.e., its lack of axial deformation) provides no additional information about the reduction of translational degrees of freedom. Such a member is an "excess member" in the context of calculating NT. Had NM been entered as 2 in Eq. 10.19 the correct result would be obtained.

As expected, Eq. 10.18 gives the same erroneous result:

$$NX = NJ - NFJ + NT$$
$$= 4 - 2 + 0$$
$$= 2$$

i.e., one less degree of freedom is suggested.

Fig. 10.13b Solution: Equation 10.19 gives

$$NT = 2NJ - (2NFJ + 2NPJ + NR + NM)$$
$$= 2(4) - [2(1) + 2(0) + 2 + 3]$$
$$= 4$$

and Eq. 10.18 gives

$$NX = NJ = NFJ + NT$$
$$= 4 - 1 + 1$$
$$= 4$$

These results are correct and the degrees of freedom are located as indicated in Fig. 10.13d. In this case, the physical explanation is the same as in Fig. 10.13c, except for member BC. Lack of axial deformation in member BC removes the horizontal degree of freedom of the roller at joint C. Thus member BC is not an excess member.

Example 10.5 emphasizes the nongenerality of the equations developed for determining NX *when axial deformation is neglected.* Any results obtained by use of equations should be carefully scrutinized by use of physical concepts. Conversely, *when axial deformation is included* Eqs. 10.5 and 10.7 are general.

PROBLEMS

Note: In each problem verify the results by a formula, if possible.

10-1.   Indicate by numbered vectors the degrees of freedom for each beam.

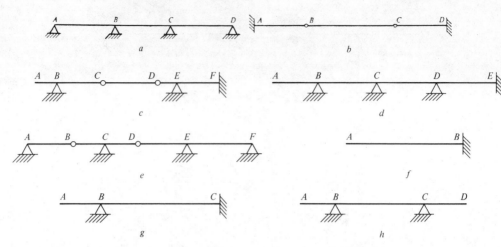

10-2. Indicate by numbered vectors the degrees of freedom for each planar truss.

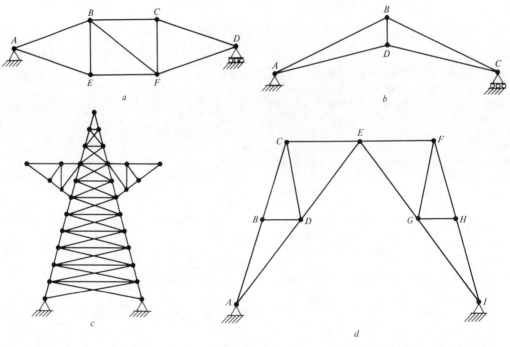

10-3. Indicate by numbered vectors the degrees of freedom for each planar frame. Include axial deformation.

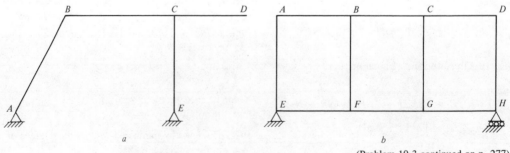

(Problem 10-3 continued on p. 277)

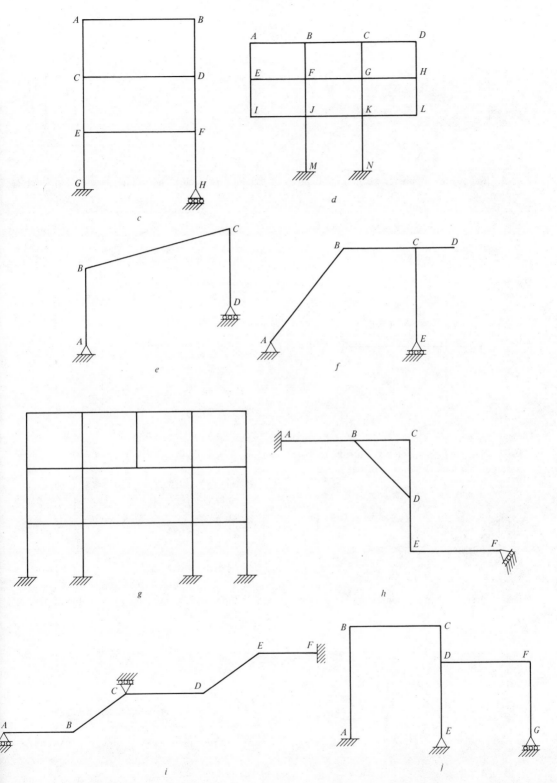

c

d

e

f

g

h

i

j

(Problem 10-3 continued on p. 278)

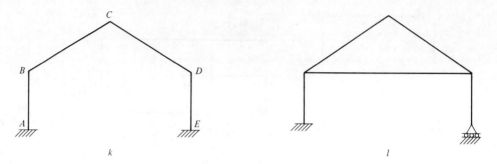

k   l

10-4.   Indicate by numbered vectors the degrees of freedom for each planar frame. Neglect axial deformation.

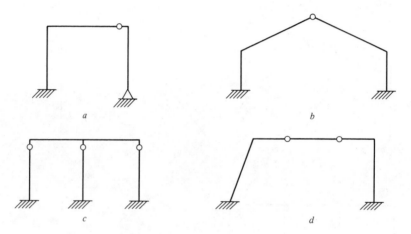

a   b

c   d

10-5.   Indicate by numbered vectors the rotational degrees of freedom that exist at each of the given joints. Assume each joint is part of a planar frame.

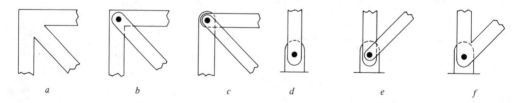

a   b   c   d   e   f

10-6.   Draw a schematic representation of each planar frame. Indicate by numbered vectors the degrees of freedom.

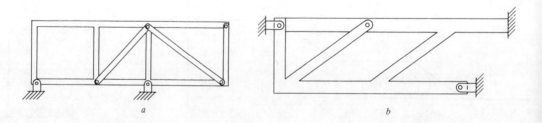

a   b

11
Flexibility Method

11.1 FUNDAMENTAL APPROACH

This chapter develops the classical *flexibility method* of analysis for framed structures. As a means to introduce the fundamental approach involved in the flexibility method, consider the three span continuous beam depicted in Fig. 11.1. From Fig. 11.1a, it is apparent that the beam is statically indeterminate to the third degree because of an excessive number of external reactions. If three of the reactions (e.g., R_3, R_4, and R_5) were known, the beam would pose no analytical difficulties. The flexibility method has the calculation of these excess reactions as its primary objective. This is accomplished as follows.

The original structure is converted to a determinate state by the removal of the three pin supports (i.e., removal of the excess reactions.) In this state, the beam can be easily analyzed for any loading using techniques described earlier in this text. Using the resulting cantilever beam, the superposition of the loadings in Fig. 11.1b through e is used to develop the response of the original structure (Fig. 11.1a).

In Fig. 11.1b, the cantilever beam is loaded with the actual loads (W_1, W_2, W_3, and W_4) taken from Fig. 11.1a. Displacements Δ_1-Δ_3 at the locations of the removed supports are of particular interest and can be calculated by established procedures. Displacements δ_{11} through δ_{33} are the displacements at the locations of the removed supports as caused by the various unit loads. Loads shown in Fig. 11.1c through e are unit values of the unknown reactions R_3, R_4, and R_5, respectively. With the displacement values determined, the condition of zero displacement at each pin support in the original structure requires (by the Principle of Superposition):

$$\Delta_1 + R_3\delta_{11} + R_4\delta_{12} + R_5\delta_{13} = 0$$

$$\Delta_2 + R_3\delta_{12} + R_4\delta_{22} + R_5\delta_{23} = 0$$

$$\Delta_3 + R_3\delta_{13} + R_4\delta_{23} + R_5\delta_{33} = 0 \quad (11.1)$$

The preceding are called either compatibility conditions or *flexibility equations*, and δ_{11}-δ_{33} are referred to as *flexibility coefficients*. Simultaneous solution of Eqs. 11.1 yields the values of the reactions R_3-R_5. Once R_3 through R_5 are known, the beam in Fig. 11.1a is statically determinate and can be analyzed directly for other quantities, e.g., the shear and moment at any location. Alternatively, such other quantities can be calculated in each of Figs. 11.1b through e and summed.

11.2 STATICAL INDETERMINACY

The primary unknown quantities in the flexibility method of analysis are the "redundant re-

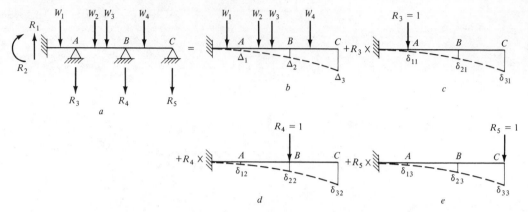

Fig. 11.1.   Flexibility analysis of a continuous beam.

actions'' (or, simply, the "redundants"). Redundant reactions are those reactions in excess of the number that can be determined from external statics. Internal forces can also be considered redundant. If so, redundants can be defined as being comprised of those reactions and internal forces whose values must be known to render a given structure to be statically determinate. Thus the number of redundants is equal to the degree of statical indeterminacy.

The set of redundants for any given structure is not unique. Any set of redundants whose removal from the structure would render it statically determinate and stable is a proper choice. It is important to master the ability to determine the degree of statical indeterminacy as a prerequisite to selecting redundants.

Determination of the degree of statical indeterminacy for planar framed structures has been described in earlier chapters. Expressions for the total degree of statical indeterminacy were derived for planar trusses, planar frames and beams. These are summarized in Table 11.1. Use of these equations serves to determine the number of redundants for any given

structure. When $I > 1$, there usually are several choices for the needed set of redundants. Use of the "method of trees" allows designation of particular redundants that might be removed in the flexibility method.

In Section 5.5.2.1 the equation

$$I = NF - NP \qquad (5.14)$$

was developed as an alternative means of determining the degree of statical indeterminacy of a planar truss. Recall, NF is equal to the number of member forces (equal to the number of members) and NP is the number of possible independent load directions (equal to NX, the number of degrees of freedom.) By redefining NF as the number of independent internal member forces, Eq. 5.14 has applicability to all types of framed structures.

Each of the structures drawn in Fig. 11.2 exemplifies a statically indeterminate beam. This is apparent from the simple observation that, in all cases, the two available equilibrium conditions are insufficient in number to enable computation of all reactions. (Horizontal loading of a beam is not considered, and horizontal reac-

Table 11.1.   Expressions for the Total Degree of Statical Indeterminancy.

Type of structure	Degree of statical indeterminancy	
Planar truss	$I = NM - (2NJ - NRF)$	(5.11)
Continuous beam	$I = NRF - NIH - 3$	(3.2)
Planar frame	$I = NRF + 3(NM - NJ) - NIH$	(6.48)

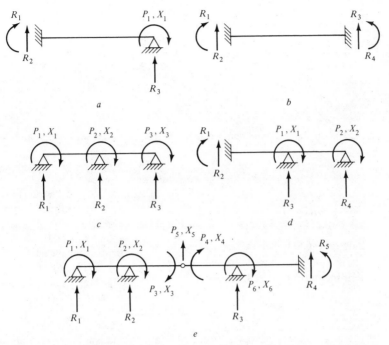

Fig. 11.2.   Statically indeterminate beams.

tions are excluded from the illustrations.) The number of excessive reactions in each structure constitutes its degree of statical indeterminacy, I. In the matrix stiffness method of analysis, presented in later chapters, the external reactions and the end moments for each span are the unknown force quantities in a continuous beam. Mathematically, the total number of unknowns is expressed as

Number of unknowns

$$= NF + NRF - 2NIH \quad (11.2)$$

in which

NF = Number of internal member end moments (two per member), including those of zero magnitude (i.e., at hinged ends).

NRF = Number of external reactions (including moments)

NIH = Number of internal hinges

The last term in Eq. 11.2 recognizes that for each internal hinge, two end moments are known to have zero magnitude and need not be computed. If the joints of a continuous beam are isolated as free-body diagrams, two equations of statics can be applied to each of them with one exception. Moment equilibrium has no meaning at an internal hinge, and only one equilibrium equation is available at such points.

The degrees of freedom and reactions that exist at each of the possible joints of a continuous beam are shown in Fig. 11.3. Examination of the conditions shown indicates that the total number of equilibrium equations in a continuous beam is given by

Number of equations

$$= NP + NRF - 2NIH \quad (11.3)$$

The last term in Eq. 11.3 recognizes that the number of degrees of freedom at an internal hinge (3) exceeds the number of available statics equations (1) by two.

For a statically determinate beam, Eqs. 11.2 and 11.3 should reflect equality and

$$NF + NRF - 2NIH = NP + NRF - 2NIH$$

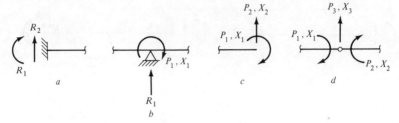

Fig. 11.3.   Common support conditions. *a*, fixed; *b*, pinned; *c*, free end; and *d*, internal hinge.

or

$$NF - NP = 0 \qquad (11.4)$$

For a statically indeterminate beam, NF always exceeds NP, and Eq. 11.4 is invalid. In such cases, the degree of indeterminancy, I, is given by

$$I = NF - NP \qquad (11.5)$$

which is identical to the relationship used for trusses, except for the underlying meaning of the term NF. NF was defined as the number of member end moments, including those of zero value. The member end moments are the independent internal force quantities in a continuous beam. This statement is verified as follows. If the end moments are established, complete shear and moment diagrams are easily plotted. Given the end moments, the end shears can be obtained by applying statics to each member (e.g., by summing moments about each end of the member). These can be used to directly plot the shear and moment diagrams for each span. Alternatively, after transferring the end moments and shears to free body diagrams of the joints, the reactions can be determined by applying statics to the joints. With these established, the shear and moment diagrams are developed directly for the complete beam.

Example 11.1.   Determine the degree of statical indeterminacy for each beam in Fig. 11.2. No horizonal loads are to be applied. Indicate reactions and/or forces that could be used as redundants.

Solution:
Fig. 11.2a: $I = NF - NP = 2(1) - 1 = 1$
R_1 or R_3 could serve as the redundant
Fig. 11.2b: $I = 2(1) - 0 = 2$
either R_1 and R_2 or R_1 and R_4 could serve as redundants
Fig. 11.2c: $I = 2(2) - 3 = 1$
any of the vertical reactions could serve as the redundant
Fig. 11.2d: $I = 2(2) - 2 = 2$
either any two vertical reactions or R_1 and any one vertical reaction could serve as redundants
Fig. 11.2e: $I = 4(2) - 6 = 2$
any combination of two reactions, except R_1 and R_2 together, could serve as redundants

---

Equation 11.5 is extended to include planar frames as follows. In Chapter 10, the number of degrees of freedom NX in a planar frame (with axial deformation neglected) is derived as

$$NX = NP = 3NJ - 3NFJ$$
$$- 2NPJ - NR - NM \qquad (10.17)$$

It is easily deduced that the expression $3NFJ + 2NPJ + NR$ in Eq. 10.17 is numerically equal to NRF. Making this substitution yields

$$NP = 3NJ - NRF - NM \qquad (11.6)$$

Equation 11.6 can be rearranged to the form

$$I = 2NM - (3NJ - NRF - NM) \qquad (11.7)$$

and substitution of Eq. 11.6 gives

$$I = 2NM - NP \qquad (11.8)$$

The total number of internal-member end moments in planar frames is $2NM$. This normally comprises the set of independent internal forces. With some exceptions, if the end moments are known, all other internal forces in a planar frame can be established by use of statics alone. In other words, a planar frame is statically determined if all the end moments are known. On this basis, the term $2NM$ can be replaced by NF. This replacement gives

$$I = NF - NP \qquad (11.5)$$

which is the desired result.

The sufficiency of the end moments as a basis for determining all unknown forces is easily justified. Usually, if it is assumed that the end moments are known for a given frame, the remaining forces can be determined by executing the following steps:

1. The members and joints are isolated as separate free-body diagrams. Longitudinal equilibrium establishes the axial force at one end of each member is equal in magnitude and opposite in direction to the axial force at the other end.
2. Each shear force is calculated by summing moments about the opposite end of the member.
3. The unknown axial forces and known shear forces and moments are transferred to the joint free-body diagrams.
4. Moment reactions are determined by enforcing moment equilibrium at each fixed support.
5. Horizontal and vertical equilibrium are enforced at each joint. This produces the axial forces and remaining reactions.

For planar frames in which axial deformation is included, Eq. 11.5 remains applicable and general, if NP and NF are properly defined. More important, whether axial deformation is neglected or not, the degree of statical indeterminacy of a planar frame is the same. Thus, by including axial deformation Eq. 11.5 can be applied without fear of exceptions.

The application and validity of Eq. 11.5 when axial deformation is considered is explained as follows. If free-body diagrams of all members and joints of a planar frame are drawn, the total number of unknowns is

$$\text{Number of unknowns} = NRF + 6NM$$

$$(11.9)$$

where $6NM$ represents the presence of moment, shear and axial force at each member end. The total number of equations obtained by applying statics to each free-body diagram is

$$\text{Number of equations} = 3NJ + 3NM$$

$$(11.10)$$

Equating Eqs. 11.9 and 11.10 and recognizing

$$3NJ = NRF + NP$$

yields

$$NRF + 6NM = 3NJ + 3NM$$

$$NRF + 6NM = NRF + NP + 3NM$$

$$3NM = NP \qquad (11.11)$$

If the axial force at one member end and the two end moments are known for each member, the remaining end forces are determined by statics. Satisfaction of equilibrium at the joints establishes the unknown reactions. On this basis, there are three independent end forces per member and knowing these values renders a planar frame to be statically determinate. These independent forces are the basic unknowns in the matrix stiffness method of analysis. Thus, the term $3NM$ in Eq. 11.11 can be replaced by NF giving

$$NF = NP$$

If the number of unknowns, Eq. 11.9, exceeds the number of equations, Eq. 11.10, the result is $NF > NP$. Thus the frame is statically indeterminate and

$$I = NF - NP$$

which is the desired result.

In the present context, a planar frame with axial deformation included, we define

NF = number of independent internal forces

= two end moments plus one axial force

per member

= $3NM$

$NP = NX$ = number of degrees of freedom

with axial deformation

included (Eq. 10.7)

Example 11.2. Determine the degree of indeterminacy for the structures shown in Fig. 11.4. Neglect axial deformation. Confirm results by including axial deformation.

Solution (using Eq. 11.4):
Without axial deformation (Fig. 11.4a)

NF = 18 end moments

$NP = NX$ = 7 rotations + 4 translations

= 11 degrees of freedom

$$\therefore I = NF - NP = 18 - 11 = 7$$

With axial deformation (Fig. 11.5a)

NF = 18 end moments + 9 axial forces = 27

NP = 20

$$\therefore I = NF - NP = 27 - 20 = 7$$

With axial deformation (Fig. 11.4b)

NF = 8 end moments

$NP = NX$ = 6 rotations + 2 translations

= 8 degrees of freedom

$$\therefore I = NF - NP = 8 - 8 = 0$$

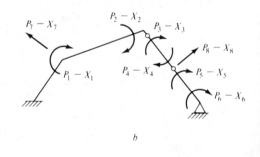

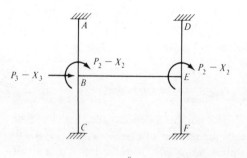

Fig. 11.4.   Planar frames—degrees of freedom with axial deformation neglected. Example 11.2.

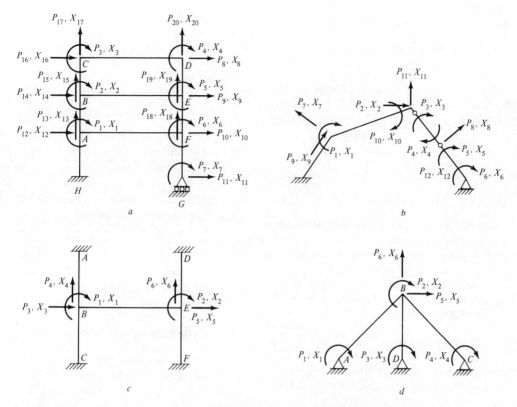

Fig. 11.5. Planar frame—degrees of freedom with axial deformation included. Example 11.2.

With axial deformation (Fig. 11.5b)

$$NF = 8 \text{ end moments} + 4 \text{ axial forces} = 12$$

$$NP = 12$$

$$\therefore I = NF - NP = 12 - 12 = 0$$

Without axial deformation (Fig. 11.4c)

$$NF = 10 \text{ end moments}$$

$$NP = 1 \text{ rotation} + 2 \text{ translations}$$

$$= 3 \text{ degrees of freedom}$$

$$\therefore I = NF - NP = 7?$$

With axial deformation (Fig. 11.5c)

$$NF = 10 \text{ end moments} + 5 \text{ axial forces} = 15$$

$$NP = 6$$

$$\therefore I = NF - NP = 15 - 6 = 9$$

In this case, an incorrect answer is obtained if axial deformation is neglected. Based on including axial deformation, the frame is statically indeterminate to the ninth degree. This observation was verified in Example 6.9c by Eq. 6.48. The source of the error lies in the use of Eq. 10.17 in deriving Eq. 11.7. The term NM implies that the knowledge that the axial deformation is zero for each member reduces the number of degrees of freedom by NM. In this case, "excess" members (as defined in Example 10.5) are present in the frame. In Fig. 11.4c, the fixed joint at A eliminates 3 degrees of freedom at that point. Also, knowing that member AB has no axial deformation eliminates the vertical degree of freedom at joint B. Member BC also has no axial deformation, but this fact is irrelevant because it does not eliminate any additional degrees of freedom. By similar reasoning, either member DE or EF can be viewed as irrelevant to the elimination of degrees of freedom. Use of Eq. 11.8 (or Eq. 11.4) implies the two irrelevant members are included in the term NM. The consequence is the calculated degree of indeterminacy is two less than the actual number. Fig. 11.4d:

Without axial deformation (Fig. 11.4d)

$$NF = 6 \text{ end moments}$$

$$NP = NX = 4 \text{ rotations}$$

$$\therefore I = NF - NP = 6 - 4 = 2?$$

With axial deformation (Fig. 11.5d)

$$NF = 6 \text{ end moments} + 3 \text{ axial forces} = 9$$

$$NP = 6$$

$$\therefore I = 9 - 6 = 3$$

Clearly, the structure is statically indeterminate (externally) to the third degree. This was verified in Example 6.9. In this case, the incorrect value given by Eq. 11.5 when axial deformation is neglected, is attributed to the concurrency of the three member axes connecting to joint B. Specifically, knowledge of zero axial deformation in any two of the members is sufficient to eliminate the two translational degrees of freedom at joint B. Recognition of zero axial deformation in the third member contributes no additional input to the determination of degrees of freedom. Hence, the calculated value of I is low by one.

In light of this example, the reader is urged to be cautious in applying Eq. 11.5 to planar frames when axial deformation is neglected. Any given structure should be examined for members whose presence is irrelevant to determining the degrees of freedom. The degree of statical indeterminacy calculated by Eq. 11.5 is then increased by one for each "excess" member. Alternatively, the analyst can apply Eq. 11.5 but *include* axial deformation. If this is done,

the axial force in each member is *independent* of the member end moments. Thus, NF is equal to $3NM$ in Eq. 11.5 instead of $2NM$. Also, NP constitutes the degrees of freedom that exist when axial deformation is not neglected.

In general, Eq. 11.5 states that the degree of statical indeterminacy is equal to the number of independent internal member forces minus the number of degrees of freedom. Stated this way, Eq. 11.5 is applicable to any planar or space frame structure, provided the independent internal member forces are properly counted.

11.3   THE FLEXIBILITY METHOD OF ANALYSIS

11.3.1   Beams with One Redundant

Details of the solution technique used in the flexibility analysis will be introduced by use of a simple example, a beam with a single redundant.

Example 11.3.   A propped cantilever beam is subjected to a single concentrated load as illustrated in Fig. 11.6a. By use of the flexibility method, calculate all reactions and determine the internal moment at the load point. The member has a constant value of EI.

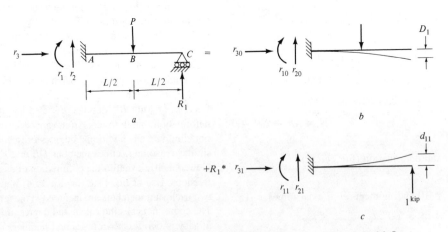

Fig. 11.6.   Alternative flexibility analysis for the beam of Example 11.3.

Solution:

Step 1: A single degree of statical indeterminacy exists. This is visually obvious as four support reactions exist, indicating an excess of one. The vertical reaction at the pinned end is chosen as the redundant force, R_1. Remaining reactions are considered nonredundant and labeled r_1 through r_3, accordingly. (Other choices for declaration of the redundant exist and some are examined in subsequent example problems.)

Step 2: Figure 11.6b depicts the released (determinate) structure subjected to the actual load. In this state, three non-redundant reactions (r_{10}, r_{20} and r_{30}) are determined as

$$r_{10} = -\tfrac{1}{2} PL = \tfrac{1}{2} PL \text{ ccw} \qquad r_{20} = P \uparrow$$
$$r_{30} = 0$$

In addition, the displacement at the location of the removed redundant is easily calculated by any suitable analysis method, e.g., the moment area method, as

$$D_1 = \frac{5}{48} \frac{PL^3}{EI} \downarrow$$

The moment at midspan is

$$F_{10} = 0$$

Step 3: In Fig. 11.6c, a unit value of the redundant is applied. Nonredundant reactions produced by this loading are.

$$r_{11} = L \text{ cw} \qquad r_{21} = 1 \downarrow \qquad r_{31} = 0$$

and the displacement in the direction of the removed redundant is

$$d_{11} = \frac{L^3}{3EI} \uparrow$$

A moment of magnitude

$$f_{11} = \frac{L}{2} \text{ (tension in bottom fibers)}$$

results at midspan. Positive signs were arbitrarily chosen to correspond to positive moments by the designer's convention.

Step 4: Compatibility of displacements at C requires

$$D_1 + R_1 d_{11} = 0$$

Substitution of previously computed values gives

$$\frac{5}{48} \frac{PL^3}{EI} \downarrow + R_1 \left(\frac{L^3}{3EI} \uparrow \right) = 0$$
$$-\frac{5}{48} \frac{PL^3}{EI} + \frac{L^3}{EI} R_1 = 0$$
$$R_1 = \frac{5}{16} P = \frac{5}{16} P \uparrow$$

Step 5: The moment at midspan, F_1, in Fig. 11.6a, is determined by superposition

$$F_1 = F_{10} + R_1 f_{11}$$
$$= 0 + \left(\frac{5}{16} P \right) \left(\frac{L}{2} \right)$$
$$F_1 = \frac{5}{16} PL \text{ (tension in bottom fibers)}$$

Step 6: Nonredundant reactions are determined by superposition

$$r_1 = r_{10} + R_1 r_{11}$$
$$r_2 = r_{20} + R_1 r_{21}$$
$$r_3 = r_{30} + R_1 r_{31}$$

Substitution of known values yields

$$r_1 = -\frac{PL}{2} + \left(\frac{5}{16} P \right) (L) = -\frac{3}{16} PL$$
$$= \frac{3}{16} PL \text{ ccw}$$
$$r_2 = P + \left(\frac{5}{16} P \right) (-1) = \frac{11}{16} PL$$
$$r_3 = 0 + \left(\frac{5}{16} P \right) (0) = 0$$

Computed results for the external reactions are shown in Fig. 11.7.

Example 11.4. Repeat Ex. 11.3 except use the moment at the fixed support as the redundant.

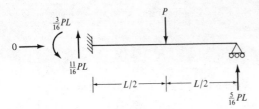

Fig. 11.7.   Results of Example 11.5.

Solution:

Step 1:   The moment resistance at the fixed-end is labeled as the redundant force R_1. Remaining non-redundant reactions are labeled r_1 through r_3, accordingly.

Step 2:   Figure 11.8b depicts the released structure subjected to the actual load. In this state, the three nonredundant support reactions, r_{10}, r_{20}, and r_{30} are

$$r_{10} = \frac{P}{2} \uparrow \qquad r_{20} = 0 \qquad r_{30} = \frac{P}{2} \uparrow$$

In addition, the displacement at the redundant is calculated as

$$D_1 = \frac{PL^2}{16EI} \text{ cw}$$

Next, the moment at midspan is computed to be

$$F_{10} = \frac{PL}{4} \text{ (tension in bottom fibers)}$$

Step 3:   A unit moment, which acts in the direction assumed for the redundant R_1 is applied at A. This auxiliary unit loading is shown in Fig. 11.8c Non-redundant support reactions produced by this loading are

$$r_{11} = -\frac{1}{L} = \frac{1}{L} \downarrow \qquad r_{21} = 0 \qquad r_{31} = \frac{1}{L} \downarrow$$

and the displacement in the direction of the released redundant is

$$d_{11} = \frac{L}{3EI} \text{ (cw)}$$

A moment of magnitude

$$f_{11} = \tfrac{1}{2} \text{ kip-ft (tension in bottom fibers)}$$

results at midspan.

Step 4:   For the single degree of indeterminacy,

$$D_1 + R_1 d_{11} = 0$$

Substitution of previously computed values gives

$$\frac{PL^2}{16EI} + R_1 \left(\frac{L}{3EI} \right) = 0$$

if clockwise rotations are treated as positive quantities (this is an arbitrary choice). The redundant force is easily calculated as

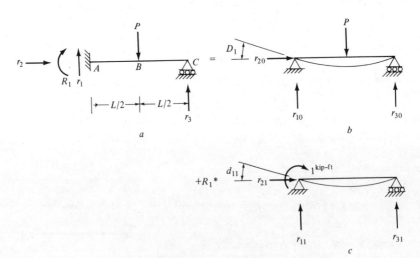

Fig. 11.8.   Indeterminate beam for Example 11.4.

$$R_1 = -\frac{3}{16} PL$$

The negative sign (to be retained in the ensuing computations) indicates R_1 is actually a counterclockwise moment.

Step 5: The moment at midspan is given by

$$F_1 = F_{10} + R_1 f_{11}$$

which, by substitution of known values, gives

$$F_1 = \frac{PL}{4} + \left(-\frac{3PL}{16}\right)\left(\frac{1}{2}\right)$$

$$= \frac{5PL}{32} \text{ (tension in bottom fibers)}$$

Step 6: Nonredundant reactions are determined from

$$r_1 = r_{10} + R_1 r_{11}$$

$$r_2 = r_{20} + R_1 r_{21}$$

$$r_3 = r_{30} + R_1 r_{31}$$

Substitution of known values gives

$$r_1 = +\frac{P}{2} + \left(-\frac{3PL}{16}\right)\left(-\frac{1}{L}\right) = \frac{11}{16} PL \uparrow$$

$$r_2 = 0 + \left(-\frac{3PL}{16}\right)(0) = 0$$

$$r_3 = +\frac{P}{2} + \left(-\frac{3PL}{16}\right)\left(+\frac{1}{L}\right) = \frac{5}{16} PL \uparrow$$

Upward reactions were arbitrarily chosen to be positive. All results agree with Example 11.3.

Examples 11.3 and 11.4 illustrate the technique employed for the analysis of a particular beam by the flexibility method. When an external reaction is used as a redundant within the restriction of creating a stable, determinate structure, the choice of which force quantity should be released is arbitrary. Use of the internal moment as the redundant quantity is a less obvious approach and is numerically demonstrated in the ensuing example problem.

Example 11.5.   Determine the moment at midspan of the indeterminate beam of Example 11.3. Use the flexibility method and the superposition procedure depicted schematically in Fig. 11.9.

Solution: Removal of the moment resistance at midspan leaves the determinate structure shown in Fig. 11.9b. Its moment diagram is shown in Fig. 11.10a. The displacement D_1 is the relative rotation of the member ends that frame into the internal hinge. Using the conjugate-beam method and the moment diagram given in Fig. 11.10a, the numerical value for D_1 can be determined in a straightforward manner. D_1 is comprised of the two rotations θ_L and θ_R shown in Fig. 11.11a. Referring to the conjugate beam drawn in Fig. 11.12a, these individual rotations correspond to the shear values to the left and right, respectively, of point B. By simple statics

$$R_B' = \frac{1}{2}\left(\frac{1}{2}\frac{PL}{EI}\right)\left(\frac{L}{2}\right)\left(\frac{5}{6}L\right) \bigg/ \left(\frac{L}{2}\right) = \frac{5}{24}\frac{PL^2}{EI}$$

$$\theta_L = V_{BL}' = -\frac{1}{2}\left(\frac{1}{2}\frac{PL}{EI}\right)\left(\frac{L}{2}\right) = -\frac{1}{8}\frac{PL^2}{EI}$$

$$= \frac{1}{8}\left(\frac{PL^2}{EI}\right) \text{cw}$$

$$\theta_R = V_{BR}' = R_B' + V_{BL}'$$

$$= \frac{5}{24}\frac{PL^2}{EI} + \left(-\frac{1}{8}\frac{PL^2}{EI}\right)$$

$$= +\frac{1}{12}\frac{PL^2}{EI} = \frac{1}{12}\frac{PL^2}{EI} \text{ccw}$$

$$D_1 = |\theta_L| + |\theta_R|$$

$$= \frac{5}{24}\frac{PL^2}{EI} \text{ cw relative to left tangent}$$

A clockwise notation is assigned to D_1 because the tangent on the right side of the hinge rotates in that sense relative to the displaced tangent on the left side of the hinge.

For a single degree of indeterminacy, compatibility requires

$$D_1 + R_1 d_{11} = 0$$

In the present case, R_1 is the unknown internal moment at point B of the indeterminate beam in Fig. 11.9a. The displacement d_{11} is the hinge rotation caused by the unit auxiliary loading shown in Fig.

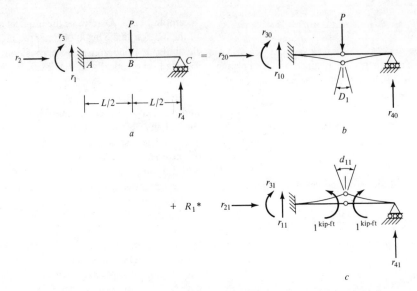

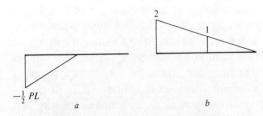

Fig. 11.9.   Flexibility analysis using an internal redundant. Example 11.5.

11.9c. The left-hand side of the compatibility equation represents the superposition of the hinge rotations indicated in Figs. 11.9b and c. The right side of the equation represents the net rotation that results from this superposition. Because no discontinuity exists at point B in the indeterminate beam (Fig. 11.9a), the net displacement must be zero.

The moment diagram created by the unit auxiliary loading is given in Fig. 11.10b. Referring to Fig. 11.11b, d_{11} is seen to consist of the two individual rotations, ϕ_L and ϕ_R. Consequently, computation of d_{11} is similar to the determination of D_1. The only difference in this computation is that the conjugate beam has the loading shown in Fig. 11.12b.

$$R_B'' = -\frac{1}{2}\left(\frac{2}{EI}\right)(L)\left(\frac{2}{3}L\right)\bigg/\left(\frac{L}{2}\right)$$

$$= -\frac{4}{3}\frac{L}{EI}$$

$$\phi_L = V_{BL}'' = +\frac{1}{2}\left(\frac{2}{EI}+\frac{1}{EI}\right)\left(\frac{L}{2}\right) = \frac{3}{4}\frac{L}{EI}$$

$$= \frac{3}{4}\frac{L}{EI}\ \text{ccw}$$

$$\phi_R = V_{BR}''$$

$$= R_B'' + V_{BL}'' = \left(-\frac{4}{3}\frac{L}{EI}\right)+\left(\frac{3}{4}\frac{L}{EI}\right)$$

$$= -\frac{7}{12}\frac{L}{EI} = \frac{7}{12}\frac{L}{EI}\ \text{cw}$$

$$d_{11} = |\phi_L| + |\phi_R|$$

$$= \frac{4}{3}\frac{L}{EI}\ \text{ccw relative to left tangent}$$

The clockwise hinge rotations are treated as positive, and the computed values are substituted into the compatibility equation

$$D_1 + R_1 d_{11} = d_{10}$$

$$\frac{5}{24}\frac{PL^2}{EI} + R_1\left(-\frac{4}{3}\frac{L}{EI}\right) = 0$$

$$R_1 = \frac{5}{32}PL^2$$

A positive value results and indicates that R_1 acts in the direction of the auxiliary unit moment and therefore is positive according to the designer's convention (i.e., it produces tension in the bottom fibers).

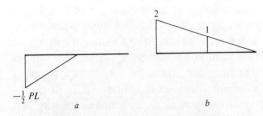

Fig. 11.10.   Moment diagrams for Example 11.5. *a*, for the actual loading; *b*, for the unit auxiliary loading.

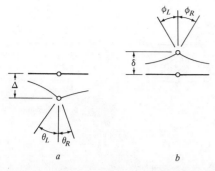

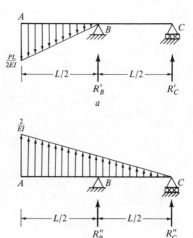

Fig. 11.11. Hinge rotations. Example 11.5 a, for the actual loading; b, for the unit auxiliary loading.

Fig. 11.12. Conjugate beams for Example 11.5. a, for the actual loading; b for the unit auxiliary loading.

This result agrees with the value of F_1 calculated in Example 11.3.

Remaining quantities such as external reactions are determined by superposition of values obtained by statics for each of the beams shown in Fig. 11.9b and c. As an example, the reaction, r_4, is calculated as follows:

From Fig. 11.9b:

$$r_{40} = 0 \, (\text{by statics})$$

From Fig. 11.9c:

$$r_{41} = \frac{2}{L} \uparrow (\text{by statics})$$

By superposition:

$$r_4 = r_{40} + R_1 r_{41}$$

$$= 0 + \left(\frac{5}{32} PL\right)\left(\frac{2}{L}\right)$$

$$= \frac{5}{16} PL \uparrow$$

and is found to be in agreement with the results of Example 11.3. The reader is urged to calculate the reactions at the fixed-end by superposition and verify their correctness by further comparison with Example 11.3.

11.3.2 Beams With Multiple Redundants

Examples 11.3 to 11.5 illustrated the flexibility method of analysis when one redundant was

present. The solution involved development of a compatibility condition involving the redundant as a single unknown. For structures with multiple degrees of statical indeterminacy, a compatibility condition must be developed for each selected redundant. The resulting compatibility equations are a simultaneous set involving the redundants as the unknowns. Solution of these equations for these unknowns allows the subsequent superposition of determinate beams to proceed as usual.

Example 11.6. Determine the end reactions and the midspan moment for the beam given in Fig. 11.13a. Use a constant EI value. Solve the problem by the flexibility method and the superposition shown in Fig. 11.13.

Solution: Horizontal reactions are not present because no horizontal loads are applied. Of the four remaining reactions, two are redundant. Moments R_1 and R_2 are chosen. The vertical reactions that remain are nonredundant forces and are designated r_1 and r_2. The superposition of Figs. 11.3b, c, and d serves to develop the needed pair of compatibility conditions. By linearly combining the end slopes from each determinate beam, one gets

$$D_1 + R_1 d_{11} + R_2 d_{12} = 0 \qquad \text{(a)}$$

$$D_2 + R_1 d_{21} + R_2 d_{22} = 0 \qquad \text{(b)}$$

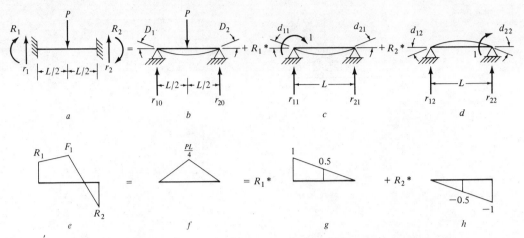

Fig. 11.13.   Beam for Example 11.6. *a, b, c, d,* superposition by the flexibility method; *e, f, g, h,* moment diagrams for superposed beams.

where d_{11} through d_{22} are due to unit values of the redundants. These compatibility equations enforce zero slope at both ends of the member.

The end rotations for the determinate beams in Fig. 11.13b, c, and d are easily calculated by the conjugate-beam method using the simple beam moment diagrams given in Fig. 11.13f, g, and h. Computations are not shown and are left for the reader to verify.

For Fig. 11.13b

$$D_1 = \frac{PL^2}{16EI} \text{ cw} \qquad D_2 = \frac{PL^2}{16EI} \text{ ccw}$$

for Fig. 11.13c

$$d_{11} = \frac{L}{3EI} \text{ cw} \qquad d_{21} = \frac{L}{6EI} \text{ ccw}$$

for Fig. 11.13d

$$d_{12} = \frac{L}{6EI} \text{ ccw} \qquad d_{22} = \frac{L}{3EI} \text{ cw}$$

Substitution into Eqs. (a) and (b) with attention to signs ("cw" is arbitrarily taken as positive) yields

$$\frac{PL^2}{16EI} + R_1 \left(+ \frac{L}{3EI} \right) + R_2 \left(- \frac{L}{6EI} \right) = 0$$

$$- \frac{PL^2}{16EI} + R_1 \left(- \frac{L}{6EI} \right) + R_2 \left(+ \frac{L}{3EI} \right) = 0$$

and simultaneous solution gives

$$\begin{Bmatrix} R_1 \\ R_2 \end{Bmatrix} = \frac{EI}{L} \begin{bmatrix} \frac{1}{3} & -\frac{1}{6} \\ -\frac{1}{6} & \frac{1}{3} \end{bmatrix}^{-1} \begin{Bmatrix} -\dfrac{PL^2}{16EI} \\ +\dfrac{PL^2}{16EI} \end{Bmatrix} = \begin{Bmatrix} -\dfrac{PL}{8} \\ +\dfrac{PL}{8} \end{Bmatrix}$$

R_1 is negative, indicating that the direction chosen in Fig. 11.13a is incorrect and the moment actually acts in a counterclockwise direction (producing compression on the bottom fibers).

Nonredundant reactions are determined by superposition. By application of statics to Figs. 11.13b through d, the simple beam reactions are found to be

for Fig. 11.13b,

$$r_{10} = +\frac{P}{2} \qquad r_{20} = +\frac{P}{2}$$

for Fig. 11.13c,

$$r_{11} = -\frac{1}{L} \qquad r_{21} = +\frac{1}{L}$$

for Fig. 11.13d,

$$r_{12} = -\frac{1}{L} \qquad r_{22} = +\frac{1}{L}$$

Therefore

$$r_1 = r_{10} + R_1r_{11} + R_2r_{12}$$

$$= \frac{P}{2} + \left(-\frac{PL}{8}\right)\left(-\frac{1}{L}\right) + \left(+\frac{PL}{8}\right)\left(-\frac{1}{L}\right)$$

$$= +\frac{P}{2} = \frac{P}{2} \uparrow \qquad\qquad (c)$$

$$r_2 = r_{20} + R_1r_{21} + R_2r_{22}$$

$$= \frac{P}{2} + \left(-\frac{PL}{8}\right)\left(+\frac{1}{L}\right) + \left(+\frac{PL}{8}\right)\left(+\frac{1}{L}\right)$$

$$= +\frac{P}{2} = \frac{P}{2} \uparrow \qquad\qquad (d)$$

The moment, F_i, at any location in the indeterminate structure is calculated by superposition. In effect, the moment diagrams for the released structure and the scaled auxiliary unit loadings are combined linearly, as shown in Fig. 11.13e, f, g, and h. At midspan, this equation gives a moment

$$F_1 = F_{10} + R_1f_{11} + R_2f_{12}$$

$$= \frac{PL}{4} + \left(-\frac{PL}{8}\right)(+0.5) + \left(+\frac{PL}{8}\right)(-0.5)$$

$$= +\frac{PL}{8}$$

The positive value indicates the sign assumed for F_1 in the moment diagram of Fig. 11.13e is correct. For clarity, the final moment diagram is shown in Fig. 11.14 with the computed numerical values shown in the usual sign convention for such diagrams.

11.3.3 Planar Frames

Consider the three planar frames depicted in Fig. 11.15. Figure 11.15a through c illustrate a three-member portal frame with 1, 2, and 3 degrees of statical indeterminacy, respectively. These structures will be used as a means of demonstrating the treatment of planar frames by the flexibility method. In particular, the effect of increasing the degree of indeterminacy will be shown by investigating the frames in Fig. 11.15 in sequential order. At the conclusion, the compatibility equations developed for

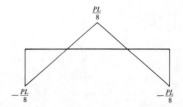

Fig. 11.14. Final moment diagram for Example 11.6.

the structures in Figs. 11.15a and b will be shown to be subsets of those developed for the structure in Fig. 11.15c.

Examine the 1-degree indeterminate frame in Fig. 11.15a. Of the four reactions that exist in this frame, the vertical reaction at D is chosen as the redundant, R_1. The superposition shown in the rest of the figure visually displays the needed compatibility condition. With the redundant force released, point D experiences a vertical displacement, D_1, and the magnitude of the redundant reaction must be sufficient to restore the joint to its proper position. Application of a vertical unit-auxiliary loading causes a vertical displacement d_{11}. After scaling this displacement by the factor R_1, the condition obtained by superposition is

$$D_1 + R_1d_{11} = 0 \qquad\qquad (11.12)$$

This equation is identical with that used earlier for beams except for the nature of the displacement quantities and the redundant force. The released structure is determinate, and the desired displacements, D_1 and d_{11}, can be computed by any of several methods discussed earlier. Solution of Eq. 11.12 produces the value of R_1.

Two choices exist for completion of the analysis of the indeterminate structure. With R_1 known, the determinate structures in Fig. 11.15a can be analyzed for desired quantities and the results combined by superposition. Alternatively, knowledge of R_1 reduces the original frame to a statically determinate system and ordinary determinate structural analysis can be performed directly on this system.

The frame in Fig. 11.15b differs from that of

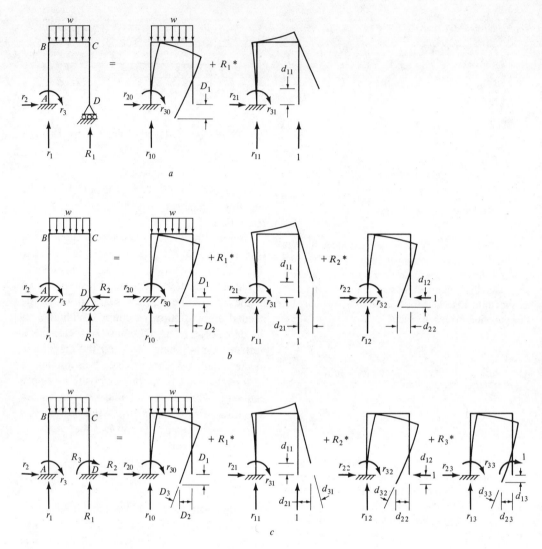

Fig. 11.15. Flexibility method analysis of planar frames. *a*, one degree of indeterminacy; *b*, two degrees of indeterminacy; *c*, three degrees of indeterminacy.

Fig. 11.15a only by the change to a pin support at joint D. Its presence increases the degree of statical indeterminacy to 2. As a consequence, two redundant forces must be selected prior to generating compatibility conditions. The pair of reactions at joint D are chosen and labeled R_1 and R_2. In this way, the vertical reaction is numbered as before, and the introduction of R_2 is an extension of the preceding problem. Both redundant reactions must be released to initiate the analysis.

Compatibility equations are generated from

the condition that the vertical displacement and the horizontal displacement created at D by release of the redundants must both be restored to a position of zero motion. This is accomplished by the separate application of the two unit-auxiliary loads shown in Fig. 11.15b. The first unit load acts in the direction of redundant R_1 and causes a vertical displacement, d_{11}, and a horizontal displacement, d_{21}, at joint D. The second unit load acts in the direction of redundant R_2 and causes a vertical displacement, d_{12}, and a horizontal displacement, d_{22}, at joint D.

Superposition of the displacements in the released structure and the unit-load cases yields the two compatibility conditions

$$D_1 + R_1d_{11} + R_2d_{12} = 0$$

$$D_2 + R_1d_{21} + R_2d_{22} = 0 \quad (11.13)$$

Displacements created by the unit loadings have been scaled by the appropriate redundant reaction. Computation of the six required displacements by ordinary determinate frame analysis and substitution into Eq. 11.13 permits a simultaneous solution for the redundant reactions. Completion of the analysis is accomplished by the superposition shown in Fig. 11.15b or by direct computation (determinate frame analysis) applied to the original structure, once R_1 and R_2 are known.

It should be noted that the second auxiliary loading and computation of the displacements at joint D produced by this load are not the only additional features beyond those included in the analysis of the 1-degree indeterminate frame in Fig. 11.15a. The computation of the horizontal displacement, D_2, in the released structure and of its counterpart, d_{21}, caused by the first auxiliary-unit load, which are unnecessary in the preceding problem, are required in the present analysis.

To increase the complexity of the preceding problem, consider the effect of changing joint D to a fixed support (see Fig. 11.15c). This alteration introduces a third redundant force, the moment resistance at D, resulting in a frame that is statically indeterminate to the third degree. The implication of this third redundant is the need for three compatibility equations. This need is accommodated by use of a third unit-auxiliary loading as indicated in Fig. 11.15c in addition to those used in Fig. 11.15b. In this loading, a unit moment is applied at joint D in the direction assumed for the redundant moment, R_3. Each of the three auxiliary loads are depicted in Fig. 11.15c. In each case, a vertical displacement, horizontal displacement, and a rotation are induced at joint D. The nine displacements are labeled d_{11} through d_{33} on the appropriate frames. When the displacements in

each of these directions are scaled by the proper redundant force and combined with the corresponding displacements from the released structure, three compatibility equations are obtained. Examining the various displacement quantities shown in Fig. 11.15c and superimposing their effects gives

$$\boxed{D_1 + R_1d_{11}} + R_2d_{12}\boxed{} + R_3d_{13} = 0$$

$$D_2 + R_1d_{21} + R_2d_{22} + R_3d_{23} = 0$$

$$D_3 + R_1d_{31} + R_2d_{31} + R_3d_{33} = 0$$

The conditions stated in Eqs. 11.14 correspond, in the order presented, to the need for zero vertical displacement, zero horizontal displacement, and zero rotation, respectively, at joint D in the indeterminate frame. After computation of the 12 displacement quantities and substitution into Eqs. 11.14, a simultaneous solution yields the values of the redundant reactions.

At this point, it is useful to examine Eqs. 11.12 and 11.13 and observe that each is a subset of Eq. 11.14. To dramatize this observation, the broken and solid rectangles in Eq. 11.14 contain the left side of Eq. 11.12 and the left side of Eq. 11.13, respectively. Indeed, solving the third-degree indeterminate problem inherently produces the displacement quantities needed in the other two problems. Earlier comments regarding the growth in the amount of computational effort are also evident in Eqs. 11.12 through 11.14.

Example 11.7.   Determine the redundant reactions for the frame depicted in Fig. 11.16a. Assume EI is constant, and use the superposition shown in Fig. 11.15c.

Solution:
Step 1:   Reactions forces at joint D are selected as the redundants and released to produce the displacements D_1, D_2, and D_3 shown in Fig. 11.15c.

Step 2:   Unit auxiliary forces are applied individually in the directions of the selected redundants. Displacements $d_{11}, d_{21} \cdots d_{33}$ are created as shown in Fig. 11.15c.

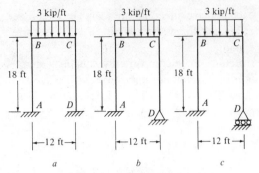

Fig. 11.16.   Indeterminate frames.

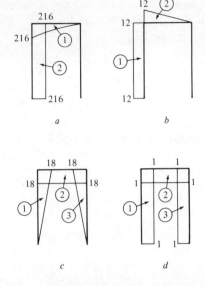

Fig. 11.17.   Moment diagrams for Example 11.7. *a, M; b, m; c, m; d, m.*

Step 3:   The 12 displacement values are calculated by a suitable method. The unit-load method, conjugate-frame method, or the angle-change method are suitable choices. Because joint A is fixed in the released structure, the angle-change principle is chosen for use. Moment diagrams corresponding to each of the determinate frames in Fig. 11.15c are drawn in Fig. 11.17. These diagrams are used as follows
From Fig. 11.17a:

Angle Change	$\phi = M/EI$ area	x	y	ϕx	ϕy
1	$\left(\dfrac{216}{EI}\right)(18) = \dfrac{3888}{EI}$ cw	12	9	$\dfrac{46{,}656}{EI}$ ↓	$\dfrac{34{,}992}{EI}$ ←
2	$\dfrac{1}{3}\left(\dfrac{216}{EI}\right)(12) = \dfrac{864}{EI}$ cw	9	18	$\dfrac{7760}{EI}$ ↓	$\dfrac{15{,}552}{EI}$ ←
	$\Sigma = D_3 = \dfrac{4752}{EI}$ cw			$\Sigma = D_1 = \dfrac{54{,}432}{EI}$ ↓	$\Sigma = D_2 = \dfrac{50{,}544}{EI}$ ←

From Fig. 11.17b:

Angle Change	$\phi = M/EI$ area	x	y	ϕx	ϕy
1	$\left(\dfrac{13}{EI}\right)(18) = \dfrac{216}{EI}$ ccw	12	9	$\dfrac{2592}{EI}$ ↑	$\dfrac{1994}{EI}$ →
2	$\dfrac{1}{2}\left(\dfrac{12}{EI}\right)(12) = \dfrac{72}{EI}$ ccw	8	18	$\dfrac{576}{EI}$ ↑	$\dfrac{1296}{EI}$ →
	$\Sigma = d_{31} = \dfrac{288}{EI}$ ccw			$\Sigma = d_{11} = \dfrac{3168}{EI}$ ↑	$\Sigma = d_{21} = \dfrac{3240}{EI}$ →

From Fig. 11.17c:

Angle Change	$\phi = M/EI$ area	x	y	ϕx	ϕy
1	$\dfrac{1}{2}\left(\dfrac{18}{EI}\right)(18) = \dfrac{162}{EI}$ cw	12	12	$\dfrac{1944}{EI}$ ↓	$\dfrac{1944}{EI}$ ←
2	$\left(\dfrac{18}{EI}\right)(12) = \dfrac{216}{EI}$ cw	6	18	$\dfrac{1296}{EI}$ ↓	$\dfrac{3888}{EI}$ ←
3	$\dfrac{1}{2}\left(\dfrac{18}{EI}\right)(18) = \dfrac{162}{EI}$ cw	0	12	0	$\dfrac{1944}{EI}$ ←
	$\Sigma = d_{32} = \dfrac{540}{EI}$ cw			$\Sigma = d_{12} = \dfrac{3240}{EI}$ ↓	$\Sigma = d_{22} = \dfrac{7776}{EI}$ ←

From Fig. 11.17d:

Angle Change	$\phi = M/EI$ area	x	y	ϕx	ϕy
1	$\left(\dfrac{1}{EI}\right)(18) = \dfrac{18}{EI}$ cw	12	9	$\dfrac{216}{EI}$ ↓	$\dfrac{162}{EI}$ ←
2	$\left(\dfrac{1}{EI}\right)(12) = \dfrac{12}{EI}$ cw	6	18	$\dfrac{72}{EI}$ ↓	$\dfrac{216}{EI}$ ←
3	$\left(\dfrac{1}{EI}\right)(18) = \dfrac{18}{EI}$ cw	0	9	0	$\dfrac{162}{EI}$ ←
	$\Sigma = d_{33} = \dfrac{48}{EI}$ cw			$\Sigma = d_{13} = \dfrac{288}{EI}$ ↓	$\Sigma = d_{23} = \dfrac{540}{EI}$ ←

Step 4: Equations 11.14 state the compatibility requirements that all displacements at joint D should be zero. Substituting the values obtained in step 3 gives

$$\underset{\downarrow}{\frac{53{,}432}{EI}} + R_1\underset{\uparrow}{\left(\frac{3168}{EI}\right)} + R_2\underset{\downarrow}{\left(\frac{3240}{EI}\right)} + R_3\underset{\downarrow}{\left(\frac{288}{EI}\right)} = 0$$

$$\underset{\leftarrow}{\frac{50{,}544}{EI}} + R_1\underset{\rightarrow}{\left(\frac{3240}{EI}\right)} + R_2\underset{\leftarrow}{\left(\frac{7776}{EI}\right)} + R_3\underset{\leftarrow}{\left(\frac{540}{EI}\right)} = 0$$

$$\underset{\substack{\text{cw}}}{\frac{4752}{EI}} + R_1\underset{\substack{\text{ccw}}}{\left(\frac{288}{EI}\right)} + R_2\underset{\substack{\text{cw}}}{\left(\frac{540}{EI}\right)} + R_3\underset{\substack{\text{cw}}}{\left(\frac{48}{EI}\right)} = 0$$

Declaring displacements that are upward or to the left as positive and clockwise rotations as positive, yields (after dropping EI from all terms):

$$\begin{Bmatrix} -54{,}432 \\ 50{,}544 \\ 4752 \end{Bmatrix} + \begin{bmatrix} 3168 & -3240 & -288 \\ -3240 & 7776 & 540 \\ -288 & 540 & 48 \end{bmatrix} \begin{Bmatrix} R_1 \\ R_2 \\ R_3 \end{Bmatrix} = \begin{Bmatrix} 0 \\ 0 \\ 0 \end{Bmatrix}$$

Values of the redundants are calculated by inversion.

$$\begin{Bmatrix} R_1 \\ R_2 \\ R_3 \end{Bmatrix} = \begin{bmatrix} 3168 & -3240 & -288 \\ -3240 & 7776 & 540 \\ -288 & 540 & 48 \end{bmatrix}^{-1} \begin{Bmatrix} 54{,}432 \\ -50{,}544 \\ -4752 \end{Bmatrix}$$

$$= \begin{bmatrix} 0.0006944 & 0 & 0.004167 \\ 0 & 0.0005879 & -0.006614 \\ 0.004167 & -0.006614 & 0.1202 \end{bmatrix} \begin{Bmatrix} 54{,}432 \\ -50{,}544 \\ -4752 \end{Bmatrix}$$

$$= \begin{Bmatrix} +18.00 \\ +1.71 \\ -10.13 \end{Bmatrix} = \begin{Bmatrix} 18.00 \text{ kips } \uparrow \\ 1.71 \text{ kips } \leftarrow \\ 10.13 \text{ kip-ft ccw} \end{Bmatrix}$$

Example 11.8. Repeat Example 11.7 for the frame of Fig. 11.16b. Use the superposition shown in Fig. 11.15b. The frame is identical to that of Fig. 11.16a except for the pin support at D.

Solution: With the reactions at joint D specified as the redundants, this problem is a subset of Example 11.7. Compatibility equations have the form of Eqs. 11.13, and numerical values are directly extracted from the preceding example.

$$\begin{Bmatrix} -54{,}432 \\ -50{,}544 \end{Bmatrix} + \begin{bmatrix} 3168 & -3240 \\ -3240 & 7776 \end{bmatrix} \begin{Bmatrix} R_1 \\ R_2 \end{Bmatrix} = \begin{Bmatrix} 0 \\ 0 \end{Bmatrix}$$

Solution by inversion yields

$$\begin{Bmatrix} R_1 \\ R_2 \end{Bmatrix} = \begin{bmatrix} 3168 & 3240 \\ -3240 & 7776 \end{bmatrix}^{-1} \begin{Bmatrix} 54{,}432 \\ 50{,}544 \end{Bmatrix}$$

$$= \begin{bmatrix} 0.0005501 & 0.0002292 \\ 0.0002292 & 0.0002241 \end{bmatrix} \begin{Bmatrix} 54{,}432 \\ -50{,}544 \end{Bmatrix}$$

$$= \begin{Bmatrix} 18.36 \\ 1.15 \end{Bmatrix} \equiv \begin{Bmatrix} 18.36 \text{ kips } \uparrow \\ 1.15 \text{ kips } \leftarrow \end{Bmatrix}$$

Example 11.9. Repeat Example 11.7 for the frame of Fig. 11.16c. Use the superposition shown in Fig. 11.15a. The frame is identical to that of Fig. 11.16b except for the roller support at c.

Solution: By simple extraction from Examples 11.6 or 11.7, the sole compatibility equation is

$$-54{,}432 + R_1(3168) = 0$$

$$R_1 = +17.18 = 17.18 \text{ kips } \uparrow$$

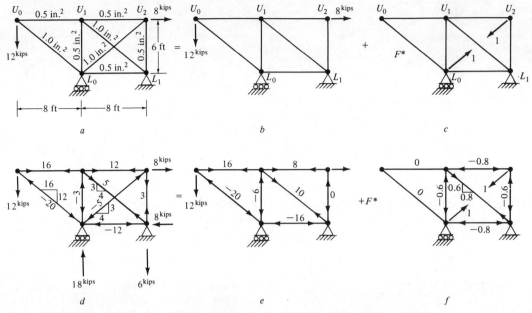

Fig. 11.18.   Indeterminate truss analysis by the flexibility method.

11.3.4   Planar Trusses

Truss analysis by the flexibility method is schematically indicated in Figs. 11.18a through c. The truss is the same as that given in Fig. 9.6a except for the addition of member L_0U_2, which makes the system statically indeterminate internally to the first degree. First, indeterminacy is eliminated by removing the redundant member L_0U_2. This results in a *relative* motion, D_1, between joints L_0 and U_2 under the action of the applied loads shown in Fig. 11.18b. Such motion would be incompatible with the condition of Fig. 11.18a, in which joints L_0 and U_2 are partially restrained by the presence of member L_0U_2. Compatibility can be restored by superposition of the deformation effects of the truss in Fig. 11.18b with the deformation effects of the truss shown in Fig. 11.18c. In the latter figure, F is the force of member L_0U_2 pulling on the joints. Because the value of F is unknown, the compatibility condition must be expressed as

$$D_1 + Fd_{11} = \frac{FL}{EA} \qquad (11.15)$$

in which d_{11} is the relative motion caused by a pair of unit loads pulling on joints L_0 and U_2.

The term on the right-hand side is the elongation of bar L_0U_2 under the action of the force F. In effect, Eq. 11.15 states that the relative motion between joints L_0 and U_2 must equal the change in length of member L_0U_2.

Example 11.10.   Determine the member forces in the truss of Fig. 11.18a. Use the flexibility method. $E = 30,000$ kip/in.^{2}.

Solution: The problem can be solved in the following steps:

1. Analyze the trusses given in Fig. 11.18b and c.
2. Determine the redundant member force, F, by applying Eq. 11.15.
3. With F determined, execute the multiplication shown in Fig. 11.18c, and combine the results with those obtained in Fig. 11.18b.

Step 1:   Member forces for the truss of Fig. 11.18b were determined in Example 9.1. Results are repeated in Fig. 11.18e. Member forces for the truss in Fig. 11.18c are shown in Fig. 11.18f.

Step 2: D_1 is determined by Eq. 9.4. The Δ_i values are the member deformations for the truss of Fig. 11.18b and were determined in Example 9.1. The f_1 values are obtained from Fig. 11.18f. Computation of D_1 was accomplished in Example 9.2, giving

$$D_1 = 0.0982 \text{ in. together}$$

Displacement d_{11} is the relative motion between L_0 and U_2 caused by the unit loading shown in Fig. 11.18f and is also computed by employing Eq. 9.4. In applying Eq. 9.4, the Δ_i values are the member deformations for the truss shown in Fig. 11.18f.

Member	F_i (kips)	L_i (in.)	A_i (in.²)	Δ_i (in.)
U_1U_2	-0.8	96	0.5	$-153.6/E$
U_1L_0	-0.6	72	0.5	$-86.4/E$
U_2L_1	-0.6	72	0.5	$-86.4/E$
L_0L_1	-0.8	96	0.5	$-153.6/E$
U_1L_1	$+1.0$	120	1.0	$+120.0/E$

In this case, the unit auxiliary load is the same as the actual loads, i.e., the f_i and F_i forces are identical. Consequently,

$$1 \cdot d_{11} = \sum_{i=1}^{7} f_i \cdot \Delta_i$$

$$d_{11} = (-0.8)\left(\frac{153.6}{E}\right) + (-0.6)\left(-\frac{86.4}{E}\right)$$

$$+ (-0.6)\left(-\frac{86.4}{E}\right)$$

$$+ (-0.8)\left(-\frac{153.6}{E}\right)$$

$$+ (+1.0)\left(+\frac{120}{E}\right)$$

$$= +\frac{469.4}{E} = \frac{469.4}{30,000}$$

$$= 0.0156 \text{ in./kip together} \qquad (a)$$

Compatibility requires:

$$D_1 + Fd_{11} = \frac{FL}{EA} \qquad (b)$$

$$\underset{\text{together}}{0.0982} + \underset{\text{together}}{F(0.0156)} = \underset{\text{apart}}{\frac{F(120)}{(30,000)(1)}}$$

Care must be taken in assessing the sense of the motion for the term on the right-hand side. Since the *redundant* member force was assumed to be in tension, this implies that the *net* motion of joints U_0 and L_2 must be apart (otherwise the member would have to be in compression). Arbitrarily calling "together" motions positive:

$$0.0982 + F(0.0156) = -0.004F$$

$$F = -5.01 \text{ kips} = 5.01 \text{ kips compression}$$

Step 3: Member forces for the truss of Fig. 11.18a are computed by the addition shown schematically in Figs. 11.18e and f, e.g.,

$$U_1U_2 = +8 + F(-0.8)$$

$$= +8 + (-5.01)(-0.8)$$

$$= +12.0 \text{ kips} = 12.0 \text{ kips tension}$$

The complete set of results is shown in Fig. 11.18d. Reactions have been computed by the visual method of joints and are in agreement with equilibrium conditions for the truss taken as a whole.

Example 11.11.   Outline the procedure for analyzing the truss shown in Fig. 11.19a by the flexibility method. Assume a linear, elastic material, and define all terms that appear in the developed equations.

Solution:
Step 1:   Three degrees of statical indeterminacy exist in the truss. It is necessary to choose three redundant quantities in a manner that renders the structure statically determinate while maintaining stability. Any three of the reactions may be chosen, and, in this case, the three interior reactions R_1, R_2, and R_3 are selected.

Step 2:   With the redundant quantities removed, the truss appears as shown in Fig. 11.19b. With the actual loads (P_1-P_{10}) from Fig. 11.19a acting on the determinate structure, the released joints will experience displacements. In Fig. 11.19b, these vertical displacements have been designated D_1, D_2, and D_3. In the actual truss, the supports do not move, which is the basis for establishing three compatibility conditions. Three auxiliary unit loadings, one per redundant, are needed to make a solution possible.

Step 3:   The first auxiliary loading is shown in Fig. 11.19c and consists of a unit force applied at the location of and in the direction of the redundant, R_1. The vertical displacements caused at the released joints by this auxiliary loading have been designated d_{11}, d_{21}, and d_{31}.

Step 4:   Figure 11.19d depicts the second auxiliary loading. Due to the unit force applied at the location of and in the direction of redundant R_2, the released

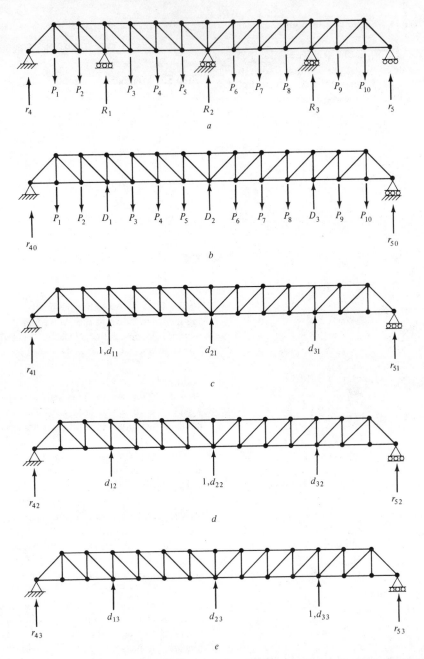

Fig. 11.19. Flexibility method of truss with multiple redundants. *a,* truss indeterminate to third degree; *b,* truss made determinate by removal of redundants; *c,* first auxiliary load; *d,* second auxiliary load; *e,* third auxiliary load.

joints will displace vertically by the amounts designated as d_{12}, d_{22}, and d_{32}.

Step 5: The third auxiliary loading is shown in Fig. 11.19e and consists of a unit force at the location of and in the direction of the third redundant, R_3. Ver-

tical displacements that result at the released joints are designated as d_{13}, d_{23}, and d_{33}.

Step 6: Compatibility conditions require that the interior supports in Fig. 11.19a do not displace vertically. Superposition allows one to obtain the response for the truss in Fig. 11.19a by linearly combining the trusses in Figs. 11.19b through e in such a way as to satisfy these compatibility conditions. After multiplying the response of the trusses in Fig. 11.19c through e by R_1, R_2, and R_3, respectively, the desired superposition is achieved by the compatibility

$$D_1 + R_1 d_{11} + R_2 d_{12} + R_3 d_{13} = 0 \qquad \text{(a)}$$

$$D_2 + R_1 d_{22} + R_2 d_{22} + R_3 d_{23} = 0 \qquad \text{(b)}$$

$$D_3 + R_1 d_{31} + R_2 d_{32} + R_3 d_{33} = 0 \qquad \text{(c)}$$

or

$$D_i + \sum_{j=1}^{3} R_j d_{ij} = 0 \qquad i = 1, 3 \quad (11.16)$$

With the 12 displacement quantities known, the redundant forces are obtained by a simultaneous solution of Eqs. 11.16. The *flexibility coefficient* d_{ij} constitutes a displacement in direction i, caused by a unit force acting in direction j.

Step 7: Each of the 12 displacement quantities can be computed by the unit-load method according to

$$1 \cdot X = \Sigma f_i \cdot \Delta_i$$

with the summation executed according to the schedule given in Table 11.2. X is a generic symbol for any of the desired joint displacements. From the computations indicated in Table 11.2, the flexibility coefficients exhibit symmetry, i.e., $d_{ij} = d_{ji}$. The symmetry of the flexibility coefficients is a *general* occurrence. Because of symmetry, only 9 displacement quantities require actual computation.

Step 8: Member forces in the actual truss can be determined by the superposition:

$$F_i = F_{i0} + R_1 f_{i1} + R_2 f_{i2} + R_3 f_{i3}$$

or

$$F_i = F_{i0} + \sum_{j=1}^{3} R_j f_{ij} \qquad i = 1, NF \quad (11.17)$$

where

NF = number of member forces
f_{ij} = force in the ith member of the determinate truss subjected to the jth auxiliary loading.

Table 11.2. Computation of Flexibility Coefficients for Ex. 11.11

To determine	Obtain Δ_i from Fig.	Obtain f_i from Fig.
D_1	11.19b	11.19c
D_2	11.19b	11.19d
D_3	11.19b	11.19e
d_{11}	11.19c	11.19c
d_{21}	11.19c	11.19d
d_{31}	11.19c	11.19e
d_{12}	11.19d	11.19c
d_{22}	11.19d	11.19d
d_{32}	11.19d	11.19e
d_{13}	11.19e	11.19c
d_{23}	11.19e	11.19d
d_{33}	11.19e	11.19e

F_i = force in the ith member of the actual truss, Fig. 11.19a.
F_{i0} = force in the ith member of the truss in Fig. 11.19b.
R_j = value of the jth redundant reaction.

Similarly, reactions other than the redundants are determined by

$$r_i = r_{i0} + R_1 r_{i1} + R_2 r_{i2} + R_3 r_{i3}$$

or

$$r_i = r_{i0} + \sum_{j=1}^{3} R_j r_{ij} \qquad i = 1, NNR \quad (11.18)$$

where

NNR = number of non-redundant reactions
r_i = value of the ith nonredundant reaction in the indeterminate truss when subjected to the actual loading (Fig. 11.19a).
r_{i0} = value of the ith nonredundant reaction in Fig. 11.19b.
r_{ij} = value of the ith nonredundant reaction in the jth auxiliary loading.

11.3.5   General Equations

Equations 11.16 through 11.18 developed in Example 11.11 are applicable to any structure that is statically determinate externally to the third degree. It is useful to introduce general equations applicable to structures that are statically indeterminate to the nth degree, regardless of the nature of the indeterminacy. General compatibility relationships for the flexibility method are given by

$$D_i + \sum_{j=1}^{I} R_j d_{ij} = d_{i0} \quad i = 1, I \quad (11.19)$$

which, when $I = 3$, is equivalent to Eq. 11.16, except the right-hand side is nonzero. This accommodates situations, such as removal of a member force as a redundant (Example 11.10), for which the net displacement at the location of a redundant is not zero. In Eq. 11.19,

I = degree of statical indeterminacy.
D_i = displacement in the direction of the ith redundant in the determinate (all redundants are released) structure when subjected to the actual loading.
R_j = unknown value of the jth redundant force in the indeterminate structure.
d_{ij} = displacement in the direction of the ith redundant in the determinate structure when subjected to the jth auxiliary loading.
d_{i0} = net displacement in the direction of the ith redundant force in the indeterminate structure.

In using Eq. 11.19, it is important that a consistent sign convention be used for each redundant and the corresponding displacement quantities in the released state and the auxiliary unit load states. Equation 11.19 serves to determine the redundant forces. With these known, the force in the i-th member is determined by the superposition of forces indicated in Eq. 11.20:

$$F_i = F_{i0} + \sum_{j=1}^{I} R_j f_{ij} \quad i = 1, NF \quad (11.20)$$

where

NF = number of member forces in the in determinate structure.
F_i = force in the ith member of the indeterminate structure when subjected to the actual loading.

F_{i0} = force in the ith member of the determinate structure when subjected to the actual loading.
F_{ij} = force in the ith member of the determinate structure when subjected to the jth auxiliary loading.

Reactions at supports that are not included in the redundants are determined from Eq. 11.21:

$$r_i = r_{i0} + \sum_{j=1}^{I} R_j r_{ij} \quad i = 1, NR - I$$

$$(11.21)$$

where

NR = number of external reactions in the indeterminate structure.
r_i = value of the ith nonredundant reaction in the indeterminate structure when subjected to the actual loading.
r_{i0} = value of the ith nonredundant reaction in the determinate structure when subjected to the actual loading.
r_{ij} = value of the ith nonredundant reaction in the determinate structure when subjected to the jth auxiliary loading.

If $I = 3$, Eqs. 11.20 and 11.21 are equivalent to Eqs. 11.17 and 11.18, respectively.

11.3.6 Selection of Redundants

Usually, several choices for selection of the redundants exist in any given problem. In selecting redundants, two conditions must be satisfied. First, the number of redundants must equal the degree of statical indeterminacy. Second, when the redundants are removed, the resulting structure must be stable.

Both conditions require a thorough understanding of the determination of statical indeterminacy and stability of framed structures. These topics have been thoroughly covered in earlier sections of this book. Example 11.10 involved a truss that was statically indeterminate internally to the first degree. Thus, no external reaction can be used as a redundant. Further, some members cannot serve as redundants. Examples 11.3, 11.4 and 11.5 illustrated the use of different redundants for the same problem. Example 11.4 is probably the preferable choice. By choosing the end moment as the redundant, it is only necessary to compute a rotation. This is a bit less effort than required to compute a displacement, as was done in Example 11.3. The choice also avoids the introduction of an internal hinge (and the discontinuity associated with it) which was necessary in Example 11.5. While all three choices produce the same results, differences in either the length or the complexity of the solution exist. These differences are compounded when multiple redundants are involved. Thus, the analyst should exercise forethought when choosing redundants.

11.4 SYMMETRY OF FLEXIBILITY COEFFICIENTS

In general, symmetry exists in the flexibility coefficients when arranged in a matrix form. This was observed for a beam in Example 11.6, for planar frames in Examples 11.7 and 11.8, and for a truss in Example 11.11. A general proof of this condition is appropriate here.

The influence coefficient d_{ij} represents a displacement in a direction i, as signified by the first subscript, due to the application of a unit force to some structure. The second subscript, j, is used to distinguish the line of action of the unit load. Similarly, the flexibility coefficient d_{ji} is a displacement in the direction j caused by a unit load acting in the direction i on the same structure. The two separate loadings, which create d_{ij} and d_{ji}, are interrelated by Maxwell's reciprocal theorem (see Section 8.5). A mathematical statement of the theorem is

$$F_i \Delta_j = F_j \Delta_i \qquad (11.22)$$

F_i is the load that causes Δ_i, and F_j is the loading that creates Δ_j. In terms of the present discussion, both the F_i and F_j loads have a magnitude of one. Taking Δ_i as the displacement d_{ij} caused by the unit load F_i, and taking Δ_j as the displacement d_{ji} caused by the unit load F_j, yields

$$d_{ij} = d_{ji} \qquad (11.23)$$

Directions i and j can be chosen arbitrarily, which demonstrates that the symmetry of the flexibility coefficients is general.

An alternative proof of the relationship given in Eq. 11.23 can be developed by the use of virtual work concepts. It is useful to do this for the specific case of planar frames.

Consider any given frame. Let M_i be the moment diagram created by a single concentrated load in the direction i. Moment diagram, M_i, leads to a specific displaced shape and, in particular, causes a displacement, D_{ji}, in the direction j. The magnitude of D_{ji} is computed by the virtual work approach (unit-load method) as

$$D_{ji} = \int_0^L \frac{M_i m_j}{EI} \, dl \qquad (11.24)$$

In accordance with the virtual work concept, m_j is the moment diagram caused by a unit-auxiliary load applied in the direction j. The integration limits in this and all succeeding integrals in this section signify integrating "over the length of all members in the frame."

Consider an alternate loading applied to the same structure. Let M_j be the moment diagram created by a single concentrated load in the direction j. Moment diagram M_j leads to a different displaced shape for the structure. In this case, the displacement, D_{ij}, in the direction i is of interest. Its magnitude is also determined by the virtual work approach as

$$D_{ij} = \int_0^L \frac{M_j m_i}{EI} \, dl \qquad (11.25)$$

The term m_i represents the moment diagram caused by a unit-auxiliary load applied in the direction i.

Equations 11.24 and 11.25 permit computation of displacements created by moment diagrams M_i and M_j, respectively. If the concentrated loads that cause these moment diagrams were to have unit value, the displacements D_{ij} and D_{ji} would constitute flexibility coefficients d_{ij} and d_{ji}, respectively. More importantly, the moment diagrams M_i and m_i would be identical. Similarly, the equality of the moment diagrams M_j and m_j would be evident. Under these circumstances, Eqs. 11.24 and 11.25 become

$$d_{ij} = \int_0^L \frac{m_j m_i}{EI} \, dl \qquad (11.26)$$

and

$$d_{ji} = \int_0^L \frac{m_i m_j}{EI} \, dl \qquad (11.27)$$

The identity of d_{ij} and d_{ji} is clearly seen by comparison of these two equations. This second proof is presented to demonstrate the applicability of the unit load method to the calculation of flexibility coefficients for frames. Equation 11.26 (or, equivalently, Eq. 11.27) is the basis of the required computations. Because the m_i and m_j moment diagrams result from single unit loads, they are simple in form, and the implied integration is not tedious.

Example 11.12. Determine the flexibility coefficients for the frame of Example 11.7 using the unit-load method. Also determine the displacements D_1, D_2, D_3.

Solution: Using the moment diagrams in Fig. 11.17 (no additional diagrams are needed), the necessary integrals are formed and evaluated below. It should be noted that the reference point for the girder moment diagrams is taken at the right end of the member.

$$d_{11} = \int_0^L \frac{m_1 m_1}{EI} \, dl$$

$$= \int_0^{18} \frac{(12)^2}{EI} \, dx + \int_0^{12} \frac{(x)^2}{EI} \, dx$$

$$= \frac{3168}{EI}$$

$$d_{22} = \int_0^L \frac{m_2 m_2}{EI} \, dl$$

$$= 2 \int_0^{18} \frac{(x)^2}{EI} \, dx + \int_0^{12} \frac{(18)^2}{EI} \, dx$$

$$= \frac{7776}{EI}$$

$$d_{33} = \int_0^L \frac{m_3 m_3}{EI} \, dl$$

$$= 2 \int_0^{18} \frac{(1)^2}{EI} \, dx + \int_0^{12} \frac{(1)^2}{EI} \, dx$$

$$= \frac{48}{EI}$$

$$d_{12} = d_{21} = \int_0^L \frac{m_1 m_2}{EI} \, dl$$

$$= \int_0^{18} \frac{(12)(-x)}{EI} \, dx$$

$$+ \int_0^{18} \frac{(x)(-18)}{EI} \, dx$$

$$= -\frac{3240}{EI}$$

$$d_{13} = d_{31} = \int_0^L \frac{m_1 m_3}{EI} \, dx$$

$$= \int_0^{18} \frac{(12)(-1)}{EI} \, dx$$

$$+ \int_0^{12} \frac{(x)(-1)}{EI} \, dx$$

$$= -\frac{288}{EI}$$

$$d_{23} = d_{32} = \int_0^L \frac{m_2 m_3}{EI} \, dl$$

$$= \int_0^{18} \frac{(-x)(-1)}{EI} \, dx$$

$$+ \int_0^{12} \frac{(-18)(-1)}{EI} \, dx$$

$$= \frac{540}{EI}$$

The second moment term in each integral is due to the applicable *auxiliary*-unit load. The directions assumed for the various unit loads determines the directions of the calculated displacement. Recognizing this and referring to Fig. 11.15, the proper directions can be established for each displacement.

$$d_{11} = \frac{3168}{EI} \uparrow \quad d_{12} = \frac{3240}{EI} \rightarrow \quad d_{13} = \frac{288}{EI} \text{ ccw}$$

$$d_{21} = \frac{3240}{EI} \downarrow \quad d_{22} = \frac{7776}{EI} \leftarrow \quad d_{23} = \frac{540}{EI} \text{ cw}$$

$$d_{31} = \frac{288}{EI} \downarrow \quad d_{32} = \frac{540}{EI} \leftarrow \quad d_{33} = \frac{48}{EI} \text{ cw}$$

Again referring to Fig. 11.17,

$$D_1 = \int_0^L \frac{Mm_1}{EI} \, dx = \int_0^{18} \frac{(-216)(12)}{EI} \, dx$$

$$+ \int_0^{12} \frac{(-1.5 \, x^2)(x)}{EI} \, dx$$

$$= -\frac{54{,}432}{EI} = \frac{54{,}432}{EI} \downarrow$$

$$D_2 = \int_0^L \frac{Mm_2}{EI} \, dx = \int_0^{18} \frac{(-216)(-x)}{EI} \, dx$$

$$+ \int_0^{12} \frac{(-1.5 \, x^2)(-18)}{EI} \, dx$$

$$= \frac{50{,}544}{EI} = \frac{50{,}544}{EI} \leftarrow$$

$$D_3 = \int_0^L \frac{Mm_3}{EI} \, dx = \int_0^{18} \frac{(-216)(1)}{EI} \, dx$$

$$+ \int_0^{12} \frac{(-1.5 \, x^2)(-1)}{EI} \, dx$$

$$= \frac{4752}{EI} = \frac{4752}{EI} \text{ cw}$$

Computed values and directions are in agreement with Example 11.7.

11.5 ALTERNATIVE APPROACH FOR PLANAR TRUSSES

In Example 11.10, it was necessary to carefully assess the direction of motion and signs for the terms in the compatibility equation (Eq. b in that example.) In particular, the term on the right side of the equality required caution. This assessment can be avoided by a modification to the analysis.

Example 11.13. Solve Example 11.10 by an alternative approach.

Solution: Consider the superposition shown in Fig. 11.20. Contrary to the previous solution the redundant member is not removed. Instead, it is cut at some point length, so as to create an infinitesimal gap.

Application of load (Fig. 11.20b) causes the gap to close by an amount

$$D_1 = .0982 \text{ in. together}$$

as calculated in Example 11.10. Since the joints move together, so does the gap.

In Fig. 11.20c, a unit value of the redundant member force is applied to the member ends adjacent to the gap (as opposed to "at the joints" as was done in Example 11.10). The movement of the gap, d_{11}, is calculated as

$$1 \cdot d_{11} = \sum_{i=1}^{8} f_i \cdot \Delta_i \qquad \text{(a)}$$

Comparison of Eq. a of the present example with Eq. a of Example 11.10 indicates that the present summation is over 8 members. Unlike Example 11.10, the redundant member is included. Thus

$$d_{11} = .0156 + (1) \left(\frac{L}{EA} \right) \text{ together}$$

where the first term is obtained from Example 11.10. The second term is for the redundant member. Compatibility requires

$$D_1 + F_1 \, d_{11} = 0$$

$$D_1 + F \left(.0156 + \frac{L}{EA} \right) = 0$$

$$D_1 + F(.0156) + \frac{FL}{EA} = 0$$

$$\underset{\text{together}}{.0982} + \underset{\text{together}}{F \, (.0156)} + \underset{\text{together}}{\frac{F(120)}{30{,}000(1)}} = 0$$

$$.0982 + .0156F + .004F = 0$$

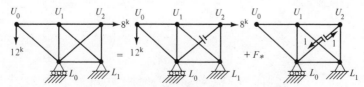

Fig. 11.20. Superposition for Example 11.13.

$$F = -5.01 \text{ kips}$$

$$= 5.01 \text{ kips tension}$$

Two points should be recognized. First, the right side of the compatibility equation is set to zero. This is because the net gap dimension must be zero, i.e., the member ends must stay connected. Second, the elongation of the redundant member is automatically on the left side and with the proper sign. This occurs because the member was included in the summation in Eq. (a) above. In other words,

$$d_{11} = .0156F + .004F = .0196F$$

11.6   SUPPORT SETTLEMENTS

Differential settlement of supports causes deformations and internal forces in the members of any statically indeterminate structure. This effect is easily accounted for in the flexibility method. No new concepts are needed. If a support settlement occurs in the direction of a redundant, Eqs. 11.19 are used, with the d_{10} value set equal to the applicable known settlement. Settlements that occur in the direction of nonredundant reactions are part of the loading effect for the statically determinate (released) structure.

Example 11.14.   Analyze the beam in Fig. 11.21a by the flexibility method. Support B settles 0.01 m. $EI = 48,000 \text{ kN m}^2$. Use support B as the redundant.

Solution: The superposition shown in Figs. 11.21b and c is applied. Reactions shown are computed by statics. In the released state subjected to the actual load (Fig. 11.21b), the settled support has been removed. Thus, nothing happens.

$$D_1 = 0$$

Due to a unit value of the redundant, the midspan displacement d_{11} is

$$d_{11} = \frac{L^3}{48EI} = \frac{(12)^3}{48\,(48,000)} = .00075 \text{ m/kN} \uparrow$$

as can be confirmed by any method (e.g., the unit-load method.)

Compatibility requires

$$D_1 + R_1\,d_{11} = d_{10}$$

$$0 + R_1(.00075 \uparrow) = (.01 \downarrow)$$

$$0 + R_1(.00075) = -.01$$

$$R = -13.33 \text{ kN}$$

$$= 13.33 \text{ kN} \downarrow$$

$$r_1 = r_{10} + R_1\,r_{11} = 0 + (-13.33)\,(-.5)$$

$$= 6.67 \text{ kN} \uparrow$$

$$r_2 = r_{20} + R_1\,r_{21} = 0 + (-13.33)\,(-.5)$$

$$= 6.67 \text{ kN} \uparrow$$

By direct analysis of Fig. 17.21a, the midspan moment is

$$M = 6r_1 = 6(6.67) = 40 \text{ kN} \cdot \text{m}$$

Example 11.15.   Repeat Example 11.14 except use the support at C as the redundant.

Solution: The superposition shown in Figs. 11.22b and c is applied. Reactions shown are computed by statics. The motion D_1 at C due to the settlement at support B is obtained by geometry. Because no external load is applied to the *statically determinate* beam, its reactions are zero and no internal moment develops. Therefore, it displaces as a rigid body and

$$D_1 = .02 \text{ m} \downarrow$$

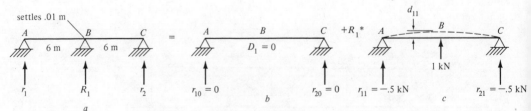

Fig. 11.21.   Beam for Example 11.14.

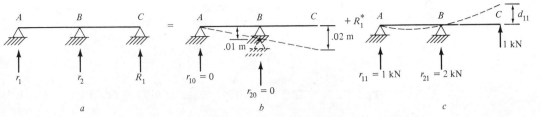

Fig. 11.22. Beam for Example 11.15.

In Fig. 11.22c, the displacement d_{11} is calculated by the moment-area method (computations not shown) as

$$d_{11} = \frac{L^3}{12EI} = .003 \text{ m/kN}$$

Compatibility requires

$$D_1 + R_1 d_{11} = d_{10}$$

$$.02 \downarrow + R_1(.003 \uparrow) = 0$$

$$-.02 + R_1(.003) = 0$$

$$R_1 = 6.67 \text{ kN} \uparrow$$

Therefore

$$r_1 = r_{10} + R_1 r_{11} = 0 + 6.67(1) = 6.67 \text{ kN} \uparrow$$

$$r_2 = r_{20} + R_1 r_{21} = 0 + 6.67(-2)$$

$$= -13.33 \text{ kN} = 13.33 \text{ kN} \downarrow$$

This confirms the results of Example 11.14.

11.7 INITIAL DEFORMATIONS

Fabrication errors and thermal expansion are sources of initial deformation in a member.

Application of the flexibility method to such problems is similar to the treatment of support settlements.

Example 11.16. Analyze the truss in Example 11.10 for the case where member $L_0 U_2$ is fabricated 0.05 in. too long.

Solution: The superposition in Fig. 11.23 is used. For Fig. 11.23b,

$$D_1 = 0.05 \text{ in. together (member ends overlap)}$$

For Fig. 11.23c, from Example 11.13,

$$d_{11} = .0196 \text{ in. together}$$

Thus, calling the redundant member force F,

$$D_1 + Fd_{11} = 0$$

$$.05 + F(.0196) = 0$$

$$F = -2.55$$

$$= 2.55 \text{ kips compression}$$

Since the member forces in Fig. 11.23b are zero, the member forces in the indeterminate truss are obtained by multiplying the forces in Fig. 11.18f by the calculated F. The results are shown in Fig. 11.24. The reactions can be computed similarly, but are known to be zero. (Why?)

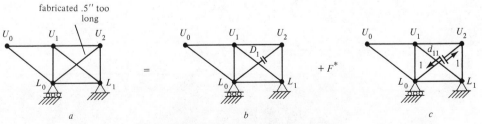

Fig. 11.23. Truss for Example 11.16.

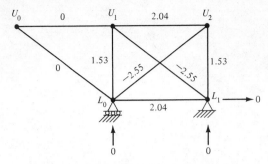

Fig. 11.24. Results for Example 11.16.

Note: The inquisitive reader should rework this problem for member U_1U_2 being fabricated 0.05 in. too long and member L_0U_2 as the redundant.

11.8 COMPOUND STRUCTURES

The preceding sections addressed the analysis of individual structures by the flexibility method. In many cases two or more structures are framed so as to act together. The flexibility method is particularly powerful for the analysis of such "compound structures" as illustrated by the following example problems.

Example 11.17. Two 12-meter-long beams cross each other at midspan, as shown in Fig. 11.25. Calculate the displacement under the 10 kN load.

Solution: An unknown reaction R, acts between the two members. It is determined by a compatibility

condition derived from the superposition shown in Figs. 11.25b and c. The two beams must displace equally. Therefore, assuming all displacements are in the same direction (despite knowing otherwise),

$$D_1 + Rd'_{11} = Rd''_{11} \qquad (a)$$

Any conventional method of deflection analysis will show the midspan displacement Δ of a beam loaded by a midspan concentrated load, P, is

$$\Delta = \frac{PL^3}{48EI}$$

Thus,

$$D_1 = \frac{10(12)^3}{48\,(3EI)} = \frac{120}{EI} \downarrow$$

$$d'_{11} = \frac{1(12)^3}{48\,(3EI)} = \frac{12}{EI} \uparrow$$

$$d''_{11} = \frac{1(12)^3}{48EI} = \frac{36}{EI} \downarrow$$

Substitution into Eq. a yields

$$\frac{120}{EI} \downarrow + R\left(\frac{12}{EI} \uparrow\right) = R\left(\frac{36}{EI} \downarrow\right)$$

$$-\frac{120}{EI} + R\left(\frac{12}{EI}\right) = R\left(-\frac{36}{EI}\right)$$

$$R = 2.5 \text{ kN pushing on both members}$$

and either

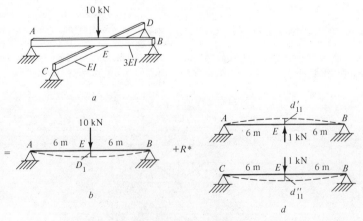

Fig. 11.25. Structure for Example 11.17.

$$\Delta_E = D_1 - Rd'_{11} = \frac{120}{EI} - (2.5)\left(\frac{12}{EI}\right) = \frac{90}{EI} \downarrow$$

or

$$\Delta_E = Rd''_{11} = 2.5\left(\frac{36}{EI}\right) = \frac{90}{EI} \downarrow$$

Example 11.18. Calculate the tension in the cable of the structure in Fig. 11.26a. For the beam, $EI = 320,000$ kip-ft². For the cable, $EA = 100,000$ kips.

(a) Neglect axial deformation in the beam.
(b) Include axial deformation in the beam with $EA = 160,000$ kips.

Solution:

Part (a): The superposition shown in Figs. 11.26b and c is used. Moment diagrams for each of these figures are given in Figs. 11.26d and e.

In Fig. 11.26b, the cable is removed. The displacement of B toward C is

$$D_1 = \int_0^{24} \frac{M(x)m(x)}{EI} dx = \int_0^{24} \frac{(-24x)(1x)}{320,000} dx$$

$$= -.06144 \text{ ft}$$

$$D_1 = .0614 \text{ ft apart}$$

Due to the unit load in Fig. 11.26c, the displacement of B toward C is

$$d_{11} = \int_0^{24} \frac{m(x)m(x)}{EI} dx$$

$$= \int_0^{24} \frac{(1\,x)^2}{320,000} dx = .001536 \text{ ft/kip together}$$

Compatibility requires

$$(D_1 + Td_{11})_{beam} = \left(\frac{TL}{EA}\right)_{cable}$$

$$\underset{\text{apart}}{.06144} + \underset{\text{together}}{T(.001536)} = \underset{\text{apart}}{\frac{T(20)}{\$00,000}}$$

$$-.06144 + .001536T = -.0002T$$

$$T = 35.39 \text{ kips tension}$$

Part (b): D_1 is unchanged, but

$$d_{11} = \int_0^{24} \frac{m(x)m(x)}{EI} dx + (f \cdot \Delta)_{beam}$$

where

f = horizontal component of the auxiliary load
$\quad = .80$ kip

$\Delta = (FL/EA)_{beam}$ due to the 1 kip value of T

$$= \frac{(.8)(16)}{160,000} = .000064 \text{ ft/kip together}$$

$d_{11} = .001536 + .000064 = .00160$ ft

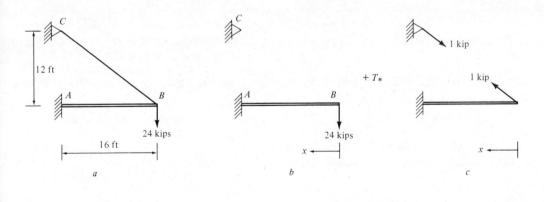

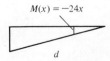

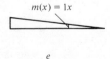

Fig. 11.26. Structure for Example 11.18.

Compatibility requires

$$-.06144 + .00160T = -.0002T$$

$$T = 34.13 \text{ kips}$$

11.9 FRAMED STRUCTURE COMBINED WITH A CONTINUUM

Sometimes structures are comprised of framed members combined with a continuum. The fluid tank depicted in Fig. 11.27 is an example. Similar to Example 11.18, the axial forces in the ties between the sidewalls could serve as redundants. The tank without the ties would need to be analyzed for the effect of the hydrostatic fluid pressure. In addition it would have to be analyzed separately for unit values of each unknown tie force. These analyses could be performed by computer using the "finite element method," which is an advanced method for analysis of continua.

Three compatibility conditions must be satisfied and are of the form

$$D_1 + T_1 d_{11} + T_2 d_{12} + T_3 d_{13} = \left(\frac{TL}{EA}\right)_1$$

$$D_2 + T_1 d_{21} + T_2 d_{22} + T_3 d_{23} = \left(\frac{TL}{EA}\right)_2$$

$$D_3 + T_1 d_{31} + T_2 d_{32} + T_3 d_{33} = \left(\frac{TL}{EA}\right)_3$$

where T_1 through T_3 are the unknown cable tensions. Displacements D_1, D_2, and D_3 are relative displacements between the sidewalls at each tie location due to hydrostatic pressure. Displacements d_{11} through d_{33} are flexibility coefficients (i.e., due to unit values of T_1. T_2,

and T_3 individually acting on the sidewalls. The terms on the right side of the equations are the cable deformations. Solution of the three simultaneous equations produces the cable tensions needed to numerically effect the superposition stage of the flexibility method.

11.10 COMPUTER PROGRAMS

Appendix B reviews matrix algebra for use in later chapters. Inversion of a matrix is useful for solving simultaneous equations. The programs, INVERSE.BAS (in BASIC) and INVERSE.F (in FORTRAN) are presented in Appendix C as tools for doing matrix inversion by the Gauss elimination method. They are valuable tools for the problem exercises in this chapter and the remainder of the text. Initially it is preferable to solve problems manually and check them by use of the computer.

PROBLEMS

Note: Do the following problems by the flexibility method. Unless stated otherwise, all truss members have the same EA value and all beam and frame members have the same EI value. If the unit-load method is used for trusses, tabulate the computations in the order of the circled numbers shown for the members.

Note: If previously solved by the reader, some of the problem exercises at the ends of earlier chapters involve determinate structures useful in the following problems. When this is the case a reference (see Problem n) is given.

11-1. Determine the degree and nature (internal and/or external) the statical indeterminacy for each of the given beams.

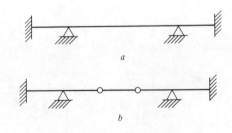

a

b

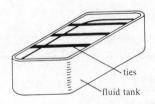

Fig. 11.27. Fluid tank with tied sidewalls.

(Problem 11-1 continued on p. 311)

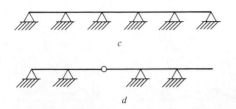

c

d

11-2. Repeat Problem 11-1 for the given planar frames.

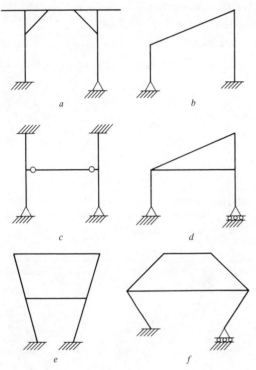

a

b

c

d

e

f

11-3. Repeat Problem 11-1 for the given planar trusses.

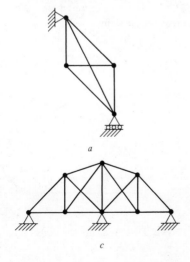

a

c

11-4. Determine the reactions for the given beam. Use the reaction at as the redundant.

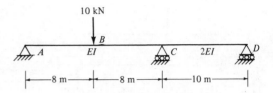

10 kN

B

A   EI   C   2EI   D

|——8 m——|——8 m——|——10 m——|

11-5. Repeat Problem 11-4 except use the reaction at *D* as the redundant.

11-6. Determine the moment diagram for each beam. Hint: Use the moments at the interior supports as redundants.

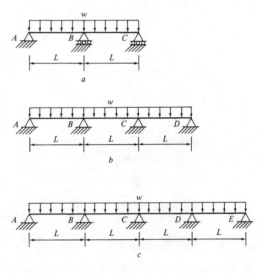

w

A   B   C

L   L

a

w

A   B   C   D

L   L   L

b

w

A   B   C   D   E

L   L   L   L

c

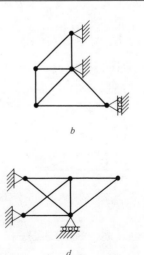

b

d

11-7. Using the reactions at B and C as redundants, determine the moment diagram for the given beam.

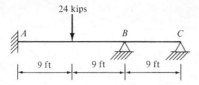

11-8. Using the reaction at C as the redundant, determine the moment diagram for the given beam.

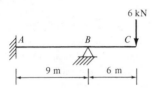

11-9. Determine the reactions for the given beam. Hint: Use the reaction at the roller as the redundant.

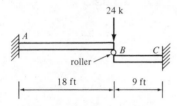

11-10. Determine the force in the cable.

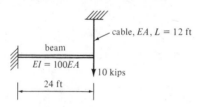

11-11. Repeat Problem 11-4 except the support at C settles 0.5 in. Use $EI = 24,000$ kip-ft^2.

11-12. Repeat Problem 11-5 except the support at C settles 0.5 in. Use $EI = 24,000$ kip-ft^2.

11-13. Determine the fixed end moments.

(a) Use the end moments as the redundants.
(b) Use the reactions at B as the redundants.

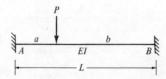

11-14. Determine the reactions for each planar frame. Also, draw the moment diagrams.

11-15. Determine the reactions and moment diagram for the given planar frames.

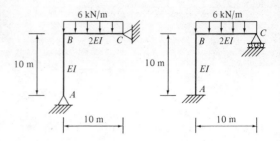

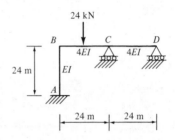

11-16. Determine the reactions for the given planar frame. Use the reactions at D as the redundants. Draw a complete moment diagram.

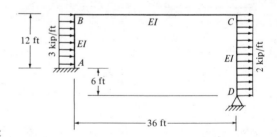

11-17. Determine the moment diagram for the given planar frame.

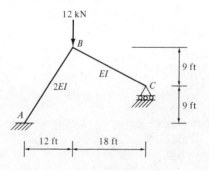

11-18. Determine the reactions for the given frame.

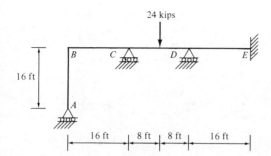

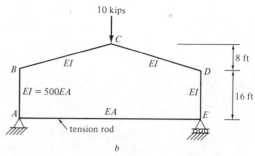

b

11-19. Determine the reactions for the given frame.

11-22. Determine the reactions for the given frame.

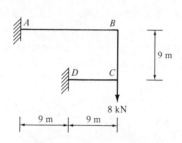

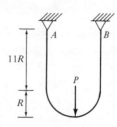

11-20. Determine the moment diagram for the given frame. Hint: Use the internal forces at a point (carefully chosen) as the redundants.

11-23. Determine the internal moment at each load point.

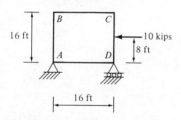

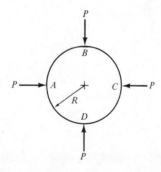

11-21. Determine the reactions for the given frames. In frame b also determine the tension in the rod. The rod has no bending resistance ($EI = 0$).

11-24. Determine the moment diagram for the frame *ABCD*. The cable is free to move around a pulley device at *D*. The cable has $EA = .025EI$.

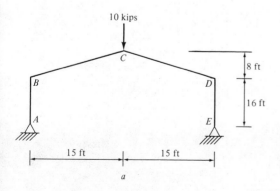

a

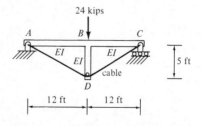

11-25. Determine the member forces in the given planar truss. Use the force in member U_1L_2 as the redundant. Member areas are given in parentheses.

(Problem 11-25 continued on p. 314)

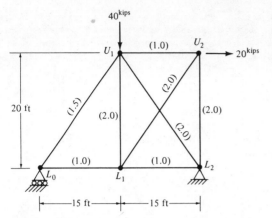

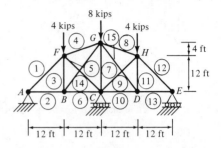

11-29. Determine the member forces in the given truss. Use the reactions at C and the force in members *CF* and *CH* as the redundants and employ symmetry. (See Problems 5.1e and 9-9.)

11-26. Determine the member forces in the given planar truss. Use the force in member *BD* as the redundant. (See Problems 5-1a and 9-8.)

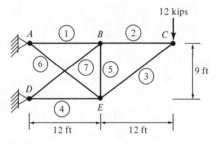

11-30. Determine the member forces in the given planar truss. Use the force in member *AD* as the redundant. (See Problems 5-1g and 9-10.)

11-27. Determine the member force in the given truss. Use the force in member *BD* and the reaction at *C* as the redundants. (See Problems 5-1a and 9-3.)

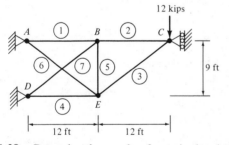

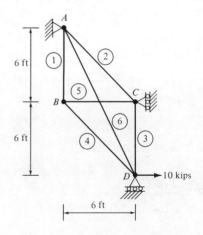

11-31. Determine the member forces in the given planar truss. Use the force in member *AD* and the reaction at *C* as the redundants. (See Problems 5.1g, 9-6 and 9-10.)

11-28. Determine the member forces in the given planar truss. Use the reaction at *C* as the redundant. (See Problems 5-1e and 9-5.)

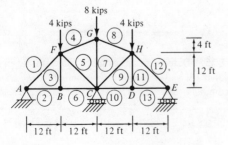

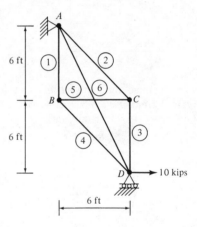

11-32. Determine the member forces in the given planar truss. Use the reaction at C and the vertical reaction at E as the redundants.

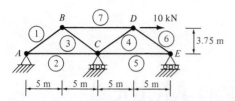

11-33. Determine the member forces in the given planar truss. Use the reaction at C as the redundant. (See Problems 5-11a and 9-7.)

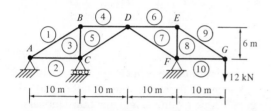

11-34. Determine the member forces in the given planar truss. Use the reaction at C and force in member CF as the redundants. (See Problems 9-12 and 11-33.)

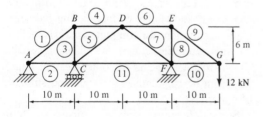

11-35. Repeat Problem 11-34 except use the reaction at C and the horizontal reaction at F as the redundants.

11-36. Determine the member forces in the given planar truss. Use the force in Member U_1L_2 and the horizontal reaction at joint L_0 as redundants. Member areas are the same as in Problem 11-25.

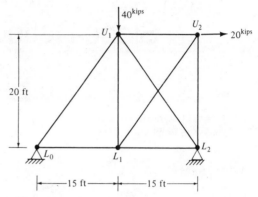

11-37. Determine the member forces in the given planar truss. Use the reactions at E and F as the redundants.

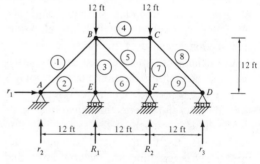

12
Slope-Deflection Method

12.1 REMARKS

Historically, several *stiffness methods* of analysis exist in classical structural analysis. The *slope-deflection method* is a stiffness approach that, at one time, had common usage. Practically speaking, this technique has limited application in contemporary structural analysis. Nevertheless, the method will be developed in this section because of its value in deriving other methods of structural analysis. In particular, the moment-distribution method, which is treated in Chapter 13, can be developed with the aid of the slope-deflection equations.

The basic approach in the slope-deflection method is to use the three conditions of structural analysis (equilibrium, material laws, and compatibility) to generate a set of equations that relates applied loads to joint displacements. In this manner, the independent generalized displacements are introduced as the unknown quantities because the loads are known. Because one such equation is developed for each independent displacement quantity, the number of equations is sufficient for a complete determination of their values. Stated differently, the slope-deflection method has the determination of the displacements, which signify the degree of kinematic indeterminacy, as its initial step.

As an introduction to the concepts of the slope-deflection method, consider the planar frame shown in Fig. 12.1a. Given the indicated support conditions, 4 degrees of freedom exist (axial deformation is neglected), and the frame is kinematically indeterminate to the fourth degree. Three rotational and one translational degrees of freedom can be defined and are depicted in Fig. 12.1b. The frame responds to arbitrary loading by developing internal forces and assuming a new position in space. Internal moments developed at the member ends are shown in Fig. 12.2a. With axial deformation neglected, knowledge of the six end moments alone is sufficient for complete analysis including the determination of the displaced shape shown in Fig. 12.2b. All other desired quantities can be calculated by statics and strength of materials.

Slope-deflection analysis consist of four basic stages. These stages consist of:

1. Use of material laws and compatibility to develop equations that relate the set of member end moments to the independent displacement quantities (degrees of freedom). Equations generated in this stage are called *slope-deflection equations*.
2. Use of equilibrium conditions and the slope-deflection equations to calculate the values of the independent displacement quantities.
3. Substitution of the calculated displacements into the slope-deflection equations to determine the member end moments.

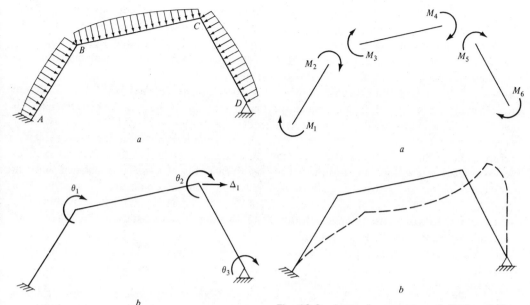

Fig. 12.1.   Three-member   planar   frame.   *a*, structure; *b*, degrees of freedom.

Fig. 12.2.   Internal moments and displaced shape for the frame of Fig. 12.1. *a*, internal moments; *b*, displaced shape.

4. Use of statics and conventional methods (such as the moment-area method) to calculate other desired quantities (e.g., internal shear forces, elastic curves).

12.2   SLOPE-DEFLECTION EQUATIONS FOR PRISMATIC MEMBERS

In developing the general slope-deflection equations, remember that if a member's loading and the displacements at its ends are known,

the internal end moments are theoretically established in value. With this knowledge, one can derive two equations needed in the first stage of the slope-deflection method. Consider the member shown in Fig. 12.3a, which is extracted from a planar frame. For reference, the ends of the member are labeled i and j. The member is loaded in some general fashion and, due to this loading and its interaction with the rest of the framework, experiences displacements at both ends. In going from its undeformed position to its displaced position, the

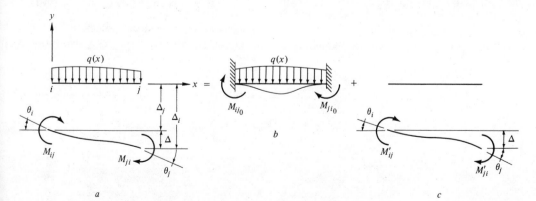

Fig. 12.3.   Deformation of a framed member.

member undergoes rotations θ_i and θ_j at its ends and experiences a differential translation, Δ, due to the motion of its ends, Δ_i and Δ_j. In this development, the displacement quantities are measured relative to the members' local coordinates and are considered to be positive when directed as shown in Fig. 12.3a. Consequently, θ_i and θ_j are measured from the original member axis location, and Δ_i and Δ_j are measured perpendicular to this axis. Clockwise rotations and downward translations are positive. The relative translation, Δ, is positive when Δ_j exceeds Δ_i, i.e., when the chord of the member rotates in a clockwise sense. These deformations and the loading, $q(x)$, create the end moments, M_{ij} and M_{ji}, shown in Fig. 12.3a. The internal moments are considered positive when they act in a clockwise sense on the member ends.

End moments shown in Fig. 12.3a, can be considered as being created by two separate effects: (1) moments, M_{ijo} and M_{jio}, developed if the ends are temporarily fixed in position and, (2) moments, M'_{ij} and M'_{ji}, caused by the displacements that occur if the temporarily fixed ends are released. The superposition of effects is shown in Fig. 12.3b and c.

Moments M_{ijo} and M_{jio} created in the fixed-end condition can be determined by the quadruple-integration method (as demonstrated in Example 3.14) or by the flexibility method described in Chapter 11. If the flexibility method is used, the restraining moments are taken as the redundants, and the compatibility conditions consist of the known zero slopes at the member ends. Example 11.6 gave an illustration of the flexibility method. Fixed-end moments will also be used in the matrix stiffness method (Chapter 15), and, for convenience

values for common loadings are listed in Table A.3. This table should be accessed in the slope-deflection method. It is important to note that Table A.3 applies only to members with constant EI.

The moments M'_{ij} and M'_{ji} in Fig. 12.3c are created by the member end displacements. For development purposes, it is useful to consider a further subdivision of these displacements. A more detailed picture of the deformed member is given in Fig. 12.4a. The displaced shape in this figure can be obtained by a superposition of the two simple beam effects shown in Figs. 12.4b and c. Consequently,

$$\theta_i = \phi_i + R \qquad (12.1)$$

$$\theta_j = \phi_j + R \qquad (12.2)$$

where ϕ_i and ϕ_j are the end rotations created if the moments M'_{ij} and M'_{ji} are applied to a simply-supported beam. Angle R is the chord rotation created by displacing end j of the simply-supported beam a distance Δ relative to end i.

Displacements for the beam in Fig. 12.4b can be determined by the conjugate-beam method. The appropriate conjugate structure is shown in Fig. 12.5. It is a simple matter to compute the desired end rotations

$$\phi_i^* = \text{shear at end } i = -\frac{L}{3EI}M'_{ij} + \frac{L}{6EI}M'_{ji} \qquad (12.3)$$

$$\phi_j^* = \text{shear at end } j = +\frac{L}{6EI}M'_{ij} - \frac{L}{3EI}M'_{ji} \qquad (12.4)$$

In the context of the slope-deflection equations, positive end rotations are clockwise motions.

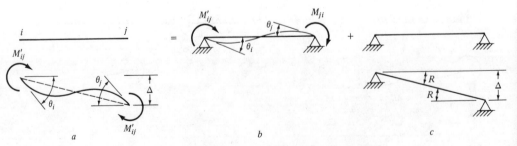

Fig. 12.4. End displacement effects by superposition.

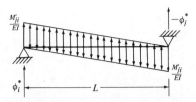

Fig. 12.5.   Conjugate beam for Fig. 12.4.

According to the sign convention established in Chapter 7 for the conjugate-beam method, the rotations given by Eqs. 12.3 and 12.4 are, instead, *counterclockwise* in sense. The notations ϕ_i^* and ϕ_j^* are employed to distinguish the *computed* counterclockwise rotations from the *desired* clockwise rotations, ϕ_i and ϕ_j. These clockwise values are given by

$$\phi_i = +\frac{L}{3EI} M'_{ij} - \frac{L}{6EI} M'_{ji} \quad (12.5)$$

$$\phi_j = -\frac{L}{6EI} M'_{ij} + \frac{L}{3EI} M'_{ji} \quad (12.6)$$

Equations 12.5 and 12.6 can be expressed in matrix form as

$$\left\{ \begin{array}{c} \phi_i \\ \phi_j \end{array} \right\} = \begin{bmatrix} \dfrac{L}{3EI} & -\dfrac{L}{6EI} \\[3mm] -\dfrac{L}{6EI} & \dfrac{L}{3EI} \end{bmatrix} \left\{ \begin{array}{c} M'_{ij} \\ M'_{ji} \end{array} \right\} \quad (12.7)$$

By direct inversion, the alternative relationship

$$\left\{ \begin{array}{c} M'_{ij} \\ M'_{ji} \end{array} \right\} = \begin{bmatrix} \dfrac{4EI}{L} & \dfrac{2EI}{L} \\[3mm] \dfrac{2EI}{L} & \dfrac{4EI}{L} \end{bmatrix} \left\{ \begin{array}{c} \phi_i \\ \phi_j \end{array} \right\} \quad (12.8)$$

is obtained. The square matrices in Eqs. 12.7 and 12.8 are termed the member flexibility and stiffness matrices, respectively.

Written in algebraic form, Eq. 12.8 becomes

$$M'_{ij} = \frac{4EI}{L} \phi_i + \frac{2EI}{L} \phi_j \quad (12.9)$$

$$M'_{ji} = \frac{2EI}{L} \phi_i + \frac{4EI}{L} \phi_j \quad (12.10)$$

Solving Eqs. 12.1 and 12.2 for ϕ_i and ϕ_j and substituting the results into Eqs. 12.9 and 12.10 yields

$$M'_{ij} = \frac{4EI}{L} (\theta_i - R) + \frac{2EI}{L} (\theta_j - R)$$

$$(12.11)$$

$$M'_{ji} = \frac{2EI}{L} (\theta_i - R) + \frac{4EI}{L} (\theta_j - R)$$

$$(12.12)$$

Recognizing that chord rotation, R, is given by

$$R = \frac{\Delta}{L} \quad (12.13)$$

Equations 12.11 and 12.12 can be written in the form

$$M'_{ij} = \frac{4EI}{L} \theta_i + \frac{2EI}{L} \theta_j - \frac{6EI}{L^2} \Delta \quad (12.14)$$

$$M'_{ji} = \frac{2EI}{L} \theta_i + \frac{4EI}{L} \theta_j - \frac{6EI}{L^2} \Delta \quad (12.15)$$

Equations 12.14 and 12.15 constitute basic material laws for an *unloaded* frame member of constant EI that can displace freely in space as shown in Fig. 12.3c. When combined with the fixed-end moments depicted in Fig. 12.3b, material laws are obtained for a *loaded* frame member that is free to move in space. These laws can be written as

$$M_{ij} = M_{ij0} + M'_{ij} \quad (12.16)$$

$$M_{ji} = M_{ji0} + M'_{ji} \quad (12.17)$$

Substitution of Eqs. 12.14 and 12.15 produces the expanded relationships

$$M_{ij} = M_{ij0} + \frac{4EI}{L} \theta_i + \frac{2EI}{L} \theta_j - \frac{6EI}{L^2} \Delta$$

$$(12.18)$$

$$M_{ji} = M_{jio} + \frac{2EI}{L}\theta_i + \frac{4EI}{L}\theta_j - \frac{6EI}{L^2}\Delta$$

$$(12.19)$$

The expressions given in Eqs. 12.18 and 12.19 are the desired relationships, i.e., the slope-deflection equations. At this point, the misnomer of the terminology *slope-deflection* should be apparent and is an important observation. Displacement quantities in these equations are rotations and the relative transverse displacement of the member ends. In global coordinates, these quantities do not represent, in general, the true "slope" of the member ends and the vertical "deflection" as one commonly thinks of these quantities. Hence, the term *rotation-displacement equations* is more appropriate but not used commonly. For members that originally lie in a horizontal position, slope and deflection are correct terms; otherwise they are misnomers. Only if one refers to displacements with respect to a member's local coordinates, can slope and deflection be viewed as general terms.

12.3 APPLICATION OF EQUILIBRIUM CONDITIONS

In the second stage of the slope-deflection method, it is necessary to apply the independent equilibrium conditions inherent in a given problem. The number of conditions needed equals the number of degrees of freedom, and the corresponding set of structural equations is obtained by applying statics in each of these directions. Equilibrium applied in the directions of the rotational degrees of freedom yields a set of *moment conditions*. Conditions of equilibrium in the directions of the translational degrees of freedom are referred to as *shear conditions*. Moment conditions are trivial to establish but shear conditions involve more effort. The basic purpose of the equilibrium conditions is to eliminate the unknown moments from the set of slope-deflection equations applicable to the problem. Subsequent simultaneous solution of the equilibrium equations produces the values of the independent joint displacements.

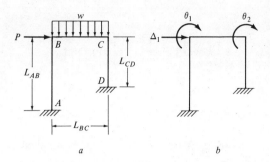

Fig. 12.6. Rectangular portal frame.

Generation of equilibrium conditions is best demonstrated by consideration of an example frame such as that shown in Fig. 12.6a. The given frame has three degrees of freedom in the directions indicated in Fig. 12.6b. Thus, three equilibrium equations must be written for use in the slope-deflection method. These can be developed with the aid of the free-body diagrams constructed in Fig. 12.7. Axial force, shear force, and moment have been shown at both ends of each member of the frame. Moments must be shown acting in a positive sense, but directions chosen for the remaining quantities are arbitrarily chosen. Equal and opposite forces are shown at each joint and, for convenience, have been left unlabeled. The necessary equations are obtained by satisfying statics at each joint *in the direction of the independent degrees of freedom*. Corresponding to the two rotational degrees of freedom is the need for the equilibrium of moments at joints B and C. This necessitates

$$M_{BA} + M_{BC} = 0 \qquad (12.20)$$

$$M_{CB} + M_{CD} = 0 \qquad (12.21)$$

which are the desired *moment conditions*. The single translational degrees of freedom in the horizontal direction at joint B has the need for horizontal equilibrium at joint B as its counterpart. Application of statics in that direction yields the shear condition:

$$V_{BA} + P_{BC} = P \qquad (12.22)$$

Equations 12.20 through 12.22 comprise the three equilibrium conditions needed in the

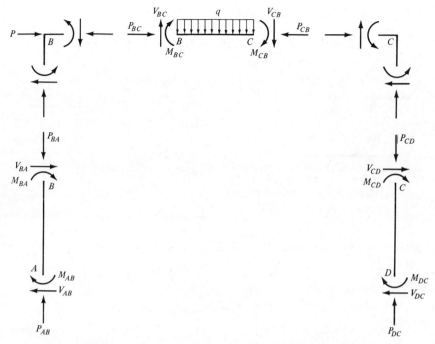

Fig. 12.7.   Free-body diagrams for the frame of Fig. 11.6.

slope-deflection analysis of the frame in Fig. 12.6. However, the latter equation is not useful in its present form. As a secondary step, V_{BA} and P_{BC} must be expressed in terms of the unknown end moments. This can be done as follows (refer to Fig. 12.7).

For member AB:

$$\Sigma M_A = M_{AB} + M_{BA} + V_{BA}(L_{AB}) = 0$$

$$\therefore V_{BA} = -\frac{1}{L_{AB}}(M_{AB} + M_{BA}) \quad (12.23)$$

For member CD:

$$\Sigma M_D = M_{CD} + M_{DC} + V_{CD}(L_{CD}) = 0$$

$$\therefore V_{CD} = -\frac{1}{L_{CD}}(M_{CD} + M_{DC}) \quad (12.24)$$

Inspection of horizontal equilibrium at joint C and in member BC indicates

$$P_{BC} = V_{CD} \quad (12.25)$$

Sequential substitution of Eqs. 12.23 through 12.25 into Eq. 12.22 yields the desired form of the shear condition:

$$\frac{M_{AB} + M_{BA}}{L_{AB}} + \frac{M_{CD} + M_{DC}}{L_{CD}} + P = 0$$

$$(12.26)$$

Equations 12.20, 12.21, and 12.26 are the equilibrium equations applicable to the frame in Fig. 12.6. Each of the moment terms that appear in these equations can be expressed in the slope-deflection format (Eqs. 12.18 and 12.19). In this way, the three displacement quantities θ_1, θ_2, and Δ_1 become the unknown quantities. Simultaneous solution for their values completes the second stage of slope-deflection analysis. Back-substitution of the computed displacements into the slope-deflection equations for each member yields the end moment values. Execution of this third step completes the major phases of slope-deflection analysis. With the end moments and displacements known for each member, further computations can be made by any of the various methods of determinate frame analysis to complete the final stage. Ideally, the analyst could draw shear and moment diagrams for each member and determine its displaced shape by the double- or quadruple-integration method.

12.4 EXAMPLES

Example 12.1. Analyze the beam given in Fig. 12.8a by the slope-deflection method. Draw the shear and moment diagram. Use $E = 30,000$ kip/in.^{2} and $I = 216$ in.^{4}

Solution: Two rotational degrees of freedom exist for the structure as indicated in Fig. 12.8b, making the beam kinematically indeterminate to the second degree. Rotations θ_1 and θ_2 at supports B and C are the unknown quantities in the slope-deflection method.

Step 1: Fixed-end moments are calculated by use of formulas obtained from Table A.3.

$$M_{AB_0} = M_{BC_0} = -\tfrac{1}{12}wL^2 = -\tfrac{1}{12}(2)(18)^2$$
$$= -54.0 \text{ kip-ft}$$

$$M_{BA_0} = M_{CB_0} = +\tfrac{1}{12}wL^2 = +\tfrac{1}{12}(2)(18)^2$$
$$= +54.0 \text{ kip-ft}$$

Step 2: Coefficients in the slope-deflection equations are computed for each member

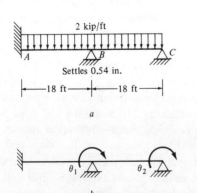

a

b

Fig. 12.8. Beam for Example 12.1. *a*, loading; *b*, degrees of freedom.

$$\left(\frac{4EI}{L}\right)_{AB} = \left(\frac{4EI}{L}\right)_{BC} = \frac{4(30,000)(216)}{216}$$
$$= 120,000 \text{ kip-in.}$$
$$= 10,000 \text{ kip-ft}$$

$$\left(\frac{6EI}{L^2}\right)_{AB} = \left(\frac{6EI}{L^2}\right)_{BC} = \frac{6(30,000)(216)}{(216)^2}$$
$$= 833.33 \text{ kips}$$

Using Eqs. 12.14 and 12.15.

Step 3: A pair of slope-deflection equations is written for each member. (Note that the slope at the left end of member AB is known to be zero.)

$$M_{AB} = M_{AB_0} + \frac{2EI}{L_{AB}}\theta_1 - \frac{6EI}{L_{AB}^2}\Delta$$

$$M_{BA} = M_{BA_0} + \frac{4EI}{L_{AB}}\theta_1 - \frac{6EI}{L_{AB}^2}\Delta$$

$$M_{BC} = M_{BC_0} + \frac{4EI}{L_{BC}}\theta_1 + \frac{2EI}{L_{BC}}\theta_2 - \frac{6EI}{L_{BC}^2}\Delta$$

$$M_{CB} = M_{CB_0} + \frac{2EI}{L_{CB}}\theta_1 + \frac{4EI}{L_{CB}}\theta_2 - \frac{6EI}{L_{CB}^2}\Delta$$

Substitution of the values calculated in Step 2 and the known Δ (be alert to signs) gives:

$$M_{AB} = -54 + 5000\theta_1 - 833.33\left(\frac{0.54}{12}\right)$$

$$M_{BA} = +54 + 10,000\theta_1 - 833.33\left(\frac{0.54}{12}\right)$$

$$M_{BC} = -54 + 10,000\theta_1 + 5000\theta_2$$
$$- 833.33\left(-\frac{0.54}{12}\right)$$

$$M_{CB} = +54 + 5000\theta_1 + 10,000\theta_2$$
$$- 833.33\left(-\frac{0.54}{12}\right)$$

which reduce to

$$M_{AB} = 5000\theta_1 - 91.5$$
$$M_{BA} = 10,000\theta_1 + 16.5$$
$$M_{BC} = 10,000\theta_1 + 5000\theta_2 - 16.5$$
$$M_{CB} = 5000\theta_1 + 10,000\theta_2 + 91.5$$

Step 4: Equilibrium consists of the two moment conditions at joints B and C

$$M_{BA} + M_{BC} = 0$$

$$M_{CB} = 0$$

Substitution of the slope-deflection equations yields

$$20,000\theta_1 + 5000\theta_2 = 0$$

$$5000\theta_1 + 10,000\theta_2 + 91.5 = 0$$

which are solved simultaneously for the rotations

$$\theta_1 = 0.002614 \text{ rad} = 0.002614 \text{ rad cw}$$

$$\theta_2 = -0.010456 \text{ rad} = 0.010456 \text{ rad ccw}$$

Step 5: Back substitutions of θ_1 and θ_2 into the slope-deflection equations gives

$$M_{AB} = 5000(0.002614) - 91.5 = -78.43 \text{ kip-ft}$$

$$M_{BA} = 10,000(0.002614) + 16.5$$

$$= +42.64 \text{ kip-ft}$$

$$M_{BC} = 10,000(0.002614) + 5000(-0.010456)$$

$$- 16.5 = -42.64 \text{ kip-ft}$$

$$M_{CB} = 5000(0.002614) + 10,000(-0.010456)$$

$$+ 91.5 = 0$$

Overly precise values of θ_1 and θ_2 have been employed to demonstrate that M_{CB} is correctly zero and M_{BA} and M_{BC} are correctly equal in magnitude and opposite direction. Usually, three significant figures are sufficient in all steps and the ensuing round-off errors are acceptable for practical answers.

Step 6: With appropriate attention to the directions of the end moments, free-body diagrams can be drawn for each member. Statics suffices to determine the end shear values. Results are shown in Fig. 12.9. These free bodies provide the information necessary to plot the shear and moment diagrams given in Fig. 12.10. Care must be taken because the sign convention for moments in the slope-deflection method differs from the "designer's convention" used in the moment diagrams.

Example 12.2. Analyze the planar frame given in Fig. 12.11 by the slope-deflection method. Draw shear and moment diagrams.

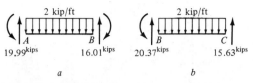

Fig. 12.9. Free-body diagram of members. a, member AB; b, member BC.

Solution: The given structure fits the geometry of the rectangular portal frame (Fig. 12.6) discussed in Section 12.3. The degrees of freedom labeled in Fig. 12.6b.

Step 1: The slope-deflection equations are

$$M_{AB} = \frac{2EI}{15}\theta_1 - \frac{EI}{75}\Delta_1 \tag{a}$$

$$M_{BA} = \frac{4EI}{15}\theta_1 - \frac{EI}{75}\Delta_1 \tag{b}$$

$$M_{BC} = -600 + \frac{EI}{3}\theta_1 + \frac{EI}{6}\theta_2 \tag{c}$$

$$M_{CB} = 600 + \frac{EI}{6}\theta_1 + \frac{EI}{3}\theta_2 \tag{d}$$

$$M_{CD} = \frac{EI}{5}\theta_2 - \frac{3EI}{200}\Delta_1 \tag{e}$$

$$M_{DC} = \frac{EI}{10}\theta_2 - \frac{3EI}{200}\Delta_1 \tag{f}$$

In formulating the above equations, it was recognized that the slope and deflection at points A and D

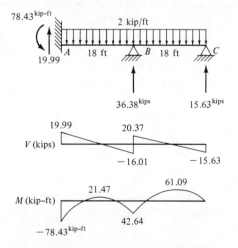

Fig. 12.10. Shear and moment diagrams.

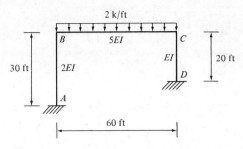

Fig. 12.11.   Planar frame for Example 12.2.

In matrix form, Eqs. g, h, and l are

$$\frac{EI}{90,000} \begin{bmatrix} 54,000 & 15,000 & -1200 \\ 15,000 & 48,000 & -1350 \\ -1200 & -1350 & 215 \end{bmatrix} \begin{Bmatrix} \theta_1 \\ \theta_2 \\ \Delta_1 \end{Bmatrix}$$

$$= \begin{Bmatrix} 600 \\ -600 \\ 0 \end{Bmatrix} \qquad (m)$$

are zero and members AB and CD both move by Δ_1 at the top end.

Step 2: Equilibrium conditions are given by Eqs. 12.20 through 12.22. Substitution of (a) through (f) into Eqs. 12.20 and 12.21 gives

$$\frac{9EI}{15}\theta_1 + \frac{EI}{6}\theta_2 - \frac{EI}{75}\Delta_1 = 600 \qquad (g)$$

$$\frac{EI}{6}\theta_1 + \frac{8EI}{15}\theta_2 - \frac{3EI}{200}\Delta_1 = -600 \qquad (h)$$

Because $P = 0$, Eq. 12.22 becomes

$$V_{BA} + V_{CD} = 0 \qquad (i)$$

and Eqs. 12.23 and 12.24 give

$$V_{BA} = -\tfrac{1}{30}(M_{AB} + M_{BA}) \qquad (j)$$

$$V_{CD} = -\tfrac{1}{20}(M_{CD} + M_{DC}) \qquad (k)$$

Substitution of Eqs. j and k into Eq. i and then Eqs. a through f into the result yields

$$-\frac{EI}{75}\theta_1 - \frac{3EI}{200}\theta_2 + \frac{215EI}{90,000}\Delta_1 = 0 \qquad (l)$$

Step 3: Using computer inversion of the coefficient matrix, Eqs. m are solved to produce

$$\theta_1 = \frac{1398.9}{EI}$$

$$\theta_2 = \frac{-1630.5}{EI}$$

$$\Delta_1 = \frac{2430.2}{EI}$$

Step 4: Substitution into Eqs. a through f gives

$M_{AB} = 218.9$ kip-ft $M_{CB} = 298.7$ kip-ft

$M_{BA} = 405.4$ kip-ft $M_{CD} = -289.7$ kip-ft

$M_{BC} = -405.4$ kip-ft $M_{DC} = -126.6$ kip-ft

Step 5: Free body diagrams of the members are shown in Fig. 12.12. The end shears are calculated by summing moments about each end. All quantities indicated in Fig. 12.12 are in their correct directions. Shear and moment diagrams follow directly, recognizing the designer's convention for signs. The assembled diagrams are shown in Fig. 12.13.

Example 12.3. A planar frame is shown in Fig. 12.14a. Express the equilibrium equations (for the

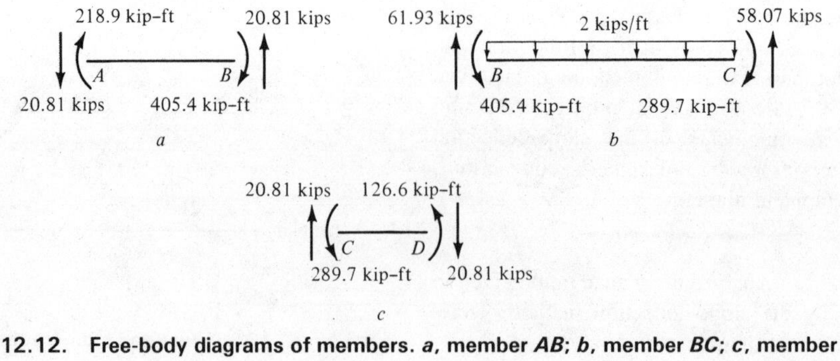

218.9 kip-ft   20.81 kips   61.93 kips   2 kips/ft   58.07 kips

20.81 kips   405.4 kip-ft   405.4 kip-ft   289.7 kip-ft

a   *b*

20.81 kips   126.6 kip-ft

289.7 kip-ft   20.81 kips

c

Fig. 12.12.   Free-body diagrams of members. *a*, member *AB*; *b*, member *BC*; *c*, member *CD*.

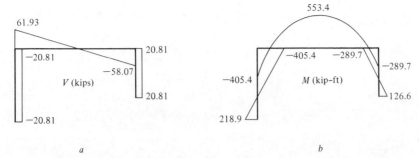

Fig. 12.13.   Results for Example 12.2. *a*, shear diagram; *b*, moment diagram.

slope-deflection method) in terms of the kinematic unknowns. Use a value of $1000L$ kN $\cdot$ m (L = the member length) for EI for each member.

Solution: Degrees of freedom for the frame are labeled in Fig. 12.14b. The single translational degree of freedom can be indicated in several ways and the horizontal motion at joint C is chosen in this solution.

Step 1:   It is first necessary to determine the relative translation of the ends of each member in terms of the declared translational degree of freedom. This is accomplished by considering the effect depicted in Fig. 12.15a. If joint B is cut and joint C allowed to move an amount Δ_1 to point C_1, end B of member BC must move an equal amount to point B_1. Because

joint B must actually remain intact, the true displaced position of the joint is established at point B_2. Point B_2 is located by recognizing that lines BB_2 and B_1B_2 must be orthogonal to their respective members in a first-order analysis. The motion of joint B is isolated in Fig. 12.15b, which also shows the inclination of lines BB_2 and B_1B_2.

Distances BB_2 and B_1B_2 are the relative displacements of the ends of members AB and BC, respectively, and can be computed using the diagram in Fig. 12.15b. Two geometric conditions can be established

$$\frac{12}{21.63} BB_1 + \frac{9}{25.63} B_1B_2 = \Delta_1$$

$$\frac{18}{21.63} BB_2 - \frac{24}{25.63} B_1B_2 = 0$$

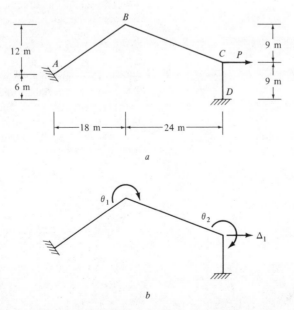

Fig. 12.14.   Frame for Example 12.3. *a*, loaded structure; *b*, degrees of freedom.

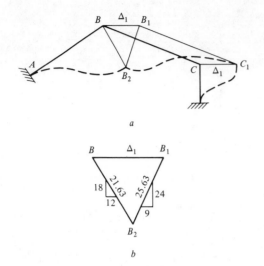

a

b

Fig. 12.15. Effect of translational motion of joint C. a, deformed shape; b, motion of joint B.

Solving these equations gives

$$BB_2 = 1.154\Delta_1$$

$$B_1B_2 = 1.025\Delta_1$$

Step 2: Slope-deflection equations are obtained by substitution into the general Eqs. 12.18 and 12.19. For the given loading, all fixed-end moments have zero value.

$$M_{AB} = \frac{2(1000L_{AB})}{L_{AB}}\theta_1$$
$$-\frac{6(1000L_{AB})}{L_{AB}^2}(1.154\Delta_1)$$

$$M_{BA} = \frac{4(1000L_{AB})}{L_{AB}}\theta_1$$
$$-\frac{6(1000L_{AB})}{L_{AB}^2}(1.154\Delta_1)$$

$$M_{BC} = \frac{4(1000L_{BC})}{L_{BC}}\theta_1 + \frac{2(1000L_{BC})}{L_{BC}}\theta_2$$
$$-\frac{6(1000L_{BC})}{L_{BC}^2}(-1.025\Delta_1)$$

$$M_{CB} = \frac{2(1000L_{BC})}{L_{BC}}\theta_1 + \frac{4(1000L_{BC})}{L_{BC}}\theta_2$$
$$-\frac{6(1000L_{BC})}{L_{BC}^2}(-1.025\Delta_1)$$

$$M_{CD} = \frac{4(1000L_{CD})}{L_{CD}}\theta_2 - \frac{6(1000L_{CD})}{L_{CD}^2}(\Delta_1)$$

$$M_{DC} = \frac{2(1000L_{CD})}{L_{CD}}\theta_2 - \frac{6(1000L_{CD})}{L_{CD}^2}(\Delta_1)$$

(Note that the chord of member BC rotates counter-clockwise causing a negative relative displacement!) These relationships reduce to

$$M_{AB} = 2000\theta_1 - 320\Delta_1$$

$$M_{BA} = 4000\theta_1 - 320\Delta_1$$

$$M_{BC} = 4000\theta_1 + 2000\theta_2 + 240\Delta_1$$

$$M_{CB} = 2000\theta_1 + 4000\theta_2 + 240\Delta_1$$

$$M_{CD} = 4000\theta_2 - 667\Delta_1$$

$$M_{DC} = 2000\theta_2 - 667\Delta_1$$

Step 3: Moment conditions of equilibrium are established by inspection

$$M_{BA} + M_{BC} = 0$$

$$M_{CB} + M_{CD} = 0$$

Step 4: A single shear condition is needed and is established from the free-body diagrams constructed in Fig. 12.16. Two orthogonal internal forces and a single flexural moment exist at each member end. Careful examination of joint equilibrium establishes that only three independent direct internal forces exist. Declaring the forces at end A of member AB as H_1 and V_1, and the horizontal reaction at end D of member CD as H_2, all other internal forces are established by satisfying the conditions $\Sigma F_x = 0$ and $\Sigma F_y = 0$ for all members and all joints. At joint C, two horizontal forces of different magnitude exist because of the presence of the force P. Coincidentally, the shear condition required in this problem is established by this very condition, that is

$$P + H_1 + H_2 = 0$$

Step 5: It is necessary to express H_1 and H_2 in terms of the member end moments. Referring to Fig. 12.16, moment equilibrium requires
    For member AB:

$$\Sigma M_A = 0$$

$$12H_1 - 18V_1 - M_{AB} - M_{BA} = 0$$

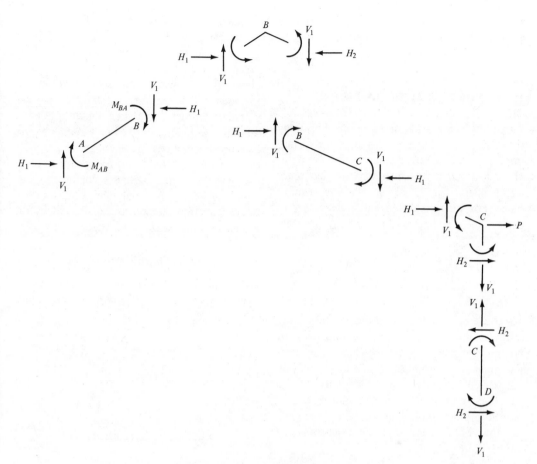

Fig. 12.16.   Free-body diagrams for Example 12.3.

For member BC:

$$\Sigma M_B = 0$$

$$9H_1 + 24V_1 + M_{BC} + M_{CB} = 0$$

For member CD:

$$\Sigma M_C = 0$$

$$9H_2 - M_{CD} - M_{DC} = 0$$

Solving the three preceding equations yields (the value of V_1 is not needed and, thus, is omitted)

$$H_1 = \tfrac{4}{75}(M_{AB} + M_{BA}) - \tfrac{1}{25}(M_{BC} + M_{CB})$$

$$H_2 = \tfrac{1}{9}(M_{CD} + M_{DC})$$

Consequently, the shear condition of equilibrium is established as

$$\tfrac{4}{75}(M_{AB} + M_{BA}) - \tfrac{1}{25}(M_{BC} + M_{CB})$$

$$+ \tfrac{1}{9}(M_{CD} + M_{DC}) + P = 0$$

Step 6: Substitution of the slope-deflection equations into the equilibrium equations yields the desired results.

$$8000\theta_1 + 2000\theta_2 - 80\Delta_1 = 0$$

$$2000\theta_1 + 8000\theta_2 - 427\Delta_1 = 0$$

$$-80\theta_1 - 427\theta_2 + 201\Delta_1 = P$$

or in matrix form

$$\begin{bmatrix} 8000 & 2000 & -80 \\ 2000 & 8000 & -427 \\ -80 & -427 & +201 \end{bmatrix} \begin{Bmatrix} \theta_1 \\ \theta_2 \\ \Delta_1 \end{Bmatrix} = \begin{Bmatrix} 0 \\ 0 \\ P \end{Bmatrix}$$

The symmetry observed in the coefficient matrix is a general condition of the slope-deflection method.

12.5 NONPRISMATIC MEMBERS

Flexibility coefficients in Eq. 12.7 apply only to members that have a constant value of moment of inertia over their entire length. Haunched and curved members are examples of members that do not satisfy this requirement. For such conditions, it is necessary to express the member flexibility relationships as

$$\begin{Bmatrix} \phi_i \\ \phi_j \end{Bmatrix} = \begin{bmatrix} d_{ii} & d_{ij} \\ d_{ji} & d_{jj} \end{bmatrix} \begin{Bmatrix} M'_{ij} \\ M'_{ji} \end{Bmatrix} \quad (12.27)$$

where d_{ii}, d_{ij}, d_{ji}, and d_{jj} are the member flexibility coefficients. Flexibility coefficients for nonuniform members can be derived by various *numerical methods* of structural analysis not treated in this textbook. Generally, this is a tedious process and tabulated values are employed in practice. Given Eq. 12.28, the general form of the member stiffness matrix is obtained by inversion as

$$\begin{Bmatrix} M'_{ij} \\ M'_{ji} \end{Bmatrix} = \begin{bmatrix} S_{ii} & S_{ij} \\ S_{ji} & S_{jj} \end{bmatrix} \begin{Bmatrix} \phi_i \\ \phi_j \end{Bmatrix} \quad (12.28)$$

Both the flexibility coefficient matrix and the stiffness coefficient matrix are symmetric in general form. Repeating the steps of the derivation of Eqs. 12.18 and 12.19 will produce the general slope-deflection equations.

$$M'_{ij} = M_{ijo} + S_{ii}\theta_i + S_{ij}\theta_j - (S_{ii} + S_{ij})\frac{\Delta}{L}$$

$$(12.29)$$

$$M'_{ji} = M_{jio} + S_{ji}\theta_i + S_{jj}\theta_j - (S_{ji} + S_{jj})\frac{\Delta}{L}$$

$$(12.30)$$

It is vital to reiterate that the fixed-end moments, M_{ijo} and M_{jio} in Eqs. 12.29 and 12.30 cannot be extracted from Table A.3. Values given in that table apply specifically to members having a constant cross section. Numerical methods can also be used to calculate fixed-end moments for members with a nonuniform cross section. In practice, fixed-end moments for common beam types and loadings are generally available in tabulated form. Except for employment of the correct stiffness coefficients and fixed-end moments, analytical stages in the slope-deflection method are unaffected by the presence of nonuniform members.

PROBLEMS

12-1. Calculate the slope and/or displacement at points A, B, and C of the given beams. Use the slope-deflection method.

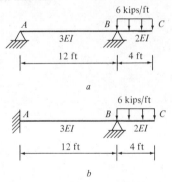

a

b

12-2. Do Problem 11-4 by the slope-deflection method.

12-3. Calculate the member end moments for the given beams. Use the slope-deflection method. Draw shear and moment diagrams.

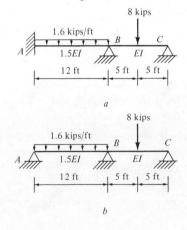

a

b

12-4. Repeat Problem 12-3 for the given beam.

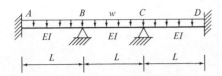

12-5. Determine the member end moments by using the slope-deflection method. Hint: Treat the third points as joints.

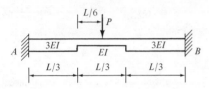

12-6. Derive slope-deflection equations for the member *AB* in Problem 12-7. The result should be the relationship between the end moments and displacements for the variable *I* member.

12-7. Determine the fixed end moment for the given beam. Use the slope-deflection method. Hint: Treat midspan as a joint.

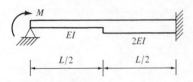

12-8. Do Problem 11-6 by the slope-deflection method.

12-9. Do Problem 11-7 by the slope-deflection method.

12-10. Do Problem 11-8 by the slope-deflection method.

12-11. Redo Problem 12-4 except include settlements of 0.5 in. at supports *B* and *C*. *EI* = 24,000 kip-ft².

12-12. Determine the spring displacement and the slope at *B* using the slope-deflection method.

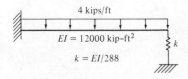

12-13. Determine the member end moments for the given planar frame. Use the slope-deflection method. Draw shear and moment diagrams.

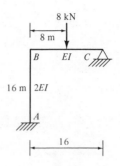

12-14. Determine the member end moments for the given planar frame. Use the slope-deflection method.

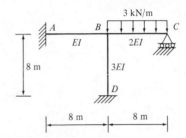

12-15. Repeat Problem 12-14 for the given planar frame.

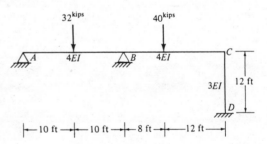

12-16. Determine the member end moments for the given planar frame by the slope-deflection method.

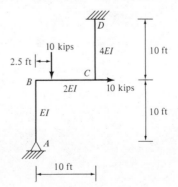

12-17. Determine the member end moments for the given planar frame by the slope-deflection method.

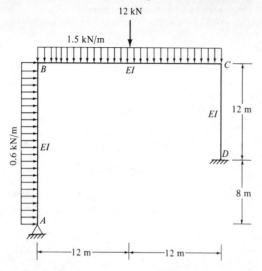

12-18. Do Problem 11-15 by the slope-deflection method.

12-19. Do Problem 11-16 by the slope-deflection method.

12-20. Do Problem 11-17 by the slope-deflection method.

12-21. Do Problem 11-19 by the slope-deflection method.

12-22. Do Problem 11-21a by the slope-deflection method.

13

Moment-Distribution Method

13.1 INTRODUCTION

The flexibility method and the slope-deflection method were developed in the preceding two chapters. Both techniques have general applicability but also suffer certain drawbacks that reduce their desirability as practical tools.

Employment of the flexibility method introduces the need for a multitude of displacement computations. As the degree of redundancy increases, the volume of computations required expands considerably. An additional constraint on the flexibility method is the need to solve the compatibility equations as a simultaneous set. This factor, by itself, suffices to eliminate this analytical method as a means of manual computation for complex frames.

The slope-deflection procedure reduces the need to employ direct geometrical analysis. Despite this advantage, the slope-deflection method is undesirable except for small frameworks. Similar to the flexibility method, there is an inherent requirement to solve several simultaneous equations. The source of the simultaneous equations is the set of equilibrium equations. One statics condition must be developed for each degree of freedom present in the structure. Few structures have less than three degrees of freedom, which severely limits the applicability of the slope-deflection procedure. The presence of translational degrees of freedom infers a need to generate shear equa-

tions. This requires caution and tedious analysis even if the framework is relatively simple in form. Example 12.3 offers evidence of this. Clearly, the presence of multiple translational degrees of freedom greatly complicates the analysis and causes the slope-deflection method to become prohibitive.

Frame analysis by the moment distribution is a numerical procedure that eliminates the lengthy displacement computations that are part of the flexibility method. This technique is essentially a cyclic operation based upon the principles used in developing the slope-deflection equations. In essence, the effects of joint rotations are examined in a repetitive, sequential manner. Each cycle results in a closer approximation to the exact solution, and the operations are continued until convergence is achieved within acceptable accuracy. Simultaneous equations can occur in the analysis but only if the number of translational degrees of freedom exceeds one. A simultaneous equation is needed for each translational degree of freedom. Before the advent of matrix methods of structural analysis, moment distribution was the predominant analytical procedure employed. Its usefulness as a general method has decreased in the face of computer methods, but applications still exist in the analysis of reinforced concrete structures. Unlike their predecessors, contemporary engineers who employ moment distribution commonly resort to computer algorithms as an alternative to the manual computations described herein.

13.2   BASIC CONCEPT

Moment distribution relies upon one elementary concept for its basic foundation. Fundamentally, one must understand the effect of a joint rotation on the internal forces at the ends of a single frame member. The structural member shown in Fig. 13.1 will serve to introduce the rudiments of moment distribution.

13.2.1   Member Stiffnesses and Carryover Factors

Two basic definitions in the moment-distribution method are "member stiffness" and "carryover factor." These terms are defined in the context of two types of support conditions for an individual member. Each is discussed in the following sections.

13.2.1.1   Pinned-Fixed Members

In Fig. 13.1, a beam element pinned at one end and fixed at the other end is subject to a clockwise moment, m_i, at the pinned end. This applied moment causes a clockwise rotation, θ_i, at the pinned end and a resisting moment, m_j, at the fixed end. The magnitude of m_i and m_j can be computed by consideration of the general member stiffness relationship

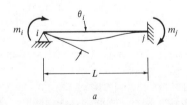

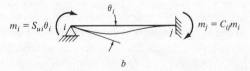

Fig. 13.1.   Effect of rotation θ_i on a pinned-fixed member.

$$\begin{Bmatrix} m_i \\ m_j \end{Bmatrix} = \begin{bmatrix} S_{ii} & S_{ij} \\ S_{ji} & S_{jj} \end{bmatrix} \begin{Bmatrix} \theta_i \\ \theta_j \end{Bmatrix} \quad (13.1)$$

In this relationship, clockwise moments and rotations at the member ends are considered positive. For the beam in Fig. 13.1a, the rotation at end j is known to be zero, which gives

$$\begin{Bmatrix} m_i \\ m_j \end{Bmatrix} = \begin{bmatrix} S_{ii} & S_{ij} \\ S_{jl} & S_{jj} \end{bmatrix} \begin{Bmatrix} \theta_i \\ 0 \end{Bmatrix}$$

The unknown end moments are determined to be

$$m_i = S_{ii}\theta_i \quad (13.2)$$

$$m_j = S_{ji}\theta_i \quad (13.3)$$

The ratio of the moment, m_j, at the fixed-end to that applied at the pinned end, m_i, is

$$C_{ij} = \frac{m_j}{m_i} = \frac{S_{ji}}{S_{ii}} \quad (13.4)$$

C_{ij} is commonly called the *carryover factor*. Figure 13.1b illustrates the end-moment values necessary to create the situation shown in Fig. 13.1a. The carryover factor is always positive, indicating that an applied clockwise moment induces a clockwise carryover moment, i.e., the direction for both moments is the same. The stiffness coefficient S_{ii} represents the value of the moment, m_i, which must be applied to cause a unit clockwise value of θ_i. In the context of moment distribution, this particular stiffness coefficient is called the *member stiffness*, S_i, and

$$S_i = S_{ii} \quad (13.5)$$

If the locations of the supports are reversed, the end moments would have the magnitude and directions shown in Fig. 13.2. The value of the member stiffness at end j would be

$$S_j = S_{jj} \quad (13.6)$$

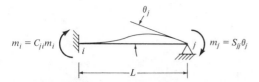

Fig. 13.2. Effect of end rotation θ_j on a pinned-fixed member.

and the carryover factor, C_{ji}, from end j to end i would have a value

$$C_{ji} = \frac{S_{ij}}{S_{jj}} \qquad (13.7)$$

13.2.1.2 Pinned-Pinned Members

Member stiffness and carryover factors derived in the preceding paragraphs are applicable only if the particular support conditions specified in their development actually exist for the beam. To complete this preliminary discussion, it is necessary to investigate the behavior of a member that is pin-supported at both ends. In Fig. 13.3a, a member that is pin-supported at both ends is subjected to a moment m_i at end i. The value of the moment, m_i, and induced rotation, θ_i, can be computed using Eq. 13.1. Substituting known values

$$\left\{ \begin{array}{c} m_i \\ 0 \end{array} \right\} = \left[\begin{array}{cc} S_{ii} & S_{ij} \\ S_{ji} & S_{jj} \end{array} \right] \left\{ \begin{array}{c} \theta_i \\ \theta_j \end{array} \right\}$$

gives

$$m_i = S_{ii}\theta_i + S_{ij}\theta_j \qquad (13.8)$$

$$0 = S_{ji}\theta_i + S_{jj}\theta_j \qquad (13.9)$$

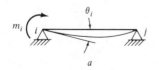

a

b

Fig. 13.3. Effect of end rotation θ_i on a pinned-pinned member.

Solving Eq. 13.9 for θ_j gives

$$\theta_j = -\frac{S_{ji}}{S_{jj}} \theta_i \qquad (13.10)$$

and substitution into Eq. 13.8 yields

$$m_i = \left(S_{ii} - \frac{S_{ji}}{S_{jj}} S_{ij} \right) \theta_i \qquad (13.11)$$

Due to the symmetry of the stiffness coefficient matrix, $S_{ij} = S_{ji}$, and Eq. 13.11 can be written

$$m_i = \left(S_{ii} - \frac{S_{ij}^2}{S_{jj}} \right) \theta_i \qquad (13.12)$$

The coefficient of θ_i in Eq. 13.12 is the "member stiffness" for a member pinned at both ends, i.e.,

$$S_i = S_{ii} - \frac{S_{ij}^2}{S_{jj}} \qquad (13.13)$$

In many texts, this term is referred to as the *modified member stiffness* but this terminology is not preferred by the author. Because no moment is induced at the other pinned end, the carry-over factor has zero value, that is

$$C_{ij} = 0 \qquad (13.14)$$

Furthermore, if the moment were applied at end j instead of end i the member stiffness, S_j, would be

$$S_j = S_{jj} - \frac{S_{ij}^2}{S_{ii}} \qquad (13.15)$$

which is shown in Fig. 13.4. The carryover factor, C_{ji}, is zero.

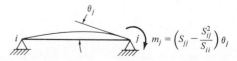

Fig. 13.4. Effect of end rotation θ_j on a pinned-pinned member.

13.2.1.3 Members with Constant *EI*

At this point, it is useful to compute the member stiffnesses and carryover factors for a member with a uniform cross section. For such a member, Eq. 13.1 is of the form

$$\begin{Bmatrix} m_i \\ m_j \end{Bmatrix} = \begin{bmatrix} \dfrac{4EI}{L} & \dfrac{2EI}{L} \\ \dfrac{2EI}{L} & \dfrac{4EI}{L} \end{bmatrix} \begin{Bmatrix} \theta_i \\ \theta_j \end{Bmatrix}$$

which is seen from Eq. 12.8. Thus

$$S_{ii} = S_{jj} = \frac{4EI}{L}$$

$$S_{ij} = S_{ji} = \frac{2EI}{L}$$

Consequently, simple substitutions into earlier equations give:

For a pinned fixed member:

$$S_i = S_j = \frac{4EI}{L} \qquad (13.16)$$

$$C_{ij} = C_{ji} = \tfrac{1}{2} \qquad (13.17)$$

For a pinned-pinned member:

$$S_i = S_j = \frac{3EI}{L} \qquad (13.18)$$

$$C_{ij} = C_{ji} = 0 \qquad (13.19)$$

It should be noted that for a nonuniform member, two member stiffnesses exist, as do two carryover factors, i.e., $S_i \neq S_j$ and $C_{ij} \neq C_{ji}$.

13.2.2 Distribution of an Unbalanced Moment

Before the complete moment-distribution process can be fully demonstrated the additional terms *unbalanced moment*, *distributed moment*, and *distribution factor* must be defined. Analysis of the arbitrary frame drawn in Fig. 13.5a will permit definition of each of these terms. This three-member frame is subjected to a moment, *M*, at the common fixed joint. If the fixity of this common joint is removed, the joint cannot maintain its position and will rotate to a new position (Fig. 13.5b). The applied moment, *M*, which causes the joint motion, is termed an *unbalanced moment*. In its new position, the structure will be in static equilibrium. Of particular interest is the common joint. As shown in Fig. 13.5c, this joint must maintain moment equilibrium. This is accomplished by the development of end moments M_1, M_2, and M_3, which must satisfy the relationship

$$M_1 + M_2 + M_3 = M \qquad (13.20)$$

Each of the necessary end moments M_1 through M_3 is termed a *distributed moment*. The magnitude of the distributed moments can be determined by the slope-deflection method.

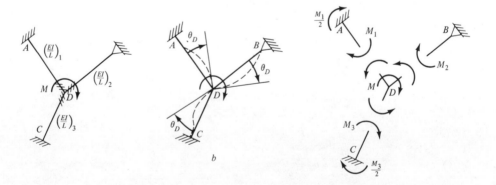

Fig. 13.5. Concept of distributed moments. *a*, common joint fixed; *b*, common joint released; *c*, distributed moments.

The free joint must rotate an amount θ_D and for compatibility each member must rotate by an equal amount at the location of the joint. After consideration of the support condition at the far end of each member vis-à-vis Figs. 13.3 and 13.4, one establishes

$$M_1 = \left(\frac{4EI}{L}\right)_1 \theta_D \qquad (13.21)$$

$$M_2 = \left(\frac{3EI}{L}\right)_2 \theta_D \qquad (13.22)$$

$$M_3 = \left(\frac{4EI}{L}\right)_3 \theta_D \qquad (13.23)$$

That is, each moment is the product of the member stiffness times the rotation of the member end. By substitution of Eqs. 13.21 through 13.23 into Eq. 13.20, the joint rotation is computed as

$$\theta_D = \frac{M}{\left(\dfrac{4EI}{L}\right)_1 + \left(\dfrac{3EI}{L}\right)_2 + \left(\dfrac{4EI}{L}\right)_3}$$

$$(13.24)$$

Back-substitution into Eqs. 13.21 through 13.23 yields the desired moment values

$$M_1 = \frac{\left(\dfrac{4EI}{L}\right)_1}{\left(\dfrac{4EI}{L}\right)_1 + \left(\dfrac{3EI}{L}\right)_2 + \left(\dfrac{4EI}{L}\right)_3} M$$

$$M_2 = \frac{\left(\dfrac{3EI}{L}\right)_2}{\left(\dfrac{4EI}{L}\right)_1 + \left(\dfrac{3EI}{L}\right)_2 + \left(\dfrac{4EI}{L}\right)_3} M$$

$$M_3 = \frac{\left(\dfrac{4EI}{L}\right)_3}{\left(\dfrac{4EI}{L}\right)_1 + \left(\dfrac{3EI}{L}\right)_2 + \left(\dfrac{4EI}{L}\right)_3} M$$

$$(13.25)$$

Inspection of the results permits the observation that the unbalanced moment at the joint is distributed into the members that frame into that joint in proportion to their relative member stiffnesses. The member stiffnesses employed must take account of the support condition that exists at the far end of each member.

The general mathematical form for the distributed moment M_m in a member k is expressed by

$$M_m = \frac{(S_i)_m}{\sum\limits_{m=1}^{NM} (S_i)_m} M \qquad (13.26)$$

where $(S_i)_m$ is the stiffness of member m, $(S_i)_m$ is the stiffness of member m, and NM is the number of members that frame into the joint. As an alternative to Eq. 13.26, the distributed moment in member m can be expressed as

$$M_m = DF_m M \qquad (13.27)$$

DF_m is termed the *distribution factor* and by inference is given by

$$DF_m = \frac{(S_i)_m}{\sum\limits_{m=1}^{NM} (S_i)_m} \qquad (13.28)$$

In other words, the distribution factor indicates what proportion of the total unbalanced moment at a joint is resisted by member m. Clearly, the addition of the distribution factors for all members that frame into a given free joint must sum to unity.

A final observation regarding the structure of Fig. 13.5b is the existence of carryover moments at the fixed ends of members AD and CD. Each of these carryover moments acts in the same direction as the corresponding distributed moment and is half its value in magnitude. The pin support, has no carryover moment. In other words, the carryover moment is the product of the carryover factor times the distributed moment.

Finally, note from Eq. 13.10, that

$$\theta_j = -\frac{\dfrac{2EI}{L}}{\dfrac{4EI}{L}} \theta_i = -\frac{1}{2}\theta_i$$

and thus

$$\theta_B = -\tfrac{1}{2}\theta_D$$

The minus sign indicates θ_B is reverse in direction to θ_D, i.e., it is counterclockwise.

Member stiffness, distribution of unbalanced moments at a joint, and carry-over moments have been explained previously. These three simple and interrelated ideas are the foundation of the moment-distribution method. Each was discussed in the context of a number of individual members that frame into a single free joint. Viewed in their pure form, these concepts constitute a closed-form solution to such a framework when the only loading is an external moment applied at the free joint. However, extrapolation of the underlying principles permits the analysis of planar frames of any size, configuration and loading. Prior to the treatment of general frameworks, it is useful to examine the application of the moment-distribution concepts to frameworks that are limited to a single free joint but subjected to transverse member loads.

Example 13.1. Analyze the two-span, continuous beam shown in Fig. 13.6 by application of the principles of moment distribution. Assume EI is the same and constant for both spans.

Solution: The configuration of the continuous beam is such that it has a single free joint and two fixed ends. Except for the apparent absence of an unbalanced moment at the free joint, the structure is perfectly tailored to the conditions maintained dur-

ing the development of the basic concepts of moment distribution. A solution will be obtained in a stepwise application of these concepts after replacing the transverse loading with a suitable unbalanced moment at joint B.

Step 1: As a preliminary operation, consider joint B to be held temporarily in a fixed position as depicted in Fig. 13.7a. In this state, each span acts as an independent beam and can be analyzed individually. A direct result is the development of resisting moments at the fixed ends of each member. Their numerical values are calculated from formulas extracted directly from Table A.3.

For member AB:

$$M_{AB_0} = -\tfrac{1}{12}wL^2 = -\tfrac{1}{12}(1)(8)^2 = -5.33 \text{ kN} \cdot \text{m}$$

$$M_{BA_0} = +\tfrac{1}{12}wL^2 = +\tfrac{1}{12}(1)(8)^2 = +5.33 \text{ kN} \cdot \text{m}$$

For member BC:

$$M_{BC_0} = -\tfrac{1}{12}wL^2 = -\tfrac{1}{12}(1)(12)^2$$
$$= -12.00 \text{ kN} \cdot \text{m}$$

$$M_{CB_0} = +\tfrac{1}{12}wL^2 = +\tfrac{1}{12}(1)(12)^2$$
$$= +12.00 \text{ kN} \cdot \text{m}$$

These moment values are shown in Fig. 13.7b. In this figure, all end moment *vectors* are shown in the positive sense (clockwise on the member ends and counterclockwise at the joints) whether they truly act in these directions or not. However, the proper mathematical sign associated with each moment value is also given so that the correct direction for the moment vectors can be surmised. This manner of presentation is employed because it corresponds to a tabulated procedure, which is to be described subsequently.

Step 2: Examination of Fig. 13.7b indicates that joint B is not in equilibrium unless a restraining moment, RP_1, is developed at joint B. The magnitude of this moment (shown clockwise) must be -6.67 kN $\cdot$ m. The implication is that if joint B is released from its fixed state, RP_1 cannot be developed, and the net internal moment will act as an unbalanced moment at joint B. The existence of this moment is shown in Fig. 13.7c. As before, the moment vector is shown in the positive sense (counterclockwise on

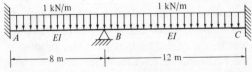

Fig. 13.6. Two-span continuous beam, Example 13.1.

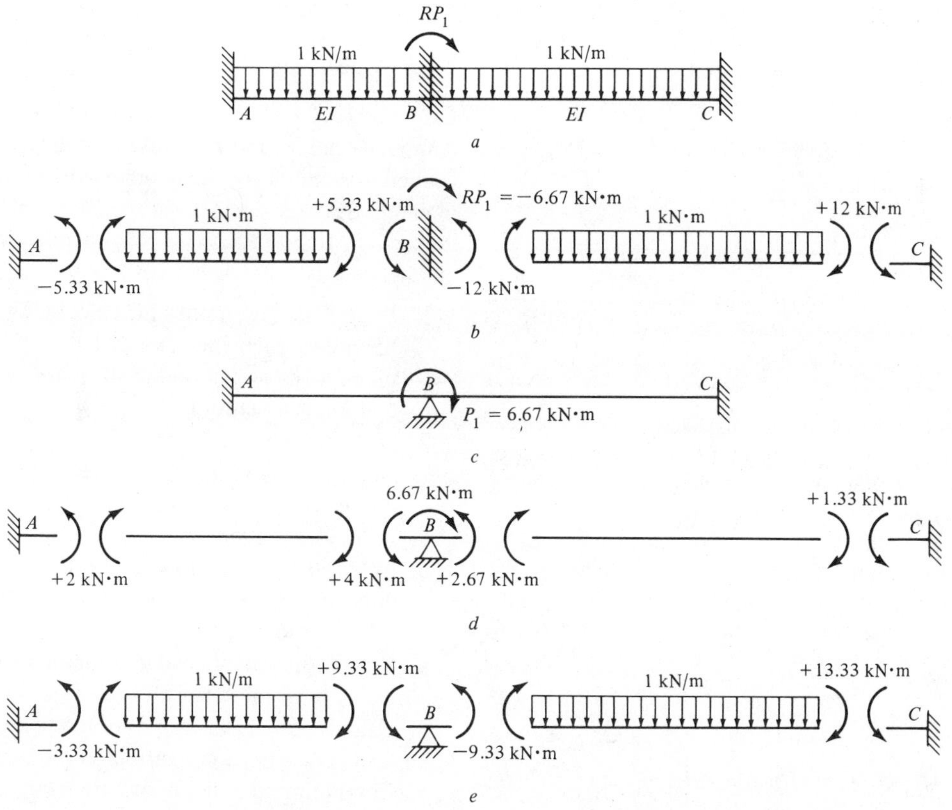

Fig. 13.7. Moment distribution applied to beam of Fig. 13.6. *a*, joint *B* held fixed temporarily; *b*, fixed-end moments; *c*, unbalanced moment at joint *B*; *d*, distributed and carryover moments; *e*, final moment values.

the joint) with the proper mathematical sign (positive in this case) given to permit interpretation of the true direction.

Step 3: The structure in Fig. 13.7c is perfectly suited to the moment distribution principles. Therefore, the unbalanced moment at joint *B* is distributed into the member ends in proportion to the relative stiffnesses of these members. The distributed moments are computed as follows (note that the far end of each member is fixed).

Stiffness of member $BA = \dfrac{4EI}{8} = 0.500EI$

Stiffness of member $BC = \dfrac{4EI}{12} = 0.333EI$

$DF_{BA} = \dfrac{0.500EI}{0.833EI} = 0.60$

$DF_{BC} = \dfrac{0.333EI}{0.833EI} = 0.40$

Distributed moment in member *BA*

$$= 0.60(+6.67)$$

$$= +4.00 \text{ kN} \cdot \text{m}$$

Distributed moment in member *BC*

$$= 0.40(+6.67)$$

$$= +2.67 \text{ kN} \cdot \text{m}$$

These moments are shown at joint *B* in Fig. 13.7d. Both moments have a positive sign because they must act in a counterclockwise direction on the joint.

Step 4: The distributed moments shown at joint *B* in Fig. 13.7d have a further influence on the structure. Because the joints at *A* and *C* are fixed in position, carryover moments are created at both locations. Each carryover moment has a magnitude equal to half the magnitude of the distributed moment (at the other end of the member which frames into the joint) and acts in the same sense (i.e., has the same mathematical sign if clockwise vectors are shown on the ends of the members). In this instance,

$$\text{Carryover moment at } A = \tfrac{1}{2}\,(+4.00)$$

$$= +2.00 \text{ kN} \cdot \text{m}$$

$$\text{Carryover moment at } C = \tfrac{1}{2}\,(+2.67)$$

$$= +1.33 \text{ kN} \cdot \text{m}$$

The carryover moments are also included in Fig. 13.7d.

Step 5: As a final stage of the solution, it is necessary to determine the end moments corresponding to the original structure of Fig. 13.7a. This is done by combining the end moments created in Fig. 13.7b with those produced in Fig. 13.7d. In other words, the temporary fixed-end moments are combined with the moments caused by releasing the joint at B.

$$M_{AB} = M_{AB_0} + \text{Carryover moment}$$

$$= -5.33 + 2.00 = -3.33 \text{ kN} \cdot \text{m}$$

$$M_{BA} = M_{BA_0} + \text{Distributed moment}$$

$$= +5.33 + 4.00 = 9.33 \text{ kN} \cdot \text{m}$$

$$M_{BC} = M_{BC_0} + \text{Distributed moment}$$

$$= -12.00 + 2.67 = -9.33 \text{ kN} \cdot \text{m}$$

$$M_{CB} = M_{CB_0} + \text{Carryover moment}$$

$$= +12.00 + 1.33 = +13.33 \text{ kN} \cdot \text{m}$$

At this point, the analysis by moment distribution is complete. The final moment results are depicted in Fig. 13.7e. As an indicator of correctness, the equilibrium of joint B should be checked. In this case, M_{BA} and M_{BC} have equal magnitude but opposite sense (as indicated by the difference in mathematical signs) and statics is, indeed, satisfied. Finally, further computations, such as constructing the shear and moment diagrams, require a recognition of the proper direction of the end monents. In this example, M_{AB} and M_{BC} are negative, and their true sense is counterclockwise on the member ends. M_{BA} and M_{CB} are positive, and their true sense is clockwise on the member ends. However, all four moments produce compression in the bottom fibers and would be negative by the "designer's convention" employed in constructing shear and moment diagrams.

13.3 MOMENT-DISTRIBUTION TABLE

13.3.1 Beams

Example 13.1 serves as an introduction to the computational process involved in analyzing framed structures by the moment-distribution method. Several distinct steps can be distinguished in that solution. These steps can be enumerated as follows:

1. Fixing the free joint and calculating the fixed-end moments associated with the given loads acting on the structure in this temporary condition.
2. Determining the restraining moment resulting at the temporarily fixed joint as a consequence of the algebraic difference in fixed-end moments at that joint.
3. Releasing the free joint and allowing the unbalanced internal moment to act, by itself, on the structure.
4. Calculating the distributed moments that result from the release of the free joint.
5. Calculation of the carryover moments associated with each distributed moment.
6. Combining the fixed-end moment with the distributed moment (or carryover moment) at each member end to obtain the total end moments.

As illustrated in Example 13.1, the computations associated with this procedure are straightforward, however, the paperwork is somewhat lengthy. Moreover, for problems involving more than one free joint, the moment-distribution procedure becomes more involved. It will be shown that all of the free joints must be temporarily fixed and then released in an alternating fashion for a number of cycles before convergence to a solution is achieved. In this fashion, a *sequence* of distributed moments and carryover moments is produced at each member end; these must be correctly summed at the conclusion of the problem and added to the fixed-end moments.

If, as a general procedure, the moment-distribution computations are recorded in the fashion of Example 13.1, the task of keeping track of the numbers becomes unwieldy. For this reason, it is common practice to assemble the moment distribution into an organized table. In this way, the various stages of computation are reduced to little more than a bookkeeping procedure. A variety of tabular arrangements for moment distribution appears in different textbooks. To acquaint the reader with the configuration adopted by the author, Example 13.1 will be repeated in a table format. All subsequent examples will be solved using the same format. Complete understanding of the computations implied in the table is essential for comprehension of these later examples.

Example 13.2. Analyze the beam of Fig. 13.7 by moment distribution using the tabular method of computation.

Solution: All necessary computations are assembled in Table 13.1. All joints are listed in the first line of the table, and each is separated by a double vertical line. In line 2, each joint heading is subdivided into a number of columns, with a column assigned for each member that frames into the joint. Line 3 contains the member stiffness values, which permit calculation of the distribution factors (*DF*) given in line 4. Joints *A* and *C* are fixed and do not require distribution factors. These first four lines constitute preliminary steps, and the remainder of the table serves to permit analysis of any given loading.

Line 5 is labeled with a 1 as it constitutes the "first cycle" of the moment-distribution process. This line is subdivided into two parts. The upper part (labeled M_0) is a tabulation of the fixed-end moments computed in Fig. 13.7b. At this point, the free joint at

B is held temporarily fixed. The initial effect of releasing joint *B* is shown in the lower part of line 5 (labeled *Dist*). Upon release, joint *B* rotates to an equilibrium position and develops the distributed moments shown in Fig. 13.7d. In the table, these are computed by first recognizing that in the fixed state the restraining moment at joint *B* is the algebraic difference in the fixed-end moments, which is $(5.33 - 12.00) = -6.67$. Upon release of the joint, the net internal moment acts as an unbalanced moment at joint *B*. This moment is a clockwise moment of equal magnitude but opposite sign to that of the restraining moment, i.e. +6.67 in the present example.

Next, this unbalanced moment is multiplied by each of the distribution factors tabulated in line 4 for joint *B* to produce the distributed moments. The results are entered in the *Dist* line. Joints *A* and *C* remain fixed, so no distributed moments are developed at these locations and no table entries are necessary. At this stage, joint *B* is in equilibrium, and the first cycle is completed.

Line 6 constitutes the "second cycle" of the moment distribution. This cycle is necessitated by the secondary effect of releasing joint *B*. Contingent with the development of distributed moments at joint *B*, is the development of carryover moments at joints *A* and *B* (see Fig. 13.7d). Numerical values for the carry-over moments are entered in the upper part of line 6 (labeled *CO*) with the appropriate signs. (Note that carry-overs occur across double vertical lines in the table.) No carry-over moments occur at joint *B*, so dashes are entered. The lower part of line 6 is labeled *Dist* and has no entries for this particular problem. An explanation for the purpose of this line is reserved for a subsequent example.

Table 13.1. Moment Distribution for Example 13.2.

Joint	A	B		C	
Member	AB	BA	BC	CB	
S_i or S_j	0.5EI	0.5EI	0.333EI	0.333EI	
DF	–	0.60	0.40	–	
1 M_0	−5.33	+5.33	−12.00	+12.00	see Fig. 13.7b
1 Dist	–	+4.00	+2.67	–	see Fig. 13.7d
2 CO	+2.00	–	–	+1.33	see Fig. 13.7d
2 Dist	–	–	–	–	
	−3.33	+9.33	−9.33	+13.33	see Fig. 13.7e

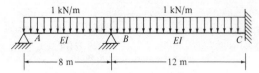

Fig. 13.8.   Beam for Example 13.3.

$$S_i = S_j = \frac{3EI}{8} = 0.375EI$$

For member BC:

$$S_i = S_j = \frac{4EI}{12} = 0.333EI$$

Final answers are obtained by summing the moment entries below each member heading. These are in agreement with values shown in Fig. 13.7e.

Example 13.3.   Analyze the beam in Fig. 13.8 by the moment-distribution method.

Solution: The given structure is identical to that examined in Example 13.2 except the support at joint A has been changed to a pin support. A solution in tabular form is presented in Table 13.2 and differs from Table 13.1 only with respect to the influence of this pin support. Several facets of this solution require comment.

The first two lines of Table 13.2 simply list the joint and member labeling. Line 3 contains the member stiffnesses and differs from Table 13.1 because of the pin support at A. In the moment distribution process, joint B will be periodically released. As a consequence, member BA behaves as a pinned-pinned span and has a member stiffness of $3EI/L$ instead of $4EI/L$. Member BC behaves as pinned-fixed span, necessitating a stiffness of $4EI/L$ for the member. Accordingly, the stiffness values are computed as follows

For member BA:

and entered in line 3. Distribution factors are computed and entered in line 4.

The first cycle of moment distribution is initiated by fixing joint B temporarily and computing the resulting fixed-end moments for each span. Because span AB is not fixed at both ends, the fixed-end moment equations cannot be extracted from Table A.3. To accommodate this need, Table A.4 is presented as a compilation of fixed-end moments for common loadings on a propped, cantilever beam. For member AB, the single fixed-end moment is computed as

$$M_{BA_0} = +\tfrac{1}{8} wL^2 = \tfrac{1}{8}(1)(8)^2 = +8.0 \text{ kN} \cdot \text{m}$$

The reader should verify the correctness of the positive sign. Fixed-end moments exist at both ends of member BC and are unchanged from Example 13.2. To complete the first cycle, the net fixed-end moment (-4.00 kN $\cdot$ m) at joint B necessitates a clockwise restraining moment of magnitude -4.00 kN $\cdot$ m. Upon release of joint B, an unbalanced moment of $+4.00$ kN $\cdot$ m (clockwise) results and is distributed into the members according to the distribution factors 0.53 and 0.47.

The second cycle of moment distribution consists of entering carryover moments. The carryover to

Table 13.2.   Moment Distribution for Example 13.3.

Joint	A	B		C
Member	AB	BA	BC	CB
S_i or S_j	0	$0.375EI$	$0.333EI$	$0.333EI$
DF	–	0.53	0.47	–
1 M_0	–	+8.00	–12.00	–12.00
1 Dist	–	+2.12	+1.88	–
2 CO	–	–	–	+0.94
2 Dist	–	–	–	–
	0	+10.12	–10.12	–11.06

joint A is zero because it is an exterior pin. Joint C receives a moment of 0.94. Similar to Example 13.2, the last line of the table labeled *Dist* is superfluous and has no entries. At this point, the columns are summed to yield the final end moments.

Moment distribution applied to multispan continuous beams is executed in exactly the same sequence as for the two-span structures illustrated in earlier examples. However, a complete solution in two cycles of distribution is not possible for such structures. The presence of several interior supports is the chief reason why the possibility of such an "immediate" solution is impossible. In truth, the first cycle of moment distribution represents an approximate solution to the given problem. In previous examples, a second cycle had to be executed to correct for the error produced by the neglect of carryover moments in the first cycle. By condition of those particular problems, convergence to a correct solution was achieved after completion of the second cycle. For larger structural systems, additional cycles beyond the first two are necessary before one achieves an acceptable degree of convergence. Reasons for the further cycling and details concerning treatment of the increased number of members and joints are best presented by use of another numerical sample problem.

Example 13.4. A multispan beam is drawn in Fig. 13.9. Analyze the structure for the given loading by employing the moment-distribution technique. In addition, the support at C settles 0.6 in. Perform a sufficient number of cycles to produce convergence to five significant figures. Use $E = 17,280$ kip/in.^{2} and $I = 400$ in.^{4}

Solution: A complete solution to the problem is contained in Table 13.3.

Step 1: Except for an increased number of columns in the table, the first four lines of Table 13.3 are entered in the usual fashion. The overhang at the right end is statically determinate and can be viewed as applying a 40 kip-ft clockwise moment on the support at D. Seen in this way, member DE can be (and is) omitted from the table. Recognizing the external pinned support at D that exists in this interim state, member stiffnesses are computed

For member AB:

$$S_i = S_j = \frac{4(2EI)}{16} = 0.500EI$$

For member BC:

$$S_i = S_j = \frac{4(2EI)}{20} = 0.400EI$$

For member CD:

$$S_i = \frac{3(4EI)}{24} = 0.500EI$$

$$S_j = 0$$

With the member stiffnesses known, distribution factors are computed *for each joint* and are based upon the relative stiffnesses of all members that frame into each particular joint. For example, at joint C,

$$DF_{CB} = \frac{0.400EI}{0.400EI + 0.500EI} = 0.444$$

$$DF_{CD} = \frac{0.500EI}{0.400EI + 0.500EI} = 0.556$$

Step 2: Fixed-end moments are computed taking into account the exterior pin support at D and the

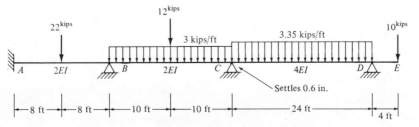

Fig. 13.9. Multispan beam for Example 13.4.

Table 13.3.　Moment Distribution for Example 13.4.

Joint	A	B		C		D
Member	AB	BA	BC	CB	CD	DC
S_i or S_j	0.500EI	0.500EI	0.400EI	0.400EI	0.500EI	0
DF	–	0.556	0.444	0.444	0.556	–
1 M_0	–44.00	+44.00	–202.00	+58.00	–171.00	+40.00
1 Dist	–	+87.85	+70.15	+50.17	+62.83	–
2 CO	+43.93	–	+25.09	+35.08	–	
2 Dist	–	–13.95	–11.14	–15.58	–19.50	
3 CO	–6.98	–	–7.79	–5.57	–	
3 Dist	–	+4.33	+3.46	+2.47	+3.10	
4 CO	+2.17	–	+1.24	+1.73	–	
4 Dist	–	–0.69	–0.55	–0.77	–0.96	
5 CO	–0.35	–	–0.39	–0.28	–	
5 Dist	–	+0.22	+0.17	+0.12	+0.16	
6 CO	+0.11	–	+0.06	+0.09	–	
6 Dist	–	–0.03	–0.03	–0.04	–0.05	
7 CO	–0.02	–	–0.02	–0.02	–	
7 Dist	–	+0.01	+0.01	+0.01	+0.01	
	–5.14	+121.74	–121.74	+125.41	–125.41	+40.00

results entered in line 5 of Table 13.3. The support settlement at joint C introduces a new aspect to moment distribution. A simple approach to the treatment of this effect is to consider the known displacement to be part of the loading. Additional fixed-end moments are calculated and combined with those created by the transverse loading. Table A.5 lists formulas needed for making the fixed-end moment calculation when a support settlement is involved. (The reader is cautioned that in the slope-deflection method, support movements are an explicit part of the slope-deflection equations and treated separately from the fixed-end moment computations. Table A.5 has no use in that method.)

For member AB (see Table A.3, Case a)

$$M_{AB_0} = -\frac{PL}{8} = -\frac{22(16)}{8} = -44 \text{ kip-ft}$$

$$M_{BA_0} = +\frac{PL}{8} = \quad = +44 \text{ kip-ft}$$

For member BC (see Table A.3, Cases a and c, and Table A.5, Case a). This member experiences the combined effects shown in Fig. 13.10.

$$M_{BC_0} = -\frac{PL}{8} - \frac{1}{12} wL^2 - \frac{6EI}{L^2}\Delta$$

$$= -\frac{12(20)}{8} - \frac{1}{12}(3)(20)^2$$

$$- \frac{6(2)(17,280)(400)}{(20)^2(1728)}(0.6)$$

$$= -30 - 100 - 72 = -202 \text{ kip-ft}$$

$$M_{CB_0} = +30 + 100 - 72 = 58 \text{ kip-ft}$$

For member CD (see Table A.4, Cases c and f, and Table A.5, Case b). This member experiences the combined effects shown in Fig. 13.11.

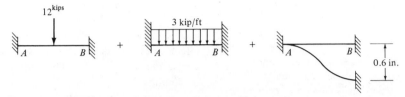

Fig. 13.10. Source of fixed-end moments for span BC.

$$M_{CD_0} = -\frac{1}{8}wL^2 + \frac{1}{2}M - \frac{3EI}{L^2}\Delta$$

$$= -\frac{1}{8}(3.35)(24)^2 + \frac{1}{2}(40)$$

$$+ \frac{3(4)(17{,}280)(400)}{(24)^2(1728)}$$

$$= -241 + 20 + 50 = -171 \text{ kip-ft}$$

$$M_{DC_0} = +40 \text{ kip-ft}$$

The 40 kip-ft moment, M_{DC_0}, is entered simply to record this known moment, and no further use is made of the last column in Table 13.3. Its effect on the structure is accounted for by the inclusion of the 20 kip-ft fixed-end moment in calculating M_{CD_0}.

Step 3: Examination of the fixed-end moments shows that, unlike previous examples, two of the joints do not satisfy equilibrium if released from the fixed position. Consequently, when the joints are released moments of magnitudes $(44 - 202) = -158$ kip-ft and $(58 - 171) = -113$ kip-ft, develop at B and C respectively. Upon release, corresponding unbalanced clockwise moments of magnitudes $+158$ kip-ft and $+113$ kip-ft act at these joints. The resulting joint rotations induce distributed moments at each location as indicated in the first *Dist* line of the table.

Step 4: If M_0 and *Dist* moments are combined after the first cycle of moment distribution, all joints are found to be "in balance" (statics is satisfied). However, this apparent solution is incorrect and only ap-

proximates the true answer because an additional influence must be considered. Clearly, the rotations at joints B and C, which occur after release from the fixed condition, do not simply influence the member ends that frame into these joints. Secondary moments are induced throughout the structure and are treated in the following manner. After the initial release of joints B and C and the development of distributed moments, carryover moments, in a sense, are "sent out" from each joint. Viewed in this way, the joints (except D) can be quickly restored to a fixed state in order to "catch" these approaching moments. This is the reason for using a $4EI/L$ stiffness for members AB and BC. Thus, all carryover moments can be entered simultaneously as the first line in the second cycle of the distribution process.

At this stage some of the joints do not satisfy equilibrium. Thus, a secondary set of restraining moments are developed at the temporarily fixed joints. A second release of the joints causes the additional distributed moments entered in the second *Dist* line of the table. With the entry of these values, the second cycle of the distribution table is complete.

Step 5: If all moments created in the first two cycles of moment distribution are combined, all joints will again be found to satisfy equilibrium. The results constitute an improved, but still unsatisfactory, approximation to the correct solution. To improve the accuracy of the solution, carry-over moments created by the second set of distributed moments must be analyzed. This is done in the next (third) moment-distribution cycle listed in Table 13.3.

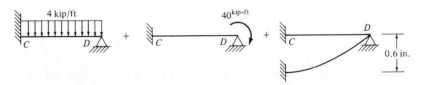

Fig. 13.11. Source of fixed-end moments for span CD.

Step 6, etc.: Each cycle of moment distribution adds a further refinement to the preceding approximate solution. Sufficient cycles must be executed to produce the desired degree of accuracy. In most textbooks, the moment-distribution process is stopped when the size of the distributed moments is 0.01 kip-ft or less. This is commonly done to compare the solution to any given problem to that obtained by alternative solution methods. Practically, distributed moments in the order of 1% of the smallest fixed-end moment entry are probably an adequate signal to halt the analysis. Seven cycles of distributions were performed for the current example. At the conclusion, all distributed moments are 0.01 kip-ft, indicating that convergence has been achieved. All moments are summed in each member column, giving the answers listed at the bottom of the table.

13.3.2   Planar Frames

Application of moment distribution to beam problems has been explained previously. Extension of the procedure to the investigation of general planar frames requires little new discussion. Tabular computation layout is the same except for the possibility of several members framing into any given joint. In fact, the treatment of any existing sidesway degrees of freedom is the only new conceptual consideration. This issue will be discussed subsequently. As in all classical methods of structural analysis, axial deformation is neglected in moment distribution.

13.3.2.1   Rotational Degrees of Freedom

When a frame has no joint translations, the execution of moment distribution table is not visibly different than for beam problems. Introduction of additional members to a joint simply increases the number of components that share in resisting any unbalanced moment that acts when the joint is released from a temporary fixed state. Such a state was used earlier as the basis for the derivation of the distribution factor concept expressed by Eq. 13.28. In earlier examples NM, the number of members framing into a joint, was equal to 1 or 2 depending upon whether the joint was an exterior or an interior

beam support, respectively. In general frames, several members may meet at a common joint and, consequently, NM would exceed 2. The effect of this situation is a simple need to increase the number of subdivisions in the corresponding joint column in the moment-distribution table. The unbalanced moment at the joint is the net moment created by the fixed-end moments induced by all members connected to the joint. In accordance with Eq. 13.26, the unbalanced moment would then be distributed among these members in proportion to their relative stiffnesses. Finally, carryover moments associated with the joint release must be transferred to the correct location in the table. Unlike beam problems, this is not always an adjacent column in the table.

Example 13.5. Analyze the frame shown in Fig. 13.12 by employing moment distribution.

Solution: With symmetrical loading, no horizontal translation sidesway is induced into the given symmetrical frame, and no special consideration of this aspect of behavior is necessary. A complete solution is compiled in Table 13.4. As clarification to the reader, a few helpful comments are made.

First, because distribution factors are based upon a ratio of stiffness terms, they are unitless, and only relative stiffness values are needed to compute their values. In this regard, stiffness values needed are:

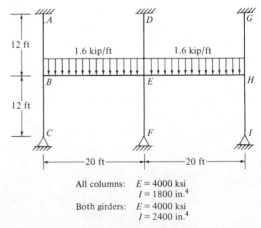

Fig. 13.12.   Two-story, two-bay frame for Example 13.5.

Table 13.4.   Moment Distribution for Transverse Loads — Example 13.5.

Joint	A	B			C	D		E			F	G	H			I
Member	AB	BA	BE	BC	CB	DE	ED	EB	EH	EF	FE	GH	HG	HE	HI	IH
S_i or S_j	2	2	1.6	1.5	1.5	2	2	1.6	1.6	1.5	1.5	2	2	1.6	1.5	1.5
DF	–	0.392	0.314	0.294	–	–	0.299	0.239	0.239	0.224	–	–	0.392	0.314	0.294	–
M_0	–	–	–53.33	–		–	–	+53.33	–53.33	–		–	–	+53.33	–	–
Dist (1)	–	+20.90	+16.75	+15.68		–	–	–	–	–		–	–20.90	–16.75	–15.68	–
CO	+10.45	–	–	–		–	–	+8.38	–8.38	–		–10.45	–	–	–	–
Dist (2)	–	–	–	–		–	–	–	–	–		–	–	–	–	–
	+10.45	+20.90	–36.58	+15.68	0	0	0	+61.71	–61.71	0	0	–10.45	–20.90	+36.58	–15.68	0

For fixed-end columns:

$$S_i = S_j = \frac{4EI}{L} = \frac{4(400,000)(1800)}{12(12)}$$

$$= 200,000 \text{ kip-in.}$$

For pinned-end columns:

$$S_i = S_j = \frac{3}{4}(200,000) = 150,000 \text{ kip-in.}$$

For girders:

$$S_i = S_j = \frac{4(4000)(2400)}{20(12)} = 160,000 \text{ kip-in.}$$

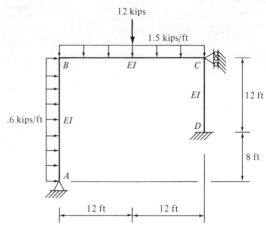

Fig. 13.13.   Planar frame for Example 13.6.

Dividing each stiffness by 100,000 gives the relative stiffness values entered in Table 13.4. Distribution factors are determined at each joint in accordance with Eq. 13.28.

Second, only two unbalanced moments, those at joints B and H, exist in the initial fixed state. Joint E experiences no unbalanced effect because the fixed-end moments from member EB and EH cancel exactly. The columns carry no load and have no fixed end moments.

Third, distributed moments are calculated at each joint, and carry-over moments produced are transferred to the opposite ends of the affected members. For example, the distributed moment of 16.75 kip-ft at end B of member BE causes a carry-over moment of 8.38 kip-ft to joint E. This moment is entered in the subcolumn of joint E, which is labeled EB. The reader should verify the remaining carry-over moments. Also, no carry-over moments are transferred to the pinned column ends, C, F, and I.

Last, after the initial carry-over moments are entered, no unbalanced moments exist (all joints are found to be in equilibrium), and no further computations are necessary. This is not typical of frames. Many more cycles are usual, as will be seen in Example 13.6. As expected, the final moments (summed at the bottom of the table) exhibit the symmetry present in the structure and its loading.

Example 13.6.   Analyze the frame given in Fig. 13.13. Use the moment-distribution method.

Solution: With axial deformation neglected, the frame has no translational degrees of freedom. The moment-distribution solution for the frame is presented in Table 13.5. Because no joint has more than

two members framed into it, the table looks much the same as for a beam problem.

Step 1:   The first four lines indicate the joints, the members framing into each joint, the member stiffnesses and the distribution factors. During the solution the interior joints B and C will be simultaneously released during the balancing of joints (distribution of unbalanced moments) and simultaneously fixed during each carry-over. Thus, a $4EI/L$ stiffness is used for all members framing to interior joints (the same as for beams.) For member BA, the exterior end is pinned and a $3EI/L$ stiffness is used. For member CD the exterior end is fixed and a $4EI/L$ stiffness is used. As usual, distribution factors reflect the relative stiffnesses of the members framing into the various joints.

Step 2:   In the restrained state (M_0 line), joints B and C are fixed. Members AB and BC develop fixed-end moments by acting as individual beams. Member CD behaves independently, too, but has no transverse loading. Thus

$$M_{AB} = 0$$

$$M_{BA} = \frac{1}{8} wL^2 = \frac{1}{8}(.6)(20)^2 = 30 \text{ kip-ft}$$

$$M_{BC} = -\frac{1}{12} wL^2 - \frac{PL}{8} = -\frac{1}{8}(1.5)(24)^2$$

$$-\frac{12(24)}{8} = -108 \text{ kip-ft}$$

$$M_{CB} = -M_{BC} = 108 \text{ kip-ft}$$

$$M_{CD} = M_{DC} = 0$$

Table 13.5.   Moment Distribution for Example 13.6.

Joint		A	B		C		D
Member		AB	BA	BC	CB	CD	DC
Si or Sj		.15EI	.15EI	.167EI	.167EI	.333EI	.333EI
DF		1	.473	.527	.333	.667	–
1	M_0	0	+30.00	−108.00	+108.00	0	0
	Dist.	–	+36.89	+41.11	−36.00	−72.C0	0
2	CO	–	–	−18.00	+20.56	–	−36.00
	Dist.		+8.51	9.49	−6.85	−13.71	–
3	CO	–	–	−3.42	+4.75	–	−6.86
	Dist.		+1.62	+1.80	−1.58	−3.17	–
4	CO	–	–	−0.79	+0.90	–	−1.59
	Dist.		+0.37	+0.42	−0.30	−0.60	–
5	CO	–	–	−0.15	+0.21	–	−0.30
	Dist.		+0.07	+0.08	−0.14	−0.07	–
6	CO	–	–	−0.07	+0.04	–	−0.04
	Dist.		+0.03	+0.04	−0.01	−0.03	–
7	CO	–	–	−0.00	+0.02	–	−0.02
	Dist.		0	+0.00	−0.01	−0.01	–
		0	77.49	−77.49	89.59	−89.59	−44.83

Step 3: Joints B and C are released and distributed moments develop to balance the joints. This completes cycle 1. Adding the column entries confirms the balanced condition.

Step 4: Joints B and C are immediately fixed to receive the carryover moments (from B to C and C to B) entered in the first line of cycle 2.

Step 5: The carryover moments represent unbalanced moments. A second release of joints B and C creates the distributed moments (entered to complete cycle 2) necessary to again balance the joints. Summing the columns will verify the restoration of a balanced state.

Step 6: The remaining cycles are a repetition of the sequence of carryover and distribution. Cycles are continued until the distributed moments in the latest cycle are sufficiently small to terminate. Seven cycles were required in the present problem.

Step 7: The columns are summed to obtain the final answers entered at the base of the table.

13.3.2.2   Translational Degrees of Freedom

In prelude to a discussion of the handling of joint translational motions in the moment-distribution method, one must recognize the difficulty that they create in this analysis technique. Fundamentally, the process of distributing unbalanced moments involves the periodic fixing and releasing of rotational resistance at free joints. All distributed moments and carryover moments that enter into the tabulated computations reflect the influence of the rotational effects. Translational motions, which might occur at released joints, also influence the structure, but no provision has been made in the derivation of the method for the direct treatment of such displacements. No joint translations occurred in the frames analyzed thus far. This is true only if such degrees of freedom are actually absent by virtue of the support conditions or if the framework and its loading exhibit geometric symmetry. In practice, this is not always the case, and some means of dealing with their effects must be developed. In most treatments of the subject, this is done through employment of "sidesway corrections," and these corrections are the approach adopted in this textbook.

Translational motions in general frames conceptually have the same influence as support settlements in a beam problem. As shown in Example 13.4, in beams these actions can be treated by including additional moments in the fixed-end moment calculations. Table A.5 was provided for this specific purpose. The difficulty with sidesway displacements is that, unlike the case of specified differential support settlements, the magnitudes of the movements are unknown at the start of the analysis. This is somewhat perplexing but, in reality, can be overcome by a simple process. The unknown displacement values constitute a barrier to the analysis, which can be bypassed by merely guessing these values. Subsequent to performing the analysis, the answers are scaled in accordance with a need to satisfy static equilibrium. This final adjustment represents the "sidesway correction" and is similar to the use of "shear conditions" in the slope-deflection method. As an alternative technique, the sidesway displacements can be retained as unknown quantities in the analysis. In essence, all moment values are expressed in terms of these quantities. At the completion of the required moment-distribution calculations, the magnitudes of the displacements are established by invoking statics conditions.

13.3.2.3 Examples with a Single Translation

Examples 13.7 and 13.8 serve to demonstrate the use of the sidesway correction.

Example 13.7.   Analyze the frame shown in Fig. 13.14 by the moment-distribution process. All member properties are identical with those given in Example 13.5.

Solution: The frame geometry is identical with that of Fig. 13.12a, but the loading differs by the addi-

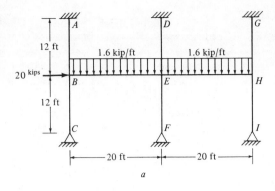

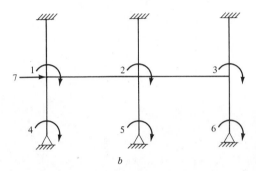

Fig. 13.14. Frame for Example 13.7. Translation included.

tion of a concentrated horizontal load at joint B. This 20-kip force has the major effect of incorporating a sidesway motion into the system. Degrees of freedom for the structure are shown in Fig. 13.14b. The unsymmetrical loading causes a nonzero translational motion in direction 7, which is unlike the situation examined in Example 13.5.

After examining Fig. 13.14 carefully, the inability to treat sidesway effects directly is quite evident. Fixed-end moments caused by the transverse loading on the two girders are easily determined and, in fact, are identical with those of Example 13.5. With the addition of the 20-kip lateral load, the frame in Fig. 13.14a is clearly different from that of Fig. 13.12. However, no direct formula exists for the inclusion of fixed-end moments caused by this load. Recognizing that the lateral force causes sidesway motion, the analysis can proceed as follows:

Step 1:   Perform a complete moment-distribution analysis for the effects of transverse member loads acting by themselves. In this particular problem, the results obtained in Example 13.5 represent this analysis, and the answers can be extracted from Table 13.4.

Step 2: Perform a moment-distribution analysis for the effects of sidesway acting by themselves. To do this, it is necessary to guess the magnitude of the sidesway displacement or simply treat it as an unknown. The latter choice will be employed in this example. Fixed-end moments can be determined by temporarily restraining the rotation at joints B, E, and H and allowing the sidesway to occur. If the magnitude of the sidesway movement is Δ, the six columns will displace as depicted in Fig. 13.15. With the column ends fixed at the girder level, formulas listed in Table A.5 apply.

For columns AB, DE, and GH:

$$M_0(\text{top}) = M_0(\text{bottom}) = \frac{6EI}{L^2} \Delta$$

$$= \frac{6(4000)(1800)}{(144)^2} \Delta$$

$$= 2083\Delta$$

For columns BC, EF, and HI:

$$M_0(\text{top}) = -\frac{3EI}{L^2} \Delta = -1042\Delta$$

$$M_0(\text{bottom}) = 0$$

These moments are entered in Table 13.6, and moment distribution is executed for a sufficient number of cycles to reduce the answers to convergence in the third decimal place. It should be noted that Δ has been factored out of the table. Realizing that its magnitude is likely to be small, the answers summed at the bottom of Table 13.6 are quite refined.

Step 3: Total moments can be obtained if the magnitude of Δ is known. This value is established by enforcing the sidesway correction. Partial free-body diagrams of each column and the continuous girder

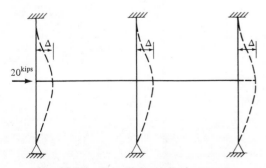

Fig. 13.15. Translational displacement.

in Fig. 13.15 are drawn in Fig. 13.16. These provide a basis for computing Δ.

For horizontal equilibrium of the girder:

$$H_{BA} + H_{ED} + H_{HG} - H_{BC} - H_{EF} - H_{HI} = 20$$

$$(a)$$

For moment equilibrium of each of the columns:

$$H_{BA} + H_{ED} + H_{HG} = \tfrac{1}{12}(M_{AB} + M_{BA} + M_{DE}$$

$$+ M_{ED} + M_{GH} + M_{HG})$$

$$H_{BC} + H_{EF} + H_{HI} = \tfrac{1}{12}(M_{BC} + M_{EF} + M_{HI})$$

Substitution into condition (a) yields

$$\tfrac{1}{12}[(M_{AB} + M_{BA} + M_{DE} + M_{ED} + M_{GH} + M_{HG})$$

$$- (M_{BC} + M_{EF} + M_{HI})] = 20 \qquad (b)$$

Extracting the moments obtained in Table 13.6 gives:

$$M_{AB} + M_{BA} + M_{DE} + M_{ED} + M_{GH} + M_{HG}$$

$$= 11,048\Delta$$

$$M_{BC} + M_{EF} + M_{HI} = -3850\Delta$$

By substitution of these known values, condition (b) becomes

$$\tfrac{1}{12}(14,898\Delta) = 20$$

which gives $\Delta = 0.0161$ in.

Final moments for the frame in Fig. 13.14a are obtained as follows:

$$M_{AB} = +10.45 + 1897\Delta = +40.99 \text{ kip-ft}$$

$$M_{BA} = +20.90 + 1711\Delta = +48.45 \text{ kip-ft}$$

$$M_{BE} = -36.58 - 389\Delta = -42.84 \text{ kip-ft}$$

$$M_{BC} = +15.68 - 1321\Delta = -5.59 \text{ kip-ft}$$

$$M_{DE} = +1972\Delta = +31.75 \text{ kip-ft}$$

$$M_{ED} = +1860\Delta = +29.95 \text{ kip-ft}$$

$$M_{EB} = +61.71 - 327\Delta = +56.45 \text{ kip-ft}$$

$$M_{EH} = -61.71 - 327\Delta = -66.97 \text{ kip-ft}$$

$$M_{EF} = -1208\Delta = -19.45 \text{ kip-ft}$$

Table 13.6.   Moment Distribution for Sidesway Effect—Example 13.7.

Joint	A	B			C	D	E				F	G	H			I
Member	AB	BA	BE	BC	CB	DE	ED	EB	EH	EF	FE	GH	HG	HE	HI	IH
S_i or S_j	2	2	1.6	1.5	1.5	2	2	1.6	1.6	1.5	1.5	2	2	1.6	1.5	1.5
DF	–	0.392	0.314	0.294	–	–	0.299	0.239	0.239	0.224	–	–	0.392	0.314	0.294	–
M_0	+2083	+2083	–	–1042		+2083	+2083	–	–	–1042		+2083	+2083	–	–1042	
1 Dist	–	–408	–327	–306		–	–312	–249	–249	–233		–	–408	–327	–306	
CO	–204	–	–125	–		–153	–	–163	–163	–		–204	–	–125	–	
2 Dist	+25	+49	+39	+39		+48	+97	+78	+78	+73		+25	+49	+39	+37	
CO	–	–	+39	–		–	–	+20	+20	–		–	–	+39	–	
3 Dist	–7	–15	–12	–11		–	–12	–10	–10	–9		–7	–15	–12	–11	
CO	–	–	–5	–		–6	–	–6	–6	–		–	–	–5	–	
4 Dist	–	+2	+2	+1		–	–4	+3	+3	+3		–	+2	+2	+1	
	+1897	+1711	–389	–1321	0	+1972	+1860	–327	–327	–1208	0	+1897	+1711	–389	–1321	0

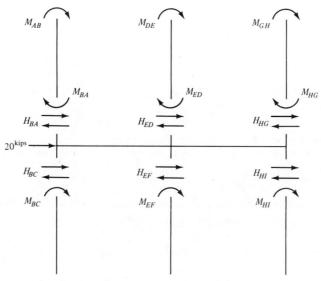

Fig. 13.16. Partial free-body diagrams for the frame in Fig. 13.15.

$M_{GH} = -10.45 + 1897\Delta = +20.90$ kip-ft

$M_{HG} = -20.90 + 1711\Delta = +6.65$ kip-ft

$M_{HE} = +36.58 - 389\Delta = +30.32$ kip-ft

$M_{HI} = -15.68 - 1321\Delta = -36.95$ kip-ft

It is important to note that the frame in Example 13.7 is symmetric as well as its transverse loading. Under this circumstance, the horizontal shears at the girder level, due to the transverse loading alone, sum to zero. Had the frame or loading been unsymmetrical, the shears due to transverse loading would not sum to zero and would have to be included when equilibrium is imposed at the girder level, e.g., in condition (a) in Example 13.7.

The alternative technique for moment distri-

bution applied to a planar frame with a translational degree of freedom is depicted in Fig. 13.17. A two-stage procedure is used.

In stage 1 (Fig. 13.17a), a temporary support is introduced (in this case at joint C) so as to prevent translation. This modified frame can be analyzed by moment distribution in the usual way. Based on statics, H_C' can be calculated subsequently.

In stage 2, the objective is to remove the effect of the reaction H_C from the stage 1 analysis. This is done by removing the temporary support and subjecting the frame to a horizontal translation of an assumed magnitude Δ'. As described in the next paragraph, the effect of Δ' can be analyzed by a second stage of moment distribution. Once done, H_C'' can be calculated by statics.

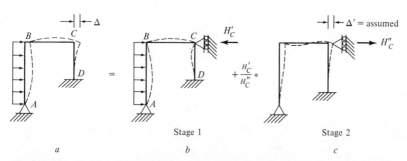

Fig. 13.17. Frame with translation-two-stage analysis by moment distribution.

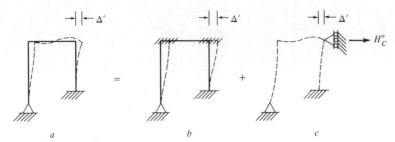

Fig. 13.18. Moment distribution for an imposed translation. *a,* actual structure; *b,* restrained state; *c,* released state.

To obtain the results for the original structure, the results of stage 2, based on an assumed Δ', are scaled by the ratio H'_C/H''_C and combined with the results of stage 1. In effect, the net horizontal force at C will be zero as required.

Moment distribution for stage 2 is performed as depicted in Fig. 13.18. In the restrained state, Fig. 13.18b, the joints (except the natural exterior pin support) are fixed against rotation and the assumed Δ' is imposed. Fixed-end moments develop in the columns and are directly calculated from equations given in Table A.5, Appendix A. In the released state, the temporary moment restraint is removed and the frame held laterally in place by the support at C. Specifically, the horizontal reaction, H''_C, keeps the joints of frame in place. Consequently, the only effect of releasing the temporary restraint is that the released joints rotate. This effect can be analyzed by the usual process of moment distribution.

Example 13.8. Analyze the planar frame (Fig. 13.19a) by moment distribution.

Solution: The two stage analysis described above is used as shown in Figs. 13.19b and c.

Stage 1 (Fig. 13.19b): This structure is identical to the one analyzed in Example 13.6. Table 13.5 contains the moment distribution results. H'_C is calculated using statics. Free-body diagrams of the columns and the girder are shown in Fig. 13.20. V'_B and V'_C are calculated by summing moments about the bottom end of each column. Horizontal equilibrium of the girder requires

$$H'_C = V'_B - V'_C = -1.33 \text{ kips} = 1.33 \text{ kips} \rightarrow$$

The negative result (H'_C to the right) indicates the original frame will sway to the left. This is opposite to the sway direction assumed in Fig. 13.19c.

Stage 2: In Fig. 13.19c, Δ is arbitrarily assumed to be -1 foot, i.e. a translation to the left. Using the general procedure shown in Fig. 13.18, the fixed-end moments for the restrained state (Fig. 13.18b) for the problem at hand are determined as shown in Fig. 13.21. From Table A.5.

$$M_{BA_0} = \frac{3EI\Delta}{L^2} = \frac{3EI(1)}{(20)^2} = \frac{3EI}{400} \, cw$$

$$M_{CD_0} = M_{DC_0} = \frac{6EI\Delta}{L^2} = \frac{6EI(1)}{(12)^2} = \frac{EI}{24} \, cw$$

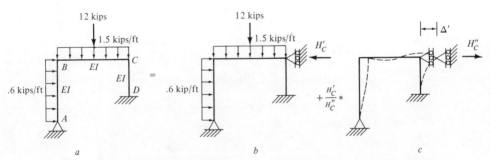

Fig. 13.19. Planar frame for Example 13.8. *a,* actual frame; *b, c,* two-stage analysis.

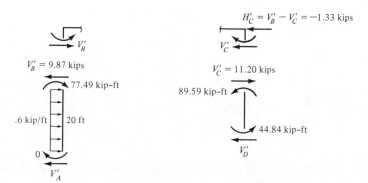

Fig. 13.20.   Free-body diagrams for stage 1 of Example 13.8.

Note the substitutions of $\Delta = +1$! Rather than substitute for EI, the assumed Δ will be scaled to

$$\Delta' = \frac{400(24)}{EI}(-1)$$

in which case the restrained state moments become

$$M_{BA_0} = 72 \text{ kip-ft}$$

$$M_{CD_0} = M_{DC_0} = 400 \text{ kip-ft}$$

The joints are then released as shown, in general, in Fig. 13.18c. The moment distribution is performed in Table 13.7. Except for the M_0 values, the table format and procedure are the same as used in Table 13.5. Extracting the end moments and using the free body diagrams drawn in Fig. 13.22,

$$H''_C = -V''_B - V''_C = -36.93 \text{ kips} = 36.93 \text{ kips}$$

Stage 1 + Stage 2: Since H''_C does not balance H'_C, a scale factor

$$k = \frac{H'_C}{H''_C} = \frac{-1.33}{-36.73} = .036$$

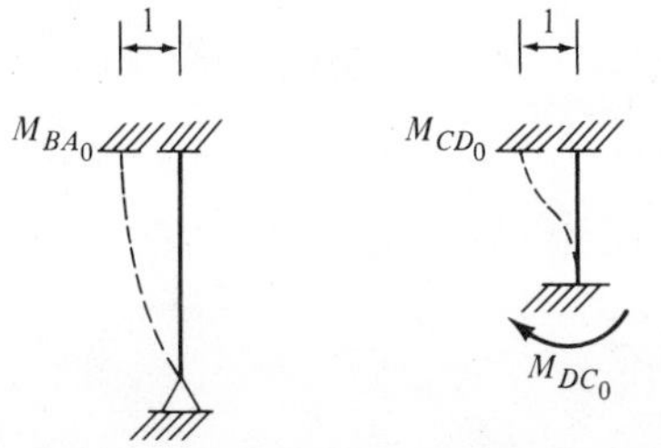

must be applied to the stage 2 results. The final results (results for the actual frame) are obtained by

$$\text{Final results} = (\text{Table 13.5 results})$$
$$+ .036 (\text{Table 13.7 results})$$

Thus

$$M_{AB} = 0 + .036(0) = 0$$

$$M_{BA} = 77.49 + .036(69.36) = 79.99 \text{ kip-ft}$$

$$M_{BC} = -77.49 + .036(-69.36)$$

$$= -79.99 \text{ kip-ft}$$

$$M_{CB} = 89.59 + .036(-134.31) = -84.75 \text{ kip-ft}$$

$$M_{CD} = -89.59 + .036(134.32) = 84.75 \text{ kip-ft}$$

$$M_{DC} = -44.83 + .036(267.18)$$

$$= -35.21 \text{ kip-ft}$$

13.3.2.4   Multiple Translations

From a conceptual point of view, frames that exhibit more than one translational degree of freedom cause little additional difficulty beyond the approach illustrated in Examples 13.7 and 13.8. Each translational degree of freedom introduces the need for a separate "sidesway correction" equation. Unfortunately, the set of corrective equations are not uncoupled and must be solved simultaneously. In addition, a separate moment distribution must be performed for each possible sidesway mode. To formulate the procedure and illustrate the nature of the analytical inconvenience, the mul-

Fig. 13.21.   Restrained state column moments, Example 13.8.

Table 13.7.   Moment Distribution for the Translational Effect—Example 13.8.

Joint		A	B		C		D
Member		AB	BA	BC	CB	CD	DC
S_t or S_j		.15EI	.15EI	.167EI	.167EI	.333EI	.333EI
CYC	DF	1	.473	.527	.333	.667	–
1	FEM	0	72.00	0	0	400.00	400.00
	BAL	–	–34.06	–37.94	–133.33	–266.67	–
2	CO	–	–	–66.67	–18.97	–	–133.33
	BAL	–	+31.53	+35.14	+6.32	+12.66	–
3	CO	–	–	3.16	+17.57	–	+6.33
	BAL	–	–1.49	–1.67	–5.85	–11.72	–
4	CO	–	–	–2.93	–0.84	–	–5.86
	BAL	–	1.39	1.64	+0.28	+.56	
5	CO	–	–	0.14	0.77	–	0.28
	BAL	–	–0.07	–0.07	–0.26	–0.51	
6	CO	–	–	–0.13	–0.03	–	.–0.25
	BAL	–	0.06	0.07	+0.01	+0.02	–
7	CO	–	–	0	0.03	–	0.01
	BAL	–	0	0	–0.01	–0.02	–
			69.36	–69.36	–134.31	134.32	267.18

tistory frame shown in Fig. 13.23 will be used as a basis for discussion.

Six translational degrees of freedom, which correspond to the six girder levels, exist for the structure of Fig. 13.23. In moment distribution, the effect of each sidesway motion is analyzed individually. As an example, to investigate the influence of a translation at the jth level, a displacement, of unknown magnitude Δ_j, is introduced while restraining all other degrees of freedom (including the joint rotations). This displacement mode is shown in Fig. 13.24. In the fashion of Example 13.7, a complete moment distribution is executed to determine the column moments and, contiguously, the column shears. In particular, at any given level i, the 10 column shears are determined and algebraically summed as inferred in Fig.

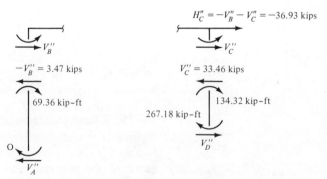

Fig. 13.22.   Free-body diagrams for stage 2 of Example 13.8.

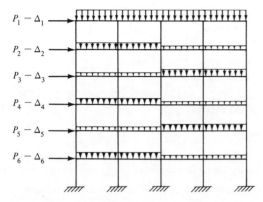

$P_1 - \Delta_1$
$P_2 - \Delta_2$
$P_3 - \Delta_3$
$P_4 - \Delta_4$
$P_5 - \Delta_5$
$P_6 - \Delta_6$

Fig. 13.23.   Translational degrees of freedom in a multistory building.

13.25. The net shear force is termed $(\Sigma H_i)_j$. The interpretation of the subscripts is that the ΣH is the net shear at level i due to the translation at level j.

If each translational displacement is treated in the manner demonstrated in Fig. 13.24, a set of sidesway correction equations is generated which equal n in number (n is the number of translational degrees of freedom). For the given frame, these equations are

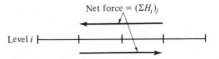

Net force $= (\Sigma H_i)_j$

Level i

Fig. 13.25.   Total shear force at level i due to jth sidesway acting alone.

girder level i, respectively, which are created by the transverse member loads acting alone. For a symmetrical frame and transverse loading pattern, each $(\Sigma H_i)_0$ term would be zero.

In essence, each equation of Eq. 13.29 represents a horizontal equilibrium condition at the corresponding girder level. Since each $\Sigma(H_i)_j$ term includes one of the n unknown translational displacements, a sufficient number of equations exists for their determination by a simultaneous solution. Once these displacement values are established and substituted into the various sidesway moment-distribution results, a linear summation (including the results obtained for the transverse loads acting alone) produces final answers. In the particular case of the frame shown in Fig. 13.23, seven separate moment-distribution analyses are required

$$(\Sigma H_1)_0 + (\Sigma H_1)_1 + \cdots + (\Sigma H_1)_j + \cdots + (\Sigma H_1)_6 = P_1$$

$$\vdots$$

$$(\Sigma H_i)_0 + (\Sigma H_i)_1 + \cdots + (\Sigma H_i)_j + \cdots + (\Sigma H_i)_6 = P_i \qquad (13.29)$$

$$\vdots$$

$$(\Sigma H_n)_0 + (\Sigma H_n)_1 + \cdots + (\Sigma H_6)_j + \cdots + (\Sigma H_n)_n = P_6$$

P_i signifies the applied horizontal load and $(\Sigma H_i)_0$ represents the summed shear forces at

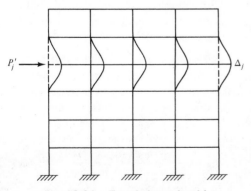

P_j'

Δ_j

Fig. 13.24.   Translation at level j.

for a complete solution. Clearly, performance of such computations is tedious and time-consuming.

In actual practice, an exact solution by moment distribution usually is not performed when the loading consists of gravity loads only. Instead, the moment distribution is performed by isolating the individual floor levels as a separate structure of the type shown in Fig. 13.26. In this way, each floor level can be analyzed in the manner demonstrated either in Example 13.7 or, possibly, as in 13.5 (if the horizontal translation is judged to be a minor effect). The number of joints at each level is considerably less than the total for the entire multistory structure, and great computational savings are

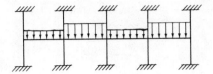

Fig. 13.26.   Girder level *i* isolated as a separate structure.

gained by this approach. Physically, this approximation assumes that gravity loads at any particular floor level have a minor effect on the other story levels. Furthermore, under certain circumstances the effect of the sidesway can be neglected. Isolation of individual floor levels and neglect of sidesway are safe assumptions if the span lengths and the upper and lower column heights are not extremely irregular and the variation in loading on individual spans is not excessive. Design codes that permit such an approximate analysis generally specify such limitations.

13.3.2.5 Torsional Stiffness

Simplification of three-dimensional structures to planar frames was discussed in Section 2.12.2. Structures that are box-shaped and exhibit a regular, lattice arrangement are suitable for substructuring. Figures 2.23 through 2.27 depict the physical manner in which this is ac-

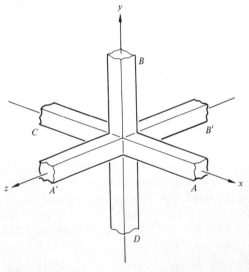

Fig. 13.27.   Typical joint in a three-dimensional frame.

complished. Analysis of the multistory, three-dimensional structure is replaced by individual analysis of the isolated frameworks. For gravity loadings, it is feasible to further reduce the size of the problem by isolating the individual floor levels and attached columns as described in the preceding paragraph.

As a further refinement of the structural model that results from the physical substructuring just described, consider the joint depicted in Fig. 13.27. As a means of reference, consider the *x-y* plane to contain a substructured planar frame extracted from a larger space frame. Members *A*, *B*, *C*, and *D* lie in the *x-y* plane and comprise part of the reduced size planar frame. Members *A'* and *B'* do not lie in the *x-y* plane and, as such, would not be included in the planar frame. However, because of their attachment to adjacent planar frames, members *A'* and *B'* do contribute to stiffness of the joint shown in Fig. 13.27. For motion within the *x-y* plane, members *A'* and *B'* primarily contribute torsional resistance to the rotation of the joint. It is a simple matter to include this type of restraint in the moment-distribution analysis. A careful review of the development of the moment-distribution method indicates that the presence of out-of-plane members has three effects:

1. Additional unbalanced moments, caused by any torsional loading on these members, are introduced at the fixed-end moment stage.
2. Torsional stiffness of these members must be combined with the flexural stiffness of the in-plane members.
3. Part of the released unbalanced moment at the joint is distributed into the out-of-plane members.

Torsional stiffness and distribution of the unbalanced moment are taken into account by a modification of Eq. 13.28 to include the torsional stiffness of any torsional members. The modified equation can be expressed as

$$DF_\kappa = \frac{(\text{Member stiffness})\kappa}{\sum_{n=1}^{NM}(S_i)_n + \sum_{p=1}^{NTM}(TS_i)_p} \quad (13.30)$$

in which $(TS_i)_p$ is the torsional stiffness of member p and *NTM* is the total number of torsional members (which is either 1 or 2). *Member stiffness* in the numerator is either a flexural stiffness or a torsional stiffness depending on the member of interest. Torsional stiffness would usually be of a form

$$TS_i = \alpha \frac{GK}{L} \qquad (13.31)$$

in which α is a constant that reflects the end connections and geometric shape of the member. As a last comment, carryover moments from one planar frame to an adjacent planar frame via torsional members is normally neglected. This statement simply reiterates the assumption that with the exception of localized torsional restraint, all substructured frames act independently. Within the bounds of regular geometry and loading, this is a safe and practical assumption.

13.3.3   General Comments

One need in the application of moment distribution to planar frames is an ability to recognize when translational degrees of freedom exist. Equation 10.19 is available for such assessment, but does not give foolproof results. An alternative means is to examine the equilibrium of the structure based on stage 1 analysis alone. If an imbalance in reactions (either horizontal, vertical or otherwise) exists in that stage without a temporary support present, then translation is involved. In such a case, a temporary support is needed and stage 2 must be executed. Provided one does not make computational mistakes, this approach is foolproof. Of course, if a frame geometry and its loading are symmetrical about any axis, translation perpendicular to that axis cannot occur. This is true, even if a translational degree of freedom exists in that perpendicular direction. This fact is useful in assessing the need for a two stage moment distribution process.

In each of the example problems executed to this point, the final answers are the moments at the member ends. With these determined, a

structure is rendered statically determinate (with some exceptions) and continued conventional analysis is possible, e.g., determining reactions, shear and moment diagrams, displacements etc. It is presumed the reader can execute such computations. It may be advisable to attempt such computations for one or more of the presented example problems. The exceptions that exist involve either frames with axial forces applied to the members or frames where some reactions parallel to a member are indeterminate. In some cases it is still possible to proceed with conventional analysis.

PROBLEMS

Do the following problems by the moment distribution method.

13-1. Determine the end moments for the beam in Problem 12-4.

13-2. Analyze the given beam to determine the end moments.

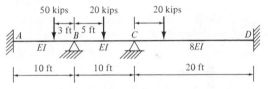

13-3. Determine the end moments for the beam in Problem 11-6.

13-4. Determine the end moments for the beam in Problem 11-7.

13-5. Determine the end moments for the given beam.

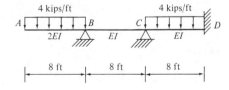

13-6. Determine the end moments for the given beam.

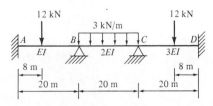

13-7. Determine the end moments for the given beams.

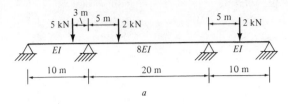

a

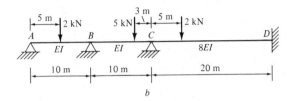

b

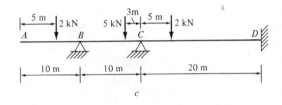

c

13-8. Determine the end moments for the beam and conditions described in Problem 12-11.

13-9. Determine the member end moments in the given planar frame.

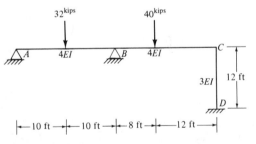

13-10. Determine the member end moments in the given planar frame.

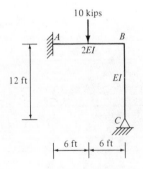

13-11. Determine the member end moments in the given planar frame.

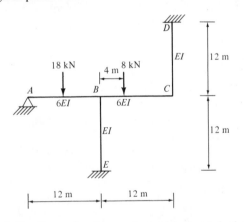

13-12. Determine the member end moments in the given planar frame.

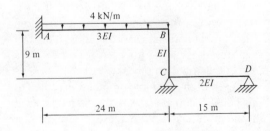

13-13.   Determine the member end moments in the given planar frame.

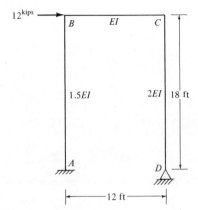

13-14.   Determine the member end moments in the given planar frame.

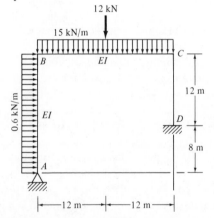

13-15.   Determine the member end moments in the given planar frame.

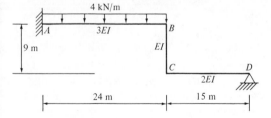

13-16.   Determine the member end moments in the given planar frame.

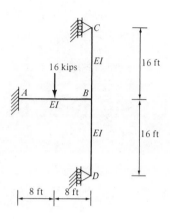

13-17.   Determine the member end moments in the given planar frame.

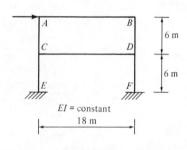

13-18.   Determine the member end moments in the given planar frame.

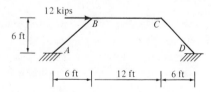

14

Basic Matrix Analysis of Trusses

14.1 LEVELS OF ANALYSIS

Computer-based structural analysis methods have the use of matrix algebra as a fundamental characteristic. Indeed, the terminology "matrix methods of structural analysis" is employed for most commonly used techniques. Several levels of sophistication can be taken to develop the structural analysis of frameworks by matrix methods. "Basic" and "semi-automated" methods are described in this text. "Fully automated" methods also exist but are beyond the scope of this text.

The descriptor "basic" infers that all needed matrices are prepared manually and the computer serves to accept these matrices as input, execute standard matrix operations, and print the output. The user must execute preparatory computations, based on conventional structural analysis principles, to prepare the input matrices and interpret the output accordingly.

"Semi-automated" refers to the use of the computer as a direct means of generating the needed matrices in addition to the execution of the matrix operations. In semi-automated procedures, the input is literal and numerical data regarding the type of structure; its geometry, member properties, and support conditions; and its loading. Data take the form of coordinates of joints, location and magnitude of loads, material constants, etc.

In fully automated methods, reliance on digital input is either minimized or eliminated. Powerful computer graphics capability is a major component in fully automated methods. "Digitizing" graphical information for automatic conversion to numerical data is a common technique. By combining or augmenting these with automated structural analysis techniques the tedium of manual computation associated with structural analysis is virtually eliminated. Pre- and post-processor software packages serve to automate the input to and implement the output from highly advanced matrix structural analysis computer algorithms. Computer-aided drafting, computer-aided design, and computer-aided manufacturing (CAD/CAM) are at the forefront of the state of the art of structural and mechanical engineering. "Artificial intelligence" and "expert systems" are growing in importance in the decision-making, fabrication, and construction processes involved in structural engineering.

In this chapter, basic matrix analysis of trusses is described and serves as the first building block toward evolving full capability in and understanding of contemporary computer-based methods of structural analysis. Chapters 15 and 16 serve to introduce and develop the principal features of semi-automated computer analysis of the various types of 2-D framed structures.

14.2 BASIC MATRIX RELATIONSHIPS

As noted in Section 1.10, all structural analysis is founded upon the need to satisfy the three general criteria:

1. *Equilibrium* of the internal forces with the applied loads.
2. Adherence to *material conditions* dictated by the stress-strain behavior of the particular material.
3. *Compatibility* of the deformations of the individual members with the support conditions and displacements of the structure as a whole.

By developing mathematical formulations to model each of these effects, a numerical determination of the response of the structure is possible. The following sections describe how to accomplish this for the planar truss structure. In the described analysis procedure, the three general criteria of equilibrium, material conditions, and compatibility are linked together in such a way as to permit a determination of the external joint displacements of the truss. Next, these displacement values are back-substituted into the compatibility equations to permit computation of the individual member deformations. Then, the computed member deformation values are back-substituted into the equations of material conditions to allow determination of the individual member forces. Finally, these member forces are back-substituted into the equilibrium equations to enable calculation of the support reactions. Matrix al-

gebra provides ready tools, and a solution procedure using matrix operations is developed in the succeeding paragraphs. (A review of matrix algebra is given in Appendix B.)

14.2.1 Equilibrium Matrices $[E_P]$ and $[E_R]$

Equilibrium conditions express the satisfaction of statics at each joint of the truss. Classical truss analysis was treated in Chapter 5. Direct solution of the equilibrium equations was described in Section 5.4.1 and illustrated in Example 5.1 for the truss in Fig. 5.9a. Equations a through h in Example 5.1 constitute the equilibrium equations. Examination of these equations indicates that they can be expressed in the two matrix relationships

$$\{P\} = [\text{EQUILIBRIUM}_P]\{F\} = [E_P]\{F\}$$
$$(14.1)$$

$$\{R\} = [\text{EQUILIBRIUM}_R]\{F\} = [E_R]\{F\}$$
$$(14.2)$$

in which

$\{P\}$ = column matrix containing the known load values

$\{R\}$ = column matrix containing the unknown reactions

$\{F\}$ = column matrix containing the unknown member forces

$[E_P]$ = rectangular matrix containing constant coefficients

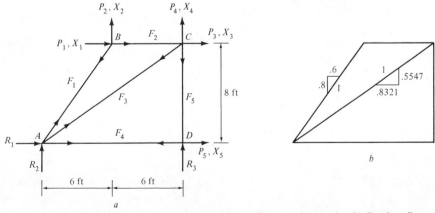

Fig. 14.1.   Truss from Example 14.1 *a*, general force diagram; *b*, member inclination diagram.

$[E_R]$ = rectangular matrix containing constant coefficients

The general force diagram and member inclination diagram for the truss of interest are repeated in Fig. 14.1. The specific loading of Example 5.1 has been replaced by the general set of loads P_1 through P_5, acting in the directions of the degrees of freedom X_1 through X_5. As a general convention in this text, loads are considered positive if acting either to the right or upward. By visual examination of the joints in Fig. 14.1a, in the order A, B, C, D, the equilibrium equations become

and

$$\{R\} = [E_R]\{F\}$$

$$\begin{Bmatrix} R_1 \\ R_2 \\ R_3 \end{Bmatrix} = \begin{bmatrix} -.6 & 0. & -.8321 & -1.0 & 0. \\ -.8 & 0. & -.5547 & 0. & 0. \\ 0 & 0. & 0 & 0. & 1.0 \end{bmatrix} \begin{Bmatrix} F_1 \\ F_2 \\ F_3 \\ F_4 \\ F_5 \end{Bmatrix}$$

$\{P\}$ is the set of known loads. For the truss at hand, $P_1 = 6$ kips and $P_2 = -12$ kips, giving

$R_1 = -.6F_1$	$- .8321F_3 - 1.0F_4$	(a)
$R_2 = -.8F_1$	$- .5547F_3$	(b)
$P_1 = .6F_1 - 1.0F_2$		(c)
$P_2 = .8F_1$		(d)
$P_3 =$	$1.0F_2 + .8321F_3$	(e)
$P_4 =$	$.5547F_3 + 1.0F_5$	(f)
$P_5 =$	$1.0F_4$	(g)
$R_3 =$	$- 1.0F_5$	(h)

These relationships should be compared with Eqs. a through h in Example 5.1. Note the change in sign of the coefficients on the right side relative to Example 5.1. The equivalent expressions in the matrix format of Eqs. 14.1 and 14.2 are

$$\{P\} = [E_P]\{F\}$$

$$\begin{Bmatrix} P_1 \\ P_2 \\ P_3 \\ P_4 \\ P_5 \end{Bmatrix} = \begin{bmatrix} .6 & -1.0 & 0. & 0. & 0. \\ .8 & 0. & 0. & 0. & 0. \\ 0. & 1.0 & .8321 & 0. & 0. \\ 0. & 0. & .5547 & 0 & 1.0 \\ 0. & 0. & 0. & 1.0 & 0. \end{bmatrix} \cdot \begin{Bmatrix} F_1 \\ F_2 \\ F_3 \\ F_4 \\ F_5 \end{Bmatrix}$$

$$\{P\} = \begin{Bmatrix} P_1 \\ P_2 \\ P_3 \\ P_4 \\ P_5 \end{Bmatrix} = \begin{Bmatrix} 6 \\ -12 \\ 0. \\ 0. \\ 0. \end{Bmatrix} \tag{14.5}$$

14.2.2 Compatibility Matrix [C]

A complete analysis of a truss includes a determination of the displaced position of the structure in addition to computation of the member forces. Of primary concern are the magnitude of the joint motions. Compatibility requires that the displacement of the joints of the truss must be consistent with the deformations (elongations or contractions) of the individual members. In effect, this requirement states that any particular member deformation can be determined as the change in length needed to keep the member connected to the displaced joints and/or supports at its ends.

Conversely, if all the member deformations are known, the new joint positions can be established on the basis of this information alone. From a physical point of view, this latter statement means that with the new length of each member known, they can be pieced together in only one admissible configuration and still satisfy the support conditions. Analytically, it is possible to use this concept to permit numerical computation of the joint displacements, but this approach is usually tedious and impractical. When the displacements of only a few selected joints are needed, the most common classical approach is to employ the unit-load method (see Chapter 9). If all the joint displacement values are desired, a matrix formulation is the most suitable method. As a part of this matrix analysis method, it is necessary to express the compatibility conditions in a matrix format. A procedure for accomplishing this will be described in the following paragraphs.

A simple example of the concept of compatibility of displacements is offered by considering the two-bar truss shown in Fig. 14.2. For the given support conditions, the direction X_1 is the sole degree of freedom. (P_1 indicates the possibility of applying load in this direction.) Compatibility for this structure infers three constraints:

1. Member 1 remains attached to joint A.
2. Member 2 remains attached to joint C.
3. Members 1 and 2 remain attached to the roller support at joint B.

These three requirements must be satisfied for all possible values of the load P_1 and all possible magnitudes of displacement X_1 if a failure is to be avoided. It should be noted that the three *geometrical* conditions must exist regardless of the material behavior and size of the members. Furthermore, it is possible to express the conditions in a quantified way by the two equations

$$\Delta_1 = X_1 \qquad (14.6)$$

$$\Delta_2 = -X_1 \qquad (14.7)$$

in which Δ_1 and Δ_2 are the internal deformations of the two members. Equations 14.6 and 14.7 are termed *compatibility equations*. They simply state that if the roller displaces an amount X_1, member 1 elongates by an equal amount and member 2 contracts by an equal amount. Compatibility with the pin supports at A and D is also inferred in Eqs. 14.6 and 14.7. If the pin supports were converted to roller supports, both members would freely move and experience no strain. In this case, Δ_1 and Δ_2 would have no value, and Eqs. 14.6 and 14.7 would give incorrect answers because of their dependence on particular boundary conditions.

Example 14.1. Determine the compatibility equations for the three-member truss given in Fig. 14.3.

Solution: To retain continuity, three conditions must be satisfied:

1. Member 1 must remain attached to supports A and B.

$$\Delta_1 = X_1$$

2. Member 2 must remain attached to supports B and C.

$$\Delta_2 = -X_1 + X_2$$

3. Member 3 must remain attached to supports C and D.

$$\Delta_3 = -X_2$$

Example 14.2. Determine the compatibility equations for the truss shown in Fig. 14.4a.

Solution: Equations of compatibility for complex frameworks are more easily determined by *individ-*

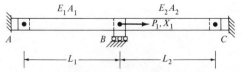

Fig. 14.2. Two-member truss.

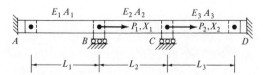

Fig. 14.3. Three-member truss.

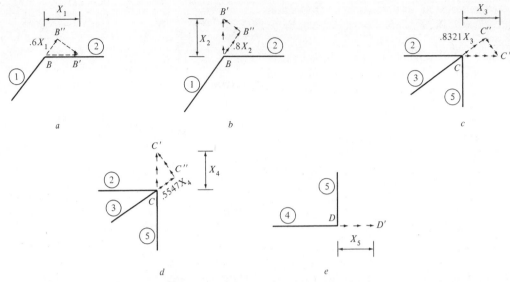

Fig. 14.4. Compatibility of a multimember truss.

ually inspecting the effects of a displacement in the direction of each of the degrees of freedom.

Step 1: Figure 14.4a depicts the member deformations that must occur if a displacement X_1 occurs by itself (i.e., all other degrees of freedom are restrained). Joint B moves from its original position to point B'. To maintain continuity, members 1 and 2 must change in length. The deformation of member 1 can be viewed as occurring in two stages. First, the member elongates by an amount $0.6X_1$ reaching point B''. Next, the member swings in a transverse direction to reach point B'. During this swing, it is assumed that the member end moves perpendicular to its *original* axis. Clearly, some additional change in length must occur if the member end is to reach point B'. It should be recalled from Chapter 3 that one assumption in a first-order analysis is to neglect this additional change in length. For small deformations, this is a safe assumption. Member 2 deforms by contracting by an amount equal to X_1. Consequently, it can be concluded that

$$\Delta_1 = 0.6X_1$$

$$\Delta_2 = -1.0X_1$$

$$\Delta_3 = \Delta_4 = \Delta_5 = 0$$

Δ_2 is negative because member 2 must contract.

Step 2: The effect of motion in the X_2 direction shown in Fig. 14.4b can be similarly formulated as

$$\Delta_1 = 0.8X_2$$

$$\Delta_2 = 0$$

$$\Delta_3 = \Delta_4 = \Delta_5 = 0$$

Δ_2 is zero because the member 2 swings transversely without a change in length.

Step 3: Motion in the X_3 direction produces the effects shown in Fig. 14.4c. Consequently,

$$\Delta_2 = 1.0X_3$$

$$\Delta_3 = .8321X_3$$

$$\Delta_1 = \Delta_4 = \Delta_5 = 0$$

Steps 4 and 5: Motion in the X_4 direction is shown in Fig. 14.4d and requires

$$\Delta_3 = .5574X_2$$

$$\Delta_5 = 1.0X_4$$

$$\Delta_1 = \Delta_2 = \Delta_4 = 0$$

Motion in the X_5 direction is shown in Fig. 14.4e and requires

$$\Delta_4 = 1.0X_5$$

$$\Delta_1 = \Delta_2 = \Delta_3 = \Delta_5 = 0$$

Step 6: Under the action of an arbitrary set of applied loads, it is likely that the five motions described in Figs. 14.4a through e would occur simultaneously. Consequently, the compatibility equations for the given truss consist of the combined effects

Note that $[C]$ in Eq. 14.13 is the transpose of $[E_P]$ in Eq. 14.3. This is not a coincidence. The general validity of $[C] = [E_P]^T$ was proven in Chapter 8.

$$\Delta_1 = \quad 0.6X_1 + 0.8X_2 \tag{14.8}$$

$$\Delta_2 = -1.0X_1 \qquad\qquad + 1.0X_3 \tag{14.9}$$

$$\Delta_3 = \qquad\qquad .8321X_3 + .5547X_4 \tag{14.10}$$

$$\Delta_4 = \qquad\qquad\qquad\qquad\qquad + 1.0X_5 \quad \tag{14.11}$$

$$\Delta_5 = \qquad\qquad\qquad\qquad + 1.0000X_4 \tag{14.12}$$

Equations 14.8 through 14.12 can be assembled into the matrix

$$\begin{Bmatrix} \Delta_1 \\ \Delta_2 \\ \Delta_3 \\ \Delta_4 \\ \Delta_5 \end{Bmatrix} = \begin{bmatrix} 0.6 & 0.8 & 0. & 0. & 0. \\ -1.0 & 0. & 1.0 & 0. & 0. \\ 0. & 0. & .8321 & .5547 & 0. \\ 0. & 0. & 0. & 0. & 1.0 \\ 0. & 0. & 0. & 1. & 0. \end{bmatrix}$$

$$\cdot \begin{Bmatrix} X_1 \\ X_2 \\ X_3 \\ X_4 \\ X_5 \end{Bmatrix} \tag{14.13}$$

Equation 14.13 has the general form

$$\{\Delta\} = [\text{COMPATIBILITY}]\{X\} = [C|\{X\}$$

$$\tag{14.14}$$

where

  $\{\Delta\}$ = column matrix containing the unknown member deformations
  $[C]$ = rectangular matrix containing constant coefficients
  $\{X\}$ = column matrix containing the unknown values of the independent joint displacements

It should be noted that for the statically determinate truss of Example 14.2, the compatibility equations permit complete determination of the member deformations, if the joint displacements are known. Conversely, if the member deformations are known, the joint displacements can be determined by solution of the compatibility equations. This is a general condition for statically determinate systems.

For statically indeterminate trusses, the number of member deformations exceeds the number of degrees of freedom. This condition is also general. Nonetheless, as in determinate trusses, knowledge of the joint displacement values establishes the member deformations. Conversely, the determination of unknown joint displacements from known member deformations is an overdefined problem. The joint displacements can be computed by solving a proper subset of the compatibility equations. The remaining compatibility conditions must be automatically satisfied by the computed displacement values and could be used as a verification of the solution.

No method for the direct (in one step) computation of the joint displacements has been described to this point in the text, and thus the joint displacements must be considered unknown quantities. Therefore, the complete set of independent joint displacements are usually termed the *kinematic* unknowns, and each one represents a "kinematic degree of indeterminacy." Consequently, all trusses are "kinematically indeterminate," and the degree of kinematic indeterminacy is equal to the number of degrees of freedom. If the kinematic unknowns can be calculated, it is then possible to establish the member deformations and, in turn, the member forces.

14.2.3   Local Member Stiffness Matrix [*S*]

Equilibrium equations for trusses were described in Section 14.2.1. These equations represent a set of statics conditions that the free-body diagrams of the individual truss joints must satisfy for any set of applied loads. For statically determinate systems, they permit a complete determination of all member forces caused by the applied loads. For statically indeterminate trusses, the equilibrium conditions are insufficient in number to perform a complete force analysis. Compatibility equations for trusses were discussed in Section 14.2.2. These equations constitute a set of boundary conditions that the truss must satisfy under all possible loadings. In effect, they permit a complete determination of all member deformations if the external joint displacements are known. Equally important, if member deformations are known, joint displacement values are established by the compatibility requirements. Clearly, if a link can be made between the internal forces of the equilibrium equations and member deformations needed in the compatibility equations, a solution is possible. Laws governing the behavior of materials constitute the needed link and thus will be termed *material conditions*.

ysis is based upon the fundamental assumption of a linear, elastic material law. For a member with constant cross-sectional area, A_i, subjected to axial force, F_i, only, Hooke's law permits calculation of the member deformation, Δ_i, as

$$\Delta_i = \frac{F_i L_i}{E_i A_i} \qquad (14.15)$$

in which L_i and E_i are the member length and modulus of elasticity, respectively. Equation 14.15 is more useful in the form

$$F_i = \frac{E_i A_i}{L_i} \Delta_i \qquad (14.16)$$

For a given truss, the set of member forces $\{F\}$ are directly related to the set of member deformations $\{\Delta\}$ according to

$$\{F\} = [\text{STIFFNESS}]\{\Delta\} = [S]\{\Delta\} \qquad (14.17)$$

in which

[*S*] = diagonal *local member stiffness matrix* containing known member properties

The descriptor "local" distinguishes [*S*] from a "global" member stiffness matrix to be defined in Chapter 15. Written in expanded form for a truss of *NF* members, Eq. 14.17 becomes

$$
\begin{Bmatrix} F_1 \\ F_2 \\ \vdots \\ F_i \\ \vdots \\ F_{NF} \end{Bmatrix}
=
\begin{bmatrix}
\dfrac{E_1 A_1}{L_1} & 0 & \cdots & 0 & \cdots & 0 \\
0 & \dfrac{E_2 A_2}{L_2} & \cdots & 0 & \cdots & 0 \\
\vdots & \vdots & & \vdots & & \vdots \\
0 & 0 & \cdots & \dfrac{E_i A_i}{L_i} & \cdots & 0 \\
\vdots & \vdots & & \vdots & & \vdots \\
0 & 0 & \cdots & 0 & \cdots & \dfrac{E_{NF} A_{NF}}{L_{NF}}
\end{bmatrix}
\begin{Bmatrix} \Delta_1 \\ \Delta_2 \\ \vdots \\ \Delta_i \\ \vdots \\ \Delta_{NF} \end{Bmatrix}
\qquad (14.18)
$$

In the context of this book, structural anal-   wherein the diagonal matrix is [*S*].

If for the truss of Fig. 14.1, all members have $E = 30000$ ksi and $A = 10$ in.^{2}, then

$$[S] = \begin{bmatrix} 2500. & 0. & 0. & 0. & 0. \\ 0. & 4167. & 0. & 0. & 0. \\ 0. & 0. & 1733. & 0. & 0. \\ 0. & 0. & 0. & 2083. & 0. \\ 0. & 0. & 0. & 0. & 3125. \end{bmatrix}$$

(14.19)

14.2.4 Interrelationship of the Basic Matrices

Matrix relationships described in Sections 14.2.1 through 14.2.3 are sufficient for developing a matrix procedure for complete analysis of any planar truss. As a set, these matrices are used in the order

$$\{P\} = [E_P]\{F\} \qquad (14.1)$$

$$\{F\} = [S]\{\Delta\} \qquad (14.17)$$

$$\{\Delta\} = [C]\{X\} \qquad (14.14)$$

$$\{R\} = [E_R]\{F\} \qquad (14.2)$$

Relationships 14.1, 14.2, 14.14, and 14.17 can be employed as follows.

Substitution of the material conditions, Eq. 14.17, into the equilibrium conditions, Eq. 14.1, yields

$$\{P\} = [E_P][S]\{\Delta\} \qquad (14.20)$$

Subsequent substitution of the compatibility conditions, Eq. 14.14 gives

$$\{P\} = [E_P][S][C]\{X\} \qquad (14.21)$$

Equation 14.21 can be written as

$$\{P\} = [K]\{X\} \qquad (14.22)$$

in which

$$[K] = [E_P][S][C] \qquad (14.23)$$

The matrix $[K]$ is generally termed the *structure stiffness matrix* and relates the applied loads, $\{P\}$, to the resulting joint displacements, $\{X\}$. It should be noted that the three basic criteria of structural analysis are embedded in $[K]$.

A complete structural analysis is accomplished by following the format outlined in Table 14.1. Determination of the unknown quantities begins at step 5. At this point, it is necessary to invert the structure stiffness matrix in order to determine the displacement values in $\{X\}$ according to

$$\{X\} = [K]^{-1}\{P\} \qquad (14.24)$$

The inverse is best obtained by an appropriate inversion routine on the computer. For very large structural systems, it is usually impractical to actually compute an inverse or to even store the matrix $[K]$ in its entirety. Efficient equation-solving routines exist and make inversion unnecessary. These solution methods are an advanced topic and are not addressed in this textbook.

14.3 EXAMPLE PROBLEMS

Example 14.3. Determine the member forces, reactions, and joint displacement for the statically determinate truss given in Fig. 14.2.

Table 14.1. Basic Matrix Analysis Procedure.

Step	Operation	Equations
1	Form the statics matrices $[E_P]$ and $[E_R]$	14.1, 14.2
2	Form the local member stiffness matrix $[S]$	14.17
3	Form the compatibility matrix $[C]$	14.14
4	Build the structure stiffness matrix $[K]$	14.23
5	Solve for the joint displacements $\{X\}$	14.24
6	Solve for the member deformations $\{\Delta\}$	14.14
7	Solve for the member forces $\{F\}$	14.17
8	Solve for the external reactions $\{R\}$	14.2

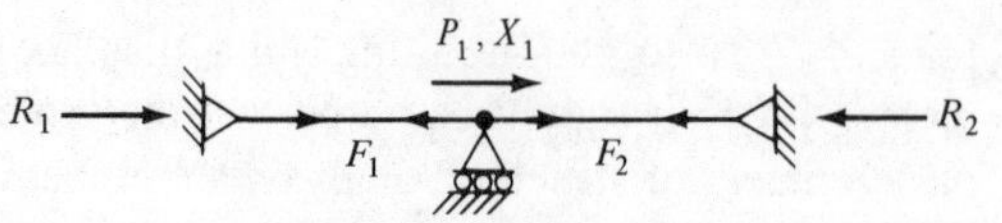

Fig. 14.5. General force diagram for the truss of Fig. 14.2.

Solution: The three sets of conditions governing structural analysis must be satisfied.

Equilibrium (from the schematic diagram in Fig. 14.5):

$$P_1 = F_1 - F_2 \qquad (14.25)$$

$$R_1 = -F_1 \qquad (14.26)$$

$$R_2 = F_2 \qquad (14.27)$$

Material conditions:

$$F_1 = \frac{E_1 A_1}{L_1} \Delta_1 \qquad (14.28)$$

$$F_2 = \frac{E_2 A_2}{L_2} \Delta_2 \qquad (14.29)$$

Compatibility (from Section 14.2.2 and Fig. 14.2):

$$\Delta_1 = X_1 \qquad (14.30)$$

$$\Delta_2 = -X_1 \qquad (14.31)$$

Substitution of Eqs. 14.28 and 14.29 into Eq. 14.25 yields

$$P_1 = \frac{E_1 A_1}{L_1} \Delta_1 - \frac{E_2 A_2}{L_2} \Delta_2$$

and upon substitution of Eqs. 14.30 and 14.31

$$P_1 = \left(\frac{E_1 A_1}{L_1} + \frac{E_2 A_2}{L_2} \right) X_1$$

which can be solved for X_1

$$X_1 = \frac{P_1}{\dfrac{E_1 A_1}{L_1} + \dfrac{E_2 A_2}{L_2}} \qquad (14.32)$$

Back-substitution into Eqs. 14.30 and 14.31 gives

$$\Delta_1 = \frac{P_1}{\dfrac{E_1 A_1}{L_1} + \dfrac{E_2 A_2}{L_2}} \qquad (14.33)$$

$$\Delta_2 = \frac{-P_1}{\dfrac{E_1 A_1}{L_1} + \dfrac{E_2 A_2}{L_2}} \qquad (14.34)$$

and upon back substitution into Eqs. 14.28 and 14.29

$$F_1 = \frac{\dfrac{E_1 A_1}{L_1}}{\dfrac{E_1 A_1}{L_1} + \dfrac{E_2 A_2}{L_2}} P_1 \qquad (14.35)$$

$$F_2 = \frac{-\dfrac{E_2 A_2}{L_2}}{\dfrac{E_1 A_1}{L_1} + \dfrac{E_2 A_2}{L_2}} P_1 \qquad (14.36)$$

Reactions are computed by substitution of F_1 and F_2 into Eqs. 14.26 and 14.27.

$$R_1 = \frac{-\dfrac{E_1 A_1}{L_1}}{\dfrac{E_1 A_1}{L_1} + \dfrac{E_2 A_2}{L_2}} P_1 \qquad (14.37)$$

$$R_2 = \frac{-\dfrac{E_2 A_2}{L_2}}{\dfrac{E_1 A_1}{L_1} + \dfrac{E_2 A_2}{L_2}} P_1 \qquad (14.38)$$

Equations 14.32 through 14.38 constitute the response of the structure to an arbitrary load P_1. For identical members, the member forces would be

$$F_1 = \frac{P}{2}$$

$$F_1 = -\frac{P}{2}$$

or, in other words, the members would share equally in resisting the applied load.

Example 14.4. Analyze the truss in Fig. 14.6a by the basic matrix analysis procedure outlined in Table 14.1.

Solution:
Equilibrium (from Figs. 14.6b,c and Eqs. 14.1 and 14.2):

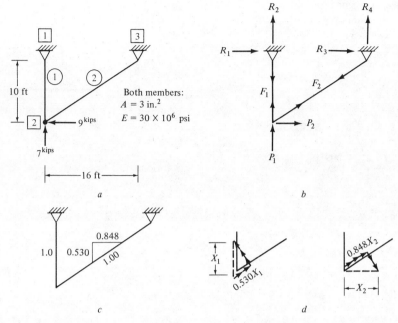

Fig. 14.6.   Truss for Example 14.5. *a*, two-member truss; *b*, general force diagram; *c*, member inclination; *d*, compatibility modes at the joints.

$$\{P\} = [E_P] \{F\}$$

$$\begin{Bmatrix} P_1 \\ P_2 \end{Bmatrix} = \begin{bmatrix} -1.0 & -0.530 \\ 0 & -8.848 \end{bmatrix} \begin{Bmatrix} F_1 \\ F_2 \end{Bmatrix}$$

$$\{R\} = [E_R] \{F\}$$

$$\begin{Bmatrix} R_1 \\ R_2 \\ R_3 \\ R_4 \end{Bmatrix} = \begin{bmatrix} 0 & 0 \\ 1.0 & 0 \\ 0 & 0.848 \\ 0 & 0.530 \end{bmatrix} \begin{Bmatrix} F_1 \\ F_2 \end{Bmatrix}$$

Material Conditions (from Eq. 14.17):

$$\{F\} = [S] \{\Delta\}$$

$$\begin{Bmatrix} F_1 \\ F_2 \end{Bmatrix} = \begin{bmatrix} \dfrac{E_1 A_1}{L_1} & 0 \\ 0 & \dfrac{E_2 A_2}{L_2} \end{bmatrix} \begin{Bmatrix} \Delta_1 \\ \Delta_2 \end{Bmatrix}$$

$$\begin{Bmatrix} F_1 \\ F_2 \end{Bmatrix} = E \begin{bmatrix} 0.0250 & 0 \\ 0 & 0.0132 \end{bmatrix} \begin{Bmatrix} \Delta_1 \\ \Delta_2 \end{Bmatrix}$$

Compatibility (from Fig. 14.6d and Eq. 14.14):

$$\{\Delta\} = [C] \{X\}$$

$$\begin{Bmatrix} \Delta_1 \\ \Delta_2 \end{Bmatrix} = \begin{bmatrix} -1.0 & 0 \\ -0.530 & -0.848 \end{bmatrix} \begin{Bmatrix} X_1 \\ X_2 \end{Bmatrix}$$

The structure stiffness matrix is obtained by multiplication

$$[S][C] = E \begin{bmatrix} 0.025 & 0 \\ 0 & 0.0132 \end{bmatrix} \begin{bmatrix} -1.0 & 0 \\ -0.53 & -0.848 \end{bmatrix}$$

$$= E \begin{bmatrix} -0.025 & 0 \\ -0.0070 & -0.0112 \end{bmatrix}$$

$$[K] = [E_P][S][C]$$

$$= E \begin{bmatrix} -1.0 & -0.530 \\ 0 & -0.848 \end{bmatrix}$$

$$\cdot \begin{bmatrix} -0.025 & 0 \\ -0.0070 & -0.0112 \end{bmatrix}$$

$$[K] = E \begin{bmatrix} 0.0287 & 0.00594 \\ 0.00594 & 0.00950 \end{bmatrix}$$

Note: Two observations should be made regarding the developed matrices. First, the equilibrium matrix $[E_P]$ is the transpose of the compatibility matrix

$[C]$ as expected. Second, the stiffness matrix $[K]$ is symmetric. The symmetry of $[K]$ is general and was proven in Chapter 8. The reader should use these conditions as indicators that the numerical work is likely to be correct.

Response is determined as follows:

$$\frac{1}{\det K} = \frac{1}{E[0.0287(0.0095) - (0.00594)^2]}$$

$$[K]^{-1} = \frac{1}{\det K} \begin{bmatrix} 0.00950 & -0.00594 \\ -0.00594 & 0.0287 \end{bmatrix}$$

$$= \frac{1}{E} \begin{bmatrix} 40.0 & -25.0 \\ -25.0 & 120.9 \end{bmatrix}$$

$$\{X\} = [K]^{-1} \{P\}$$

$$= \frac{1}{E} \begin{bmatrix} 40.0 & -25.0 \\ -25.0 & 120.9 \end{bmatrix} \begin{Bmatrix} 7.0 \\ -9.0 \end{Bmatrix}$$

$$= \begin{Bmatrix} 505/E \\ -1263/E \end{Bmatrix} = \begin{Bmatrix} +0.0168 \\ -0.0421 \end{Bmatrix}$$

$$\therefore X_1 = 0.0168 \text{ in. upward}$$

$$X_2 = 0.0421 \text{ in. to the left}$$

$$\{\Delta\} = [C] \{X\}$$

$$= \begin{bmatrix} -1.0 & 0 \\ -0.530 & -0.848 \end{bmatrix} \begin{Bmatrix} 0.0168 \\ -0.0421 \end{Bmatrix}$$

$$= \begin{Bmatrix} -0.0168 \\ +0.0268 \end{Bmatrix}$$

$$\therefore \Delta_1 = 0.0168 \text{ in. contraction}$$

$$\Delta_2 = 0.0268 \text{ in. elongation}$$

$$\{F\} = [S] \{\Delta\}$$

$$= E \begin{bmatrix} 0.025 & 0 \\ 0 & 0.0132 \end{bmatrix} \begin{Bmatrix} -0.0168 \\ 0.0268 \end{Bmatrix}$$

$$\begin{Bmatrix} F_1 \\ F_2 \end{Bmatrix} = \begin{Bmatrix} -12.6 \\ +10.6 \end{Bmatrix} = \begin{Bmatrix} 12.6 \text{ kips compression} \\ 10.6 \text{ kips tension} \end{Bmatrix}$$

$$\{R\} = [E_R] \{F\} = \begin{bmatrix} 0 & 0 \\ 1.0 & 0 \\ 0 & 0.848 \\ 0 & 0.530 \end{bmatrix} \begin{Bmatrix} -12.6 \\ +10.6 \end{Bmatrix}$$

$$\begin{pmatrix} R_1 \\ R_2 \\ R_3 \\ R_4 \end{pmatrix} = \begin{pmatrix} 0.00 \\ -12.60 \\ 8.99 \\ 5.62 \end{pmatrix} = \begin{Bmatrix} 0 \\ 12.6 \text{ kips downward} \\ 8.99 \text{ kips to the right} \\ 5.62 \text{ kips upward} \end{Bmatrix}$$

Example 14.5. Analyze the statically indeterminate truss in Fig. 14.7a by the basic matrix analyses procedure.

Solution:
Equilibrium (from Fig. 14.7b):

$$\{P\} = [E_P] \{F\}$$

$$\begin{pmatrix} P_1 \\ P_2 \end{pmatrix} = \begin{bmatrix} -.923 & -1. & -.8 & -.6 \\ .385 & 0. & -.6 & -.8 \end{bmatrix} \begin{pmatrix} F_1 \\ F_2 \\ F_3 \\ F_4 \end{pmatrix}$$

$$\{R_1\} = [E_R] \{F\}$$

$$\begin{pmatrix} R_1 \\ R_2 \\ R_3 \\ R_4 \\ R_5 \\ R_6 \\ R_7 \\ R_8 \end{pmatrix} = \begin{bmatrix} -.385 & 0. & 0. & 0. \\ .923 & 0. & 0. & 0. \\ 0. & 0. & 0. & 0. \\ 0. & 1. & 0. & 0. \\ 0. & 0. & .6 & 0. \\ 0. & 0. & .8 & 0. \\ 0. & 0. & 0. & .8 \\ 0. & 0. & 0. & .6 \end{bmatrix} \begin{pmatrix} F_1 \\ F_2 \\ F_3 \\ F_4 \end{pmatrix}$$

Material conditions:

$$\{F\} = [S] \{\Delta\}$$

$$\begin{pmatrix} F_1 \\ F_2 \\ F_3 \\ F_4 \end{pmatrix} = \begin{bmatrix} 288.5 & 0. & 0. & 0. \\ 0. & 312.5 & 0. & 0. \\ 0. & 0. & 250.0 & 0. \\ 0. & 0. & 0. & 187.5 \end{bmatrix} \begin{pmatrix} \Delta_1 \\ \Delta_2 \\ \Delta_3 \\ \Delta_4 \end{pmatrix}$$

Compatibility (from Figs. 14.7 c,d):

$$\{\Delta\} = [C]\{X\}$$

$$\begin{pmatrix} \Delta_1 \\ \Delta_2 \\ \Delta_3 \\ \Delta_4 \end{pmatrix} = \begin{bmatrix} -.923 & .385 \\ -1. & 0. \\ -.8 & -.6 \\ -.6 & -.8 \end{bmatrix} \begin{Bmatrix} X_1 \\ X_2 \end{Bmatrix}$$

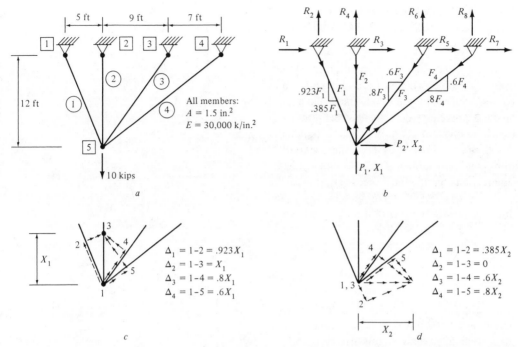

Fig. 14.7. Truss for Example 14.8. *a*, geometry; *b*, general force diagram; *c*, *d*, compatibility modes at joint 5.

Thus

$$[K] = [E_P][S][C]$$

$$= \begin{bmatrix} 785.7 & 107.7 \\ 107.7 & 252.7 \end{bmatrix}$$

The load matrix is

$$\{P\} = \left\{ \begin{array}{c} -10 \\ 0 \end{array} \right\}$$

The response is

$$\{X\} = [K]^{-1}\{P\}$$

$$\left\{ \begin{array}{c} X_1 \\ X_2 \end{array} \right\} = \left\{ \begin{array}{c} -.0135 \\ .0058 \end{array} \right\} = \left\{ \begin{array}{l} .0135 \text{ in. down} \\ .0058 \text{ in. to the right} \end{array} \right\}$$

$$\{\Delta\} = [C]\{X\}$$

$$\left\{ \begin{array}{c} \Delta_1 \\ \Delta_2 \\ \Delta_3 \\ \Delta_4 \end{array} \right\} = \left\{ \begin{array}{c} .0147 \\ .0135 \\ .00732 \\ .00346 \end{array} \right\} = \left\{ \begin{array}{l} .0147 \text{ in. elongation} \\ .0135 \text{ in. elongation} \\ .00732 \text{ in. elongation} \\ .00346 \text{ in. elongation} \end{array} \right\}$$

$$\{F\} = [S]\{\Delta\}$$

$$\left\{ \begin{array}{c} F_1 \\ F_2 \\ F_3 \\ F_4 \end{array} \right\} = \left\{ \begin{array}{c} 4.24 \\ 4.22 \\ 1.83 \\ 0.65 \end{array} \right\} = \left\{ \begin{array}{l} 4.24 \text{ kips tension} \\ 4.22 \text{ kips tension} \\ 1.83 \text{ kips tension} \\ 0.65 \text{ kips tension} \end{array} \right\}$$

$$\{R\} = [E_R]\{F\}$$

$$\left\{ \begin{array}{c} R_1 \\ R_2 \\ R_3 \\ R_4 \\ R_5 \\ R_6 \\ R_7 \\ R_8 \end{array} \right\} = \left\{ \begin{array}{c} -1.63 \\ 3.91 \\ 0.00 \\ 4.22 \\ 1.10 \\ 1.46 \\ 0.52 \\ 0.39 \end{array} \right\} = \left\{ \begin{array}{l} 1.63 \text{ kips to the left} \\ 3.91 \text{ kips upward} \\ 0.00 \text{ kips} \\ 4.22 \text{ kips upward} \\ 1.10 \text{ kips to the right} \\ 1.46 \text{ kips upward} \\ 0.52 \text{ kips to the right} \\ 0.39 \text{ kips upward} \end{array} \right\}$$

14.4   ALTERNATIVE MATRIX ANALYSIS OF DETERMINATE TRUSSES

If a truss is statically determinate, it is, by definition, possible to determine all member forces by the method of joints. In the basic matrix analysis procedure described in Section 14.2, this is equivalent to stating that the member forces can be obtained by invoking the equilibrium conditions alone. The correctness of this statement is verified by recognizing that the statics matrix, $[E_P]$, is $NP \times NF$ in size and thus is a square matrix for a determinate truss ($NP = NF$). This condition permits computation of the member forces according to

$$\{F\} = [E_P]^{-1} \{P\} \qquad (14.39)$$

without the necessity of constructing the stiffness matrix $[K]$. If it is necessary to determine the kinematic unknowns, this also can be done. Since $[S]$ is always square it is invertible. The compatibility matrix $[C]$ is $NF \times NP$ in size. Hence, for a statically determinate structure it also is square and invertible. Therefore

$$\{\Delta\} = [S]^{-1} \{F\} \qquad (14.40)$$

$$\{X\} = [C]^{-1} \{\Delta\} \qquad (14.41)$$

The reactions follow in the usual manner:

$$\{R\} = [E_R] \{F\} \qquad (14.42)$$

Example 14.6.   Analyze the determinate truss solved earlier in Example 14.4 by the alternative matrix analysis.

Solution: Using the matrices developed in Example 4.5

$$\{F\} = [E_p]^{-1} \{P\}$$

$$\begin{Bmatrix} F_1 \\ F_2 \end{Bmatrix} = \begin{bmatrix} -1.0 & -0.530 \\ 0 & -0.848 \end{bmatrix}^{-1} \begin{Bmatrix} 7 \\ -9 \end{Bmatrix}$$

$$= \frac{1}{(-1)(-0.848) - (0)} \begin{bmatrix} -0.848 & 0.530 \\ 0 & -1 \end{bmatrix}$$

$$\cdot \begin{Bmatrix} 7 \\ -9 \end{Bmatrix}$$

$$= \begin{bmatrix} -1 & 0.625 \\ 0 & -1.179 \end{bmatrix} \begin{Bmatrix} 7 \\ -9 \end{Bmatrix}$$

$$\begin{Bmatrix} F_1 \\ F_2 \end{Bmatrix} = \begin{Bmatrix} -12.6 \\ +10.6 \end{Bmatrix}$$

$$\{\Delta\} = [S]^{-1} \{F\}$$

$$\begin{Bmatrix} \Delta_1 \\ \Delta_2 \end{Bmatrix} = E \begin{bmatrix} .0250 & 0 \\ 0 & .0132 \end{bmatrix}^{-1} \begin{Bmatrix} -12.6 \\ -10.6 \end{Bmatrix}$$

$$= \frac{1}{E} \begin{bmatrix} 40 & 0 \\ 0 & 75.8 \end{bmatrix} \begin{Bmatrix} -12.6 \\ 10.6 \end{Bmatrix}$$

$$\begin{Bmatrix} \Delta_1 \\ \Delta_2 \end{Bmatrix} = \begin{Bmatrix} -.0168 \\ .0268 \end{Bmatrix}$$

$$\{X\} = [C]^{-1} \{\Delta\}$$

$$\begin{Bmatrix} X_1 \\ X_2 \end{Bmatrix} = \begin{bmatrix} -1.0 & 0 \\ -0.53 & -0.848 \end{bmatrix}^{-1} \begin{Bmatrix} -.0168 \\ .0268 \end{Bmatrix}$$

$$\begin{Bmatrix} X_1 \\ X_2 \end{Bmatrix} = \begin{Bmatrix} .0168 \\ -.0421 \end{Bmatrix}$$

The preceding numerical results agree with the Example 14.4. $\{R\}$ is computed by

$$\{R\} = [E_R] \{F\}$$

as was done in Example 14.4. Units and directions associated with the above numerical values are identical to those in Example 14.4.

14.5   TREATMENT OF SUPPORT SETTLEMENT

Uneven settlement of supports is a condition potentially encountered in structural systems. If a truss is indeterminate, a differential displacement of adjacent supports can introduce strains that are additional to those caused by "applied" joint loads. These strains develop as the structure adjusts to the new position. Determinate trusses can accommodate the motions by rigid-body translation and rotation of its members, and no strain is induced. However, in indeterminate trusses, the strains usually do not occur freely, and internal forces are developed in the members. (A more general statement is that all trusses, whether determinate or not, that can accommodate the differential settlements by rigid body rotations of their members, will not develop internal forces.) Internal forces created by the support motions can be very large in comparison with those attributed to the loading itself and cannot be neglected in the analysis of the structure. Inclusion of support settlements in the matrix analysis of framed structure can be accomplished in one of several ways. Adjustment of the $\{P\}$ matrix is described in this section. Alternatives that are more suitable for advanced computer programs are discussed in Chapter 16.

Movement of a joint in a truss is best viewed as a load type that is not directly applied to the members. Such effects are termed *actions*. The action of displacing a support can create internal forces despite the absence of joint loads. In the basic matrix analysis method, support settlements can be treated by considering a two-stage analysis. Example 14.7 serves to introduce the concept and demonstrate the computations.

Example 14.7. Determine the member forces induced in the truss shown in Fig. 14.8a if support D settles 0.05 cm. $EA = 200000$ kN for all members. Use the basic matrix analysis method.

Solution: Indeterminancy of the truss is evident from Eq. 5.14

$$I = NF - NP = 5 - 3 = 2$$

and suggests that the imposed displacement will induce force into the members.

Step 1: Analysis begins with the usual formation of the basic matrices. Detail formation of the matrices is omitted, and the resulting matrices are simply presented for later use.

$$[E_p] = \begin{bmatrix} 0 & -1.0 & 0 & -0.707 & 0 \\ 1.0 & 0 & 0 & 0.707 & 0 \\ 0 & 0 & 1.0 & 0 & 0.707 \end{bmatrix}$$

$$[E_R] = \begin{bmatrix} 0 & 0 & 0 & 0 & -0.707 \\ -1.0 & 0 & 0 & 0 & -0.707 \\ 0 & 0 & 0 & 0.707 & 0 \\ 0 & 0 & -1.0 & -0.707 & 0 \\ 0 & 1.0 & 0 & 0 & 0.707 \end{bmatrix}$$

$$[S] = EA \begin{bmatrix} 0.01 & 0 & 0 & 0 & 0 \\ 0 & 0.01 & 0 & 0 & 0 \\ 0 & 0 & 0.01 & 0 & 0 \\ 0 & 0 & 0 & 0.00707 & 0 \\ 0 & 0 & 0 & 0 & 0.00707 \end{bmatrix}$$

$$[C] = [E_p]^T$$

The preceding matrices are based upon the member number system indicated in Fig. 14.8a.

Step 2: The first stage in treating the support settlement is illustrated in Fig. 14.8b. In this stage, all joints are held in position ("restrained state") except joint D, which is allowed to experience the known displacement. To prevent motion of the other supports restraining forces, RP_1 through RP_3 must be developed as shown in Fig. 14.8b. In addition, the reactions R_{01} through R_{05} are created. The magnitudes of the restraining forces are established by a consideration of compatibility. For joint D to settle 0.05 cm, members 3 and 4 must deform as shown in the figure. In this instance, both members undergo *elongation*. Member 4 elongates an amount

$$\Delta_4 = DD'' = 0.05 \sin 45° = 0.03535 \text{ cm}$$

and member 3 elongates an amount

$$\Delta_3 = DD'' = 0.05 \text{ cm}$$

Forces corresponding to the induced elongations are obtained from Hooke's law

$$F_4 = \frac{EA}{L} \Delta_4 = \frac{200000}{141.4} (0.03535)$$

$$= 50.0 \text{ kN tension}$$

$$F_3 = \frac{EA}{L} \Delta_3 = \frac{200000}{100} (0.05)$$

$$= 100.0 \text{ kN tension}$$

displaying these results in a member force diagram (Fig. 14.8c) permits calculation of the required re-

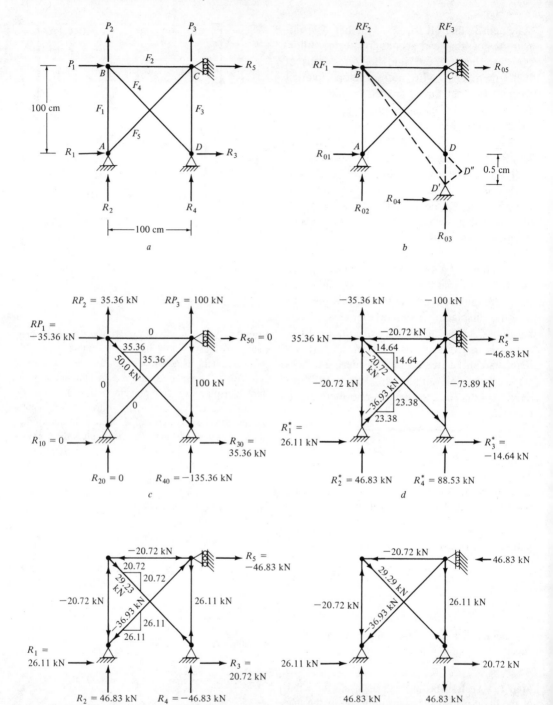

Fig. 14.8.   Matrix analysis of support settlement. a, truss and degrees of freedom; b, settlement of support D; c, restraining forces; d, released state; e, combined forces; f, final results.

straining forces and reactions by application of statics to the free joints. It is convenient to assemble an INITIAL FORCE matrix, $\{F_0\}$, and an INITIAL REACTION matrix, $\{R_0\}$, which contain the internal member forces and external reactions, respectively, in the restrained state.

$$\{F_0\} = \begin{Bmatrix} F_{01} \\ F_{02} \\ F_{03} \\ F_{04} \\ F_{05} \end{Bmatrix} = \begin{Bmatrix} 0 \\ 0 \\ 100 \\ 50 \\ 0 \end{Bmatrix} \text{kN}$$

and

$$\{R_0\} = \begin{Bmatrix} R_{01} \\ R_{02} \\ R_{03} \\ R_{04} \\ R_{05} \end{Bmatrix} = \begin{Bmatrix} 0 \\ 0 \\ 35.36 \\ -135.36 \\ 0 \end{Bmatrix} \text{kN}$$

Step 3: The restraining forces shown in Fig. 14.8c represent loads that are not actually present in the real structure. To restore the joints to their originally unloaded state, these restraining forces must be removed. Removal of the restraining forces is accomplished in the second stage of analysis. Analytically, it is simply a matter of applying loads equal in magnitude but opposite in direction to the restraining forces to cancel their presence. This concept is depicted in Fig. 14.8d, which shows the *released structure*. In this structure, joints B and C are free to move (unlike Fig. 14.8b and c), and hence the terminology *released state* is used.

The released state truss is analyzed by the usual basic matrix analysis procedure. The load matrix is

$$\{P\} = \begin{Bmatrix} P_1 \\ P_2 \\ P_3 \end{Bmatrix} = \begin{Bmatrix} 35.36 \\ -35.36 \\ -100.00 \end{Bmatrix} \text{kN}$$

Step 4: Remaining computations are executed as follows:

$$[K] = [E_p][S][C]$$

$$= EA \begin{bmatrix} 0.01353 & -0.00353 & 0 \\ -0.00353 & 0.01353 & 0 \\ 0 & 0 & 0.01353 \end{bmatrix}$$

$$\{X*\} = [K]^{-1} \{P\}$$

$$= \frac{1}{EA} \begin{bmatrix} 79.29 & 20.71 & 0 \\ 20.71 & 79.29 & 0 \\ 0 & 0 & 73.89 \end{bmatrix}$$

$$\cdot \begin{Bmatrix} 35.36 \\ -35.36 \\ -100.00 \end{Bmatrix}$$

$$= \frac{1}{EA} \begin{Bmatrix} 2072 \\ -2072 \\ -7389 \end{Bmatrix} \text{cm}$$

$$\{\Delta*\} = \begin{Bmatrix} \Delta_1^* \\ \Delta_2^* \\ \Delta_3^* \\ \Delta_4^* \\ \Delta_5^* \end{Bmatrix} = [C]\{X\}$$

$$= \frac{1}{EA} \begin{Bmatrix} -2072 \\ -2072 \\ -7389 \\ -2929 \\ -5224 \end{Bmatrix} \text{cm}$$

$$\{F*\} = \begin{Bmatrix} F_1^* \\ F_2^* \\ F_3^* \\ F_4^* \\ F_5^* \end{Bmatrix} = [S]\{\Delta\} = \begin{Bmatrix} -20.72 \\ -20.72 \\ -73.89 \\ -20.72 \\ -36.93 \end{Bmatrix}$$

$$= \begin{Bmatrix} 20.72\ C \\ 20.72\ C \\ 73.89\ C \\ 20.72\ C \\ 36.93\ C \end{Bmatrix} \text{kN}$$

$$\{R*\} = \begin{Bmatrix} R_1^* \\ R_2^* \\ R_3^* \\ R_4^* \\ R_5^* \end{Bmatrix} = [E_R]\{F\}$$

$$= \begin{Bmatrix} 26.11 \\ 46.83 \\ -14.64 \\ 88.53 \\ -46.83 \end{Bmatrix} = \begin{Bmatrix} 26.11 \rightarrow \\ 46.83 \uparrow \\ 14.64 \leftarrow \\ 88.53 \downarrow \\ 46.63 \leftarrow \end{Bmatrix} kN$$

Asterisks indicate that, unlike previous examples, these are not final answers. Internal forces and reactions for this stage are shown in Fig. 14.8d.

Step 5: Final answers are obtained by combining the forces shown in Fig. 14.8c and d. In matrix operations,

$$\{F\} = \{F_0\} + \{F*\} = \begin{Bmatrix} -20.72 \\ -20.72 \\ 26.11 \\ 29.29 \\ -36.93 \end{Bmatrix} kN$$

$$\{R\} = \{R_0\} + \{R*\} = \begin{Bmatrix} 26.11 \\ 46.83 \\ 20.72 \\ -46.83 \\ -46.83 \end{Bmatrix} kN$$

The outcome is presented in Fig. 14.8e. Visual inspection of the equilibrium of each joint is an indication of the correctness of the solution. Final results, with reactions shown in their correct directions, are given in Fig. 14.8f.

The two-stage analysis technique introduced in Example 14.7 has importance beyond the particular application in that problem. Replacement of the *real state* with a *restrained state* combined with a *released state* is used as a general procedure in subsequent subject matter

concerning matrix methods. Indeed, it will be seen that, regardless of type (beam, planar frame, etc.), unless a structure is subjected only to "applied" loads that act directly at the joints, the two-stage analysis is a necessity. A more extensive presentation of this concept is contained in Chapters 15 and 16. (Note: If the roller support at joint C were not present, no forces would develop despite the indeterminacy of the resulting truss. This is apparent because the support motion would induce only a rigid body rotation of the entire truss. However, if the bottom supports were not at the same level or if large deflection theory were employed, the absence of internal member forces would not be the case.)

14.6 TREATMENT OF INITIAL MEMBER DEFORMATIONS

Fabrication of structural members is done within the constraint of prescribed tolerances. Cross-sectional dimensions and length may not be equal to the values indicated on drawings. When the length of a truss member is different from the schematic dimension used in the structural analysis model, an unanticipated response will occur after erection is complete. A statically indeterminate truss will develop both member forces and displacements. A statically determinate truss will displace to a new position without developing member forces.

Temperature change, creep of materials, and other phenomena also cause a response in an otherwise unloaded truss. Treatment of these types of analysis problems is accomplished by the two stage solution procedure employed for treatment of support settlements. The procedure will be presented by means of an example problem.

Example 14.8. Analyze the truss in Fig. 14.7a for the effect of member 2 being fabricated $\frac{1}{16}''$ too long.

Solution:
Restrained state: With joint 5 restrained, member 2 must be precompressed to properly fit. The precompression force is obtained from Eq. 14.16

$$F_{02} = \frac{EA}{L} \Delta = 312.5 \left(-\frac{1}{16} \right) = -19.53 \text{ kips}$$

Thus

$$\{F_0\} = \begin{Bmatrix} F_{01} \\ F_{02} \\ F_{03} \\ F_{04} \end{Bmatrix} = \begin{Bmatrix} 0 \\ -19.53 \\ 0 \\ 0 \end{Bmatrix}$$

Released state: When joint 5 is released, the pre-compressed bar pushes on joints 2 and 5 with a force of magnitude 19.53 kips. The force at joint 2 acts as if it were an upward load on the support, producing an initial reaction. Thus

$$\{R_0\} = \begin{Bmatrix} R_{01} \\ R_{02} \\ R_{03} \\ R_{04} \\ R_{05} \\ R_{06} \\ R_{07} \\ R_{08} \end{Bmatrix} = \begin{Bmatrix} 0 \\ 0 \\ 0 \\ 19.53 \\ 0 \\ 0 \\ 0 \\ 0 \end{Bmatrix}$$

At joint 5, the force acts as if it were a downward load on the joint. Thus

$$\{P\} = \begin{Bmatrix} P_1 \\ P_2 \end{Bmatrix} = \begin{Bmatrix} -19.53 \\ 0 \end{Bmatrix}$$

A complete matrix analysis would be executed to obtain the response. In this case, the results can be obtained expeditiously since the load matrix is proportionate to the load matrix for Example 14.5.

$$\{X\} = \begin{Bmatrix} X_1 \\ X_2 \end{Bmatrix} = \frac{19.53}{10.00} \begin{Bmatrix} -.0135 \\ .0058 \end{Bmatrix} = \begin{Bmatrix} -.0264 \\ .0113 \end{Bmatrix}$$

$$\{\Delta^*\} = \begin{Bmatrix} \Delta_2^* \\ \Delta_2^* \\ \Delta_3^* \\ \Delta_4^* \end{Bmatrix} = \frac{19.53}{10.00} \begin{Bmatrix} .0147 \\ .0135 \\ .00732 \\ .00346 \end{Bmatrix}$$

$$= \begin{Bmatrix} .0287 \\ .0264 \\ .0143 \\ .00676 \end{Bmatrix}$$

$$\{F^*\} = \begin{Bmatrix} F_1^* \\ F_2^* \\ F_3^* \\ F_4^* \end{Bmatrix} = \frac{19.53}{10.00} \begin{Bmatrix} 4.24 \\ 4.22 \\ 1.83 \\ 0.65 \end{Bmatrix}$$

$$= \begin{Bmatrix} 8.28 \\ 8.24 \\ 3.57 \\ 1.27 \end{Bmatrix}$$

$$\begin{Bmatrix} R_1^* \\ R_2^* \\ R_3^* \\ R_4^* \\ R_5^* \\ R_6^* \\ R_7^* \\ R_8^* \end{Bmatrix} = \frac{19.53}{10.00} \begin{Bmatrix} -1.63 \\ 3.91 \\ 0. \\ 4.22 \\ 1.10 \\ 1.46 \\ 0.52 \\ 0.39 \end{Bmatrix} = \begin{Bmatrix} 3.18 \\ 7.64 \\ 0. \\ 8.24 \\ 2.15 \\ 2.85 \\ 1.02 \\ 0.76 \end{Bmatrix}$$

Combined state (real state):

$$\{F\} = \{F_0\} + \{F^*\}$$

$$\begin{Bmatrix} F_1 \\ F_2 \\ F_3 \\ F_4 \end{Bmatrix} = \begin{Bmatrix} 0 \\ -19.53 \\ 0. \\ 0. \end{Bmatrix} + \begin{Bmatrix} 8.28 \\ 8.24 \\ 3.57 \\ 1.27 \end{Bmatrix}$$

$$= \begin{Bmatrix} 8.28 \\ -11.29 \\ 3.57 \\ 1.27 \end{Bmatrix}$$

$$= \begin{Bmatrix} 8.28 \text{ kips tension} \\ 11.29 \text{ kips compression} \\ 3.57 \text{ kips tension} \\ 1.27 \text{ kips tension} \end{Bmatrix}$$

$$\{R\} = \{R_0\} + \{R^*\}$$

$$
\begin{Bmatrix} R_1 \\ R_2 \\ R_3 \\ R_4 \\ R_5 \\ R_6 \\ R_7 \\ R_8 \end{Bmatrix}
=
\begin{Bmatrix} 0. \\ 0. \\ 0. \\ 19.53 \\ 0. \\ 0. \\ 0. \\ 0. \end{Bmatrix}
+
\begin{Bmatrix} 3.18 \\ 7.64 \\ 0. \\ 8.24 \\ 2.15 \\ 2.85 \\ 1.02 \\ 0.76 \end{Bmatrix}
$$

$$
=
\begin{Bmatrix} 3.18 \text{ kips} \rightarrow \\ 7.64 \text{ kips} \uparrow \\ 0. \\ 47.77 \text{ kips} \uparrow \\ 4.15 \text{ kips} \rightarrow \\ 4.85 \text{ kips} \uparrow \\ 1.04 \text{ kips} \rightarrow \\ 0.76 \text{ kips} \uparrow \end{Bmatrix}
$$

14.7 COMPUTER PROGRAMS

Execution of the basic matrix analysis method involves a limited number of operations in matrix algebra. Development of a rudimentary computer algorithm to perform the analysis is not a complex task. Some useful computer programs are presented in Appendix C. Programs AMATRIX.BAS and AMATRIX.F perform the alternate basic matrix analysis for determinate trusses as described in Section 14.5. Programs BMATRIX.BAS and BMATRIX.F execute the basic matrix analysis enumerated in Table 14.1 and includes features necessary for the analysis of trusses that are subjected to special actions (e.g., support settlements) in addition to applied loads.

PROBLEMS

Note: The "basic matrix analysis method" refers to Section 14.2 and Programs AMATRIX in Appendix C. The "alternate matrix analysis method" refers to Section 14.4 and Programs BMATRIX in Appendix C. Unless stated otherwise, employ these programs in solving the given problems. Use the indicated numbering for members, loads and degree of freedom.

Note: Unless stated otherwise, use $EA = 300,000$ kips for U.S. customary units and $EA = 300,000$ kN for SI units.

14-1.   Analyze the given trusses by the basic matrix analysis method. Do the analysis manually, i.e. without resort to a computer. See Appendix B for a review of matrix inversion.

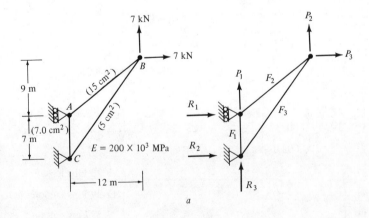

a

(Problem 14-1 continued on p. 379)

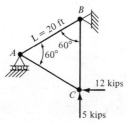

All members: $E = 30,000$ ksi
$A = 10$ in^2

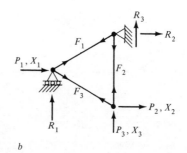

b

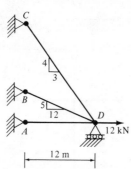

12 m

$EA = 20000$ kN for all members

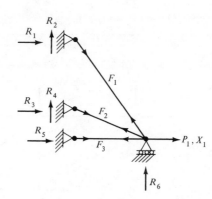

c

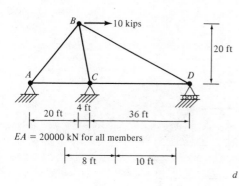

$EA = 20000$ kN for all members

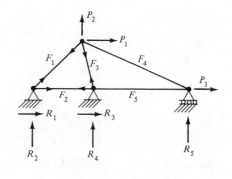

d

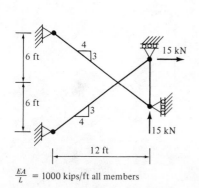

$\frac{EA}{L} = 1000$ kips/ft all members

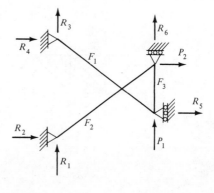

e

14-2.   Analyze the given trusses by the basic matrix analysis method.

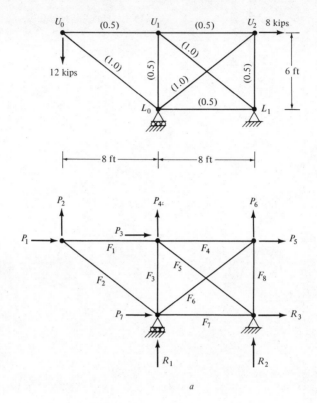

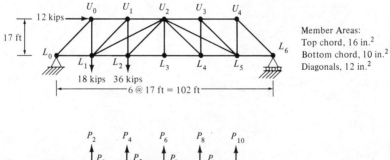

a

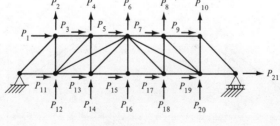

Member Areas:
Top chord, 16 in.^{2}
Bottom chord, 10 in.^{2}
Diagonals, 12 in.^{2}

b

(Problem 14-2 continued on p. 381)

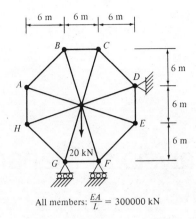

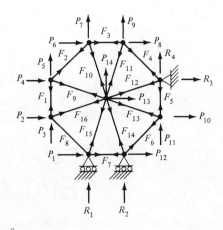

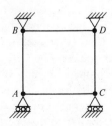

6 m | 6 m | 6 m

All members: $\frac{EA}{L} = 300000$ kN

14-3. Analyze the trusses given in Problem 5-1 by the basic matrix analysis method.

14-4. Repeat Problem 14-3 using the alternate matrix analysis method.

14-5. Analyze the trusses in Problem 5-4 by the basic matrix analysis method.

14-6. Repeat Problem 14-5 using the alternate matrix analysis method.

14-7. Analyze the trusses in Problem 5-5 by the basic matrix analysis method.

14-8. Analyze the stable trusses given in Problem 5-11 by the basic matrix analysis method.

14-9. Repeat Problem 14-8 by the alternate matrix analysis method.

14-10. Do Problem 5-6 by the alternative matrix analysis method.

14-11. Do Problem 5-7 by the alternative matrix analysis method.

14-12. Do Problem 5-8 by the alternative matrix analysis method.

14-13. Analyze the truss in Problem 14-1d for a settlement of 0.10 in. at joint C. Remove the applied loads!

14-14. Analyze the truss in Problem 14-2a for a settlement of 0.25 in. at joint L_1. Remove the applied loads!

14-15. Analyze the truss in Problem 14-2a for the case of member $L_0 U_2$ having been fabricated too short by $\frac{1}{16}$ in. and forced to fit.

14-16. Prove that multiplying all member areas in a truss by a constant factor k does not change the member forces developed for a given loading. (This must be shown to be true, even if the truss is statically indeterminate).

14-17. For a given loading, the truss in Problem 14-1a has the member forces

$$F_1 = 6 \text{ kN} \qquad F_2 = -12 \text{ kN} \qquad F_3 = 0$$

Determine the loads by matrix analysis using manual computation.

14-18. The given truss is unstable. Attempt to build $[K]$ and obtain its inverse. Comment on the outcome. The dimensions are 6 ft by 8 ft.

15

Direct Stiffness Analysis of Planar Structures

15.1 INTRODUCTION

15.1.1 Disadvantages of the Basic Matrix Analysis Method

A fundamental approach to matrix analysis of planar trusses was developed in Chapter 14. The procedure is called the "basic matrix analysis method" because of its rudimentary nature. To employ the method, the analyst must manually construct a set of basic matrices prior to the execution of any matrix algebra. From a conceptual point of a view, the method has merit for a beginning student because of its direct correlation with the three criteria of structural analysis: equilibrium, material laws, and compatibility. The use of separate matrices to represent each of these conditions gives the reader direct and clear understanding of how the structural criteria are incorporated into the matrix formulation. Conveyance of this correspondence between the concepts and the basic matrices was the principal goal in Chapter 14.

A key step in the basic matrix analysis method is the generation of the structure stiffness matrix according to the operation

$$[K] = [E_P][S][C]$$

Thus, three matrices must be developed prior to performing the indicated matrix multiplica-

tion. The difficulties associated with this task are dependent upon the nature of the structure. For planar trusses, the construction of the matrices $[E_P]$, $[S]$, and $[C]$ is straightforward. In the case of beams or planar frames with simple geometry (i.e., a rectangular member configuration), the generation of the matrices is also possible and the basic matrix analysis method could be developed for these structures. Unfortunately, if the geometry of a frame is more complex than a simple rectangular arrangement of members, the simplicity of the task quickly disintegrates. The mere presence of a single inclined member results in a tedious sequence of computations to build both the $[E_P]$ and $[C]$ matrices. Recognition that $[C] = [E_P]^T$ does not eliminate the difficulties. More importantly, an extension of the concepts to three-dimensions (space structures) is unwieldy. For general application, the basic method for developing $[K]$ is unsuitable.

One aim of this chapter is to describe alternative techniques for generating the structure stiffness matrix. By introducing one new concept, it will be shown that $[K]$ can be formed "directly" without need for any matrix operations. Elimination of the need to form $[E_P]$, $[S]$, and $[C]$ and of the two subsequent multiplications formerly needed is an obvious advantage in time and effort. Because of the manner in which $[K]$ is constructed, the procedure is termed the *direct stiffness method*. Presentation of the direct stiffness method constitutes a first progression toward an automated matrix analysis procedure. Treatment of space structures remains a prohibitive undertaking, and in this chapter applications will be limited to planar structures. Chapter 16 will describe further refinements to matrix structural analysis and create a basis for an in-depth understanding of contemporary computer methods.

15.1.2 The $[S]_m$ Matrix for a Flexural Member.

Despite the unsuitability of the basic matrix analysis method for beams and frames, it is pertinent to evolve the $[S]_m$ matrix for a flexural member. This is done as follows.

15.1.2.1 Unsupported Member

Fig. 15.1 shows an individual flexural member. If no transverse loading is present on a member—i.e., $q(x) = 0$—integration of Eq. 3.27 yields

$$y = C_1 \frac{x^3}{6} + C_2 \frac{x^2}{2} + C_3 x + C_4 \quad (15.1)$$

and determining the constants C_1–C_4 (by applying boundary conditions) establishes a solution. In local coordinates, the general boundary conditions are the four displacements at the member ends. For general purposes, the nomenclature and directions shown in Fig. 15.1b

will be employed. Positive translations are upward, and positive rotations are clockwise. (This differs from the sign convention used previously for beams and is more convenient in the later development of a matrix analysis procedure.) It must be noted that mathematically a positive rotation by this definition corresponds to a *negative* slope!

Using the terminology of Fig. 15.1b and Eq. 15.1,

$$@x = 0 \ y \ = C_4 = +\delta_L$$

$$@x = 0 \ y' = C_3 = -\Phi_L$$

$$@x = L \ y \ = C_1 \frac{L^3}{6} + C_2 \frac{L^2}{2}$$

$$+ C_3 L + C_4 = +\delta_R$$

$$@x = L \ y' = C_1 \frac{L^2}{2} + C_2 L + C_3 = -\Phi_R$$

$$(15.2)$$

Substituting the values for C_3 and C_4 from the first and second equations into the latter pair of equations and solving simultaneously yields the values of C_1 and C_2. The complete set of constants that results is

$$C_1 = -\frac{6}{L^2} \Phi_L - \frac{6}{L^2} \Phi_R + \frac{12}{L^3} \delta_L - \frac{12}{L^3} \delta_R$$

$$C_2 = \frac{4}{L} \Phi_L + \frac{2}{L} \Phi_R - \frac{6}{L^2} \delta_L + \frac{6}{L^2} \delta_R$$

$$C_3 = -\Phi_L$$

$$C_4 = +\delta_L \quad (15.3)$$

Substitution into Eq. 15.1 gives the general equation of the elastic curve for an unloaded planar frame member subjected to end deformations.

Next, consider the computation of the corresponding member end forces shown in Fig. 15.1a. These forces are determined by appropriate differentiation of Eq. 15.1. The sign convention implied in Fig. 15.1a assumes upward shears, and clockwise end moments are positive. This is contrary to the designer's sign con-

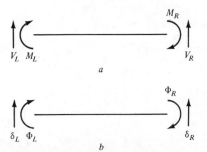

Fig. 15.1. General local end forces and deformations. *a,* local end forces; *b,* local deformations.

ventions used in deriving Eq. 3.27, and caution must be exercised. For a constant value of EI,

$$M_L = +EI \left(\frac{d^2y}{dx^2}\right)_{x=0} = EI\, C_2$$

$$M_R = -EI \left(\frac{d^2y}{dx^2}\right)_{x=L} = -EI(C_1 L + C_2)$$

$$V_L = +EI \left(\frac{d^3y}{dx^3}\right)_{x=0} = EI(C_1)$$

$$V_R = -EI \left(\frac{d^3y}{dx^3}\right)_{x=L} = -EI(C_1) \quad (15.4)$$

or, by substitution of Eqs. 15.3,

$$M_L = \frac{4EI}{L}\Phi_L + \frac{2EI}{L}\Phi_R$$
$$- \frac{6EI}{L^2}\delta_L + \frac{6EI}{L^2}\delta_R$$

$$M_R = \frac{2EI}{L}\Phi_L + \frac{4EI}{L}\Phi_R$$
$$- \frac{6EI}{L^2}\delta_L + \frac{6EI}{L^2}\delta_R$$

$$V_L = -\frac{6EI}{L^2}\Phi_L - \frac{6EI}{L^2}\Phi_R$$
$$+ \frac{12EI}{L^3}\delta_L - \frac{12EI}{L^3}\delta_R$$

$$V_R = \frac{6EI}{L^2}\Phi_L + \frac{6EI}{L^2}\Phi_R$$
$$- \frac{12EI}{L^3}\delta_L + \frac{12EI}{L^3}\delta_R \quad (15.5)$$

Written in matrix form, Eq. 15.5 becomes

The square matrix that relates the end forces to the local end deformations is the "local member stiffness matrix" $[S]_m$ of an individual flexural member. In formulating the matrix, axial deformations were neglected. Signs associated with the column matrices in Eq. 15.6 should be reiterated: (1) clockwise end moments and rotations are positive, and (2) upward shear and displacements are positive.

15.1.2.2 Supported Member

A simple beam is drawn in Fig. 15.2a with the local x-axis aligned with the member. Unlike a general planar-frame member, the beam element is restricted from experiencing displacements at its ends. If the loading is limited to the application of concentrated moments, M_L and M_R, at the supports, the local member stiffness matrix can be obtained as a degenerated form of Eq. 15.6.

At its ends, a beam element will develop internal moments and shears and experience rotational deformation as indicated in Fig. 15.2b and c, respectively. The local member stiffness matrix relates the internal end moments, M_L and M_R, to the internal rotations, Φ_L and Φ_R. By setting δ_L and δ_R equal to zero in Eq. 15.6, it is apparent that

$$\left\{\begin{array}{c} M_L \\ M_R \end{array}\right\} = \begin{bmatrix} \dfrac{4EI}{L} & \dfrac{2EI}{L} \\[2ex] \dfrac{2EI}{L} & \dfrac{4EI}{L} \end{bmatrix} \left\{\begin{array}{c} \Phi_L \\ \Phi_R \end{array}\right\} \quad (15.7)$$

The square matrix in Eq. 15.7 is the desired local member stiffness matrix $[S]_m$ for this supported member.

$$\left\{\begin{array}{c} M_L \\ M_R \\ V_L \\ V_R \end{array}\right\} = \begin{bmatrix} \dfrac{4EI}{L} & \dfrac{2EI}{L} & -\dfrac{6EI}{L^2} & \dfrac{6EI}{L^2} \\[2ex] \dfrac{2EI}{L} & \dfrac{4EI}{L} & -\dfrac{6EI}{L^2} & \dfrac{6EI}{L^2} \\[2ex] -\dfrac{6EI}{L^2} & -\dfrac{6EI}{L^2} & \dfrac{12EI}{L^3} & -\dfrac{12EI}{L^3} \\[2ex] \dfrac{6EI}{L^2} & +\dfrac{6EI}{L^2} & -\dfrac{12EI}{L^3} & \dfrac{12EI}{L^3} \end{bmatrix} \left\{\begin{array}{c} \Phi_L \\ \Phi_R \\ \delta_L \\ \delta_R \end{array}\right\} \quad (15.6)$$

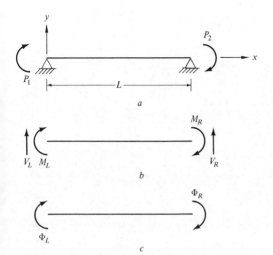

Fig. 15.2. End forces and deformations for a supported beam element.

15.1.2.3 Overhanging Members

Consider the overhanging element shown in Fig. 15.3. Examination of the internal forces and deformations indicates that the local member stiffness matrix can be obtained by setting $\delta_L = 0$ in Eq. 15.6. The result is

$$
\begin{Bmatrix} M_L \\ M_R \\ V_R \end{Bmatrix} =
\begin{bmatrix}
\dfrac{4EI}{L} & \dfrac{2EI}{L} & \dfrac{6EI}{L^2} \\[2mm]
\dfrac{2EI}{L} & \dfrac{4EI}{L} & \dfrac{6EI}{L^2} \\[2mm]
\dfrac{6EI}{L^2} & \dfrac{6EI}{L^2} & \dfrac{12EI}{L^3}
\end{bmatrix}
\begin{Bmatrix} \Phi_L \\ \Phi_R \\ \delta_R \end{Bmatrix}
$$

(15.8)

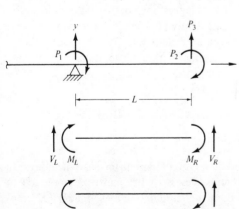

Fig. 15.3.   Overhanging beam element.

Equation 15.8 contains the desired 3×3 local member stiffness matrix. However, this matrix applies only if the overhang is at the right end of the member. If the overhang is at the left end, it is necessary to set $\delta_R = 0$ in Eq. 15.6. The result is

$$
\begin{Bmatrix} M_L \\ M_R \\ V_L \end{Bmatrix} =
\begin{bmatrix}
\dfrac{4EI}{L} & \dfrac{2EI}{L} & -\dfrac{6EI}{L^2} \\[2mm]
\dfrac{2EI}{L} & \dfrac{4EI}{L} & -\dfrac{6EI}{L^2} \\[2mm]
-\dfrac{6EI}{L^2} & -\dfrac{6EI}{L^2} & \dfrac{12EI}{L^3}
\end{bmatrix}
\begin{Bmatrix} \Phi_L \\ \Phi_R \\ \delta_L \end{Bmatrix}
$$

(15.9)

The rectangular matrices in Eqs. 15.8 and 15.9 are local member stiffness matrices for overhanging members. Thus for a beam overhanging to the right

$$
[S]_m =
\begin{bmatrix}
\dfrac{4EI}{L} & \dfrac{2EI}{L} & \dfrac{6EI}{L^2} \\[2mm]
\dfrac{2EI}{L} & \dfrac{4EI}{L} & \dfrac{6EI}{L^2} \\[2mm]
\dfrac{6EI}{L^2} & \dfrac{6EI}{L^2} & \dfrac{12EI}{L^3}
\end{bmatrix}
$$
(15.10)

For a beam overhanging to the left

$$
[S]_m =
\begin{bmatrix}
\dfrac{4EI}{L} & \dfrac{2EI}{L} & -\dfrac{6EI}{L^2} \\[2mm]
\dfrac{2EI}{L} & \dfrac{4EI}{L} & -\dfrac{6EI}{L^2} \\[2mm]
-\dfrac{6EI}{L^2} & -\dfrac{6EI}{L^2} & \dfrac{12EI}{L^3}
\end{bmatrix}
$$

(15.11)

15.1.2.4   Deformations Referenced to the Displaced Chord

In Eq. 15.8, the matrix $[S]_m$ was constructed with reference to the undisplaced member axis. As a consequence, the 3×3 matrix $[S]_m$ of Eq. 15.8, resulted in which Φ_L and Φ_R are true slopes. Consider transferring the reference for

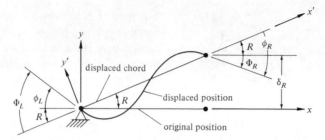

Fig. 15.4.   Positive deformations of an overhanging beam.

deformations to the displaced chord of the member

Figure 15.4 shows an overhanging member in its original and displaced positions. ϕ_L and ϕ_R are the rotations of the member ends measured from the displaced chord to the tangent to the displaced member. Because the chord rotates by an angle R, it is evident from Fig. 15.4, that

$$\Phi_L = \phi_L - R = \phi_L - \frac{\delta_R}{l}$$

$$\Phi_R = \phi_R - R = \phi_R - \frac{\delta_R}{l} \quad (15.12)$$

and substitution into Eq. 15.8 gives

$$\begin{Bmatrix} M_L \\ M_R \end{Bmatrix} = \begin{bmatrix} \dfrac{4EI}{L} & \dfrac{2EI}{L} \\ \dfrac{2EI}{L} & \dfrac{4EI}{L} \end{bmatrix} \begin{Bmatrix} \phi_L \\ \phi_R \end{Bmatrix}$$

$$= [S]'_m \begin{Bmatrix} \phi_L \\ \phi_R \end{Bmatrix} \quad (15.13)$$

The reader should note the similarity of Eqs. 15.13 and 15.7. Despite the similarity, it is critical to realize that the square matrix in Eq. 15.13 is a *modified* local stiffness matrix $[S]'_m$ and relates the clockwise end moments to the clockwise end rotations *measured relative to the displaced chord*.

Equation 15.13 can be abbreviated

$$\begin{Bmatrix} M_L \\ M_R \end{Bmatrix}_m = [S]'_m \begin{Bmatrix} \phi_L \\ \phi_R \end{Bmatrix} \quad (15.14)$$

or

$$\{F\}_m = [S]_m \{\Delta\}_m \quad (15.15)$$

where

$$[S]'_m = \begin{bmatrix} \dfrac{4EI}{L} & \dfrac{2EI}{L} \\ \dfrac{2EI}{L} & \dfrac{4EI}{L} \end{bmatrix} \quad (15.16)$$

The important observation is that, for a continuous beam with overhangs, $[S]_m$ can be written in either of two ways. In the first choice, deformations (Φ's) are measured relative to the horizontal (i.e., true slopes are used.) In the second choice, the deformations (ϕ's) are measured relative to the displaced chord.

The advantage of the second choice is that no special thought process is needed for building $[S]_m$ when overhangs are involved. The tradeoff for this simplicity is the matrix output for overhanging members would give deformations relative to the displaced chord. If true slopes are needed, Eqs. 15.12 must be implemented either within the computer algorithm or external to it.

15.2   GENERAL CONCEPTS

The matrices used in the basic matrix analysis method represent various physical meanings. In particular, each of the matrices employed to produce the structure stiffness matrix $[K]$ represents a criterion of structural analysis. Indeed, these matrices are constructed by directly applying the three criteria. It is reasonable to expect that $[K]$ itself has a physical counterpart, and this is the case.

The structure stiffness matrix relates the set of loads, $\{P\}$, applied in the direction of a structure's degrees of freedom to the corresponding set of displacements, $\{X\}$. For a structure with 4 degrees of freedom, the matrix stiffness relationship is:

$$\begin{Bmatrix} P_1 \\ P_2 \\ P_3 \\ P_4 \end{Bmatrix} = \begin{bmatrix} K_{11} & K_{12} & K_{13} & K_{14} \\ K_{21} & K_{22} & K_{23} & K_{24} \\ K_{31} & K_{32} & K_{33} & K_{34} \\ K_{41} & K_{42} & K_{43} & K_{44} \end{bmatrix} \begin{Bmatrix} X_1 \\ X_2 \\ X_3 \\ X_4 \end{Bmatrix}$$

$$(15.17)$$

A study of Eq. 15.17 suggests a useful interpretation of the stated relationship. Matrix $[K]$ gives the analyst a means to determine the unique set of loads $\{P\}$ needed to create a specific set of displacements $\{X\}$. The greater meaning of this observation is that the loads corresponding to a given displaced shape can be determined by use of the structure stiffness matrix, provided none of the members is subjected to transverse loading. This capability provides the basis for an alternative method for generating $[K]$.

Several displaced shapes—and, more importantly, the loads that cause them—are of specific interest. Consider a displaced shape in which any chosen displacement, X_i, is set equal to unity, while all other displacements in $\{X\}$ are assigned zero value. For example, if X_2 in Eq. 15.17 is chosen, then

$$\begin{Bmatrix} X_1 \\ X_2 \\ X_3 \\ X_4 \end{Bmatrix} = \begin{Bmatrix} 0 \\ 1 \\ 0 \\ 0 \end{Bmatrix}$$

By execution of Eq. 15.17 for these displacements,

$$\begin{Bmatrix} P_1 \\ P_2 \\ P_3 \\ P_4 \end{Bmatrix} = \begin{Bmatrix} K_{12} \\ K_{22} \\ K_{32} \\ K_{42} \end{Bmatrix}$$

The important observation is that the magnitudes of the calculated loads are equal to the values of the entries in the second column of $[K]$. This demonstration applies to the general displacement X_i and permits the following statement:

> The displaced shape in which X_i equals unity and all other displacements in $\{X\}$ are equal to zero requires the application of a specific set of loads $\{P\}$. The magnitude of the loads are equal in value to the entries in the ith column of structure stiffness matrix. Load P_j corresponds in magnitude to the entry K_{ji}.

Because each entry in the stiffness matrix is associated with a particular unit displacement, the entries are commonly referred to as *stiffness coefficients*.

A displaced shape of the nature described in the preceding paragraph is a *structure-compatibility mode*. $[K]$ can be constructed one column at a time, if one can establish the loads that correspond to each of the compatibility modes inherent in a given structure. Entries in the ith column of $[K]$ correspond to the applied loads of the ith compatibility mode. A planar frame (Fig. 15.5a) that has 3 degrees of freedom is used to exemplify the general principles.

Figure 15.5b depicts the compatibility mode that corresponds to the *first* column of the structure stiffness matrix. Specifically, a unit rotation is indicated at the first degree of freedom, but all other degrees of freedom are held in a restrained position. This displaced state cannot exist without the presence of a unique set of joint loads. The load values, as indicated in the figure, correspond to the stiffness coefficients in the first column of $[K]$. Determination of load values for a given displaced state constitutes the basis of direct stiffness method. In effect, one must perform structural analysis "in reverse."

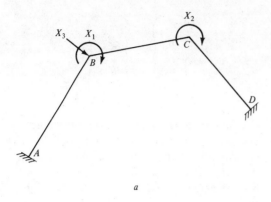

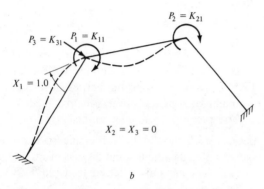

Fig. 15.5. Direct generation of a column of [*K*]. *a,* planar frame; *b,* compatibility mode 1.

15.3 DIRECT GENERATION OF [*K*] CONTINUOUS BEAMS

It is convenient to select the continuous beam as a model in an introductory discussion of the *direct stiffness matrix*. From a geometric viewpoint, a continuous beam is the simplest type of structure. All members lie along a common axis, and, by definition, axial deformation is not an issue. Normally, translational degrees of freedom are present only if overhangs exist, but, in any case, they cause no great analytical difficulties.

15.3.1 Generation by Physical Meaning

As a first consideration, the concept of *member-compatibility modes* (Fig. 15.6) must be discussed. Figure 15.6a illustrates a single beam element for which all degrees of freedom are restrained except the rotation at the right end. A unit rotation, ϕ_R, measured relative to the displaced chord (which corresponds to the original member axis in this case) is induced at that point by application of load P. This condition represents the first of two member-com-

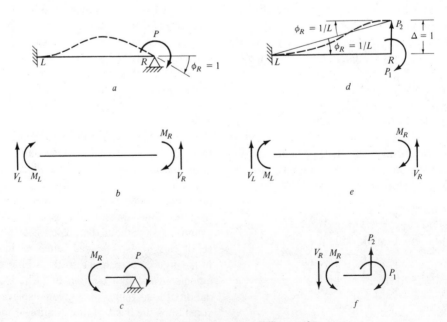

Fig. 15.6. Member compatibility modes.

patibility modes that will be examined. End moments M_L and M_R, and shear forces, V_L and V_R, are also created in this mode shape (Fig. 15.6b). The magnitude of each of these force quantities can be easily determined. By substitution of the known boundary conditions ($\Phi_R = \phi_R$, $\Phi_L = \delta_L = \delta_R = 0$) into Eq. 15.6,

$$M_L = \frac{4EI}{L}(0) + \frac{2EI}{L}(1) - \frac{6EI}{L^2}(0)$$

$$+ \frac{6EI}{L^2}(0) = +\frac{2EI}{L} \qquad (15.18)$$

$$M_R = \frac{2EI}{L}(0) + \frac{4EI}{L}(1) - \frac{6EI}{L^2}(0)$$

$$+ \frac{6EI}{L^2}(0) = \frac{4EI}{L} \qquad (15.19)$$

V_L and V_R are determined by applying moment equilibrium to the beam element

$$V_L = -\frac{M_L + M_R}{L} = -\frac{6EI}{L^2} \qquad (15.20)$$

$$V_R = +\frac{M_L + M_R}{L} = +\frac{6EI}{L^2} \qquad (15.21)$$

It is of interest to compute the load P required to produce the compatibility mode. From Fig. 15.6c,

$$P = M_R = +\frac{4EI}{L} \text{ cw}$$

A second member-compatibility mode is given in Fig. 15.6d. In this mode, a unit translation is induced at the right end while restraining all other degrees of freedom. Although the member ends do experience deformation relative to the displaced chord, *they do not rotate relative to the original member axis*. This observation is of vital importance. From Eq. 15.6

$$M_L = \frac{4EI}{L}(0) + \frac{2EI}{L}(0) - \frac{6EI}{L^2}(0)$$

$$+ \frac{6EI}{L^2}(1) = +\frac{6EI}{L^2} \qquad (15.22)$$

$$M_R = \frac{2EI}{L}(0) + \frac{4EI}{L}(0) - \frac{6EI}{L^2}(0)$$

$$+ \frac{6EI}{L^2}(1) = +\frac{6EI}{L^2} \qquad (15.23)$$

and by statics (Fig. 15.6e)

$$V_L = -\frac{M_L + M_R}{L} = -\frac{12EI}{L^3} \qquad (15.24)$$

$$V_R = +\frac{M_L + M_R}{L} = +\frac{12EI}{L^3} \qquad (15.25)$$

Joint loads required to create the mode shape are computed by satisfying equilibrium at the right joint. In this instance, two forces are necessary as determined from Fig. 15.6f.

$$P_1 = +M_R = +\frac{6EI}{L^2} = \frac{6EI}{L^2} \text{ cw}$$

$$P_2 = +V_R = +\frac{12EI}{L^3} = \frac{12EI}{L^3} \text{ upward}$$

The usefulness of the *member*-compatibility modes lies in their general presence in any *structure*-compatibility mode. Any given structure-compatibility mode is comprised of the members that deform like either of the two modes just described. Consequently, the *magnitudes* of all member end forces in a structure compatibility mode can be calculated by use of Eqs. 15.18 through 15.25. With end forces known, the equilibrium conditions at each joint can be employed to establish the corresponding joint loads. The computed loads represent entries in a particular column of the *structure* stiffness matrix. Each structure compatibility mode can be examined in like fashion to complete the building of $[K]$ one column at a time.

Example 15.1.   Build the structure stiffness matrix for the continuous beam shown in Fig. 15.7a. Use the physical meaning. Assume EI is constant and is the same value for both members.

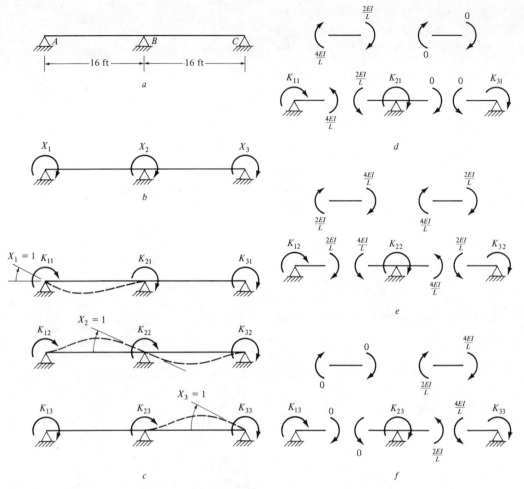

Fig. 15.7. Beam for Example 15.1. *a*, beam; *b*, degrees of freedom; *c*, compatibility modes; *d*, mode 1 forces; *e*, mode 2 forces; *f*, mode 3 forces.

Solution: The beam is kinematically indeterminate to the third degree. A rotational degree of freedom exists at each pinned support (Fig. 15.7b), and the structure-compatibility modes corresponding to them are sketched in Fig. 15.7c. Internal moments created in each mode are shown in Figs. 15.7d through f. Their magnitudes can either be calculated from Eq. 15.6 or deduced from knowledge of the results of Eqs. 15.18 and 15.19. This is done on the basis of the temporary fixed state of all but one joint in each mode. In the latter case, signs for the moment quantities are best determined from a physical investigation of the curvature in each mode shape. In each of the mode shapes in this example, all moments must act in a clockwise sense on the member ends and thus are positive. Columns of $[K]$ follow from application of statics to each of the joints.

Mode 1	Mode 2	Mode 3
$K_{11} = \dfrac{4EI}{L}$	$K_{12} = \dfrac{2EI}{L}$	$K_{13} = 0$
$K_{12} = \dfrac{2EI}{L}$	$K_{22} = \dfrac{8EI}{L}$	$K_{23} = \dfrac{2EI}{L}$
$K_{31} = 0$	$K_{32} = \dfrac{2EI}{L}$	$K_{33} = \dfrac{4EI}{L}$

Assembling the results and substituting $L = 16$ ft:

$$[K] = EI \begin{bmatrix} \frac{1}{4} & \frac{1}{8} & 0 \\ \frac{1}{8} & \frac{1}{2} & \frac{1}{8} \\ 0 & \frac{1}{8} & \frac{1}{4} \end{bmatrix}$$

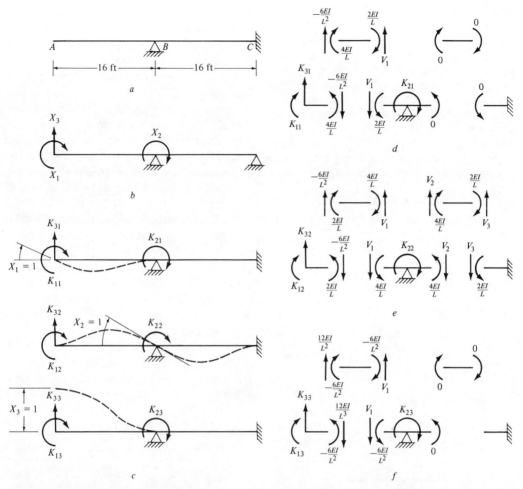

Fig. 15.8. Beam for Example 15.2. *a*, beam; *b*, degrees of freedom; *c*, compatibility modes; *d*, mode 1 forces; *e*, mode 2 forces; *f*, mode 3 forces.

Example 15.2. Follow the instructions in Example 15.1 except use the beam given in Fig. 15.8a.

Solution: The given beam has the three degrees of freedom shown in Fig. 15.8b. Corresponding compatibility modes are depicted in Fig. 15.8c. Modes 1 and 2 involve joint rotations only and produce the internal forces shown in Fig. 15.8d and e, respectively. The magnitudes of the end moments are deduced from Eqs. 15.18 and 15.19. The directions are deduced from the sense of the member curvatures. Loads associated with each mode are determined by applying statics to each joint. Prior to this, the shear force at the left end of member AB, in each case, is calculated either by statics (applied to the member itself) or by use of Eq. 15.20. From statics

of the joints shown in Fig. 15.8d and e, stiffness coefficients for each mode are determined as

Mode 1	Mode 2
$K_{11} = \dfrac{4EI}{L}$	$K_{12} = \dfrac{2EI}{L}$
$K_{21} = \dfrac{2EI}{L}$	$K_{22} = \dfrac{8EI}{L}$
$K_{31} = -\dfrac{6EI}{L^2}$	$K_{32} = -\dfrac{6EI}{L^2}$

Mode 3 is a translational mode in which only member AB experiences deformation. The magnitudes of the internal moments created by the relative motion of its two ends are deduced from previous results in Eqs. 15.22 and 15.23. However, it should be recalled that Eqs. 15.22 and 15.23 were derived for a case in which the right end moves upward relative to the left end. This is not the case in the present example. Equations 15.22 and 15.23 give the approprite magnitudes of the moments. By physical inspection of the shape of mode 3, it is apparent that the end moments must act in a counterclockwise sense to prevent rotation of the member ends. Consequently, the moments are negative. Shear forces follow by using Eqs. 15.24 and 15.25 and changing to the signs of the results. Applying statics to the joints in Fig. 15.8f gives the stiffness coefficients:

Mode 3

$$K_{13} = -\frac{6EI}{L^2}$$

$$K_{23} = -\frac{6EI}{L^2}$$

$$K_{33} = \frac{12EI}{L^3}$$

Substituting the known member lengths and assembling the results give

$$[K] = EI \begin{bmatrix} \frac{1}{4} & \frac{1}{8} & -\frac{3}{128} \\ \frac{1}{8} & \frac{1}{2} & -\frac{3}{128} \\ -\frac{3}{128} & -\frac{3}{128} & \frac{3}{1024} \end{bmatrix}$$

15.3.2   Generation by Expansion of Member Stiffness Matrix

In the preceding section, it was shown that by recognition of the physical meaning of $[K]$, the matrix can be assembled one column at a time. Although employment of this concept is an improvement over the basic matrix analysis method, some drawbacks still exist. The major deficiency is the need to investigate the structure "taken as whole" to determine the entries in any given column of $[K]$. This poses no real difficulty for continuous beams but is an unattractive task for other framed structures. In the present section, it will be shown that the use of structure-compatibility modes (the "structure approach") can be discarded in favor of the use of member-compatibility modes (the "member

approach"). In addition to its importance here, this discussion serves as an introduction to the more elegant and contemporary techniques that will be described in Chapter 16.

Entries in any given column of the structure stiffness matrix are directly attributed to a particular structure-compatibility mode. In Section 15.3.1, it was emphasized that any structure-compatibility mode can be viewed as an assemblage of member-compatibility modes. Consequently, it must be possible to develop the structure stiffness matrix by examining all possible member-compatibility modes individually. Each member compatibility mode has a numerical counterpart in the structure stiffness matrix. The total contribution made by any given member to the structure stiffness matrix is the combined effect of all its possible member-compatibility modes. It will be shown now that the combined effects are represented by the local member stiffness matrix.

The continuous beam of Example 15.1 can be reexamined in this light. From Fig. 15.7d, it is apparent that, if only member *AB* was considered, the load values in mode 1 would be

$$K_{11} = \frac{4EI}{L}$$

$$K_{21} = \frac{2EI}{L}$$

$$K_{31} = 0$$

If member *AB* is viewed alone in mode 2 (Fig. 15.7e), the load values would be

$$K_{12} = \frac{2EI}{L}$$

$$K_{22} = \frac{4EI}{L}$$

$$K_{32} = 0$$

Finally, in mode 3 (Fig. 15.7f), member *AB* has no influence, and

$$K_{13} = K_{22} = K_{33} = 0$$

The combined influence $[K]_{AB}$ of member AB on the structure stiffness matrix is the superposition of the three results or

$$[K]_{AB} = \begin{bmatrix} \dfrac{4EI}{L} & \dfrac{2EI}{L} & 0 \\[2mm] \dfrac{2EI}{L} & \dfrac{4EI}{L} & 0 \\[2mm] 0 & 0 & 0 \end{bmatrix} \quad (15.26)$$

If the isolated effects of member BC on the mode shapes depicted in Fig. 15.7d through f are examined separately, the individual results are:

Mode 1	Mode 2	Mode 3
$K_{11} = 0$	$K_{12} = 0$	$K_{13} = 0$
$K_{21} = 0$	$K_{22} = \dfrac{4EI}{L}$	$K_{23} = \dfrac{2EI}{L}$
$K_{31} = 0$	$K_{32} = \dfrac{2EI}{L}$	$K_{33} = \dfrac{4EI}{L}$

The combined influence, $[K]_{BC}$, on the structure stiffness matrix is

$$[K]_{BC} = \begin{bmatrix} 0 & 0 & 0 \\[2mm] 0 & \dfrac{4EI}{L} & \dfrac{2EI}{L} \\[2mm] 0 & \dfrac{2EI}{L} & \dfrac{4EI}{L} \end{bmatrix} \quad (15.27)$$

If the matrices in Eqs. 15.26 and 15.27 are combined, the result is

$$[K]_{AB} + [K]_{BC} = \begin{bmatrix} \dfrac{4EI}{L} & \dfrac{2EI}{L} & 0 \\[2mm] \dfrac{2EI}{L} & \dfrac{8EI}{L} & \dfrac{2EI}{L} \\[2mm] 0 & \dfrac{2EI}{L} & \dfrac{4EI}{L} \end{bmatrix}$$

$$(15.28)$$

which is identical to the structure stiffness matrix obtained in Example 15.1. This result suggests an alternative method for building $[K]$ and requires a more detailed explanation.

Member AB in Fig. 15.9 is pin-supported at both ends and has a local member stiffness matrix relationship

$$\begin{Bmatrix} M_1 \\ M_2 \end{Bmatrix} = \begin{bmatrix} \dfrac{4EI}{L} & \dfrac{2EI}{L} \\[2mm] \dfrac{2EI}{L} & \dfrac{4EI}{L} \end{bmatrix} \begin{Bmatrix} \phi_1 \\ \phi_2 \end{Bmatrix} = [S]_{AB} \begin{Bmatrix} \phi_1 \\ \phi_2 \end{Bmatrix}$$

Note that the symbol ϕ implies that rotations ϕ_1 and ϕ_2 are measured relative to the displaced chord. For a supported member, the chord does not displace at its ends, and the symbols Φ_1 and Φ_2 (implying rotations measured relative to the original member axis) could be used in place of ϕ_1 and ϕ_2. For consistency, it is preferable to use ϕ_1 and ϕ_2 and to recognize that the displaced chord merely aligns with the original member axis. The reader is alerted that the presence of a subscript distinguishes $[S]_{AB}$, the local member stiffness matrix for an individual member, from $[S]$, the member stiffness matrix used in the basic matrix analysis method.

Matrix $[K]_{AB}$ in Eq. 15.26 is merely an expanded version of matrix $[S]_{AB}$. Enough zeros are added in the appropriate location to increase the size of the matrix $[S]_{AB}$ to that of the structure stiffness matrix $[K]$. In effect, matrix $[K]_{AB}$ states the relationship

$$\begin{Bmatrix} M_1 \\ M_2 \\ 0 \end{Bmatrix} = [K]_{AB} \begin{Bmatrix} \phi_1 \\ \phi_2 \\ 0 \end{Bmatrix} = \begin{bmatrix} \dfrac{4EI}{L} & \dfrac{2EI}{L} & 0 \\[2mm] \dfrac{2EI}{L} & \dfrac{4EI}{L} & 0 \\[2mm] 0 & 0 & 0 \end{bmatrix} \begin{Bmatrix} \phi_1 \\ \phi_2 \\ 0 \end{Bmatrix}$$

$$(15.29)$$

The zeros are placed in the third row and in the third column to facilitate a conversion to an alternate form.

Member end rotations, ϕ_1, and ϕ_2, have a direct correlation with the joint rotations, X_1 and X_2, respectively. Specifically, compatibility requires

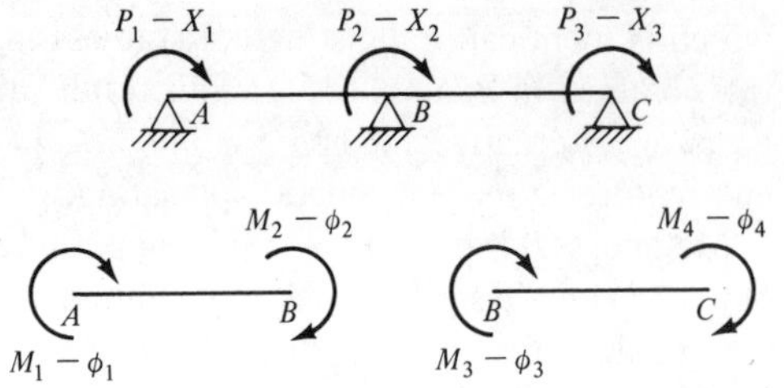

Fig. 15.9.   Subdivided continuous beam.

$$\phi_1 = X_1$$

$$\phi_2 = X_2$$

which permits Eq. 15.29 to be written as

$$\begin{Bmatrix} M_1 \\ M_2 \\ 0 \end{Bmatrix} = [K]_{AB} \begin{Bmatrix} X_1 \\ X_2 \\ X_3 \end{Bmatrix} = \begin{bmatrix} \dfrac{4EI}{L} & \dfrac{2EI}{L} & 0 \\ \dfrac{2EI}{L} & \dfrac{4EI}{L} & 0 \\ 0 & 0 & 0 \end{bmatrix} \begin{Bmatrix} X_1 \\ X_2 \\ X_3 \end{Bmatrix}$$

$$(15.30)$$

In this way, the column matrix on the right side contains the three joint displacements instead of the two member deformations. The presence of the zeros also denotes that member AB is not directly connected to joint C and moments M_1 and M_2 are not dependent upon X_3.

Member BC is also pin-supported at both ends and governed by the local member stiffness relationship

$$\begin{Bmatrix} M_3 \\ M_4 \end{Bmatrix} = \begin{bmatrix} \dfrac{4EI}{L} & \dfrac{2EI}{L} \\ \dfrac{2EI}{L} & \dfrac{4EI}{L} \end{bmatrix} \begin{Bmatrix} \phi_3 \\ \phi_4 \end{Bmatrix} = [S]_{BC} \begin{Bmatrix} \phi_3 \\ \phi_4 \end{Bmatrix}$$

Matrix $[S]_{BC}$ can be expanded in size to give the relationship

$$\begin{Bmatrix} 0 \\ M_3 \\ M_4 \end{Bmatrix} = [K]_{BC} \begin{Bmatrix} 0 \\ \phi_3 \\ \phi_4 \end{Bmatrix} = \begin{bmatrix} 0 & 0 & 0 \\ 0 & \dfrac{4EI}{L} & \dfrac{2EI}{L} \\ 0 & \dfrac{2EI}{L} & \dfrac{4EI}{L} \end{bmatrix} \begin{Bmatrix} 0 \\ \phi_3 \\ \phi_4 \end{Bmatrix}$$

$$(15.31)$$

In this instance, zeros are placed in the first row and first column to facilitate a conversion to the form

$$\begin{Bmatrix} 0 \\ M_3 \\ M_4 \end{Bmatrix} = [K]_{BC} \begin{Bmatrix} X_1 \\ X_2 \\ X_3 \end{Bmatrix} = \begin{bmatrix} 0 & 0 & 0 \\ 0 & \dfrac{4EI}{L} & \dfrac{2EI}{L} \\ 0 & \dfrac{2EI}{L} & \dfrac{4EI}{L} \end{bmatrix} \begin{Bmatrix} X_1 \\ X_2 \\ X_3 \end{Bmatrix}$$

$$(15.32)$$

This conversion recognizes the compatibility requirements

$$\phi_3 = X_2$$

$$\phi_4 = X_3$$

and the lack of direct connectivity between member BC and joint A.

Matrix addition of Eqs. 15.30 and 15.32 yields

$$\begin{Bmatrix} M_1 \\ M_2 + M_3 \\ M_4 \end{Bmatrix} = \begin{bmatrix} \dfrac{4EI}{L} & \dfrac{2EI}{L} & 0 \\ \dfrac{2EI}{L} & \dfrac{8EI}{L} & \dfrac{2EI}{L} \\ 0 & \dfrac{2EI}{L} & \dfrac{4EI}{L} \end{bmatrix} \begin{Bmatrix} X_1 \\ X_2 \\ X_3 \end{Bmatrix}$$

$$(15.33)$$

The rectangular matrix is recognized as the structure stiffness matrix $[K]$. This fact is verified by the realization that the left side of the equation is convertible to the load matrix by use of the equilibrium conditions evident in Fig. 15.9.

$$\begin{Bmatrix} P_1 \\ P_2 \\ P_3 \end{Bmatrix} = \begin{Bmatrix} M_1 \\ M_2 + M_3 \\ M_4 \end{Bmatrix} \qquad (15.34)$$

At face value, this alternate derivation of the $[K]$ matrix does not offer any particular advantage or efficiency in comparison with the earlier approaches. However, it is not the author's intent to prescribe the detailed sequence of operations just performed for general usage as an analytical procedure. A stepwise presentation has been given to emphasize the underlying concepts. The major outcome of the presentation is an understanding of the implied meaning of the "expanded member stiffness matrices" $[K]_{AB}$ and $[K]_{BC}$. $[K]_{AB}$ specifies the relationship between the joint loads and the joint displacements of the complete structure *as dictated by member AB taken alone*. $[K]_{BC}$ indicates a similar relationship between the same quantities *as dictated by member BC taken alone*. In effect, the relationship

$$\{P\} = [K]\{X\}$$

has been subdivided into

$$\{P\} = [K]_{AB}\{X\} + [K]_{BC}\{X\}$$

or in the expanded form

$$
\begin{Bmatrix} P_1 \\ P_2 \\ P_3 \end{Bmatrix} =
\begin{bmatrix}
\dfrac{4EI}{L} & \dfrac{2EI}{L} & 0 \\[2mm]
\dfrac{2EI}{L} & \dfrac{4EI}{L} & 0 \\[2mm]
0 & 0 & 0
\end{bmatrix}
\begin{Bmatrix} X_1 \\ X_2 \\ X_3 \end{Bmatrix}
$$

$$
+
\begin{bmatrix}
0 & 0 & 0 \\[2mm]
0 & \dfrac{4EI}{L} & \dfrac{2EI}{L} \\[2mm]
0 & \dfrac{2EI}{L} & \dfrac{4EI}{L}
\end{bmatrix}
\begin{Bmatrix} X_1 \\ X_2 \\ X_3 \end{Bmatrix}
$$

$$(15.35)$$

Figure 15.10 presents a pictorial description of the superposition implied in Eq. 15.35 and constitutes a visual demonstration of the general procedure referred to as the direct stiffness method. The steps in this method are:

1. Construct the local member stiffness matrix $[S]_m$ for each member of the beam.
2. Expand the local member stiffness matrices to full-size, i.e., to the size of the structure stiffness matrix $[K]$. This is done by an insertion of zero values to fill out the matrices.
3. Add the expanded member stiffness matrices by matrix algebra.

Example 15.3. Develop the structure stiffness matrix for the continuous beam given earlier in Fig. 15.8a. Use the concept of *expanded member stiffness matrices*. *EI* is constant and the same for both spans.

Solution:
Step 1: Because the structure is kinematically indeterminate to the third degree (Fig. 15.8b), the structure stiffness matrix has a 3 × 3 size. Each member stiffness matrix must be expanded to this size.

Member *AB*: This member is free at the left end and pin-supported at the right end. Its local member stiffness relationship is given by

$$
[S]_{AB} =
\begin{bmatrix}
\dfrac{4EI}{L} & \dfrac{2EI}{L} & -\dfrac{6EI}{L^2} \\[3mm]
\dfrac{2EI}{L} & \dfrac{4EI}{L} & -\dfrac{6EI}{L^2} \\[3mm]
-\dfrac{6EI}{L^2} & -\dfrac{6EI}{L^2} & \dfrac{12EI}{L^3}
\end{bmatrix}
$$

which is already 3 × 3 in size and equals $[K]_{AB}$.

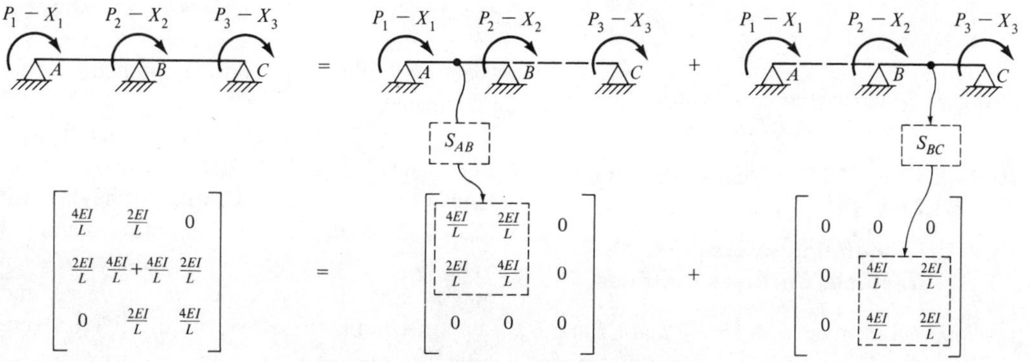

Fig. 15.10.   Direct stiffness method.

Member *BC*: With a pin-support at the left end and a fixed support at the right end, the local member stiffness relationship is $\{F\} = [S]_{BC}\{\Delta\}$ in which

$$[S]_{BC} = \left[\frac{4EI}{L}\right]$$

(Although this member stiffness matrix has not been given earlier, it is easily deduced from the fact that member *BC* is supported at both ends and free to rotate only at the left end. A unit rotation at point *B* requires the application of a moment of magnitude $4EI/L$). Expansion of the stiffness matrix $[S]_{BC}$ to full size gives

$$[K]_{BC} = \begin{bmatrix} 0 & 0 & 0 \\ 0 & \dfrac{4EI}{L} & 0 \\ 0 & 0 & 0 \end{bmatrix}$$

The reader should ponder the reason for placing the zeros in the indicated locations.

Step 2: Addition of the expanded stiffness matrices gives

$$[K] = [K]_{AB} + [K]_{BC}$$

$$= \begin{bmatrix} \dfrac{4EI}{L} & \dfrac{2EI}{L} & -\dfrac{6EI}{L} \\ \dfrac{2EI}{L} & \dfrac{8EI}{L} & -\dfrac{6EI}{L} \\ -\dfrac{6EI}{L} & -\dfrac{6EI}{L^2} & \dfrac{12EI}{L^3} \end{bmatrix}$$

Substituting $L = 16$ ft yields

$$[K] = EI \begin{bmatrix} \frac{1}{4} & \frac{1}{8} & -\frac{3}{128} \\ \frac{1}{8} & \frac{1}{2} & -\frac{3}{128} \\ -\frac{3}{128} & -\frac{3}{128} & \frac{3}{1024} \end{bmatrix}$$

which agrees with the result of Example 15.2.

15.4   DIRECT STIFFNESS METHOD: BEAMS

15.4.1   Generating by Assembly of Member Stiffness Matrices

It is apparent from Section 15.3.2 and Example 15.3 that formation of the stiffness matrix for continuous beams can be reduced to a semiautomatic procedure. By knowing the support conditions and connectivity of the members, it is a routine matter to build the structure stiffness matrix. Proceeding from left to right, the local member stiffness matrix (with end support conditions taken into account) is constructed for the first member and expanded to full (structure) size. Formation of the expanded member stiffness matrix is repeated, in turn, for each succeeding member and added to the preceding matrices. The accumulated results give the complete structure stiffness matrix.

Despite its simplicity, the stated procedure is marked by two noticeable inefficiencies, which inhibit the development of a suitable computer program. First, the size of the member stiffness must be expanded by the addition of numerous zero-valued entries. As the number of individual members increases, the size of the structure stiffness matrix grows slowly, but the number of zero-valued entries increases very rapidly. If each expanded matrix were to be stored in the computer memory, considerable storage space would be exhausted merely to "retain all the zeros." An additional (but less wasteful) inconvenience is that a set of local member stiffness matrices (i.e., one for each possible end support condition) must be stored and subsequently the appropriate matrix for each member must be extracted. With suitable modifications to the procedure, both inefficiencies can be bypassed in the programming stage. Figure 15.10 gave a pictorial representation of the use of expanded size member stiffness matrix, $[K]_m$, in developing $[K]$. In reality, generation of the full-size $[K]_m$ matrices is unnecessary and can be avoided by the use of a standard local member stiffness matrix and a reference number system for "feeding" the entries into the structure stiffness matrix.

Consider the general member illustrated in Fig. 15.11. The *P-X* vectors shown are intended to indicate the *possibility* of having two loads and two degrees of freedom at each member end. In reality, the actual number of each may be less than two depending upon the end support conditions. However, if all four were to exist, the local member stiffness matrix

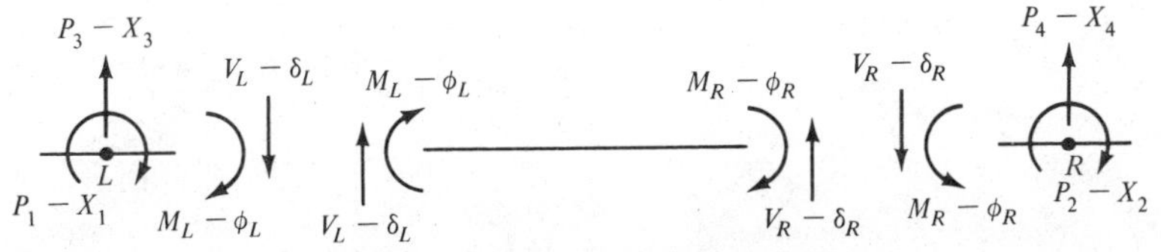

Fig. 15.11. General framed member.

would have the form presented in Eq. 15.6, which is repeated for convenience:

$$\begin{Bmatrix} M_L \\ M_R \\ V_L \\ V_R \end{Bmatrix} = \begin{bmatrix} +\dfrac{4EI}{L} & +\dfrac{2EI}{L} & -\dfrac{6EI}{L^2} & +\dfrac{6EI}{L^2} \\ +\dfrac{2EI}{L} & +\dfrac{4EI}{L} & -\dfrac{6EI}{L^2} & +\dfrac{6EI}{L^2} \\ -\dfrac{6EI}{L^2} & -\dfrac{6EI}{L^2} & +\dfrac{12EI}{L^3} & -\dfrac{12EI}{L^3} \\ +\dfrac{6EI}{L^2} & +\dfrac{6EI}{L^2} & -\dfrac{12EI}{L^3} & +\dfrac{12EI}{L^3} \end{bmatrix} \begin{Bmatrix} \phi_L \\ \phi_R \\ \delta_L \\ \delta_R \end{Bmatrix}$$

(15.6)

Note that true slopes are used in Eq. 15.6.

If the member existed by itself (no other members framing into joints L and R), equilibrium of the joints would require

$$P_1 = M_L$$

$$P_2 = M_R$$

$$P_3 = V_L$$

$$P_4 = V_R$$

and compatibility would require

$$X_1 = \Phi_L$$

$$X_2 = \Phi_R$$

$$X_3 = \delta_L$$

$$X_4 = \delta_R$$

or equivalently

$$\begin{Bmatrix} P_1 \\ P_2 \\ P_3 \\ P_4 \end{Bmatrix} = \begin{bmatrix} +\dfrac{4EI}{L} & +\dfrac{2EI}{L} & -\dfrac{6EI}{L^2} & +\dfrac{6EI}{L^2} \\ +\dfrac{2EI}{L} & +\dfrac{4EI}{L} & -\dfrac{6EI}{L^2} & +\dfrac{6EI}{L^2} \\ -\dfrac{6EI}{L^2} & -\dfrac{6EI}{L^2} & +\dfrac{12EI}{L^3} & -\dfrac{12EI}{L^3} \\ +\dfrac{6EI}{L^2} & +\dfrac{6EI}{L^2} & -\dfrac{12EI}{L^3} & +\dfrac{12EI}{L^3} \end{bmatrix} \begin{Bmatrix} X_1 \\ X_2 \\ X_3 \\ X_4 \end{Bmatrix}$$

(15.36)

Equation 15.36 can be expressed in abbreviated form

$$[K]_m = \begin{array}{c} \\ (1) \\ (2) \\ (3) \\ (4) \end{array} \begin{array}{cccc} \overset{(1)}{} & \overset{(2)}{} & \overset{(3)}{} & \overset{(4)}{} \end{array} \begin{bmatrix} +\dfrac{4EI}{L} & +\dfrac{2EI}{L} & -\dfrac{6EI}{L^2} & +\dfrac{6EI}{L^2} \\ +\dfrac{2EI}{L} & +\dfrac{4EI}{L} & -\dfrac{6EI}{L^2} & +\dfrac{6EI}{L^2} \\ -\dfrac{6EI}{L^2} & -\dfrac{6EI}{L^2} & +\dfrac{12EI}{L^3} & -\dfrac{12EI}{L^3} \\ +\dfrac{6EI}{L^2} & +\dfrac{6EI}{L^2} & -\dfrac{12EI}{L^3} & +\dfrac{12EI}{L^3} \end{bmatrix}$$

(15.37)

in which the circled numbers indicate the connectivity of the member; i.e., in this case, the local member stiffness matrix is associated with the particular loads, $P_1 - P_4$, and degrees of freedom, $X_1 - X_4$, that exist at the ends of the member. With this simple notation, a direct feeding of the member stiffness matrices into the structure stiffness matrix is facilitated and will be described in Example 15.4. The procedure presented represents the usual technique for executing the direct stiffness method.

Example 15.4. Build the structure stiffness matrix for the continuous beam given in Fig. 15.12. Use the direct stiffness method.

Solution:
Step 1: The 6 degrees of freedom are numbered (in arbitrary order) in Fig. 15.12. Figure 15.13 indicates the numbering system for the degrees of freedom of a general member. The local member stiffness matrix in Eq. 15.37 was formulated and coded with reference to this numbering pattern. Each

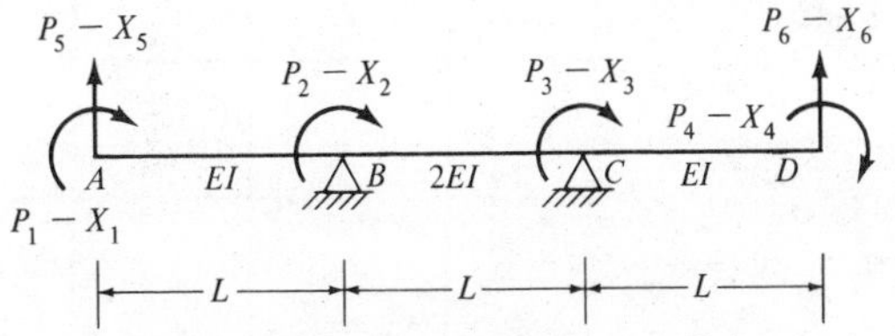

Fig. 15.12. Beam for Example 15.4.

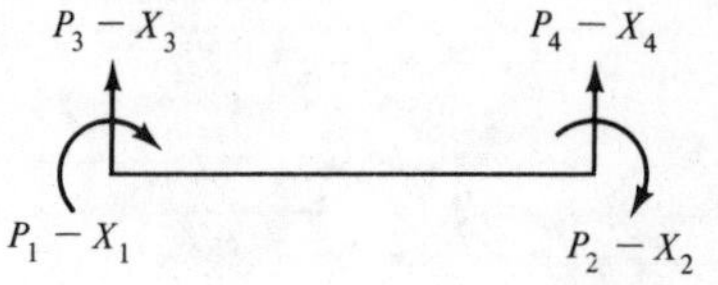

$P_3 - X_3$ $P_4 - X_4$

$P_1 - X_1$ $P_2 - X_2$

Fig. 15.13.   General member numbering.

member in Fig. 15.12 is a part of the larger structure, and the numbering of the degrees of freedom at the ends of each member is predetermined by the overall numbering system. In addition, none of the members has 4 degrees of freedom because of the existing support conditions. However, the matrix in Eq. 15.37 is still applicable provided the connectivity numbers are altered to reflect the true condition of each member. As an example

$$[K]_{AB} = \begin{array}{c} \text{①} \\ \text{②} \\ \text{⑤} \\ \text{®} \end{array} \begin{bmatrix} +\dfrac{4EI}{L} & +\dfrac{2EI}{L} & -\dfrac{6EI}{L^2} & +\dfrac{6EI}{L^2} \\[2mm] +\dfrac{2EI}{L} & +\dfrac{4EI}{L} & -\dfrac{6EI}{L^2} & +\dfrac{6EI}{L^2} \\[2mm] -\dfrac{6EI}{L^2} & -\dfrac{6EI}{L^2} & +\dfrac{12EI}{L^3} & -\dfrac{12EI}{L^3} \\[2mm] +\dfrac{6EI}{L^2} & +\dfrac{6EI}{L^2} & -\dfrac{12EI}{L^3} & +\dfrac{12EI}{L^3} \end{bmatrix} \begin{array}{c} \text{①} \quad \text{②} \quad \text{⑤} \quad \text{®} \end{array}$$

in which the actual degrees of freedom, ①, ②, ⑤, and ® (R implies restraint), replace the reference numbers ① through ④ indicated in Eq. 15.37. In this way, the connectivity of the member to joints A and B is indicated in shorthand. By similar treatment,

$$[K]_{BC} = \begin{array}{c} \text{②} \\ \text{③} \\ \text{®} \\ \text{®} \end{array} \begin{bmatrix} +\dfrac{8EI}{L} & +\dfrac{4EI}{L} & -\dfrac{12EI}{L^2} & +\dfrac{12EI}{L^2} \\[2mm] +\dfrac{4EI}{L} & +\dfrac{8EI}{L} & -\dfrac{12EI}{L^2} & +\dfrac{12EI}{L^2} \\[2mm] -\dfrac{12EI}{L^2} & -\dfrac{12EI}{L^2} & +\dfrac{24EI}{L^3} & -\dfrac{24EI}{L^3} \\[2mm] +\dfrac{12EI}{L^2} & +\dfrac{12EI}{L^2} & -\dfrac{24EI}{L^3} & +\dfrac{24EI}{L^3} \end{bmatrix} \begin{array}{c} \text{②} \quad \text{③} \quad \text{®} \quad \text{®} \end{array}$$

$$[K]_{CD} = \begin{array}{c} \text{③} \\ \text{④} \\ \text{®} \\ \text{⑥} \end{array} \begin{bmatrix} +\dfrac{4EI}{L} & +\dfrac{2EI}{L} & -\dfrac{6EI}{L^2} & +\dfrac{6EI}{L^2} \\[2mm] +\dfrac{2EI}{L} & +\dfrac{4EI}{L} & -\dfrac{6EI}{L^2} & +\dfrac{6EI}{L^2} \\[2mm] -\dfrac{6EI}{L^2} & -\dfrac{6EI}{L^2} & +\dfrac{12EI}{L^3} & -\dfrac{12EI}{L^3} \\[2mm] +\dfrac{6EI}{L^2} & +\dfrac{6EI}{L^2} & -\dfrac{12EI}{L^3} & +\dfrac{12EI}{L^3} \end{bmatrix} \begin{array}{c} \text{③} \quad \text{④} \quad \text{®} \quad \text{⑥} \end{array}$$

Step 2: Having properly referenced the connectivity of each member stiffness matrix, it is a straightforward matter to feed the entries of the $[K]_m$ matrices directly into the appropriate locations of $[K]$. Element $K_m(\text{①}, \text{①})$ of the given $[K]_m$ matrix is transferred to the location $K(\text{①},\text{①})$. Columns and rows of $[K]_m$ that are marked with an ® are bypassed in the transfer process. The member stiffness matrices are treated one at a time, and each time the contents are entered into $[K]$. The sequential results are presented below.

After feeding $[K]_{AB}$:

$$[K] = \begin{array}{c} \text{①} \\ \text{②} \\[20mm] \text{⑤} \end{array} \begin{bmatrix} \dfrac{4EI}{L} & \dfrac{2EI}{L} & & & -\dfrac{6EI}{L^2} & \\[2mm] \dfrac{2EI}{L} & \dfrac{4EI}{L} & & & -\dfrac{6EI}{L^2} & \\[2mm] & & & & & \\[2mm] & & & & & \\[2mm] -\dfrac{6EI}{L^2} & -\dfrac{6EI}{L^2} & & & +\dfrac{12EI}{L^3} & \\[2mm] & & & & & \end{bmatrix} \begin{array}{c} \text{①} \quad\;\; \text{②} \quad\quad\quad\quad \text{⑤} \end{array}$$

After feeding $[K]_{BC}$:

$[K] =$

	①	② ②	③ ②		⑤
①	$\dfrac{4EI}{L}$	$\dfrac{2EI}{L}$			$\dfrac{-6EI}{L^2}$
② ②	$\dfrac{2EI}{L}$	$\dfrac{4EI}{L}+\dfrac{8EI}{L}$	$\dfrac{4EI}{L}$		$\dfrac{-6EI}{L^2}$
③		$\dfrac{4EI}{L}$	$\dfrac{8EI}{L}$		
⑤	$\dfrac{-6EI}{L^2}$	$\dfrac{-6EI}{L^2}$			$\dfrac{12EI}{L^3}$

After feeding $[K]_{CD}$:

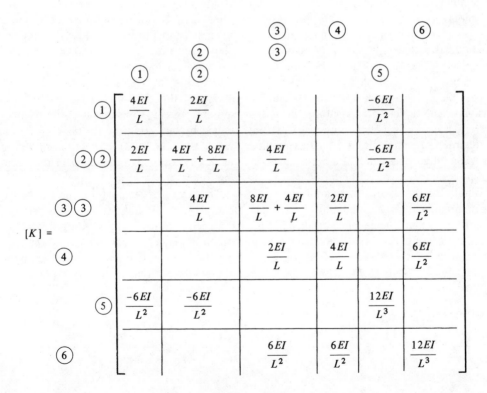

$[K] =$

	①	② ②	③ ③	④ ③	⑤	⑥
①	$\dfrac{4EI}{L}$	$\dfrac{2EI}{L}$			$\dfrac{-6EI}{L^2}$	
② ②	$\dfrac{2EI}{L}$	$\dfrac{4EI}{L}+\dfrac{8EI}{L}$	$\dfrac{4EI}{L}$		$\dfrac{-6EI}{L^2}$	
③ ③		$\dfrac{4EI}{L}$	$\dfrac{8EI}{L}+\dfrac{4EI}{L}$	$\dfrac{2EI}{L}$		$\dfrac{6EI}{L^2}$
④			$\dfrac{2EI}{L}$	$\dfrac{4EI}{L}$		$\dfrac{6EI}{L^2}$
⑤	$\dfrac{-6EI}{L^2}$	$\dfrac{-6EI}{L^2}$			$\dfrac{12EI}{L^3}$	
⑥			$\dfrac{6EI}{L^2}$	$\dfrac{6EI}{L^2}$		$\dfrac{12EI}{L^3}$

Step 3: Summing the contents gives the structure stiffness matrix.

$$
[K] =
\begin{array}{c}
\begin{array}{cccccc}
\text{①} & \text{②} & \text{③} & \text{④} & \text{⑤} & \text{⑥}
\end{array} \\
\begin{array}{c}
\text{①} \\ \text{②} \\ \text{③} \\ \text{④} \\ \text{⑤} \\ \text{⑥}
\end{array}
\begin{bmatrix}
\dfrac{4EI}{L} & \dfrac{2EI}{L} & 0 & 0 & \dfrac{-6EI}{L^2} & 0 \\[2ex]
\dfrac{2EI}{L} & \dfrac{12EI}{L} & \dfrac{4EI}{L} & 0 & \dfrac{-6EI}{L^2} & 0 \\[2ex]
0 & \dfrac{4EI}{L} & \dfrac{12EI}{L} & \dfrac{2EI}{L} & 0 & \dfrac{6EI}{L^2} \\[2ex]
0 & 0 & \dfrac{2EI}{L} & \dfrac{4EI}{L} & 0 & \dfrac{6EI}{L^2} \\[2ex]
\dfrac{-6EI}{L^2} & \dfrac{-6EI}{L^2} & 0 & 0 & \dfrac{12EI}{L^3} & 0 \\[2ex]
0 & 0 & \dfrac{6EI}{L^2} & \dfrac{6EI}{L^2} & 0 & \dfrac{12EI}{L^3}
\end{bmatrix}
\end{array}
$$

15.4.2 The Complete Direct Stiffness Analysis Procedure

Several alternatives have been described for the generation of the structures stiffness matrix for a continuous beam. The primary emphasis in Section 15.3.1 was placed upon physical meaning of the structure stiffness matrix. In contrast, the direct stiffness method illustrated the ability to develop more automated procedures, which bypass the physical concepts but expedite the analysis. The direct stiffness procedure is the preferred approach to the solution of general problems. Generation of $[K]$ by this method has been fully discussed. The complete analytical procedure consists of the following.

Within the limits of the principle of superposition the analysis of continuous beams can be performed in the manner exemplified in Fig. 15.14. This figure serves to introduce the general concept of the stiffness methods of structural analysis. Figure 15.14a depicts the beam that is the subject of the analysis. In Fig. 15.14b (the restrained state), the pin supports are considered to be temporarily held fixed in position, i.e., all degrees of freedom are restrained. If this is to be possible, restraining moments RM_1, RM_2, and RM_3 must be developed as shown in Fig. 15.14b. The directions shown are in accordance with a sign convention, which assumes that positive restraining moments act in a clockwise sense. In the actual structure (Fig. 15.14a), the restraining forces are not present. Consequently, the results obtained by analyzing the restrained state alone would be in error. To correct the results, it is necessary to remove the effect of the restraining forces.

The structure depicted in Fig. 15.14c is the released-state. In this structure, the restraint introduced in Fig. 15.14b is removed, and loads, which are equal in magnitude but opposite in direction to the restraining forces, are applied at the released joints. The sign convention for

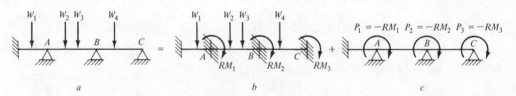

Fig. 15.14. Stiffness analysis of a continuous beam. *a*, actual state; *b*, restrained state; *c*, released state.

the external loads P_1–P_3 in the released state is identical to that used for restraining moments, namely, positive moments act in a clockwise sense. Superimposing the structures shown in Fig. 15.14b and c produces the structure shown in Fig. 15.14a. The stiffness method of analysis consists of executing this superposition. Determination of the restrained state restraining forces and analysis of the corresponding released state constitutes two distinct stages of the analysis, which are described in a matrix formulation in the next section.

15.4.3 Analysis of the Restrained State

15.4.3.1 Restraining Moments

With all degrees of freedom of a continuous beam held fixed, each span behaves as an independent beam—i.e., the deformations of one span have no influence on the other spans. By recognition of this fact, the restraining moments indicated in Fig. 15.14b can be computed by the procedure shown in Fig. 15.15. Taken by itself, the loading on the first span produces end moments M_{01} and M_{02}, which are assumed to act clockwise on the member ends. Similarly, the second and third span loadings induce the clockwise end moments M_{03}, M_{04}, M_{05}, and M_{06}. When transferred to the joints, the directions of these six end moments are reversed. The restraining moments, RM_1, RM_2, and RM_3, shown at the joints in Fig. 15.15, are needed to satisfy equilibrium. It is these equilibrating moments that are shown in Fig. 15.14b. When the restrained degrees of freedom are released, the restraining forces cannot be developed. Under this circumstance, the unbalanced moments at the joints (e.g., the resultant of M_{02} and M_{03}) behave as if they were loads applied to the joints. The net unbalanced moment at any given joint is equal in magnitude but opposite in direction to the restraining moment. This explains the negative signs shown in Fig. 15.14c. Analysis of the effects of these residual joint loads is the purpose implied in the released structure shown in Fig. 15.14c.

Determination of the end moments M_{01}–M_{06} in Fig. 15.15 requires the analysis of a particular type of beam, namely, a single span beam that is fixed at both ends and subjected to a transverse loading. The necessary background for the analysis of such beams has already been developed in earlier chapters. For the present purpose, it is only necessary to access the formulas presented in Appendix A, Table A.3.

In the basic matrix analysis procedure, it is usual to place the results of the restrained state analysis into a suitable matrix form. The fixed-end moment values for all spans comprise the pertinent output and can be arranged in a column matrix $\{F_0\}$. For a three-span system, the vector has the form

$$\{F_0\} = \left\{ \begin{array}{c} M_{01} \\ M_{02} \\ M_{03} \\ M_{04} \\ M_{05} \\ M_{06} \end{array} \right\} \qquad (15.38)$$

in which M_{01}–M_{06} are the six fixed-end moments numbered from left to right. The nomenclature $\{F_0\}$ was introduced for trusses in Chapter 14 (see Example 14.7) and is termed the INITIAL FORCE matrix. However, the natures of the contents differ. For trusses, the contents are axial forces, and for beams the contents are end moments.

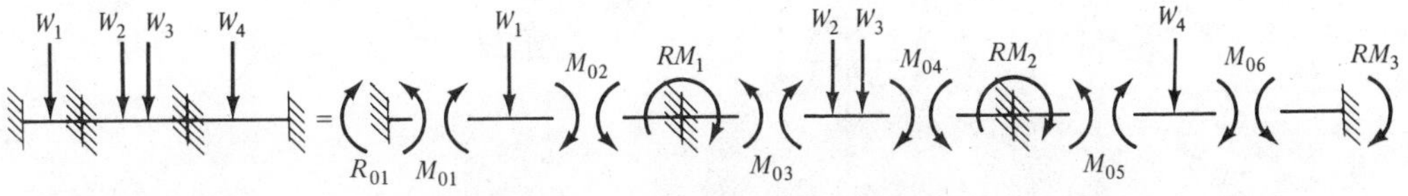

Fig. 15.15. Calculation of restraining forces.

15.4.3.2   Reactions

In addition to the temporary restraining forces, a number of actual reactions are developed in the restrained state. These reactions exist because some supports existed prior to the introduction of restraint at the degrees of freedom— i.e., if the restraints were released, certain supports would still exist. Forces that develop in the directions of the actual support reactions in the restrained state are the initial reactions and are calculated by direct application of statics.

Examine the restrained state as depicted in Fig. 15.16a. The five forces R_{01}–R_{05} are the initial reactions just described. Because the fixed-end moments and the transverse loading are known, the end shear forces, V_{01}–V_{06}, are easily determined by applying moment equilibrium to the member ends. Isolation of the joints as separate free-body diagrams (Fig. 15.16b) enables computation of the initial reactions. With all forces known at the joints, use of statics produce the desired reaction forces—e.g., $R_{03} = V_{02} + V_{03}$. As a final step, the results are entered into the INITIAL REACTION matrix

$$\{R_0\} = \begin{Bmatrix} R_{01} \\ R_{02} \\ R_{03} \\ R_{04} \\ R_{05} \end{Bmatrix} \qquad (15.39)$$

15.4.4   Analysis of the Released State

Reexamination of Fig. 15.14c reveals a special characteristic regarding the released state of a continuous beam. In the released state, the only loads that exist are concentrated moments that are applied directly at the joints. Transverse loading is absent from the structure. This feature permits a matrix formulation for the analysis of the released structure. Essentially, the matrix formulation consists of the development of the structure stiffness equations

$$\{P\} = [K]\{X\} \qquad (15.40)$$

and their subsequent solution for $\{X\}$. $[K]$ is assembled by the procedure described in Section 15.4.1. The calculated $\{X\}$ is then used in conjunction with Eq. 15.6 to get member end forces. The load matrix $\{P\}$ contains the unbalanced joint loads that develop when the restrained state is released. In the context of Fig. 15.14c, $\{P\}$ contains the loads $-RM_1$, $-RM_2$, and $-RM_3$.

15.4.5   Summary

The following steps comprise the direct stiffness method.

Step 1: Analyze the restrained state in the manner described in the basic matrix analysis method.

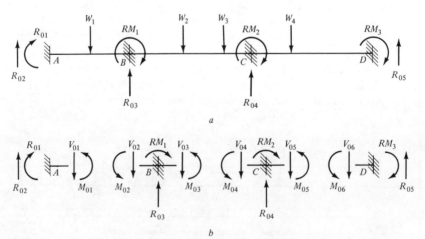

a

b

Fig. 15.16.   Calculation of restrained state reactions.

Step 2: Build the load matrix, $\{P\}$, from the results of step 1.

Step 3: Build $[K]$ by the direct-feeding technique.

Step 4: Determine the joint displacements, $\{X\}$, in the released state according to

$$\{X\} = [K^{-1}]\{P\} \qquad (15.41)$$

Step 5: Use the member stiffness relationship (Eq. 15.6) to calculate the end forces for each member. Member end deformations needed in the relationship are extracted from the results of step 4.

Step 6: Combine the results of steps 1 and 5.

A complete illustration of the use of this procedure is given in Example 15.5. It should be noted that the manner of executing step 3 is not essential. $[K]$ can be constructed by any of the described methods, and the remaining steps of the direct stiffness method could still be applied. Also, no matrix format means for calculating external reactions in the released state has been described. With the member end forces known, reactions are easily calculated by manually applying statics to each joint. It is possible to include the computation of reactions within the matrix format but this subject is deferred to Chapter 16. If the released state reactions $\{R*\}$ were to be calculated, they would be combined with the restrained state reactions according to

$$\{R\} = \{R_0\} + \{R*\} \qquad (15.42)$$

Because there are no degrees of freedom in the fixed state, the final joint displacements (or equivalently, by compatibility, the member end deformations) are those values, $\{X*\}$, obtained in the analysis of the released state i.e., $\{X\} = \{X*\}$. Conceptually, knowledge of $\{F\}$ and $\{X\}$ is sufficient to determine any other response quantity by classical structural analysis procedures. For example, knowledge of $\{X\}$ establishes the boundary conditions needed for the quadruple integration method. Alternatively, the matrix $\{F\}$ permits writing moment equations for each span and applying

the double integration method. In practice, these mathematical methods are seldom used. Instead, statics and the moment-area method are employed to produce any desired results (e.g., reactions, shear and moment diagrams, displacements within the spans). Also, unless support settlements, fabrication errors etc. occur, $\{X*\}$ always constitutes the final joint displacements and $\{\Delta*\}$ constitutes the final member deformations.

15.4.6 Sign Conventions

The sign conventions employed in the matrix analysis of beams are:

1. Positive joint loads act in a clockwise direction when externally applied to the joints.
2. Positive reactions act either upward (for forces) or in a clockwise direction (for moments).
3. Positive internal moments act in a clockwise direction on the member ends (counterclockwise on the joints).
4. Positive shears act upward at the member ends (downward on the joints).
5. Positive deformations represent clockwise rotations of the member ends.
6. Positive displacements represent clockwise joint rotations.

These conventions apply to both stages of the matrix analysis and to the combined results (the actual beam).

Example 15.5. Perform a complete structural analysis of the structure shown in Fig. 15.17a using the direct stiffness method. Assume EI is the same constant value for both members.

Solution:

Step 1: Member end forces in the restrained state are shown in Fig. 15.17b. The values were obtained by use of Table A.3.

$$\{F_0\} = \begin{Bmatrix} M_{01} \\ M_{02} \\ M_{03} \\ M_{04} \end{Bmatrix} = \begin{Bmatrix} -64.0 \\ 64.0 \\ -32.0 \\ 32.0 \end{Bmatrix}$$

Step 2: Restraining forces needed at the joints are shown in Fig. 15.17c. With the restraint released, these forces cannot be developed, and the unbalanced forces shown in Fig. 15.17d act as joint loads giving

$$\{P\} = \begin{Bmatrix} P_1 \\ P_2 \\ P_3 \end{Bmatrix} = -\begin{Bmatrix} RP_1 \\ RP_2 \\ RP_3 \end{Bmatrix} = \begin{Bmatrix} 64.0 \\ -32.0 \\ -24.0 \end{Bmatrix}$$

Step 3: Member stiffness matrices, in shorthand form, are determined from Eq. 15.6.

$$[K]_{AB} = \begin{matrix} & \overset{①}{} & \overset{②}{} & \overset{③}{} & \overset{Ⓡ}{} \\ ① \\ ② \\ ③ \\ Ⓡ \end{matrix} \begin{bmatrix} 0.250 & 0.125 & -0.02343 & 0.02343 \\ 0.125 & 0.250 & -0.02343 & 0.02343 \\ -0.02343 & -0.02343 & 0.00293 & -0.00293 \\ 0.02343 & 0.02343 & -0.00293 & 0.00293 \end{bmatrix} EI$$

$$[K]_{BC} = \begin{matrix} & \overset{②}{} & \overset{Ⓡ}{} & \overset{Ⓡ}{} & \overset{Ⓡ}{} \\ ② \\ Ⓡ \\ Ⓡ \\ Ⓡ \end{matrix} \begin{bmatrix} 0.25 & 0.125 & -0.02343 & 0.02343 \\ 0.125 & 0.25 & -0.02343 & 0.02343 \\ -0.02343 & -0.02343 & 0.00293 & -0.00293 \\ 0.02343 & 0.02343 & -0.00293 & 0.00293 \end{bmatrix} EI$$

Step 4: Direct feeding into $[K]$ gives

$$[K] = \begin{matrix} & \overset{①}{} & \overset{②}{} & \overset{③}{} \\ ① \\ ② \\ ③ \end{matrix} \begin{bmatrix} 0.25 & 0.125 & -0.02343 \\ 0.125 & 0.500 & -02343 \\ -0.02343 & -0.02343 & 0.00293 \end{bmatrix} EI$$

Step 5: Displacements in the released state are calculated

$$\{X^*\} = \begin{Bmatrix} X_1^* \\ X_2^* \\ X_3^* \end{Bmatrix} = [K]^{-1}\{P\}$$

$$= \frac{1}{EI} \begin{bmatrix} 19.92 & 3.973 & 191.1 \\ 3.973 & 3.991 & 63.69 \\ 191.1 & 63.69 & 2378.0 \end{bmatrix} \begin{Bmatrix} 64 \\ -32 \\ -24 \end{Bmatrix}$$

$$= \frac{1}{EI} \begin{Bmatrix} -3437 \text{ rad} \\ -1402 \text{ rad} \\ -46,890 \text{ ft} \end{Bmatrix}$$

which are the total displacements since all X's have zero value in the restrained state.

Step 5: Equation 15.6 is executed for each member.

Member AB:

$$\Phi_L = X_1^* \qquad \delta_L = X_3^*$$
$$\Phi_R = X_2^* \qquad \delta_R = 0$$

$$\begin{Bmatrix} M_1^* \\ M_2^* \\ V_1^* \\ V_2^* \end{Bmatrix} = [K]_{AB} \begin{Bmatrix} -\dfrac{3437}{EI} \\ -\dfrac{1402}{EI} \\ -\dfrac{46,890}{EI} \\ 0 \end{Bmatrix}$$

$$= \begin{Bmatrix} 64.1 \text{ kip-ft} \\ -318.5 \text{ kip-ft} \\ -24.0 \text{ kips} \\ +24.0 \text{ kips} \end{Bmatrix}$$

Member BC:

$$\Phi_L = X_2^* \qquad \delta_L = 0$$
$$\Phi_R = 0 \qquad \delta_R = 0$$

$$\begin{Bmatrix} M_3^* \\ M_4^* \\ V_3^* \\ V_4^* \end{Bmatrix} = [K]_{BC} \begin{Bmatrix} -\dfrac{1402}{EI} \\ 0 \\ 0 \\ 0 \end{Bmatrix} = \begin{Bmatrix} -350.5 \\ -175.3 \\ +32.85 \\ -32.85 \end{Bmatrix}$$

Step 6: Combining the results of steps 1 and 5 produces the final end moments:

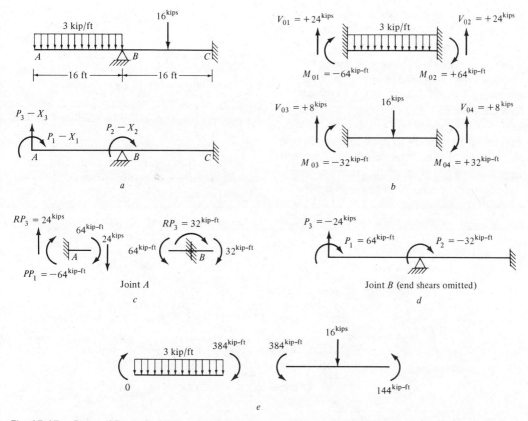

Fig. 15.17. Beam of Example 15.5. *a*, continuous beam and degrees of freedom; *b*, restrained-state internal forces; *c*, external restraining forces; *d*, released-state loads; *e*, final end moments (shears omitted).

$$\begin{Bmatrix} M_1 \\ M_2 \\ M_3 \\ M_4 \end{Bmatrix} = \begin{Bmatrix} M_{01} \\ M_{02} \\ M_{03} \\ M_{04} \end{Bmatrix} + \begin{Bmatrix} M_1^* \\ M_2^* \\ M_3^* \\ M_4^* \end{Bmatrix}$$

$$= \begin{Bmatrix} -64.0 \\ +64.0 \\ -32.0 \\ +32.0 \end{Bmatrix} + \begin{Bmatrix} 64.1 \\ 318.5 \\ -350.5 \\ -175.3 \end{Bmatrix}$$

$$= \begin{Bmatrix} +0.1 \\ 382.5 \\ -382.5 \\ -143.3 \end{Bmatrix} \text{kip-ft}$$

Note: Numbers employed in the stiffness matrices and all subsequent computations have been rounded to five significant figures. As a consequence, some round-off error is inherent in the answers. Exact moment values are $M_1 = 0$, $M_2 = 384$ kip-ft, $M_3 =$

-384 kip-ft, and $M_4 = -144$ kip-ft and can be obtained if more precise values are employed in the computations. Free-body diagrams of the members are given in Fig. 15.17e with the precise values and correct directions shown for the end moments.

15.5 DIRECT GENERATION OF [K]: PLANAR TRUSSES

Methods described in Section 15.3 reduce the effort required to generate the structure stiffness method for continuous beams. Except for the nature of the degrees of freedom, the methods can be applied to planar trusses.

15.5.1 Generation by Physical Meaning

All degrees of freedom for a truss are joint translations. The structure stiffness matrix relates the displacements at the joints to the applied joint loads. Each column of $[K]$ can be interpreted as containing entries that correspond in magnitude to the unique set of joint loads, which produces a state in which all joint displacement components have zero value except one. The ith column of $[K]$ contains the loads that cause a unit translation in the direction of the ith degree of freedom, with all other degrees of freedom held fixed in position. Determination of these loads establishes the stiffness coefficients in the ith column and can be done by the "reverse" structural analysis technique illustrated in Section 15.3.1. Repetition for all other degrees of freedom fills the remaining columns. Application of the concept is best illustrated by a numerical example.

Example 15.6.   Build the structure stiffness matrix for the truss given in Fig. 15.18a by use of its physical meaning.

Solution: The three translational degrees of freedom for the given truss are shown in Fig. 15.18b.

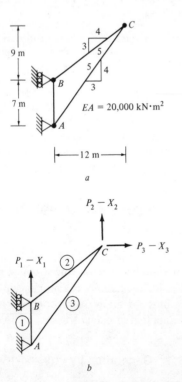

Fig. 15.18. Planar truss for Example 15.6. a, configuration; b, degrees of freedom.

Structure compatibility modes, introduced in Chapter 14 as a means of developing $[C]$ are the fundamental source of the entries in $[K]$. One such mode exists for each degree of freedom as shown in Figs. 15.19 (1a, 2a, and 3a). External loads corresponding to each mode give the entries in $[K]$.

Mode 1:   Figure 15.19 (1a): A unit translation is imposed in the direction of X_1, while joint C is held in position ($X_2 = X_3 = 0$). Joint B moves to B' and causes deformation in the attached members. Member AB must elongate by an amount, Δ_1, equal to 1.0 unit as represented by the distance BB'. Member BC must shorten by an amount, Δ_2, equal to 0.6 units as represented by the distance BB''. The elongations (a negative sign implies a contraction) are indicated in Fig. 15.19 (1b). Internal forces created by the member deformations are shown in Fig. 15.19 (1c) and are calculated by use of Hooke's law

$$F_{11} = \left(\frac{EA}{L}\right)_1 \Delta_1 = \frac{EA}{7}(1) = \frac{EA}{7}$$

$$F_{21} = \left(\frac{EA}{L}\right)_2 \Delta_2 = \frac{EA}{15}(-0.6) = -\frac{EA}{25}$$

$$F_{31} = \left(\frac{EA}{L}\right)_3 \Delta_3 = \frac{EA}{20}(0) = 0$$

Load values—shown in Fig. 15.19 (1d)—correspond to the calculated internal forces and are established by examining joint equilibrium in Fig. 15.19 (1c). (Recall that the force vectors shown in Fig. 15.19 (1c) indicate the directions of the member forces *acting on the joints*.) Hence,

$$P_1 = K_{11} = F_{11} - 0.6F_{21}$$

$$= \frac{EA}{7} - \frac{6}{10}\left(-\frac{EA}{25}\right)$$

$$= 0.167EA \tag{a}$$

$$P_2 = K_{21} = 0.6F_{21} + 0.8F_{31}$$

$$= \frac{6}{10}\left(-\frac{EA}{25}\right) + 0.8(0)$$

$$= -0.024EA \tag{b}$$

$$P_3 = K_{31} = 0.8F_{21} + 0.6F_{31}$$

$$= \frac{8}{10}\left(-\frac{EA}{25}\right) + 0.6(0)$$

$$= -0.032EA \tag{c}$$

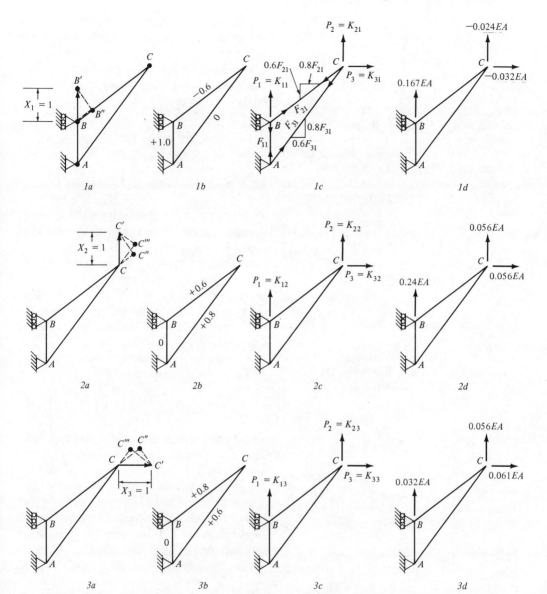

Fig. 15.19. Structure compatibility modes. *1a*, mode 1; *1b*, elongations; *1c*, forces; *1d*, loads; *2a*, mode 2; *2b*, elongations; *2c*, forces; *2d*, loads; *3a*, mode 3; *3b*, elongations; *3c*, forces; *3d*, loads.

Mode 2: Figure 15.19 (2a): A unit translation is imposed at X_2 with X_1 and X_3 restrained against movement. Joint C moves to C' and causes deformation in members BC and AC. Member BC elongates 0.6 units (distance CC''), and member AC elongates 0.8 units (distance CC''') as indicated in Fig. 15.19 (2a, b). The induced member forces (Fig. 15.19, 2c) are

$$F_{12} = \left(\frac{EA}{L}\right)_1 \Delta_1 = \frac{EA}{7}(0) = 0$$

$$F_{22} = \left(\frac{EA}{L}\right)_2 \Delta_2 = \frac{EA}{15}(0.6) = \frac{EA}{25}$$

$$F_{32} = \left(\frac{EA}{L}\right)_3 \Delta_3 = \frac{EA}{20}(0.8) = \frac{EA}{25}$$

Load values are obtained by substitution into the right side of Eqs. a through c developed above; i.e., F_{12} replaces F_{11}, etc., giving

$$K_{12} = 0 - \frac{6}{10}\left(\frac{EA}{25}\right) = -0.024EA$$

$$K_{22} = \frac{6}{10}\left(\frac{EA}{25}\right) + \frac{8}{10}\left(\frac{EA}{25}\right) = 0.056EA$$

$$K_{32} = \frac{8}{10}\left(\frac{EA}{25}\right) + \frac{6}{10}\left(\frac{EA}{25}\right) = 0.056EA$$

Mode 3: Figure 15.19 (3a): A unit translation is imposed at X_3 with $X_1 = X_2 = 0$. Joint C moves to C' causing deformation in members AC and BC. Calculation of the deformations, internal forces, and load values is left as an exercise for the reader. The final results are given in Fig. 15.19 (3d).

Assembling the results of the preceding analysis yields

$$[K] = \begin{bmatrix} 0.167 & -0.024 & -0.032 \\ -0.024 & 0.056 & 0.056 \\ -0.032 & 0.056 & 0.061 \end{bmatrix} EA$$

$$= \begin{bmatrix} 3340 & -480 & -640 \\ -480 & 1120 & 1120 \\ -640 & 1120 & 1120 \end{bmatrix}$$

The units associated with each entry are kN/m.

15.5.2 Generation by Expansion of Member Stiffness Matrix

Expansion of local member stiffness matrices to the size of the full structure stiffness matrix was discussed for beams in Section 15.3.2. The insertion of zeros to expand $[S]_m$ to structure size $[K]_m$, while trivial in execution, was shown to be an inefficient and unnecessary procedure. Application of the concept to planar trusses only compounds the inefficiency. The difference in the size between the local member stiffness matrix for a single beam element and for an entire continuous beam is generally modest. Nevertheless, numerous zeros must be processed in expanding the size of the local member stiffness matrices. The local member stiffness matrix for an individual planar truss member, $[S]_m$, has not been introduced to this point in the text. A general local member stiffness matrix will be introduced in Chapter 16. However, the difference in the size of $[S]_m$ and $[K]_m$ is generally very large because a member has no more than 4 degrees of freedom, while

the truss has 2 for every free joint plus those that exist at roller supports. Consequently, the expansion of the local member stiffness matrix to structure size is a very impractical technique when applied to planar trusses and is not recommended for use.

15.6 DIRECT STIFFNESS METHOD: PLANAR TRUSSES

15.6.1 Global Member Stiffness Matrix

Translations of the joints constitute the only degrees of freedom in a truss. Each free joint experiences two components of motion, and each roller experiences a single motion. Figure 15.20 depicts a portion of a truss in which the member AB frames into two free joints. It is useful to derive a member stiffness relationship for such a member. The load quantities, P_1–P_4, and displacement quantities, X_1–X_4, shown in Fig. 15.20 are interrelated in some way. Specifically, the relationship is of the form

$$\{P\} = [K']\{X\} \qquad (15.43)$$

where $[K']$ is a *portion* of the structure stiffness matrix. Furthermore, only part of $[K']$ is directly attributable to the presence of member AB, i.e., all members framing into joints A and B contribute entries to $[K']$ (and ultimately to $[K]$). The influence of member AB (i.e., its contribution to $[K']$) is expressible in the form of a member stiffness matrix, $[K]_m$. Member AB and joints A and B are shown as an isolated

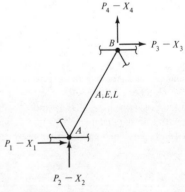

Fig. 15.20. Member in a planar truss.

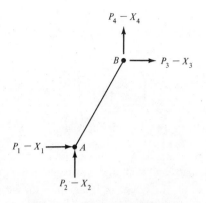

Fig. 15.21. Individual truss member.

configuration in Fig. 15.21. Matrix $[K]_m$ expresses the relationship between the joint loads and joint displacements that exist for this system. Because the directions of these quantities relate to the overall coordinate system used for the entire truss, $[K]_m$ is termed the *global* member stiffness matrix.

Derivation of the global member stiffness matrix evolves from a consideration of the member compatibility modes for the individual planar truss member. Four such modes are possible as shown in Fig. 15.22. Each of the member compatibility modes can be used to produce

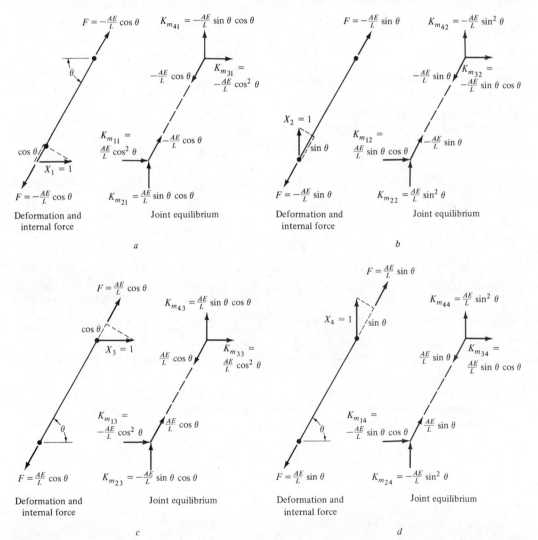

Fig. 15.22. Compatibility modes for the individual truss member. *a,* mode 1; *b,* mode 2; *c,* mode 3; *d,* mode 4.

a column of $[K]_m$ according to the physical meaning described in Section 15.5.1. The joint loads needed to constrain the member into a particular mode shape represent entries in a column of $[K]_m$. Loads corresponding to mode 1 are entered in the first column and will be calculated as an illustration to the general procedure.

In basic compatibility mode 1, a unit displacement is imposed in the X_1 direction with all other end motions restrained. To conform with the motion, the bar must deform as shown in Fig. 15.22a resulting in contraction of magnitude $\cos\theta$. This contraction produces force of magnitude

$$F = -\frac{EA}{L}\cos\theta$$

in the member. The minus sign signifies a compressive force. The joint loads, which must be present to maintain equilibrium, are easily calculated by applying statics to the joints. Using the nomenclature of Fig. 15.21

$$\begin{aligned} P_1 &= F\cos\theta = \frac{EA}{L}\cos^2\theta \\ P_2 &= F\sin\theta = \frac{EA}{L}\sin\theta\cos\theta \\ P_3 &= F\cos\theta = -\frac{EA}{L}\cos^2\theta \\ P_4 &= F\sin\theta = -\frac{EA}{L}\sin\theta\cos\theta \end{aligned} \qquad (15.44)$$

These loads constitute the entries K_{m11}, K_{m21}, K_{m31}, and K_{m41}, respectively, and are shown in the joint equilibrium diagram (Fig. 15.22a).

The remaining compatibility modes are examined in the same manner with the results summarized in Figs. 15.21a through d. Compiling the results gives

15.6.2 Assembly of [K]

The underlying relationship in Eq. 15.45 is

$$\begin{Bmatrix} P_1 \\ P_2 \\ P_3 \\ P_4 \end{Bmatrix}_m = [K]_m \begin{Bmatrix} X_1 \\ X_2 \\ X_3 \\ X_4 \end{Bmatrix}_m \qquad (15.46)$$

i.e., it expresses the association between the joint loads and joint displacements in the direction of the 4 degrees of freedom of the *individual* truss member. The circled numbers are used in Eq. 15.45 to represent Eq. 15.46. This shorthand notation was described in Section 15.4 and is an integral part of the direct stiffness method. Its general intent is also unchanged, i.e., circled members are used to indicate the assignment of the member stiffness entries to the appropriate location in the structure stiffness matrix.

Figure 15.23 illustrates a section of a much larger truss. Numbering of the particular degrees of freedom at the ends of member *AB* is entirely dependent upon the order in which the analyst numbers the complete set. The connectivity of member *AB* in the given truss is indicated by changing the numbers ① through ④ in Eq. 15.24 to the particular numerical value dictated by the overall numbering system. The elements of $[K]_m$ are then inserted into corresponding locations in the structure stiffness matrix using the direct feeding process introduced in Example 15.4. The general scheme conveyed in Fig. 15.24 is executed for each member in turn until all have been treated. The resulting matrix is the structure stiffness matrix.

Example 15.7. Build the structure stiffness matrix for the given truss (Fig. 15.25) by the direct stiffness method. Assume EA/L is the same for all members.

$$[K]_m = \frac{EA}{L}\begin{array}{c} ① \\ ② \\ ③ \\ ④ \end{array}\overset{\begin{array}{cccc} ① & ② & ③ & ④ \end{array}}{\begin{bmatrix} \cos^2\theta & \sin\theta\cos\theta & -\cos^2\theta & -\sin\theta\cos\theta \\ \sin\theta\cos\theta & \sin^2\theta & -\sin\theta\cos\theta & -\sin^2\theta \\ -\cos^2\theta & -\sin\theta\cos\theta & \cos^2\theta & \sin\theta\cos\theta \\ -\sin\theta\cos\theta & -\sin^2\theta & \sin\theta\cos\theta & \sin^2\theta \end{bmatrix}} \qquad (15.45)$$

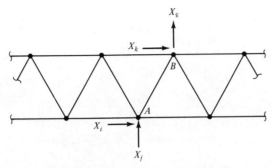

Fig. 15.23.   Segment of a planar truss.

Solution:

Step 1: Trigonometric values needed in the solution are tabulated below.

Member	$\sin\theta$	$\cos\theta$	$\sin^2\theta$	$\cos^2\theta$	$\sin\theta\cos\theta$
AB	1.00	0	1.00	0	0
BC	0	1.00	0	1.00	0
AC	0.80	0.60	0.64	0.36	0.48
DC	0.60	-0.80	0.36	0.64	-0.48
AD	0	1.00	0	1.00	0

Member stiffness matrices are generated by use of

Eq. 15.45. Connectivity (based upon the degrees of freedom indicated and numbered in Fig. 15.25) is indicated in each case by the usual shorthand notation.

$$[K]_{AB} = \begin{array}{c} \text{\small ®} \\ \text{\small ®} \\ \text{\small ①} \\ \text{\small ②} \end{array} \begin{bmatrix} 0 & 0 & 0 & 0 \\ 0 & 1.00 & 0 & -1.00 \\ 0 & 0 & 0 & 0 \\ 0 & -1.00 & 0 & 1.00 \end{bmatrix} \frac{EA}{L}$$

$$\begin{array}{cccc} \text{\small ⓘ} & \text{\small ⓙ} & \text{\small ⓚ} & \text{\small ⓛ} \end{array}$$
$$\begin{array}{c} \text{\small ⓘ} \\ \text{\small ⓙ} \\ \text{\small ⓚ} \\ \text{\small ⓛ} \end{array} \begin{bmatrix} K_{m_{11}} & K_{m_{12}} & K_{m_{13}} & K_{m_{14}} \\ K_{m_{21}} & K_{m_{22}} & K_{m_{23}} & K_{m_{24}} \\ K_{m_{31}} & K_{m_{32}} & K_{m_{33}} & K_{m_{34}} \\ K_{m_{41}} & K_{m_{42}} & K_{m_{43}} & K_{m_{44}} \end{bmatrix}$$

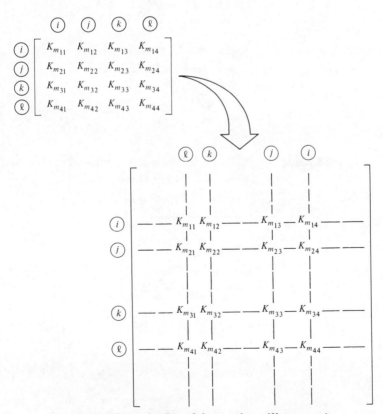

Fig. 15.24.   Direct feeding of the member stiffness matrix.

$$[K]_{BC} = \begin{array}{c} \\ ① \\ ② \\ ③ \\ ④ \end{array} \begin{array}{cccc} ① & ② & ③ & ④ \\ \left[\begin{array}{cccc} 1.00 & 0 & -1.00 & 0 \\ 0 & 0 & 0 & 0 \\ -1.00 & 0 & 1.00 & 0 \\ 0 & 0 & 0 & 0 \end{array}\right] & & & \end{array} \frac{EA}{L}$$

$$[K]_{AC} = \begin{array}{c} ⓡ \\ ⓡ \\ ③ \\ ④ \end{array} \begin{array}{cccc} ⓡ & ⓡ & ③ & ④ \\ \left[\begin{array}{cccc} 0.36 & 0.48 & -0.36 & -0.48 \\ 0.48 & 0.64 & -0.48 & -0.64 \\ -0.36 & -0.48 & 0.36 & 0.48 \\ -0.48 & -0.64 & 0.48 & 0.64 \end{array}\right] & & & \end{array} \frac{EA}{L}$$

$$[K]_{DC} = \begin{array}{c} ⑤ \\ ⓡ \\ ③ \\ ④ \end{array} \begin{array}{cccc} ⑤ & ⓡ & ③ & ④ \\ \left[\begin{array}{cccc} 0.64 & -0.48 & -0.64 & 0.48 \\ -0.48 & 0.36 & 0.48 & -0.36 \\ -0.64 & 0.48 & 0.64 & -0.48 \\ 0.48 & -0.36 & -0.48 & 0.36 \end{array}\right] & & & \end{array} \frac{EA}{L}$$

$$[K]_{AD} = \begin{array}{c} ⓡ \\ ⓡ \\ ⑤ \\ ⓡ \end{array} \begin{array}{cccc} ⓡ & ⓡ & ⑤ & ⓡ \\ \left[\begin{array}{cccc} 1.00 & 0 & -1.00 & 0 \\ 0 & 0 & 0 & 0 \\ -1.00 & 0 & 1.00 & 0 \\ 0 & 0 & 0 & 0 \end{array}\right] & & & \end{array} \frac{EA}{L}$$

Step 2:   The structure stiffness matrix is 5 × 5 in size and constructed by feeding the member stiffness matrices into the appropriate locations. Completion of the transfers yields

$$[K] = \begin{array}{c} ① \\ ② \\ ③ \\ ④ \\ ⑤ \end{array} \begin{array}{ccccc} ① & ② & ③ & ④ & ⑤ \\ \left[\begin{array}{ccccc} 1.00 & 0 & -1.00 & 0 & 0 \\ 0 & 1.00 & 0 & 0 & 0 \\ -1.00 & 0 & 2.00 & 0 & -0.64 \\ 0 & 0 & 0 & 1.00 & 0.48 \\ 0 & 0 & -0.64 & 0.48 & 1.64 \end{array}\right] & & & & \end{array} \frac{EA}{L}$$

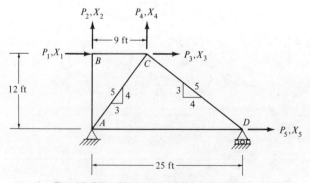

Fig. 15.25.   Planar truss, Example 15.7.

15.6.3 Calculation of Joint Displacements and Member Forces

Analysis for the joint displacements is straightforward in the direct stiffness method. The structure stiffness matrix is constructed as described in Section 15.6.2. After inverting the stiffness matrix, the joint displacements follow from the usual operation

$$\{X\} = [K^{-1}]\{P\}$$

for the given joint loads, $\{P\}$. Because loads are applied only at the joints, it is a simple matter to develop $\{P\}$. Each load is merely inserted (fed) into the appropriate location in $\{P\}$. However, subsequent to this computation, it is necessary to calculate the internal member forces. In the basic matrix analysis, the relationship

$$\{F\} = [S][C]\{X\}$$

was inferred in the determination of the internal forces. This computation is of little use in the direct stiffness method because of the absence of the compatibility matrix, $[C]$. Although this matrix could be generated (by a further consideration of the compatibility modes), this approach is undesirable. Indeed, the avoidance of building $[C]$ is one objective of the direct stiffness method. Instead, two simpler alternatives will be described.

15.6.3.1 Employment of Material Law

Hooke's law constitutes the basic material relationship for the truss member. If the elongation (contraction), Δ, of a member is known, the member force is determined as

$$F = \frac{EA}{L}\Delta$$

With the displaced position of the joints of the truss known, it is a simple matter to determine Δ for each member.

Consider the compatibility mode shapes presented earlier in Fig. 15.22. By examination of the deformation diagrams in Figs. 15.22a through d, the elongations (contractions) due to a unit value of each joint displacement component are apparent. Changing the unit displacements in these figures to displacements of magnitudes X_1 through X_4, respectively, and summing the elongation values gives

$$\Delta = -X_1 \cos \theta - X_2 \sin \theta + X_3 \cos \theta$$
$$+ X_4 \sin \theta$$

or, equivalently,

$$\Delta = (X_3 - X_1) \cos \theta + (X_4 - X_2) \sin \theta$$
$$(15.47)$$

Equation 15.47 is a general expression for the elongation of a truss member. For each member, the needed joint displacement quantities are extracted from the previously determined $\{X\}$ matrix and substituted into Eq. 15.47 to give the elongation. Hooke's law is then employed to calculate the member force.

15.6.3.2 Employment of Global Member Stiffness Matrix

The physical meaning of the member stiffness is evident in Eq. 15.46. The end force components $\{P_1, P_2, P_3, P_4\}$, needed to maintain a given set of end displacement components $\{X_1, X_2, X_3, X_4\}$ for an *individual* truss member, can be determined if $[K]_m$ is known. In the direct stiffness method, the $[K]_m$ matrix is available, and the joint displacements are the first quantities determined. Consequently, direct substitution into Eq. 15.46 is possible. Subsequently, the actual member force is determined as the resultant force at either member end

$$F = -(P_1 \cos \theta + P_2 \sin \theta)$$
$$= P_3 \cos \theta + P_4 \sin \theta \qquad (15.48)$$

where F is positive if it is a tensile force. This two-step computation is repeated for each member after extracting the needed displacement values $\{X\}_m$ from the $\{X\}$ matrix for the entire truss.

Example 15.8. Calculate the joint displacements and member forces for the truss given in Fig. 15.26a. Use $EA/L = 2000$ k/ft for all members.

Solution: Joint displacements are determined by employing the structure stiffness matrix developed in Example 15.7.

$$\{X\} = [K]^{-1}\{P\}$$

$$\begin{Bmatrix} X_1 \\ X_2 \\ X_3 \\ X_4 \\ X_5 \end{Bmatrix} = \begin{bmatrix} 2.4096 & 0 & 1.4096 & -0.3072 & 0.64 \\ 0 & 1 & 0 & 0 & 0 \\ 1.4096 & 0 & 1.4096 & -0.3072 & 0.64 \\ -0.3072 & 0 & -0.3072 & 1.2304 & -0.48 \\ 0.64 & 0 & 0.64 & -0.48 & 1 \end{bmatrix} \begin{Bmatrix} 25 \\ -12 \\ 0 \\ 0 \\ 0 \end{Bmatrix}$$

$$= \frac{L}{EA} \begin{Bmatrix} 60.24 \\ -12.00 \\ 35.24 \\ -7.68 \\ 16.00 \end{Bmatrix} = \begin{Bmatrix} 0.03012 \\ -0.00600 \\ 0.01762 \\ 0.00384 \\ 0.00800 \end{Bmatrix}$$

$$= \begin{Bmatrix} 0.03012 \rightarrow \\ 0.00600 \downarrow \\ 0.01762 \rightarrow \\ 0.00384 \downarrow \\ 0.00800 \rightarrow \end{Bmatrix}$$

Equation 15.47 is used to calculate the member deformations with needed joint displacements extracted from $\{X\}$.

AB: $\Delta_1 = (X_1 - 0) \cos (90°)$
$+ (X_2 - 0) \sin (90°)$
$= (0) + (-0.0060)(1)$
$\Delta_1 = -0.0060$ in.

BC: $\Delta_2 = (X_3 - X_1) \cos (0)$
$+ (X_4 - X_2) \sin (0)$
$= (0.01762 - 0.03012)(1) + 0$
$\Delta_2 = -0.0125$ in.

AC: $\Delta_3 = (X_3 - 0) \cos \theta + (X_4 - 0) \sin \theta$
$= (0.01762)(0.6) + (-0.00384)(0.8)$
$\Delta_3 = 0.00750$ in.

DC: $\Delta_4 = (X_3 - X_5) \cos \theta + (X_4 - 0) \sin \theta$
$= (0.01762 - 0.008)(-0.80)$
$+ (-0.00384)(0.60)$
$\Delta_4 = -0.010$ in.

AD: $\Delta_5 = (X_5 - 0) \cos (0) + (0 - 0) \sin (0)$
$= (0.008)(1)$
$\Delta_5 = 0.008$ in.

Member forces are determined from Hooke's law

$$F = \frac{EA}{L} \Delta = 2000\Delta$$

which gives

$$\begin{Bmatrix} F_1 \\ F_2 \\ F_3 \\ F_4 \\ F_5 \end{Bmatrix} = \begin{Bmatrix} -12.0 \\ -25.0 \\ +15.0 \\ -20.0 \\ +16.0 \end{Bmatrix} = \begin{Bmatrix} 12.0 \text{ kips C} \\ 25.0 \text{ kips C} \\ 15.0 \text{ kips T} \\ 20.0 \text{ kips C} \\ 16.0 \text{ kips T} \end{Bmatrix}$$

The member forces are shown in Fig. 15.26b.

Alternate solution: As an illustration of the use of Eq. 15.48, the force in member DC is determined as follows:

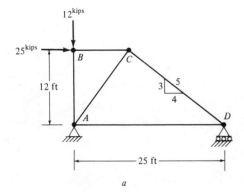

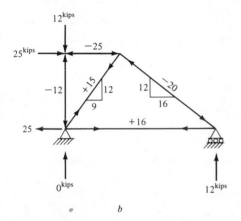

Fig. 15.26.   Results for Example 15.8. *a*, loading; *b*, internal forces.

$F_4 = -20.00$ kips

$\qquad = 20.00$ kips compression

This technique is repeated for all other members and again produces the results shown in Fig. 15.26b.

15.7   GENERATION OF [K] BY PHYSICAL MEANING: PLANAR FRAMES

Use of the physical meaning of the structure stiffness matrix to obtain its contents is a general concept. Application to planar frames is the subject matter of the present section. A planar frame is composed of individual beam elements, and development of $[K]$ is little different in execution from the approach described for continuous beams. However, the more complex geometry of planar frames leads to more extensive computations.

Compatibility modes are the fundamental tools used to quantify the physical meaning of $[K]$. As shown for beams and trusses, each structure-compatibility mode is the source of the entries in a particular column of $[K]$. The joint loads needed to maintain the frame in the particular structure-compatibility mode constitute these entries.

$$\begin{Bmatrix} P_1 \\ P_2 \\ P_3 \\ P_4 \end{Bmatrix}_{DC} = [K]_{DC} \begin{Bmatrix} X_5 \\ 0 \\ X_3 \\ X_4 \end{Bmatrix}$$

$$= 2000 \begin{bmatrix} 0.64 & -0.48 & -0.64 & 0.48 \\ -0.48 & 0.36 & 0.48 & -0.36 \\ -0.64 & 0.48 & 0.64 & -0.48 \\ 0.48 & -0.36 & -0.48 & 0.36 \end{bmatrix} \begin{Bmatrix} 0.00800 \\ 0 \\ 0.01762 \\ -0.00384 \end{Bmatrix}$$

$$= \begin{Bmatrix} -16.00 \text{ kips} \\ +12.00 \text{ kips} \\ +16.00 \text{ kips} \\ -12.00 \text{ kips} \end{Bmatrix}$$

From Eq. 15.48

$$F_4 = -(P_1 \cos \theta + P_2 \cos \theta)$$

$$= -16.00(-0.80) - (-12)(+0.60)$$

Example 15.9.   Construct the second and fourth columns of the structure stiffness matrix for the frame in Fig. 15.27a. Use their physical meanings, and assume EI is constant and the same for both members.

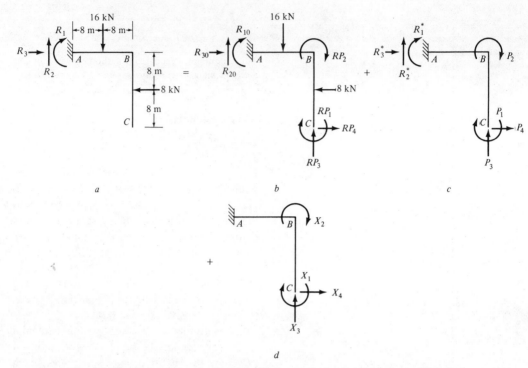

Fig. 15.27. Planar frame for Example 15.9. *a*, actual state; *b*, degrees of freedom.

Solution: Degrees of freedom for the structure are shown in Fig. 15.27b. Compatibility modes are shown in Fig. 15.28. Compatibility modes 2 and 4 are needed in the present example.

Second column of $[K]$: In mode 2, joint B experiences a unit rotation. This rotation affects both members, resulting in the internal forces shown in Fig. 15.29. Because joints A and C are restrained, the end moment values are computed as

$$M_1 = M_4 = \frac{2EI}{L} \qquad M_2 = M_3 = \frac{4EI}{L}$$

The remaining internal forces and the joint loads are calculated by sequential application of statics.

Members AB and BC:

$$V_1 = -\frac{M_1 + M_2}{L} = -\frac{6EI}{L^2}$$

$$V_2 = +\frac{M_1 + M_2}{L} = +\frac{6EI}{L^2}$$

$$V_3 = -\frac{M_3 + M_4}{L} = -\frac{6EI}{L^2}$$

$$V_4 = +\frac{M_3 + M_4}{L} = +\frac{6EI}{L^2}$$

Joint B:

$$A_2 = V_3 = -\frac{6EI}{L^2}$$

$$A_3 = V_2 = +\frac{6EI}{L^2}$$

$$P_2 = K_{22} = M_2 + M_3 = +\frac{8EI}{L}$$

Members AB and BC:

$$A_1 = A_2 = -\frac{6EI}{L^2}$$

$$A_4 = A_3 = +\frac{6EI}{L^2}$$

Joint C:

$$P_3 = K_{32} = A_4 = +\frac{6EI}{L^2}$$

$$P_4 = K_{42} = V_4 = +\frac{6EI}{L^2}$$

$$P_1 = K_{12} = M_4 = \frac{2EI}{L}$$

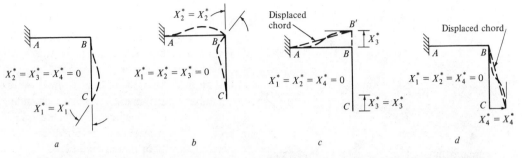

Fig. 15.28.   Compatibility modes, Example 15.9. *a,* mode 1; *b,* mode 2; *c,* mode 3; *d,* mode 4.

Fourth column of $[K]$*:* The unit displacement of joint C (with joints A and B restrained) produces the moments

$$M_1 = M_2 = 0$$

$$M_3 = M_4 = +\frac{6EI}{L^2}$$

Free-body diagrams shown in Fig. 15.29 also apply to mode 4, except $P_1 = K_{14}$, $P_2 = K_{24}$, $P_3 = K_{34}$, and $P_4 = K_{44}$. Therefore, the determination of these load values follows from the same sequence of statics as applied to mode 2. To avoid repetition, details of the numerical computation are omitted, and only the resulting load values will be given.

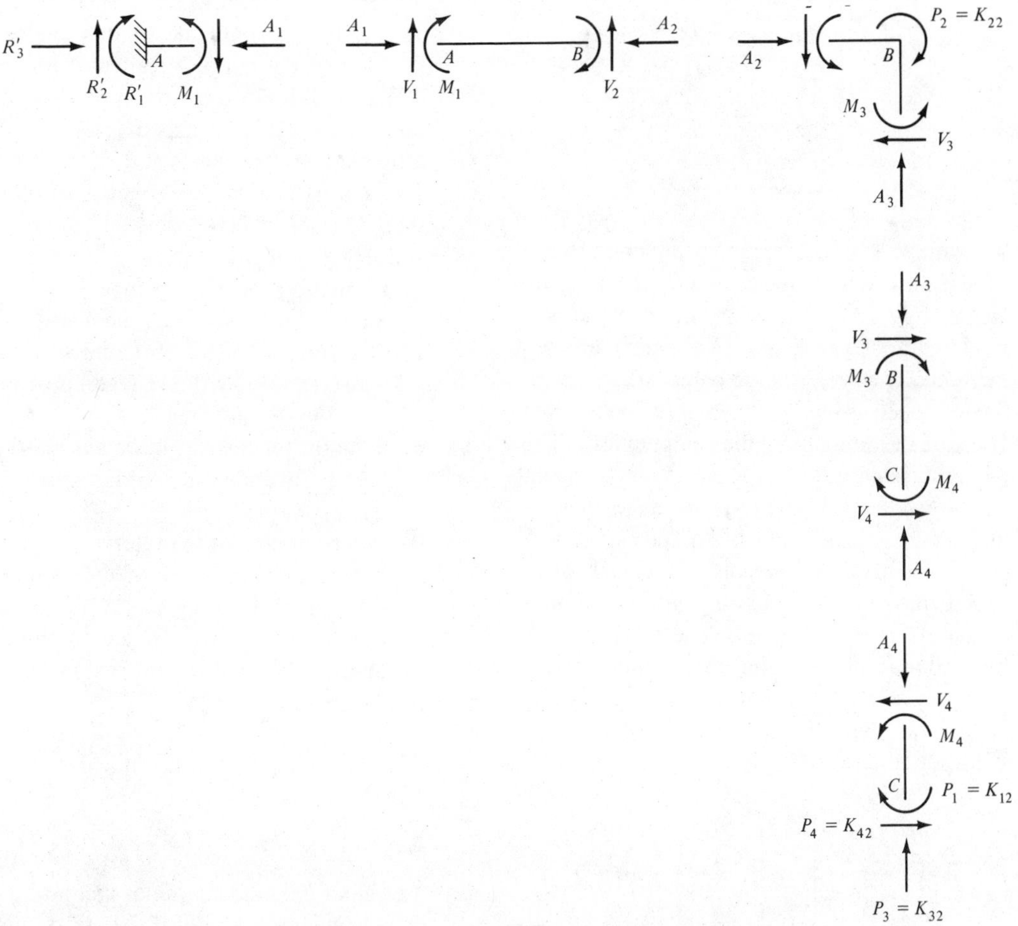

Fig. 15.29.   Free-body diagrams—Ex. 15.9.

$$P_1 = K_{14} = +\frac{6EI}{L^2}$$

$$P_2 = K_{24} = +\frac{6EI}{L^2}$$

$$P_3 = K_{34} = 0$$

$$P_4 = K_{44} = +\frac{12EI}{L^3}$$

Compatibility modes 1 and 3 are analyzed in a similar fashion to modes 2 and 4, respectively. Computations are left as an exercise for the reader. Results obtained give the first and third columns of $[K]$. The complete results for all modes give

$$[K] = \begin{bmatrix} \dfrac{4EI}{L} & \dfrac{2EI}{L} & 0 & \dfrac{6EI}{L^2} \\[2ex] \dfrac{2EI}{L} & \dfrac{8EI}{L} & \dfrac{6EI}{L^2} & \dfrac{6EI}{L^2} \\[2ex] 0 & \dfrac{6EI}{L^2} & \dfrac{12EI}{L^3} & 0 \\[2ex] \dfrac{6EI}{L^2} & \dfrac{6EI}{L^2} & 0 & \dfrac{12EI}{L^3} \end{bmatrix}$$

Example 15.9 demonstrates the construction of the structure stiffness matrix for a planar frame by a consideration of compatibility modes introduced in the basic matrix analysis method. However, minor complexities in geometry, such as inclined members, have a great effect on the difficulty of the computations. This inherent feature makes use of the physical meaning of $[K]$ unattractive as a general method for planar frames. An additional difficulty, not readily apparent, is the lack of an expedient method for subsequent computation of the member end forces. (The most straightforward procedure is, after calculating $\{X\}$, to manually compute the end rotations, ϕ_1 and ϕ_2, of each member *relative to the displaced chord* and apply the relationship

$$\begin{Bmatrix} M_1 \\ M_2 \end{Bmatrix} = \begin{bmatrix} \dfrac{4EI}{L} & \dfrac{2EI}{L} \\[2ex] \dfrac{2EI}{L} & \dfrac{4EI}{L} \end{bmatrix} \begin{Bmatrix} \phi_1 \\ \phi_2 \end{Bmatrix}$$

However, determination of ϕ_1 and ϕ_2 is cumbersome even for simple frames.) Use of the concept of expanded member stiffness matrices is not a viable alternative for generating $[K]$ because of the inefficiency of processing zero valued entries. Also, the difficulty of determining member end moment remains a problem. For these reasons, the concepts to be described in Section 15.8 represent a preferred approach to matrix analysis of planar frames.

15.8 DIRECT STIFFNESS METHOD: PLANAR FRAMES

Direct stiffness analysis of continuous beams (Section 15.4) and planar trusses (Section 15.6) proved to be an efficient approach to matrix analysis of those structural types. Steps in the general procedure are:

1. Analysis of the restrained state of the structure. The outcome of this analysis consists of the matrix of internal forces, $\{F_0\}$, and the load matrix, $\{P\}$, for the second stage of the analysis.
2. Complete the second stage of analysis; i.e., matrix analysis of the released state. This phase involves several computations, which include:
   a. Formation of the member stiffness matrix, $[K]_m$, for each member.
   b. Construction of the structure stiffness matrix, $[K]$, by direct feeding of the member stiffness matrices.
   c. Solution for the displacement matrix, $\{X^*\}$. Inversion of $[K]$ is implied in this operation.
   d. Determination of the internal member forces, $\{F^*\}$, by use of the member stiffness relationship.
3. Results from steps 1 and 2 are combined to produce the final end forces and joint displacements

$$\{F\} = \{F_0\} + \{F^*\}$$

and

$$\{X\} = \{X^*\}$$

The latter operation assumes that no joint displacements occur in the restrained state.

The feasibility of applying the direct stiffness method to beams and planar trusses is due to the subtle characteristics of both these structural types. A continuous beam and a planar truss both have a *unique* set of degrees of freedom. The analyst has no choice when declaring the degrees of freedom for these structures. As a result, the structure stiffness matrix is also unique. If axial deformation is neglected in planar frames, the kinematic unknowns do not always occur in a unique set of directions. This is because several choices are often available for the declaration of the translational degrees of freedom. In such cases, the structure stiffness matrix is *not unique* and instead is dependent upon the directions actually chosen for the translational degrees of freedom.

The significant fact concerning the non-uniqueness of a structure stiffness matrix is that it must be constructed by using the structure "as a whole." In other words, $[K]$ is dependent upon the compatibility modes associated with the structure taken in its entirety. Consequently, $[K]$ cannot be produced for a planar frame by any member-by-member approach, such as the direct stiffness method described in Section 15.6 for trusses, if axial deformation is neglected. Therefore, it can be concluded that a direct stiffness method for frames is possible *only if* axial deformation *is included* in the formulation.

The main reason for the feasibility of applying a direct stiffness technique in this circumstance is that *all* possible joint translations must be included when declaring the degrees of freedom. *When axial deformation is neglected*, all joint translations are *not independent*; therefore, all member end deformations are *not independent*. Conversely, *when axial effects are included*, all joint translations *are independent*; therefore, all member end deformations are *independent*. The latter condition makes a member-by-member construction of $[K]$ possible. The remainder of this section addresses this procedure. The reader should realize that the need to include axial deformation is not simply a feature that is adopted to enable a direct stiffness approach. Rather, its inclusion represents a more rigorous, accurate, and desirable analytical approach that was not possible (in a

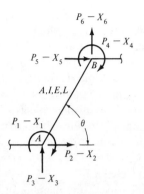

Fig. 15.30.   Member in a planar frame.

practical sense) prior to the advent of computer technology.

15.8.1   Global Member Stiffness Matrix

Derivation of a member stiffness matrix begins with a consideration of the individual framed member as a component of the overall framed system. In the most general situation, both ends of a member are connected to a free joint in the structure. In this way, 3 degrees of freedom exist at each end (see Fig. 15.30). The particular numbering system for these degrees of freedom depends upon the sequence chosen for the entire frame. In Fig. 15.30, they are numbered X_1 through X_6. The stiffness of any joint depends upon the number of members that connect to the joint, and their material and geometric properties. The contribution of any particular member to the stiffnesses of the two connecting joints reflected in its member stiffness matrix, $[K]_m$. A general $[K]_m$ matrix will be formulated with the aid of Fig. 15.31 and the technique described in Section 15.6.

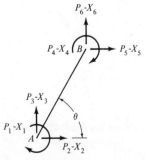

Fig. 15.31.   Individual frame member.

Equation 15.43 expresses the underlying concept of viewing the member stiffness matrix as a part of the whole structure stiffness matrix. In the present context, $[K']$ is a portion of the structure stiffness matrix and characterizes the interrelationship of the six joint loads, P_1-P_6, and the six joint displacements, X_1-X_6, depicted in Fig. 15.30. Furthermore, part of $[K']$ itself is directly due to the presence of member AB. This part of the $[K']$ is the global member stiffness matrix $[K]_m$.

Viewed by itself, the member in Fig. 15.31 has six characteristic compatibility modes. Each one is representative of a column in $[K]_m$. The entries in a particular column are determined from its physical meaning. As an initial step, consider the first member-compatibility mode depicted in Fig. 15.32. With a unit rotation induced at end A and all other degrees of freedom restrained from motion, the internal forces shown in Fig. 15.32a will be developed at the member ends. These forces are deter-

mined by substitution of the known end deformations ($\Phi_L = 1$, $\Phi_R = \delta_L = \delta_R = 0$) into Eq. 15.6. If these forces are transferred to the joints, it is clear that joint loads shown in Fig. 15.32b must exist to satisfy equilibrium. Consequently, the first column of the member stiffness matrix has the entries.

$$[K_{11}]_m = \frac{4EI}{L} \qquad [K_{41}]_m = \frac{2EI}{L}$$

$$[K_{21}]_m = \frac{6EI}{L^2} \sin \theta \qquad [K_{51}]_m = -\frac{6EI}{L^2} \sin \theta$$

$$[K_{31}]_m = -\frac{6EI}{L^2} \cos \theta \qquad [K_{61}]_m = \frac{6EI}{L^2} \cos \theta$$

Figure 15.33a presents a similar illustration of the fourth compatibility mode. In this mode, a unit rotation is induced at end B, while all other degrees of freedom are restrained from motion. Again, by substitution of the known

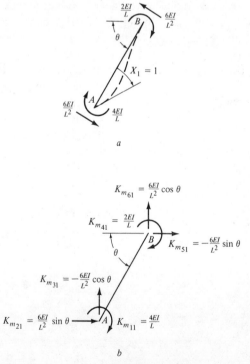

a

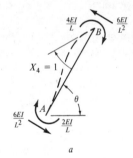

$X_4 = 1$

a

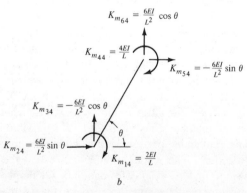

b

Fig. 15.32. First compatibility mode for a frame member ($X = 1$). a, deformation and internal forces; b, joint loads.

Fig. 15.33. Fourth compatibility mode for a frame member ($X = 1$). a, deformation and internal forces; b, joint loads.

end deformations ($\Phi_R = 1$, $\Phi_L = \delta_L = \delta_R = 0$) into Eq. 15.6, the joint loads shown in Fig. 15.33b are established and represent the entries in the fourth column of $[K]_m$.

Remaining basic compatibility modes associated with Fig. 15.31 are translational modes. Figure 15.34a depicts mode 2 in which $X_2 = 1$ and other degrees of freedom are restrained from motion. The unit translation produces both a transverse (component creating flexure) and a longitudinal component (component creating

ilar manner, and the details are presented in Fig. 15.35. Loads shown in Fig. 15.35c are the entries in the third column of $[K]_m$. Modes 5 and 6 are omitted from the discussion to permit the reader to investigate them as a self-learning process.

A compilation of the results obtained for all six modes gives the global member stiffness matrix. The complete global member stiffness relationship is given in Eq. 15.49 in shorthand format.

$$[K]_m = \begin{array}{c c} & \begin{matrix} \;\;① & \;\;② & \;\;③ & \;\;④ & \;\;⑤ & \;\;⑥ \end{matrix} \\ \begin{matrix} ① \\ ② \\ ③ \\ ④ \\ ⑤ \\ ⑥ \end{matrix} & \begin{bmatrix} \dfrac{4EI}{L} & \dfrac{6EI}{L^2}\sin\theta & -\dfrac{6EI}{L^2}\cos\theta & \dfrac{2EI}{L} & -\dfrac{6EI}{L^2}\sin\theta & \dfrac{6EI}{L^2}\cos\theta \\[2mm] & \dfrac{EA}{L}\cos^2\theta + \dfrac{12EI}{L^3}\sin^2\theta & \dfrac{EA}{L}\sin\theta\cos\theta - \dfrac{12EI}{L^3}\sin\theta\cos\theta & \dfrac{6EI}{L^2}\sin\theta & -\dfrac{EA}{L}\cos^2\theta - \dfrac{12EI}{L^3}\sin^2\theta & -\dfrac{EA}{L}\sin\theta\cos\theta + \dfrac{12EI}{L^3}\sin\theta\cos\theta \\[2mm] & & \dfrac{EA}{L}\sin^2\theta + \dfrac{12EI}{L^3}\cos^2\theta & -\dfrac{6EI}{L^2}\cos\theta & -\dfrac{EA}{L}\sin\theta\cos\theta + \dfrac{12EI}{L^3}\sin\theta\cos\theta & -\dfrac{EA}{L}\sin^2\theta - \dfrac{12EI}{L^3}\cos^2\theta \\[2mm] & & & \dfrac{4EI}{L} & -\dfrac{6EI}{L^2}\sin\theta & \dfrac{6EI}{L^2}\cos\theta \\[2mm] & \text{Symmetry} & & & \dfrac{EA}{L}\cos^2\theta + \dfrac{12EI}{L^3}\sin^2\theta & \dfrac{EA}{L}\sin\theta\cos\theta - \dfrac{12EI}{L^3}\sin\theta\cos\theta \\[2mm] & & & & & \dfrac{EA}{L}\sin^2\theta + \dfrac{12EI}{L^3}\cos^2\theta \end{bmatrix} \end{array} \quad (15.49)$$

axial deformation). The transverse component implies $\delta_L = -\sin\theta$ and $\delta_R = \Phi_L = \Phi_R = 0$ in Eq. 15.6. The longitudinal component implies an axial deformation of magnitude $\cos\theta$. Internal forces caused by these motions are shown in Fig. 15.34b. Equilibrium at the joints requires the application of the loads shown in Fig. 15.34c, which are the entries in the second column of $[K]_m$. Mode 3 is analyzed in a sim-

It is useful to separate Eq. 15.49 into two parts according to

$$[K]_m = [K_f]_m + [K_a]_m \quad (15.50)$$

Terms attributed to flexural and axial effects separately are placed in $[K_f]_m$ and $[K_a]_m$, re-

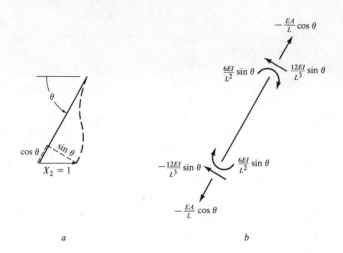

$$K_{m_{62}} = -\frac{EA}{L}\sin\theta\cos\theta + \frac{12EI}{L^3}\sin\theta\cos\theta$$

$$K_{m_{42}} = \frac{6EI}{L^2}\sin\theta \qquad K_{m_{52}} = -\frac{EA}{L}\cos^2\theta - \frac{12EI}{L^3}\sin^2\theta$$

$$K_{m_{32}} = \frac{EA}{L}\sin\theta\cos\theta - \frac{12EI}{L^2}\sin\theta\cos\theta$$

$$K_{m_{22}} = \frac{EA}{L}\cos^2\theta + \frac{12EI}{L^3}\sin^2\theta \qquad K_{m_{12}} = \frac{6EI}{L^2}\sin\theta$$

c

Fig. 15.34. Second compatibility mode for a frame member ($X = 1$). a, deformation: b, internal forces: c, joint loads.

spectively. In general terms,

$$[K_f]_m =$$

	①	②	③	④	⑤	⑥
①	$\dfrac{4EI}{L}$	$\dfrac{6EI}{L^2}\sin\theta$	$-\dfrac{6EI}{L^2}\cos\theta$	$\dfrac{2EI}{L}$	$-\dfrac{6EI}{L^2}\sin\theta$	$\dfrac{6EI}{L^2}\cos\theta$
②		$\dfrac{12EI}{L^2}\sin^2\theta$	$-\dfrac{12EI}{L^3}\sin\theta\cos\theta$	$\dfrac{6EI}{L^2}\sin\theta$	$-\dfrac{12EI}{L^3}\sin^2\theta$	$\dfrac{12EI}{L^3}\sin\theta\cos\theta$
③			$\dfrac{12EI}{L^3}\cos^2\theta$	$-\dfrac{6EI}{L^2}\cos\theta$	$\dfrac{12EI}{L^3}\sin\theta\cos\theta$	$-\dfrac{12EI}{L^3}\cos^2\theta$
④				$\dfrac{4EI}{L}$	$-\dfrac{6EI}{L^2}\sin\theta$	$\dfrac{6EI}{L^2}\cos\theta$
⑤		Symmetry			$\dfrac{12EI}{L^3}\sin^2\theta$	$-\dfrac{12EI}{L^3}\sin\theta\cos\theta$
⑥						$\dfrac{12EI}{L^3}\cos^2\theta$

(15.51)

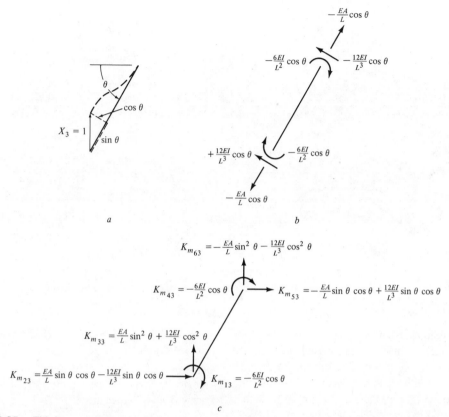

Fig. 15.35. Third compatibility mode for a frame member ($X = 1$). *a*, deformation; *b*, internal forces; *c*, joint loads.

$$[K_a]_m =$$

(15.52)

	①	②	③	④	⑤	⑥
①	0	0	0	0	0	0
②		$\dfrac{EA}{L}\cos^2\theta$	$\dfrac{EA}{L}\sin\theta\cos\theta$	0	$-\dfrac{EA}{L}\cos^2\theta$	$-\dfrac{EA}{L}\sin\theta\cos\theta$
③			$\dfrac{EA}{L}\sin^2\theta$	0	$-\dfrac{EA}{L}\sin\theta\cos\theta$	$-\dfrac{EA}{L}\sin^2\theta$
④				0	0	0
⑤		Symmetry			$\dfrac{EA}{L}\cos^2\theta$	$\dfrac{EA}{L}\sin\theta\cos\theta$
⑥						$\dfrac{EA}{L}\sin^2\theta$

Equation 15.53 represents the axial contribution to $[K]_m$ and should be compared with Eq. 15.45, which was developed for trusses earlier in the text.

15.8.2 Assembly of $[K]$

With Eq. 15.49 available, construction of the structure stiffness matrix is built by using the

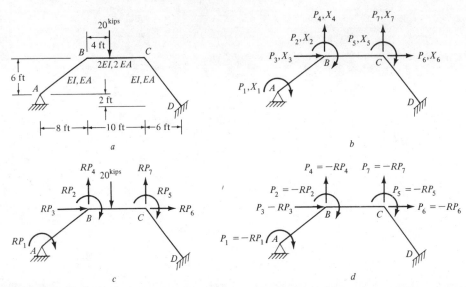

Fig. 15.36. Planar frame, Example 15.10. *a*, structure and loading; *b*, degrees of freedom; *c*, restrained state; *d*, released state.

direct feeding concept. $[K]_m$ is formulated for each member in turn and entered into the appropriate locations of $[K]$. Except for the size of $[K]_m$ and the context of a planar frame, the execution of this process is unchanged from that illustrated in Sections 15.4.1 and 15.6.2.

Example 15.10.   Construct the structure stiffness matrix for the frame in Fig. 15.36a. Use EI =

20,000 kip-ft^2 and EA = 200,000 kips. Employ the direct stiffness method.

Solution:

Step 1:   Form the member stiffness matrices. Trigonometric data needed are:

Member	$\sin \theta$	$\cos \theta$	$\sin^2 \theta$	$\cos^2 \theta$	$\sin \theta \cos \theta$
AB	0.60	0.80	0.36	0.64	0.48
BC	0.00	1.00	0.00	1.00	0.00
CD	−0.80	0.60	0.64	0.36	−0.48

Flexural effects (Eq. 15.51):

$$
[K_f]_{AB} =
\begin{array}{c}
 \\
① \\
® \\
® \\
② \\
③ \\
④
\end{array}
\begin{array}{cccccc}
① & ® & ® & ② & ③ & ④ \\
8000 & 720 & -960 & 4000 & -720 & 960 \\
720 & 86.4 & -115.2 & 720 & -86.4 & 115.2 \\
-960 & -115.2 & 153.6 & -960 & 115.2 & -153.6 \\
4000 & 720 & -960 & 8000 & -720 & 960 \\
-720 & -86.4 & 115.2 & -720 & 86.4 & -115.2 \\
960 & 115.2 & -153.6 & 960 & -115.2 & 153.6
\end{array}
$$

$$
[K_f]_{BC} =
\begin{array}{c}
② \\
③ \\
④ \\
⑤ \\
⑥ \\
⑦
\end{array}
\begin{array}{ccccccc}
② & ③ & ④ & ⑤ & ⑥ & ⑦ \\
16,000 & 0 & -2400 & 8000 & 0 & 2400 \\
0 & 0 & 0 & 0 & 0 & 0 \\
-2400 & 0 & 480 & -2400 & 0 & -480 \\
8000 & 0 & -2400 & 16,000 & 0 & 2400 \\
0 & 0 & 0 & 0 & 0 & 0 \\
2400 & 0 & -480 & 2400 & 0 & 480
\end{array}
$$

$$
[K_f]_{CD} = \begin{array}{c} \\ \text{⑤} \\ \text{⑥} \\ \text{⑦} \\ \text{⑧} \\ \text{⑧} \\ \text{⑧} \end{array}
\begin{array}{cccccc}
\text{⑤} & \text{⑥} & \text{⑦} & \text{⑧} & \text{⑧} & \text{⑧} \\
\begin{bmatrix}
8000 & -960 & -720 & 4000 & 960 & 720 \\
-960 & 153.6 & 115.2 & -960 & -153.6 & -115.2 \\
-720 & 115.2 & 86.4 & -720 & -115.2 & -86.4 \\
4000 & -960 & -720 & 8000 & 960 & 720 \\
-960 & -153.6 & -115.2 & 960 & 153.6 & 115.2 \\
720 & -115.2 & -86.4 & 720 & 115.2 & 86.4
\end{bmatrix}
\end{array}
$$

Axial effects (Eq. 15.52):

$$
[K_a]_{AB} = \begin{array}{c} \\ \text{①} \\ \text{⑧} \\ \text{⑧} \\ \text{②} \\ \text{③} \\ \text{④} \end{array}
\begin{array}{cccccc}
\text{①} & \text{⑧} & \text{⑧} & \text{②} & \text{③} & \text{④} \\
\begin{bmatrix}
0 & 0 & 0 & 0 & 0 & 0 \\
0 & 12,800 & 9600 & 0 & -12,800 & -9600 \\
0 & 9600 & 7200 & 0 & -9600 & -7200 \\
0 & 0 & 0 & 0 & 0 & 0 \\
0 & -12,800 & -9600 & 0 & 12,800 & 9600 \\
0 & -9600 & -7200 & 0 & 9600 & 7200
\end{bmatrix}
\end{array}
$$

$$
[K_a]_{BC} = \begin{array}{c} \\ \text{②} \\ \text{③} \\ \text{④} \\ \text{⑤} \\ \text{⑥} \\ \text{⑦} \end{array}
\begin{array}{cccccc}
\text{②} & \text{③} & \text{④} & \text{⑤} & \text{⑥} & \text{⑦} \\
\begin{bmatrix}
0 & 0 & 0 & 0 & 0 & 0 \\
0 & 40,000 & 0 & 0 & -40,000 & 0 \\
0 & 0 & 0 & 0 & 0 & 0 \\
0 & 0 & 0 & 0 & 0 & 0 \\
0 & -40,000 & 0 & 0 & 40,000 & 0 \\
0 & 0 & 0 & 0 & 0 & 0
\end{bmatrix}
\end{array}
$$

$$
[K_a]_{CD} = \begin{array}{c} \\ \text{⑤} \\ \text{⑥} \\ \text{⑦} \\ \text{⑧} \\ \text{⑧} \\ \text{⑧} \end{array}
\begin{array}{cccccc}
\text{⑤} & \text{⑥} & \text{⑦} & \text{⑧} & \text{⑧} & \text{⑧} \\
\begin{bmatrix}
0 & 0 & 0 & 0 & 0 & 0 \\
0 & 7200 & -9600 & 0 & -7200 & 9600 \\
0 & -9600 & 12,800 & 0 & 9600 & -12,800 \\
0 & 0 & 0 & 0 & 0 & 0 \\
0 & -7200 & 9600 & 0 & 7200 & -9600 \\
0 & 9600 & -12,800 & 0 & -9600 & 12,800
\end{bmatrix}
\end{array}
$$

Step 2: Feed the member stiffness matrices into $[K]$.

Flexural effects:

	①	②	③	④	⑤	⑥	⑦
①	8000	4000	−720	960	0	0	0
②	4000	24,000	−720	−1440	8000	0	2400
③	−720	−720	86.4	−115.2	0	0	0
④	960	−1440	−115.2	633.6	−2400	0	−480
⑤	0	8000	0	−2400	24,000	−960	1680
⑥	0	0	0	0	−960	153.6	115.2
⑦	0	2400	0	−480	1680	115.2	566.4

$[K_f] = $ (rows ① through ⑦)

Axial effects:

	①	②	③	④	⑤	⑥	⑦
①	0	0	0	0	0	0	0
②	0	0	0	0	0	0	0
③	0	0	52,800	9600	0	−40,000	0
④	0	0	9600	7200	0	0	0
⑤	0	0	0	0	0	0	0
⑥	0.	0	−40,000	0	0	47,200	−9600
⑦	0	0	0	0	0	−9600	12,800

$[K_a] = $ (rows ① through ⑦)

Combined effects:

	①	②	③	④	⑤	⑥	⑦
①	8000	4000	−720	960	0	0	0
②	4000	24,000	−720	−1440	8000	0	2400
③	−720	−720	52,886.4	−9484.8	0	−40,000	0
④	960	−1440	−9484.8	7833.6	−2400	0	−480
⑤	0	8000	0	−2400	24,000	−960	1680
⑥	0	0	−40,000	0	−960	47,353.6	−9484.8
⑦	0	2400	0	−480	1680	−9484.8	13,366.4

$[K] = $ (rows ① through ⑦)

15.8.3 Transverse Loads and Equivalent Joint Loads

In the direct stiffness method, the actual frame loading is replaced by a two-stage analysis sequence. In the first stage, all degrees of freedom are held in a restrained position. The second stage consists of releasing the restraint and allowing the restraining forces to act as joint loads. Calculation of these forces necessitates isolating all members and joints of the frame as free-body diagrams. A sequential application of statics to each free body produces the desired forces. Restraining forces are calculated by applying statics to the members and joints.

Example 15.11. Build the load matrix for the frame of Example 15.10.

Solution:
Step 1: A general representation of the restrained state forces is shown in Fig. 15.36c. The load values

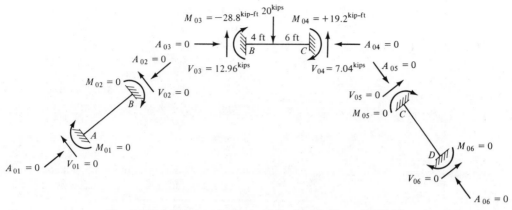

Fig. 15.37. Members in the restrained state, Example 15.11.

in the released state are of equal magnitude but opposite direction (Fig. 15.36d). Analysis of the restrained state begins with the free-body diagrams drawn in Fig. 15.37. With all degrees of freedom restrained, each member behaves as a fixed-end beam. Members AB and CD have no transverse loading, and the end moments, shears, and axial forces have zero value. Member BC is subjected to a 20 kips concentrated load. End moments for this loading are calculated by accessing Appendix A. The results are

$$M_{03} = -28.8 \text{ kip-ft}$$

$$M_{04} = +19.2 \text{ kip-ft}$$

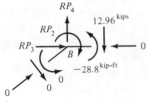

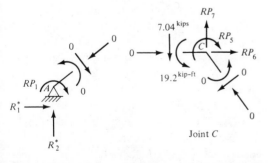

Joint A

Fig. 15.38. Joints in the restrained state, Example 15.11.

End shears are calculated by satisfying moment equilibrium at each end of the member. Results are

$$V_{03} = +12.96 \text{ kips}$$

$$V_{04} = +7.04 \text{ kips}$$

Finally, because both ends are fixed and no longitudinal loads exist, no axial forces are developed and

$$A_{03} = A_{04} = 0$$

Step 2: Figure 15.38 depicts free-body diagrams of the joints that were omitted in Fig. 15.37. Restraining forces are determined by straightforward application of statics:

Joint A:

$$RP_1 = 0$$

Joint B:

$$-RP_2 + (-28.8) = 0 \qquad \therefore RP_2 = -28.8 \text{ kip-ft}$$

$$RP_3 = 0$$

$$-RP_4 + 12.96 = 0 \qquad \therefore RP_4 = +12.96$$

Joint C:

$$-RP_5 + (19.2) = 0 \qquad \therefore RP_5 = +19.2 \text{ kip-ft}$$

$$RP_6 = 0$$

$$-RP_7 + 7.04 = 0 \qquad \therefore RP_7 = +7.04 \text{ kips}$$

The load matrix corresponding to Fig. 15.36d is

$$\{P\} = \begin{Bmatrix} P_1 \\ P_2 \\ P_3 \\ P_4 \\ P_5 \\ P_6 \\ P_7 \end{Bmatrix} = -\begin{Bmatrix} RP_1 \\ RP_2 \\ RP_3 \\ RP_4 \\ RP_5 \\ RP_6 \\ RP_7 \end{Bmatrix}$$

$$= \begin{Bmatrix} 0 \\ +28.8 \\ 0 \\ -12.96 \\ -19.2 \\ 0 \\ -7.04 \end{Bmatrix} = \begin{Bmatrix} 0 \\ 28.8 \text{ kip-ft cw} \\ 0 \\ 12.96 \text{ kips} \downarrow \\ 19.2 \text{ kip-ft ccw} \\ 0 \\ 7.04 \text{ kips} \downarrow \end{Bmatrix}$$

and the entries are shown in Fig. 15.39.

In the Example 15.11, execution of statics to determine the load matrix was straightforward. All axial forces in the restrained state had zero value (this is usually so when axial deformation is included in the analysis). Consequently, load values were obtained by observing equilibrium of each member by itself and then examining equilibrium of each joint.

Replacement of transverse loads by direct forces applied at the degrees of freedom is a feature common to all matrix methods of analysis. A common term used to denote the resulting forces is *equivalent joint loads*. When axial deformation is included in the analysis, calculation of the equivalent joint loads is trivial. Fixed-end forces (moments and shears) are computed for all members and transferred to the joints of the frame. Next, the restraining forces at any given joint are calculated. Normally, these are computed and then changed in

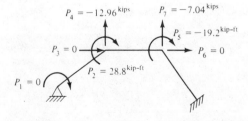

$P_4 = -12.96^{\text{kips}}$ $P_7 = -7.04^{\text{kips}}$
$P_5 = -19.2^{\text{kip-ft}}$
$P_3 = 0$ $P_6 = 0$
$P_2 = 28.8^{\text{kip-ft}}$
$P_1 = 0$

Fig. 15.39.   Joint loads for the released state, Example 15.11.

sign to produce the load matrix $\{P\}$. The end result is a set of forces corresponding in magnitude and direction to the resultant internal forces at the joints themselves. This observation makes computation of the restraining forces and changing their signs an unnecessary exercise. Instead, one can compute the resultant of the member end forces at each joint and directly feed them in appropriate locations in P. If the structure is a continuous beam, axial forces and deformations are not an issue. Nonetheless, the same concept applies when computing equivalent joint levels.

15.8.4   Computation of Internal Forces

Internal forces in a planar frame member consist of the axial force, shear force, and flexural moment at each end. Member forces in the restrained state must be combined with those of the released state to produce the complete results. The former set of forces are readily available as an inherent feature of that analytical stage because the member end forces are needed in the process of building the load matrix. The latter set of forces are not readily available, and several analytical steps are necessary to determine them.

Initially, the member end forces in the released state are determined in the directions of the degrees of freedom designated in Fig. 15.31. These forces, $\{P^*\}$, are related to the displacements experience by the member ends according to

$$\{P^*\}_{m6 \times 1} = [K]_{m6 \times 6} \{X^*\}_{m6 \times 6} \quad (15.53)$$

Subscripts are given to indicate the size of the matrices and to emphasize that $\{X^*\}_m$ is not the full array of joint displacements. Entries in $\{X^*\}_m$ are extracted from the larger array $\{X^*\}$ as needed for the particular member at hand.

Forces obtained by application of Eq. 15.53 do not align with the directions used in establishing $\{F_0\}_m$. Forces in $\{F_0\}_m$ are moments and axial force components at the member ends. $\{P^*\}_m$ contains forces whose directions align with particular degrees of freedom. A subsequent calculation is necessary to convert the forces in $\{P^*\}_m$ to the proper directions.

With the orientation of the member known, this is a matter of simple trigonometry. The adjusted values are entered in the array $\{F*\}_m$ and combined with $\{F_0\}_m$ to yield the final answers. It is of interest to note that if only the end moments are needed, no adjustment to the moment values extracted from $\{P*\}_m$ is necessary because moment values are not coordinate-dependent.

Example 15.12. Compute the member end moments for the frame in Fig. 15.36a. Use the results of Examples 15.10 and 15.11.

Solution:
Step 1: Joint displacements are calculated by using $[K]$ from Ex. 12.10 and $\{P\}$ from Example 15.11. Thus,

$$\{X\} = \{X*\} = [K]^{-1}\{P\}$$

$$
\begin{Bmatrix} X_1^* \\ X_2^* \\ X_3^* \\ X_4^* \\ X_5^* \\ X_6^* \\ X_7^* \end{Bmatrix} =
\begin{bmatrix}
8000 & 4000 & -720 & 960 & 0 & 0 \\
4000 & 24{,}000 & -720 & -1440 & 0 & 2400 \\
-720 & -720 & 52{,}866.4 & 9484.8 & -40{,}000 & 0 \\
960 & -1440 & 9484.8 & 0 & 0 & -480 \\
0 & 8000 & 0 & -2400 & -960 & 1680 \\
0 & 0 & -40{,}000 & -960 & 47{,}353.8 & -9484.8 \\
0 & 2400 & 0 & 1680 & -9484.8 & 13{,}366.4
\end{bmatrix}^{-1}
\begin{Bmatrix} 0 \\ 28.8 \\ 0 \\ -12.96 \\ -19.2 \\ 0 \\ -7.04 \end{Bmatrix}
$$

$$
= \begin{Bmatrix}
0.001094 \text{ rad} \\
0.001020 \text{ rad} \\
0.005778 \text{ ft} \\
-0.009032 \text{ ft} \\
-0.002042 \text{ rad} \\
0.005459 \text{ ft} \\
0.003096 \text{ ft}
\end{Bmatrix}
$$

Step 2: End forces are determined from the member stiffness relationships. In Example 15.10, $[K]_m$ was separated into flexural and axial components, and these must be combined prior to the succeeding calculations.

Member AB:

$$\{P*\}_{AB} = [K]_{AB}\{X*\}_{AB}$$

$$
\begin{Bmatrix} P_1^* \\ P_2^* \\ P_3^* \\ P_4^* \\ P_5^* \\ P_6^* \end{Bmatrix} =
\begin{array}{c} (1) \\ (R) \\ (R) \\ (2) \\ (3) \\ (4) \end{array}
\begin{bmatrix}
8000 & 720 & 720 & 4000 & -720 & 960 \\
720 & 12{,}886.4 & 12{,}886.4 & 720 & -12{,}886.4 & -9484.8 \\
-960 & 9484.8 & 9484.8 & -960 & -9484.8 & -7353.6 \\
4000 & 720 & 720 & 8000 & -720 & 960 \\
-720 & -12{,}886.4 & -12{,}886.4 & -720 & 12{,}886.4 & 9484.8 \\
960 & -9484.8 & -9484.8 & 960 & 9484.8 & 7353.6
\end{bmatrix}
\begin{Bmatrix} 0.001094 \\ 0 \\ 0 \\ 0.001020 \\ 0.005778 \\ -0.009032 \end{Bmatrix}
$$

$$
= \begin{Bmatrix}
0.00 \text{ ft-kips} \\
12.73 \text{ kips} \\
9.58 \text{ kips} \\
-0.29 \text{ ft-kips} \\
-12.73 \text{ kips} \\
-9.58 \text{ kips}
\end{Bmatrix}
$$

where the top of the matrix columns are labeled (1) (R) (R) (2) (3) (4).

Note that displacement values are set equal to zero at points of support.
Member BC:

$$\{P^*\}_{BC} = [K]_{BC}\{X^*\}_{BC}$$

$$
\begin{Bmatrix} P_1^* \\ P_2^* \\ P_3^* \\ P_4^* \\ P_5^* \\ P_6^* \end{Bmatrix}_{BC}
=
\begin{array}{c}
② \\ ③ \\ ④ \\ ⑤ \\ ⑥ \\ ⑦
\end{array}
\begin{bmatrix}
16{,}000 & 0 & -2400 & 8000 & 0 & 2400 \\
0 & 40{,}000 & 0 & 0 & -40{,}000 & 0 \\
-2400 & 0 & 480 & -2400 & 0 & -480 \\
8000 & 0 & -2400 & 16{,}000 & 0 & 2400 \\
0 & -40{,}000 & 0 & 0 & 40{,}000 & 0 \\
2400 & 0 & -480 & 2400 & 0 & 480
\end{bmatrix}
\begin{Bmatrix}
0.001020 \\ 0.005778 \\ -0.009032 \\ -0.002042 \\ 0.005459 \\ 0.003096
\end{Bmatrix}
$$

(column headers: ② ③ ④ ⑤ ⑥ ⑦)

$$
=
\begin{Bmatrix}
29.09 \text{ ft-kips} \\
12.76 \text{ kips} \\
-3.36 \text{ kips} \\
4.60 \text{ ft-kips} \\
-12.76 \text{ kips} \\
3.36 \text{ kips}
\end{Bmatrix}
$$

Member CD:

$$
\begin{Bmatrix} P_1^* \\ P_2^* \\ P_3^* \\ P_4^* \\ P_5^* \\ P_6^* \end{Bmatrix}_{CD}
=
\begin{array}{c}
⑤ \\ ⑥ \\ ⑦ \\ Ⓡ \\ Ⓡ \\ Ⓡ
\end{array}
\begin{bmatrix}
8000 & 960 & 720 & 4000 & -960 & -720 \\
960 & 7353.6 & -9484.8 & 960 & -7353.6 & 9484.8 \\
720 & -9484.8 & 12{,}886.4 & 720 & 9484.8 & -12{,}886.4 \\
4000 & 960 & 720 & 8000 & -960 & -720 \\
-960 & -7353.6 & 9484.8 & -960 & 7353.6 & -9484.8 \\
-720 & 9484.8 & -12{,}886.4 & -720 & -9484.8 & 12{,}886.4
\end{bmatrix}
\begin{Bmatrix}
-0.002042 \\ 0.005459 \\ 0.003096 \\ 0 \\ 0 \\ 0
\end{Bmatrix}
$$

(column headers: ⑤ ⑥ ⑦ Ⓡ Ⓡ Ⓡ)

$$
=
\begin{Bmatrix}
-23.81 \text{ ft-kips} \\
12.73 \text{ kips} \\
-10.41 \text{ kips} \\
-15.64 \text{ ft-kips} \\
-12.73 \text{ kips} \\
10.41 \text{ kips}
\end{Bmatrix}
$$

Step 3: Final end moments are obtained by combining the restrained forces (Example 15.10) with the results of Step 2.

$$
\begin{Bmatrix} M_1 \\ M_2 \\ M_3 \\ M_4 \\ M_5 \\ M_6 \end{Bmatrix}
=
\begin{Bmatrix}
0 \\ 0 \\ -28.8 \\ 19.2 \\ 0 \\ 0
\end{Bmatrix}
+
\begin{Bmatrix}
0 \\ -0.29 \\ 29.09 \\ 4.60 \\ -23.81 \\ -15.64
\end{Bmatrix}
=
\begin{Bmatrix}
0 \\ -0.29 \\ +0.29 \\ 23.80 \\ -23.81 \\ -15.64
\end{Bmatrix}
=
\begin{Bmatrix}
0 \\ 0.29 \text{ ccw} \\ 0.29 \text{ cw} \\ 23.80 \text{ cw} \\ 23.81 \text{ ccw} \\ 15.64 \text{ ccw}
\end{Bmatrix}
\text{ ft-kips}
$$

End axial and shear forces can be combined in similar fashion, except the end forces calculated in step 2 are horizontal and vertical components. These latter forces would usually have to be converted to axial and shear forces before the addition can be performed. As an alternative direct application of statics to free-body diagrams of the members and joints of the frame is possible.

Note: The influence of including axial deformation can be observed. This is partially done in the following. Results for a greatly increased axial stiffness ($EA = 2,000,000$ kips) are included to simulate the elimination of axial deformation. Although not developed in this textbook, the exact answers for a solution without axial deformation are also shown.

All listed rotations and end moments are in rad and kip-ft, respectively.

8.5 COMPUTER PROGRAMS

Several computer algorithms for executing the direct stiffness method are presented in Appendix C. Programs BEAM.BAS, TRUSS.BAS, and FRAME.BAS solve beam, planar truss, and planar frame problems, respectively, in BASIC. Programs BEAM.F, TRUSS.F and FRAME.F are counterparts in FORTRAN. These are useful in some of the PROBLEMS exercises.

	$EA = 200,000$ kips	$EA = 200,000,000$ kips	$EA = \infty$
X_1	0.001094	0.000941	0.000924
X_2	0.001020	0.000992	0.000989
X_3	-0.002042	-0.001966	-0.001957
M_1	0.00	0.00	0.00
M_2	-0.29	0.20	0.26
M_3	0.29	-0.20	-0.26
M_4	23.80	24.13	24.17
M_5	-23.81	-24.13	-24.17
M_6	-15.64	-16.26	-16.34

PROBLEMS

15-1. Generate the applied forces necessary to create the member compatibility modes shown in a and b. Assume EI is constant, and see Section 15.3.1.

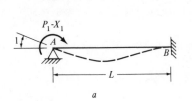

a

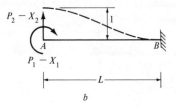

b

15-2. Generate each continuous beam by its physical meaning (see Examples 15.1 and 15.2). Use the given numbering for the degrees of freedom.

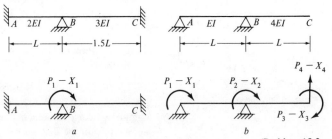

a

b

(Problem 15-2 continued on p. 432)

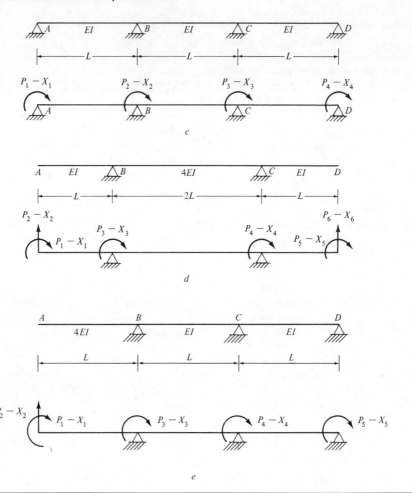

c

d

e

15-3. Generate $[K]$ for the given beams by its physical meaning.

15-4. Generate $[K]$ for the given beam by its physical meaning.

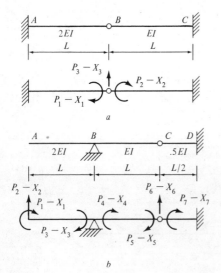

a

b

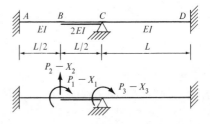

15-5. Do Problem 15-2 by the direct stiffness feeding method.

15-6. Do Problem 15-3 by the direct stiffness feeding method.

15-7. Perform a complete structural analysis of the given beams using the direct stiffness method. Draw complete shear and moment diagrams. Use the given numbering for the degrees of freedom.

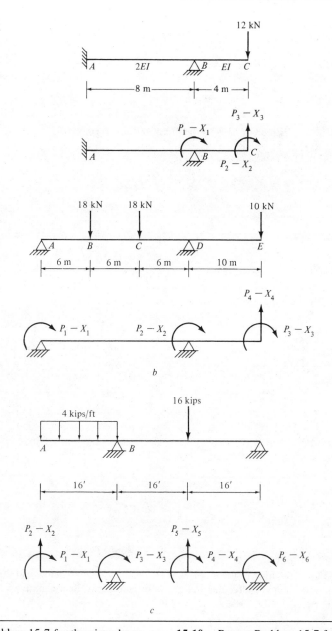

12 kN

$$A \quad 2EI \quad B \quad EI \quad C$$

|← 8 m →|← 4 m →|

$P_3 - X_3$

$P_1 - X_1$

$A \qquad B \qquad C$

$P_2 - X_2$

18 kN   18 kN   10 kN

$A \quad B \quad C \quad D \quad E$

|← 6 m →|← 6 m →|← 6 m →|← 10 m →|

$P_4 - X_4$

$P_1 - X_1 \qquad P_2 - X_2 \qquad\qquad P_3 - X_3$

b

16 kips

4 kips/ft

$A \qquad\qquad B$

|← 16' →|← 16' →|← 16' →|

$P_2 - X_2 \qquad\qquad P_5 - X_5$

$P_1 - X_1 \quad P_3 - X_3 \quad P_4 - X_4 \quad P_6 - X_6$

c

15-8. Repeat Problem 15-7 for the given beam.   **15-10.** Repeat Problem 15-7 for the given beam.

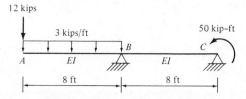

12 kips

3 kips/ft   50 kip-ft

$A \qquad EI \qquad B \qquad EI \qquad C$

|← 8 ft →|← 8 ft →|

2 kips/ft

$A \qquad\qquad B$

$4EI \qquad\qquad EI \qquad C$

|← 20 ft →|← 20 ft →|

15-9. Analyze the beam in Problem 12-4 by the direct stiffness method. Use $w = 2$ kip/ft and $L = 12$ ft.

15-11. Repeat Problem 15-7 for the given beam.

(Problem 15.11 continued on p. 434)

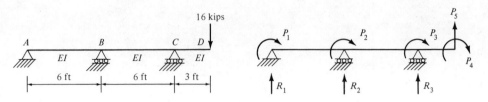

15-12.   Analyze the beam in Problem 13-2 by the direct stiffness method.

15-13.   Repeat Problem 15-7 for the given beam.

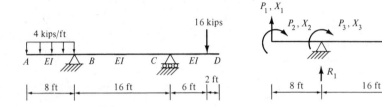

15-14.   Repeat Problem 15-7 for the given beams.

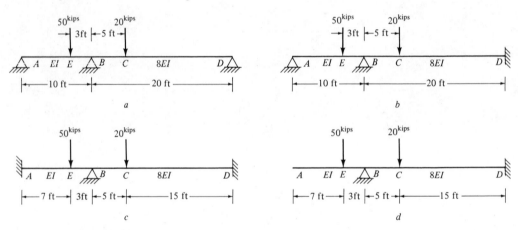

15-15.   Analyze the beams in Problem 13-7 by the direct stiffness method.

15-16.   Determine what proportion of the moment M is resisted by each member end that frames into joint B. Use the direct stiffness method. Solve the problem in general terms, and then give numerical values for each of the following cases.

a. $(EI)_1 = (EI)_2$ and $L_1 = L_2$
b. $(EI)_1 = 3(EI)_2$ and $L_1 = L_2$
c. $(EI)_1 = 3(EI)_2$ and $L_1 = 0.5L_2$

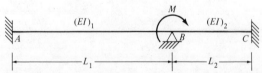

15-17.   Build the structure stiffness matrix for the given trusses by use of its physical meaning. Use $EA = 24,000$ kips for all members. Number the degrees of freedom and members as shown in the accompanying figure.

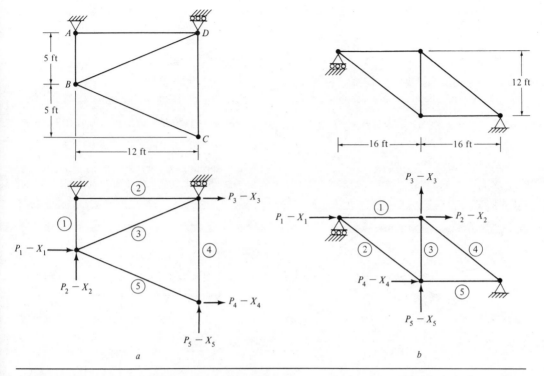

a

b

15-18. Build the structure stiffness matrix for the given truss by use of its physical meaning. Use EA = 20,000 kips and the indicated numbers for the members and degrees of freedom. Member areas (in.^{2}) are shown in parentheses.

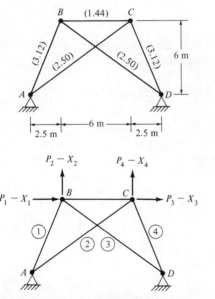

15-19. Repeat Problem 15-18 for the given truss. Member areas (in^2) are given in parentheses. Use EA = 200 × 10^3 kip.

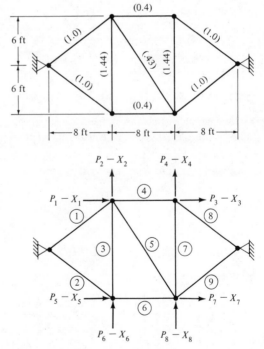

15-20. Build $[K]$ for each planar truss in Problem 14-1 by use of its physical meaning.

15-21. Do Problem 15-17 by use of the direct stiffness feeding method.

15-22. Do Problem 15-18 by use of the direct stiffness feeding method.

15-23. Do Problem 15-19 by use of the direct stiffness feeding method.

15-24. A matrix analysis of the truss in Problem 15-17a for a particular loading (not shown) produces the results:

$$X_1 = 0.0512 \text{ in.} \qquad X_4 = -0.0075 \text{ in.}$$

$$X_2 = 0.0912 \text{ in.} \qquad X_5 = 0.0010 \text{ in.}$$

$$X_3 = -0.1100 \text{ in.}$$

Determine the member forces corresponding to the given joint displacements. Also calculate the joint loads that create the given displacements.

15-25. Repeat Problem 15-24 for the truss in Problem 15-17b and the values:

$$X_1 = 0.0551 \text{ in.} \qquad X_4 = -0.0011 \text{ in.}$$

$$X_2 = 0.1150 \text{ in.} \qquad X_5 = -0.1020 \text{ in.}$$

$$X_3 = -0.1171 \text{ in.}$$

15-26. Repeat Example 15-8 for the given trusses. Use $A = 36 \text{ cm}^2$ and $E = 200 \times 10^3$ MPa for all members.

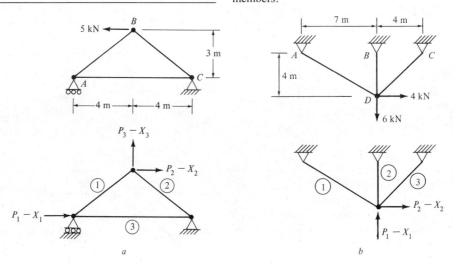

15-27. Analyze the planar trusses in Problem 14-2 by the direct stiffness method.

15-28. Build the structures stiffness matrix for each frame by the direct stiffness method. Use $EI = 20,000$ kip-ft^2 and $EA = 20,000$ kips. Number the degrees of freedom in the order shown.

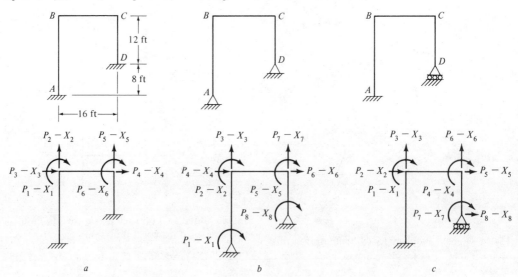

15-29. Repeat Problem 15-28 for each given planar frames. Use $EI = 200 \times 10^3$ kN · m^2 and $EA =$ 8000 kN. Use the same member lengths in each frame.

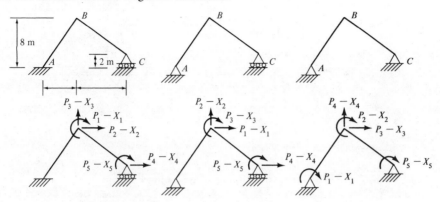

15-30. Repeat Problem 15-28 for the given frame. Use the same member properties.

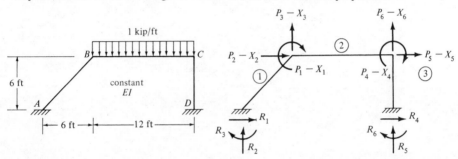

15-31. Analyze the given frame by the direct stiffness method. Number the degrees of freedom in the order shown. Use $EI = 200 \times 10^2$ kip-ft^2 and $EA = 2000$ kips for all members.

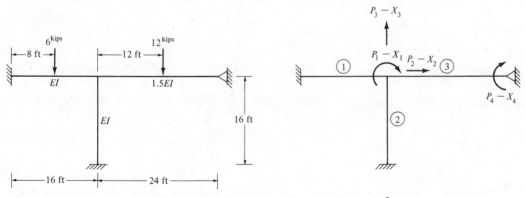

15-32. Repeat Problem 15-31 for the given frame. Use $EI = 20,000$ kN · m^2 and $EA = 2000$ kN.

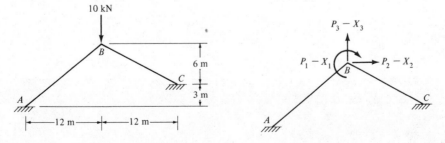

15-33.   Repeat Problem 15-31 for the given frame. Use $EI = 24{,}000$ kip-ft^2 and $EA = 24{,}000$ kips for all members.

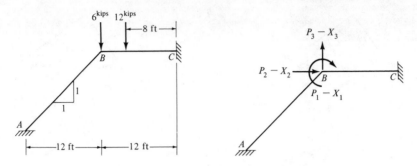

15-34.   Reanalyze the frame in Problem 15-33 for the given loadings.

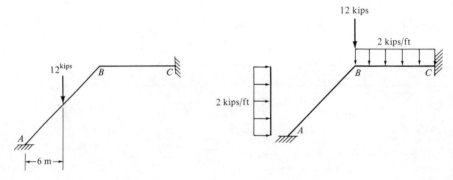

15-35.   Determine the structure stiffness matrix and the load matrix for the given planar frames. Use the direct stiffness method and the indicated numbering for the degrees of freedom. Both members have $EI = 20{,}000$ kN · m^2 and $EA = 20{,}000$ kN.

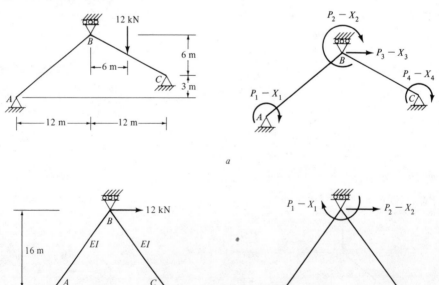

a

b

15-36. Repeat Problem 15-35 for the given planar frames. Use $EI = 20,000$ kip-ft^2 and $EA = 20000$ kips.

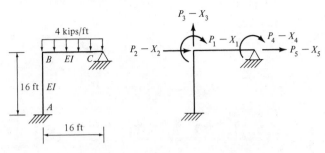

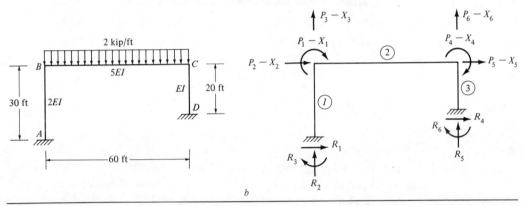

15-37. Analyze the frame given in Problem 15-33 by the direct stiffness method. The loading is a uniform downward load of 1 kip/ft on member BC. Use $[K]$ developed in Problem 15-33.

15-38. Explain the physical meaning of a column of $[K]^{-1}$.

15-39. Explain why the contents of any column of the matrix in Eq. 15.45 sum to zero. Also explain why the same fact is true for the contents of any row. Why don't these conditions exist in Eq. 15.49.

15-40. Can a local member stiffness matrix (e.g., Eq. 15.45 or Eq. 15.49) be inverted? Give a mathematical reason for your answer and a physical interpretation.

15-41. Is it possible to derive a structure stiffness matrix for a planar frame by the operation.

$$[K] = [E_p][S][C]$$

when axial deformation is not neglected? If so, do this for the frame in Problem 15-28a. If not, explain why.

15-42. What happens when you attempt to solve a planar frame problem by using Eq. 15.51 for the member stiffness matrix, and you exclude the terms in Eq. 15.52.

15-43. Calculate the fixed-end moments for the given beam. Use matrix stiffness analysis. Hint: Use the degrees of freedom shown.

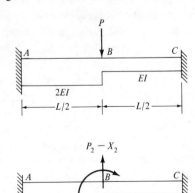

16
Transformation Matrices for Planar Structures

16.1 INTRODUCTION

Chapter 14 presented the basic matrix analysis method as applied to planar trusses. The term *basic* reflects the use of separate matrices to represent each of the three fundamental criteria for structural analysis. Employment of these concepts requires separate consideration of equilibrium, material laws, and compatibility of the complete structure. Although the concepts involved are general and applicable to all types of structures, the method has inherent drawbacks.

In the execution of the basic matrix analysis method, the analyst is required to make a careful examination of the particular structure of interest. No two structures (even of the same type) are treated in exactly the same fashion. In other words, the numerical procedures followed in building $[E_p]$, $[S]$, and $[C]$ is "problem dependent." Prior to performing any computer operations, some manual computations are necessary in the development of the numerical input. This necessity eliminates the possibility of using automated computer procedures and commits the engineer to a period of preparatory computations. Consequently, the basic matrix analysis method is impractical for everyday usage. Its primary merit lies in the direct correlation of specific matrices with each

of the three structural analysis criteria and its usefulness in verifying solutions obtained by classical analysis methods. Also, the direct use of phsyical concepts is essential in developing the beginning student's "feel" for structural behavior.

Some of the tedious features of the basic matrix approach were removed in the treatment presented in Chapter 15. First, it was shown that a stiffness matrix could be constructed by consideration of its physical meaning. Each column of $[K]$ has as its counterpart a set of equilibrating loads that exist in a particular structure-compatibility mode. By directly computing these loads and building each column of $[K]$ one at a time, the matrix operations

$$[K] = [E_p][S][C]$$

could be avoided entirely. However, this process does not remove the tedious task of performing a time-consuming statics analysis of each required compatibility mode. In addition, the determination of internal member forces and deformations requires additional manual computations, which is an inconvenience.

As a second major topic in Chapter 15, the direct stiffness method was introduced and featured a member-by-member approach. Basically, the mathematical representation of the structure was formed in a manner that models the actual physical erection of a structure. Physically, each member is placed in the system one at a time and connected to the appropriate joints. Analytically, the structure stiffness matrix was obtained by assembling member stiffness matrices and directly feeding their contents into the compartments of $[K]$. The specific locations in $[K]$ that assimilate the elements of a particular member stiffness matrix are determined by the connectivity of the member, i.e., the correlation between the degrees of freedom of the structure and those of the individual member. This process of assembling $[K]$ is advantageous because it eliminates the need to examine the complete structure "as a whole." Indeed, with a simple description of the member properties, geometry, and connectivity, the direct feeding can be performed automatically by the computer. Furthermore, the geometric complexity of any particular structure poses no complications and does not compromise the direct feeding process.

The direct stiffness method is the contemporary approach to matrix structural analysis (and to the finite-element method, which is used in the analysis of continua). In the current state of the art, many fully automated computer routines are available on the commercial market. These packaged programs permit rapid, efficient analysis of very large and complex structures with a minimal amount of computer input. The intent of the present chapter is to give an elementary introduction to techniques that are fundamental to most automated computer programs. Principles and techniques that underlie automated procedures are emphasized, and little attention is directed toward the formulation of specific computer algorithms. An adequate basis is developed to allow the competent reader to produce such programs. However, it is advisable to reference other textbooks that concentrate solely on matrix methods of structural analysis. Many are available in the general literature. A complete and detailed treatment of the direct stiffness method and the "nuts and bolts" of developing pol-

ished computer programs is subject matter for advanced study.

16.2 MEMBER STIFFNESS MATRICES

For clarity of development, it is advantageous to begin by limiting the discussion to the study of planar structures. The planar frame is the most general type in this category of structures and is examined first. Planar trusses and continuous beams are simpler types and are subsequently shown to be solvable by a degeneration of the planar frame procedure.

16.2.1 Structure Coordinates vs. Member Coordinates

A source of difficulty in the analysis of structures (whether by classical or matrix methods) is the inherently complex geometry of even simple structures. Geometrical barriers are most evident in the quantification of the equilibrium and compatibility conditions. If the structure is viewed as a whole, these relationships reflect the combined interaction of the many members that comprise the structure. Prediction of the response of the complete structure is the objective of structural analysis. Conventionally, this is done by examining the full structure first and subsequently investigating the component members. Cast in this sequence, structural analysis is an imposing proposition. It is preferable (and highly advantageous) to view structural analysis in the reverse order—behavior of the individual member followed by a consideration of its effect on the overall structure. The direct stiffness method of matrix analysis exemplifies this latter approach.

The geometry of a structure is normally described in terms of orthogonal Cartesian coordinates such as shown in Fig. 16.1a. Positions of the joints are specified by X-Y coordinates, and motion of any point in the structure is described by its two components of motion in the X-Y directions. Because a common origin and coordinate axes are used for all members, this system of reference is termed the *structure coordinate system*. Computation of the displaced shape (motion in the X-Y directions of each

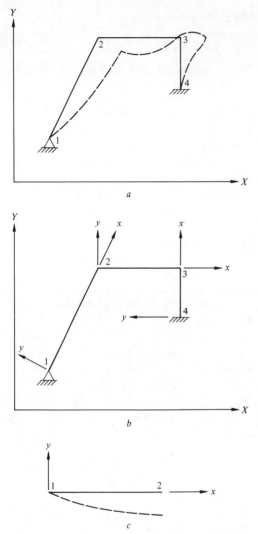

Fig. 16.1. Structure vs. member coordinates. *a*, structure coordinate system; *b*, member coordinate systems; *c*, member coordinates for member 1-2.

dinate system, and thus each is also called a *member coordinate system*. The member coordinate system is better viewed by reorienting the member, as shown (Fig. 16.1c) for member 1-2. Visually, it would appear far easier to describe the deformation of a member at the local level than at the structure level shown in Fig. 16.1a. This is indeed the case and is the basic reason for defining such coordinates. In some textbooks, the member coordinate system is referred to as the *local coordinate system*. Also, the terminology *global member stiffness matrix* was introduced in Chapter 15. The terminology implies a member stiffness relationship developed in the structure coordinate system. The descriptions *structure* and *global* are interchangeable, as are the descriptors *member* and *local*. The choice of the terminology is simply a matter of personal preference. The interrelationship between the two basic coordinates systems is the important issue, and it is discussed in the following section.

16.2.2   Transformation of Coordinates

Two types of vector quantities are important in structural analysis, those representing forces (internal or external) and those representing displacements (internal or external). In general, external loads can be applied in any direction at any point in the structure (Fig. 16.2) relative to the structure coordinates. Internal forces, instead, are usually specified in particular directions with respect to the member coordinates; i.e., axial and shear force are, respectively, coincident with and perpendicular

point *relative* to its original position) is an important task in structural analysis. As shown in previous chapters, the determination of this displaced position (broken line) is a tedious effort by classical analysis methods.

Consider the alternative depicted in Fig. 16.1b. As described in Chapter 10, it is possible to specify the displacements of a point on any member in terms of a local *x-y* coordinate system. Specifically, the *x*-axis aligns with the member and originates at one of its ends. In this way, each member has its own *x-y* coor-

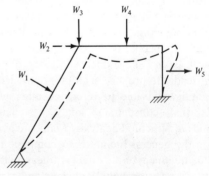

Fig. 16.2.   Loaded planar frame.

to the member x-axis. The same inconsistency is true of external displacements (usually specified as movements in the directions of the structure coordinate axes) and the internal deformations, (usually specified as movements in the directions of the member coordinate axes). To facilitate the mathematics, it is desirable to describe all vector quantities with respect to a common set of reference axes. Figure 16.3 will be used to derive a general method for accomplishing a transformation to common reference coordinates.

In structure coordinates, any load can be prescribed in terms of two orthogonal components, P_1 and P_2, acting in the directions of the structure X and Y coordinates, respectively. The equivalent components F_1 and F_2 in the member x and y coordinate directions, respectively, are obtained (see Fig. 16.3a) as

$$F_1 = P_1 \cos \theta + P_2 \sin \theta$$

$$F_2 = -P_1 \sin \theta + P_2 \cos \theta \quad (16.1)$$

and act as shown in Fig. 16.3c. Angle θ is the inclination of the member x-axis with respect to the structure X-axis. The matrix version of

Eq. 16.1 is

$$\begin{Bmatrix} F_1 \\ F_2 \end{Bmatrix} = \begin{bmatrix} \cos \theta & \sin \theta \\ -\sin \theta & \cos \theta \end{bmatrix} \begin{Bmatrix} P_1 \\ P_2 \end{Bmatrix} \quad (16.2)$$

or, in abbreviated form,

$$\{F\}_m = [T']\{P\}_m \quad (16.3)$$

in which

$$[T'] = \begin{bmatrix} \cos \theta & \sin \theta \\ -\sin \theta & \cos \theta \end{bmatrix} \quad (16.4)$$

is a *transformation matrix*.

A similar discussion can be made regarding the transformation from displacement components, X_1 and X_2, in the structure coordinate directions to components, Δ_1 and Δ_2, in the member coordinate directions. Examination of Fig. 16.3b gives the resultant displacements (Fig. 16.3d) as

$$\begin{Bmatrix} \Delta_1 \\ \Delta_2 \end{Bmatrix} = \begin{bmatrix} \cos \theta & \sin \theta \\ -\sin \theta & \cos \theta \end{bmatrix} \begin{Bmatrix} X_1 \\ X_2 \end{Bmatrix} \quad (16.5)$$

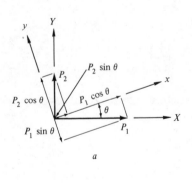

a

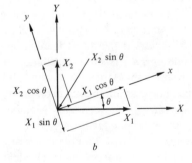

b

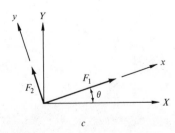

c

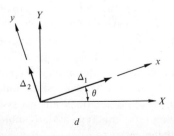

d

Fig. 16.3.   Transformation of forces and displacements. *a*, forces; *b*, displacements; *c*, resultant forces; *d*, resultant displacements.

or, in abbreviated form,

$$\{\Delta\}_m = [T']\{X\}_m \qquad (16.6)$$

Equations 16.3 and 16.6 are basic transformation relationships and are key features of semi-automated matrix stiffness analysis. Employment of these relationships is described in the remainder of this chapter.

16.2.3 Member Stiffness Matrices in Global Coordinates

The concept of a local member stiffness matrix $[S]_m$ was introduced earlier. In Chapter 15, $[S]_m$ matrices were developed for individual members in continuous beams. In that application, different $[S]_m$ matrices are used to accommodate the various end conditions that are common for any individual member. It was also shown that a single member stiffness relationship could be used to accommodate all three possibilities provided that the deformations were measured *relative to the displaced chord of the member*.

Member stiffness matrices in global coordinates have also been employed. Chapter 15 addressed this concept within the development of the direct stiffness method. Use of the member stiffness matrix in structure coordinates $[K]_m$ is an essential feature if the direct feeding approach to building $[K]$ is to be feasible. For continuous beams, the structure X-axis and member x-axis normally align with each other, and $[S]_m$ and $[K]_m$ are identical. Member stiffnesses matrices oriented to the structure coordinate systems were derived for planar trusses and planar frames by use of their physical meaning. In the case of the planar frames, the inclusion of axial deformations was shown to be a prerequisite. Generation of member stiffness matrices in global coordinates for grids and space structures is possible by an extension of the physical concepts employed for planar structures. However, tedious three-dimensional geometry considerations are necessary and are a deterrent to this type of derivation. It is preferable and easier to introduce the use of coordinate transformation matrices represented

by Eqs. 16.3 and 16.6 into the matrix analysis procedures. This chapter describes the process for the various planar structures. Extension to space structures is a topic for advanced study.

16.2.3.1 Planar Frame Member

A general planar frame member develops the internal end forces and deformations shown in Fig. 16.4a. The relationships between the forces, F_1–F_6, and the deformations, Δ_1–Δ_6, constitutes the local member stiffness matrix, $[S]_m$. Six basic compatibility modes exist for the planar frame member, one for each end deformation quantity. Internal forces corresponding to each mode are displayed in Fig. 16.4b. These forces are the entries in $[S]_m$ and are calculated by the principles used in developing Eq. 15.37 (in this case, the member inclination angle, θ, is zero). The complete matrix relationship is

$$\{F\}_m = [S]_m\{\Delta\}_m \qquad (16.7)$$

in which

$$\{F\}_m = \begin{Bmatrix} F_1 \\ F_2 \\ F_3 \\ F_4 \\ F_5 \\ F_6 \end{Bmatrix} \quad \{\Delta\}_m = \begin{Bmatrix} \Delta_1 \\ \Delta_2 \\ \Delta_3 \\ \Delta_4 \\ \Delta_5 \\ \Delta_6 \end{Bmatrix} \qquad (16.8)$$

and

$$[S]_m = \begin{bmatrix} \dfrac{4EI}{L} & 0 & -\dfrac{6EI}{L^2} & \dfrac{2EI}{L} & 0 & \dfrac{6EI}{L^2} \\[2ex] 0 & \dfrac{EA}{L} & 0 & 0 & -\dfrac{EA}{L} & 0 \\[2ex] -\dfrac{6EI}{L^2} & 0 & \dfrac{12EI}{L^3} & -\dfrac{6EI}{L^2} & 0 & -\dfrac{12EI}{L^3} \\[2ex] \dfrac{2EI}{L} & 0 & -\dfrac{6EI}{L^2} & \dfrac{4EI}{L} & 0 & \dfrac{6EI}{L^2} \\[2ex] 0 & -\dfrac{EA}{L} & 0 & 0 & \dfrac{EA}{L} & 0 \\[2ex] \dfrac{6EI}{L^2} & 0 & -\dfrac{12EI}{L^3} & \dfrac{6EI}{L^2} & 0 & \dfrac{12EI}{L^2} \end{bmatrix} \qquad (16.9)$$

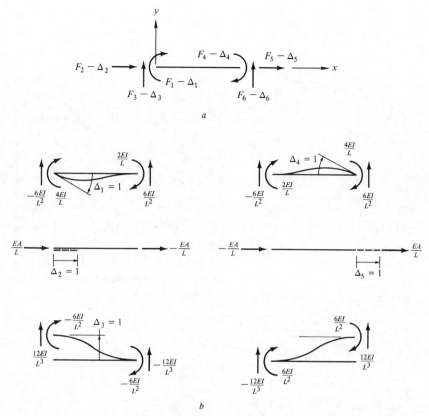

Fig. 16.4. Planar frame member stiffness in member coordinates. *a*, member; *b*, compatibility modes.

In many frameworks, most of the members are not oriented in the horizontal position shown in Fig. 16.4a. Consider a general member (Fig. 16.5), which is inclined at an angle θ relative to the X-axis of the structure coordinate system. To accommodate direct feeding into the

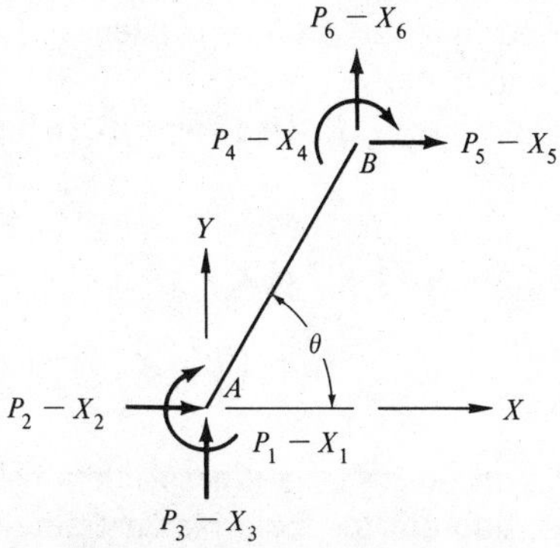

Fig. 16.5. Planar frame member in structure coordinates.

structure stiffness matrix, the member stiffness matrix in structure coordinates (global stiffness matrix) $[K]_m$ must be known. Consequently, Eq. 16.7 by itself is of no direct use. Development of $[K]_m$ for a planar frame member by use of its physical meaning was described in Section 15.8.1. An alternative derivation based upon the use of coordinate transformations is now presented. The value of this alternate approach lies in its general application to all structural types, including continua.

The member stiffness relationship in local coordinates is obtained by use of Eq. 16.7 and transformation laws. Forces and displacements at both ends of the general member must be transformed to member coordinates. Figure 16.6 shows the member end forces (F_2, F_3, F_5, and F_6) in member coordinates and their counterparts (P_2, P_3, P_5, and P_6) in structure coordinates. The corresponding displacement quantities are also shown. It is evident that the

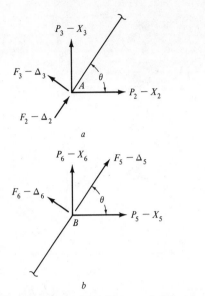

Fig. 16.6. Transformation of forces at the ends of member AB.

local quantities are related to the global quantities according to

$$\begin{Bmatrix} F_2 \\ F_3 \\ F_5 \\ F_6 \end{Bmatrix} = \begin{bmatrix} \cos\theta & \sin\theta & 0 & 0 \\ -\sin\theta & \cos\theta & 0 & 0 \\ 0 & 0 & \cos\theta & \sin\theta \\ 0 & 0 & -\sin\theta & \cos\theta \end{bmatrix} \begin{Bmatrix} P_2 \\ P_3 \\ P_5 \\ P_6 \end{Bmatrix}$$

$$(16.10)$$

and

$$\begin{Bmatrix} \Delta_2 \\ \Delta_3 \\ \Delta_5 \\ \Delta_6 \end{Bmatrix} = \begin{bmatrix} \cos\theta & \sin\theta & 0 & 0 \\ -\sin\theta & \cos\theta & 0 & 0 \\ 0 & 0 & \cos\theta & \sin\theta \\ 0 & 0 & -\sin\theta & \cos\theta \end{bmatrix} \begin{Bmatrix} X_2 \\ X_3 \\ X_5 \\ X_6 \end{Bmatrix}$$

$$(16.11)$$

When the end moments, F_1 and F_4, are included, Eq. 16.10 expands to

$$\begin{Bmatrix} F_1 \\ F_2 \\ F_3 \\ F_4 \\ F_5 \\ F_6 \end{Bmatrix} = \begin{bmatrix} 1 & 0 & 0 & 0 & 0 & 0 \\ 0 & \cos\theta & \sin\theta & 0 & 0 & 0 \\ 0 & -\sin\theta & \cos\theta & 0 & 0 & 0 \\ 0 & 0 & 0 & 1 & 0 & 0 \\ 0 & 0 & 0 & 0 & \cos\theta & \sin\theta \\ 0 & 0 & 0 & 0 & -\sin\theta & \cos\theta \end{bmatrix} \begin{Bmatrix} P_1 \\ P_2 \\ P_3 \\ P_4 \\ P_5 \\ P_6 \end{Bmatrix}$$

$$(16.12)$$

or, in symbolic form,

$$\{F\}_m = [T]\{P\}_m \qquad (16.13)$$

in which

$$[T] = \begin{bmatrix} 1 & 0 & 0 & 0 & 0 & 0 \\ 0 & & & 0 & 0 & 0 \\ 0 & [T'] & & 0 & 0 & 0 \\ 0 & 0 & 0 & 1 & 0 & 0 \\ 0 & 0 & 0 & 0 & & \\ 0 & 0 & 0 & 0 & [T'] & \end{bmatrix} \qquad (16.14)$$

and $[T']$ is given by Eq. 16.4. When end rotations Δ_1 and Δ_4 are included, Eq. 13.11 expands to

$$\begin{Bmatrix} \Delta_1 \\ \Delta_2 \\ \Delta_3 \\ \Delta_4 \\ \Delta_5 \\ \Delta_6 \end{Bmatrix} = [T] \begin{Bmatrix} X_1 \\ X_2 \\ X_3 \\ X_4 \\ X_5 \\ X_6 \end{Bmatrix} \qquad (16.15)$$

or, symbolically,

$$\{\Delta\}_m = [T]\{X\}_m \qquad (16.16)$$

Unit values in two diagonal locations within $[T]$ are in recognition of the fact that the magnitude and direction of the member end moments and rotations are independent of the reference coordinates.

Substitution of Eqs. 16.13 and 16.16 into Eq. 13.7 yields

$$[T]\{P\}_m = [S]_m[T]\{X\}_m$$

$$\{P\}_m = [T]^{-1}[S]_m[T]\{X\}_m \qquad (16.17)$$

which is the member stiffness relationship in structure coordinates for a planar frame member. The member stiffness matrix is extracted as

$$[K]_m = [T]^{-1}[S]_m[T] \quad (16.18)$$

It is trivial to show that

$$[T]^T[T] = [I]$$

and thus

$$[T]^{-1} = [T]^T \quad (16.19)$$

Thus, the final form of $[K]_m$ is obtained as

$$[K]_m = [T]^T[S]_m[T] \quad (16.20)$$

Execution of Eq. 16.20 in general terms for a planar frame member yields the member stiffness matrix given earlier as Eq. 15.37.

The merit of Eq. 16.20 is its general form. With appropriate changes in the content of $[T]$ and $[S]_m$, it can be used for all structural types. In this way, difficult physical concepts inherent in the formulation of $[K]_m$ are bypassed. The transformation matrix $[T]$ is also referred to as a *rotation matrix* because it is used to account for the effects of rotating the reference coordinates through an angle, θ. In space structures, the reference coordinates must be rotated more than once, and additional rotation matrices are required. Finally, Eqs. 16.13 and 16.16 are direct means of converting the forces and displacements from structure to member coordinates. In particular, calculation of member end shears and axial forces (which proved to be awkward in Example 15.12) is greatly facilitated.

16.2.3.2   Planar Truss Member

Members in a planar truss differ from those in a planar frame in several distinct ways. First, no flexural moments exist in the members. Second, the absence of flexural moment and transverse member loads infers the absence of shear forces. Third, joint displacements consist of two orthogonal displacements but no rotation (rigid body rotation of the member itself occurs but has no influence on behavior). By recognizing these differences, a suitable member

stiffness matrix can be extracted from Eq. 16.9. Figure 16.7a and b depict the general planar truss member in the structure coordinate and the member coordinate systems, respectively. Using the indicated nomenclature and referring to Eqs. 16.7 through 16.9,

$$\begin{Bmatrix} F_1 \\ F_2 \\ F_3 \\ F_4 \end{Bmatrix} = [S]_m \begin{Bmatrix} \Delta_1 \\ \Delta_2 \\ \Delta_3 \\ \Delta_4 \end{Bmatrix} \quad (16.21)$$

in which

$$[S]_m = \begin{bmatrix} \dfrac{EA}{L} & 0 & -\dfrac{EA}{L} & 0 \\ 0 & 0 & 0 & 0 \\ -\dfrac{EA}{L} & 0 & \dfrac{EA}{L} & 0 \\ 0 & 0 & 0 & 0 \end{bmatrix} \quad (16.22)$$

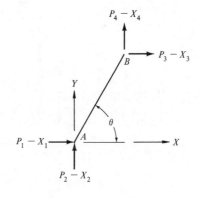

a

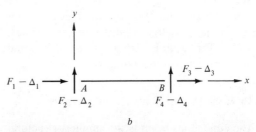

b

Fig. 16.7.   General planar truss member. *a*, structure coordinate system; *b*, member coordinate system.

Because no moments exist at both ends of the member, the transformation of forces between Fig. 16.7a and b is expressed by

$$\begin{Bmatrix} F_1 \\ F_2 \\ F_3 \\ F_4 \end{Bmatrix} = \begin{bmatrix} \cos\theta & \sin\theta & 0 & 0 \\ -\sin\theta & \cos\theta & 0 & 0 \\ 0 & 0 & \cos\theta & \sin\theta \\ 0 & 0 & -\sin\theta & \cos\theta \end{bmatrix} \begin{Bmatrix} P_1 \\ P_2 \\ P_3 \\ P_4 \end{Bmatrix}$$

(16.23)

or, symbolically,

$$\{F\}_m = [T]\{P\}_m \qquad (16.24)$$

in which

$$[T] = \begin{bmatrix} \begin{bmatrix} T' \end{bmatrix} & \begin{bmatrix} 0 & 0 \\ 0 & 0 \end{bmatrix} \\ \begin{bmatrix} 0 & 0 \\ 0 & 0 \end{bmatrix} & \begin{bmatrix} T' \end{bmatrix} \end{bmatrix}$$

(16.25)

Transformation of the displacement quantities is accomplished by

$$\begin{Bmatrix} \Delta_1 \\ \Delta_2 \\ \Delta_3 \\ \Delta_4 \end{Bmatrix} = [T] \begin{Bmatrix} X_1 \\ X_2 \\ X_3 \\ X_4 \end{Bmatrix} \qquad (16.26)$$

or, symbolically,

$$\{\Delta\}_m = [T]\{X\}_m \qquad (16.27)$$

Substituting of Eqs. 16.22 and 16.25 into the general expression, Eq. 16.20, gives the member stiffness matrix in structure coordinates, $[K]_m$. The result is identical to Eq. 15.45 and is not repeated at this point.

16.2.3.3 Continuous Beam Member

The continuous beam is the simplest structural form. By normal convention, axial forces are omitted as a consideration, and flexural action

is the primary source of deformation. Shear stresses contribute to the deformation of the member, but the effect is usually small. Thus, shear deformation is neglected in this discussion. Of more significance is the unidirectional orientation of all members along the structure X-axis. By nature of this orientation, the structure and member coordinate directions align (see Fig. 16.8), and no transformation is necessary. Viewed in another way, the transformation matrix is an identity matrix (4×4 in size). Consequently, the equalities

$$\begin{Bmatrix} P_1 \\ P_2 \\ P_3 \\ P_4 \end{Bmatrix} = \begin{Bmatrix} F_1 \\ F_2 \\ F_3 \\ F_4 \end{Bmatrix} \qquad (16.28a)$$

and

$$\begin{Bmatrix} X_1 \\ X_2 \\ X_3 \\ X_4 \end{Bmatrix} = \begin{Bmatrix} \Delta_1 \\ \Delta_2 \\ \Delta_3 \\ \Delta_4 \end{Bmatrix} \qquad (16.28b)$$

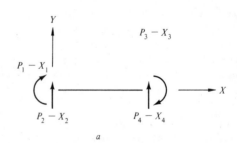

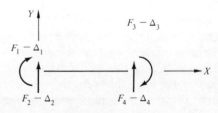

Fig. 16.8.   General continuous beam member. *a,* structure coordinate system; *b,* member coordinate system.

exist and

$$[K]_m = [S]_m$$

Extracting the appropriate entries from Eq. 16.9 yields

$$[K]_m = \begin{bmatrix} \dfrac{4EI}{L} & -\dfrac{6EI}{L^2} & \dfrac{2EI}{L} & \dfrac{6EI}{L^2} \\[2ex] -\dfrac{6EI}{L^2} & \dfrac{12EI}{L^3} & -\dfrac{6EI}{L^2} & -\dfrac{12EI}{L^3} \\[2ex] \dfrac{2EI}{L} & -\dfrac{6EI}{L^2} & \dfrac{4EI}{L} & \dfrac{6EI}{L^2} \\[2ex] \dfrac{6EI}{L^2} & -\dfrac{12EI}{L^3} & \dfrac{6EI}{L^2} & \dfrac{12EI}{L^3} \end{bmatrix}$$

$$(16.29)$$

which, if rearranged, is identical to Eq. 15.37

16.2.3.4 Grid Member

Grids are composed of members that lie in a common plane and are subject to loads that act transverse to that plane. Resistance to loads is provided by the internal forces shown in Fig. 16.9. It is seen that a grid member is simply a beam element that is subjected to torsional moment in addition to the normal flexural moments. In Fig. 16.9, the grid member is depicted in its local coordinate system. The internal forces at end A are the torsional moment F_1, the flexural moment F_2, and the shear force F_3. The corresponding deformations consist of a twist, Δ_1, a rotation, Δ_2, and a displacement, Δ_3. Similar quantities (P_4-P_6 and Δ_4-Δ_6) exist at end B. The local stiffness matrix defines the relationship between the six internal forces and their corresponding deformations. This relationship is given by

$$\begin{Bmatrix} F_1 \\ F_2 \\ F_3 \\ F_4 \\ F_5 \\ F_6 \end{Bmatrix} = [S]_m \begin{Bmatrix} \Delta_1 \\ \Delta_2 \\ \Delta_3 \\ \Delta_4 \\ \Delta_5 \\ \Delta_6 \end{Bmatrix} \qquad (16.30)$$

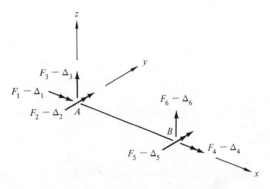

Fig. 16.9.   Planar grid member in local coordinates.

in which

$$[S]_m = \begin{bmatrix} \dfrac{GK}{L} & 0 & 0 & -\dfrac{GK}{L} & 0 & 0 \\[2ex] 0 & \dfrac{4EI}{L} & -\dfrac{6EI}{L^2} & 0 & \dfrac{2EI}{L} & \dfrac{6EI}{L^2} \\[2ex] 0 & -\dfrac{6EI}{L^2} & \dfrac{12EI}{L^3} & 0 & -\dfrac{6EI}{L^2} & \dfrac{12EI}{L^3} \\[2ex] -\dfrac{GK}{L} & 0 & 0 & \dfrac{GK}{L} & 0 & 0 \\[2ex] 0 & \dfrac{2EI}{L^2} & -\dfrac{6EI}{L_2} & 0 & \dfrac{4EI}{L} & \dfrac{6EI}{L^2} \\[2ex] 0 & \dfrac{6EI}{L_2} & -\dfrac{12EI}{L^3} & 0 & \dfrac{6EI}{L^2} & \dfrac{12EI}{L^3} \end{bmatrix}$$

$$(16.31)$$

Equation 16.31 can be viewed as follows: Flexural effects in a member are independent of its torsional behavior, and each can be examined one at a time. The vector quantities F_2-Δ_2, F_3-Δ_3, F_5-Δ_5, and F_6-Δ_6 in Fig. 16.9 are the flexural effects at the ends of the grid members. The interrelationship of these quantities is identical to that of a beam element, and corresponding entries in Eq. 16.31 are simply extracted from the local stiffness matrix for a continuous beam member given in Eq. 16.29. Examination of Eqs. 16.29 and 16.31 shows that the former expression is indeed contained in the latter expression. Remaining entries in Eq. 16.31 represent the torsional effects. The first column of $[K]_m$ is easily verified by use of a compatibility mode (see Fig. 16.10). If a unit twist ($\Delta_1 = 1$) is induced at end A while end B is restrained, torsional moments are cre-

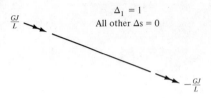

Fig. 16.10.   Compatibility   mods   1   for   a   grid member.

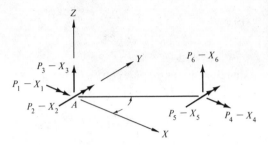

Fig. 16.11.   Planar   grid   member   in   global coordinates.

ated at both ends. For the relative twist of unity, the torsional moment at end A is obtained as

$$F_1 = \frac{GK}{L} \qquad (16.32)$$

where K is the torsional constant for the member cross section. For equilibrium to be maintained, the magnitude of the torsional moment at end B must be

$$F_2 = -\frac{GK}{L} \qquad (16.33)$$

No other forces are created, and the values given in Eqs. 16.32 and 16.33 are the sole entries in the first column of $[S]_m$. Entries in the fourth column are obtained by similar reasoning, which is left to the reader's initiative.

Next, consider the grid member in Fig. 16.11 whose location has been specified in terms of a global coordinate system. To obtain the global member stiffness matrix, it is necessary to develop a transformation matrix, $[T]$, that relates the local force (or deformation) quantities in Fig. 16.9 to the global force (or deformation) quantities in Fig. 16.11. Because of the nature of the vectors, this is done easily.

All members of a grid lie in a common plane. Consequently, the only possible change of reference coordinates is a rotation, θ, within the plane of the grid. Figure 16.11 is employed to obtain the transformation matrix corresponding to such a rotation. Vectors F_3–Δ_3 and F_6–Δ_6 in Fig. 16.9 are unaffected by the in-plane rotation, θ, which implies

$$\begin{Bmatrix} F_3 \\ F_6 \end{Bmatrix} = \begin{Bmatrix} P_3 \\ P_6 \end{Bmatrix} \qquad (16.34a)$$

$$\begin{Bmatrix} \Delta_3 \\ \Delta_6 \end{Bmatrix} = \begin{Bmatrix} X_3 \\ X_6 \end{Bmatrix} \qquad (16.34b)$$

The remaining vectors can be transformed by use of Eqs. 16.2 and 16.5. At end A,

$$\begin{Bmatrix} F_1 \\ F_2 \end{Bmatrix} = \begin{bmatrix} \cos\theta & \sin\theta \\ -\sin\theta & \cos\theta \end{bmatrix} \begin{Bmatrix} P_1 \\ P_2 \end{Bmatrix} \qquad (16.35a)$$

$$\begin{Bmatrix} \Delta_1 \\ \Delta_2 \end{Bmatrix} = \begin{bmatrix} \cos\theta & \sin\theta \\ -\sin\theta & \cos\theta \end{bmatrix} \begin{Bmatrix} X_1 \\ X_2 \end{Bmatrix} \qquad (16.35b)$$

and at end B

$$\begin{Bmatrix} F_4 \\ F_5 \end{Bmatrix} = \begin{bmatrix} \cos\theta & \sin\theta \\ -\sin\theta & \cos\theta \end{bmatrix} \begin{Bmatrix} P_4 \\ P_5 \end{Bmatrix} \qquad (16.36a)$$

$$\begin{Bmatrix} \Delta_4 \\ \Delta_5 \end{Bmatrix} = \begin{bmatrix} \cos\theta & \sin\theta \\ -\sin\theta & \cos\theta \end{bmatrix} \begin{Bmatrix} X_4 \\ X_5 \end{Bmatrix} \qquad (16.36b)$$

Assembling Eqs. 16.34 through 16.36 yields

$$\begin{Bmatrix} F_1 \\ F_2 \\ F_3 \\ F_4 \\ F_5 \\ F_6 \end{Bmatrix} = [T] \begin{Bmatrix} P_1 \\ P_2 \\ P_3 \\ P_4 \\ P_5 \\ P_6 \end{Bmatrix} \qquad (16.37a)$$

$$\begin{Bmatrix} \Delta_1 \\ \Delta_2 \\ \Delta_3 \\ \Delta_4 \\ \Delta_5 \\ \Delta_6 \end{Bmatrix} = [T] \begin{Bmatrix} X_1 \\ X_2 \\ X_3 \\ X_4 \\ X_5 \\ X_6 \end{Bmatrix} \qquad (16.37b)$$

in which the transformation matrix, $[T]$, is similar in content to that developed earlier for planar frame members and is given by

$$[T] = \begin{bmatrix} \cos\theta & \sin\theta & 0 & 0 & 0 & 0 \\ -\sin\theta & \cos\theta & 0 & 0 & 0 & 0 \\ 0 & 0 & 1 & 0 & 0 & 0 \\ 0 & 0 & 0 & \cos\theta & \sin\theta & 0 \\ 0 & 0 & 0 & -\sin\theta & \cos\theta & 0 \\ 0 & 0 & 0 & 0 & 0 & 1 \end{bmatrix}$$

Substitution of $[T]$ and Eq. 16.31 into Eq. 16.20 produces the global member stiffness matrix for a grid member (see Problem 16-36 at the end of this chapter). With an appropriate reordering of the numbers assigned to the degrees of freedom in Fig. 16.9, it is possible to make the transformation matrices for planar frames and planar grids identical. This has some merit in preparing computer algorithms. However, the author prefers that the student understand the process for developing $[T]$ in accordance with any given numbering scheme instead of relying on a general "formula."

16.3 ASSEMBLY OF THE STRUCTURE STIFFNESS MATRIX

Building the structure stiffness matrix is accomplished most expeditiously by direct feeding of the member stiffness matrices. Formerly, this required that the $[K]_m$ matrices be available in explicit form.

In light of the preceding discussion, it is clear that this is no longer a requirement. A modified procedure for assembling $[K]$ can be stated:

1. Form the local member stiffness matrix $[S]_m$.
2. Transform each to structure coordinates using the appropriate $[T]$ matrix and the matrix operation

$$[K]_m = [T]^T [S]_m [T]$$

3. Feed each $[K]_m$ matrix into $[K]$ based upon its connectivity to the joints of the complete structure.

In earlier applications of the direct feeding process, entries in $[K]_m$ associated with directions of restraint were simply discarded in the solution process. In reality, the discarded elements are valuable in the analysis and should be retained in some orderly fashion. The merits of retaining all elements of the $[K]_m$ matrices will be described.

As a basis of discussion, consider the planar truss drawn in Fig. 16.12a. Six degrees of free-

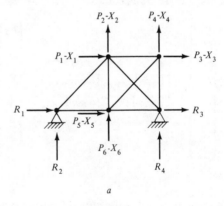

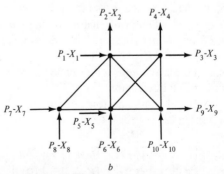

Fig. 16.12. Planar truss. a, supported case; b, unsupported case.

dom and four external reactions are features of this particular structure. By replacing the actual truss with the unsupported truss (Fig. 16.12b), an orderly retention of all entries in each $[K]_m$ matrix is possible. With knowledge of the member properties and geometry, the $[K]_m$ matrices are obtainable by use of Eq. 16.20. If the truss in Fig. 16.12b is used as a reference, the direct feeding process will yield the larger, fictitious, stiffness relationship

$$\begin{Bmatrix} P_1 \\ \vdots \\ P_{10} \end{Bmatrix} = [K]_{10 \times 10} \begin{Bmatrix} X_1 \\ \vdots \\ X_{10} \end{Bmatrix} \qquad (16.38)$$

Support conditions in the actual truss are then entered into Eq. 16.38 by simple assignment. It is known that the vectors P_7-P_{10} are, in fact, the four reactions, R_1-R_4, whose values are unknown. Likewise, the vectors X_7-X_{10} have known zero values. By making these replacements, Eq. 16.38 becomes

$$\begin{Bmatrix} P_1 \\ \vdots \\ \vdots \\ P_6 \\ R_1 \\ \vdots \\ R_4 \end{Bmatrix} = \begin{bmatrix} [K_{PP}]_{6 \times 6} & [K_{PR}]_{6 \times 4} \\ [K_{RP}]_{4 \times 6} & [K_{RR}]_{4 \times 4} \end{bmatrix} \begin{Bmatrix} X_1 \\ \vdots \\ \vdots \\ X_6 \\ 0 \\ \vdots \\ 0 \end{Bmatrix}$$

$$(16.39)$$

The matrix $[K_{PP}]$ is the structure stiffness matrix of the supported truss but in the context of Chapter 15 wherein the nature of the support conditions is not evident. The additional matrices $[K_{PR}]$, $[K_{RP}]$, and $[K_{RR}]$ in Eq. 16.39. expand the size to reflect the support conditions present in the structure.

A general shorthand form of Eq. 16.39 can be written as

$$\begin{Bmatrix} \{P\} \\ \{R\} \end{Bmatrix} = \begin{bmatrix} [K_{PP}] & [K_{PR}] \\ [K_{RP}] & [K_{RR}] \end{bmatrix} \begin{Bmatrix} \{X\} \\ \{0\} \end{Bmatrix}$$

$$(16.40)$$

Equation 16.40 signifies the two relationships

$$\{P\} = [K_{PP}]\{X\} \qquad (16.41)$$

$$\{R\} = [K_{RP}]\{X\} \qquad (16.42)$$

The usefulness of this formulation of $[K]$ is discussed in subsequent sections.

16.4 GENERATION OF THE LOAD MATRIX

16.4.1 Treatment of Transverse Loads

The load matrix $\{P\}$ has been defined and used in earlier applications of matrix analysis in this book. $\{P\}$ is a column matrix containing the joint loads for the released state. For trusses, $\{P\}$ contains the global components of the loads applied at the joints. Generation of $\{P\}$ for beams and planar frames has been previously described. The joint loads have the same magnitudes but act in directions opposite to the restraining forces of the restrained state. However, it is advantageous to alter the procedure to apply the transformation concept in lieu of the earlier use of statics. It also possible to incorporate the direct computation of support reactions in an efficient manner. In Chapter 15, no procedure was described. The inference was the support reactions could be subsequently calculated by applying statics to the supported joints.

16.4.1.1 Planar Frames

Consider the planar frame member drawn in Fig. 16.13a. The initial step in establishing $\{P\}$ consists of a computation of the restrained-state forces, $F_{01}-F_{06}$. This is easily done with the use of Tables A.3 and A.7 in Appendix A. Unfortunately, the calculated forces are aligned with the local x-y coordinate system and cannot be directly used in this form. This inconvenient form can be avoided by a simple transformation to the global orientation shown in Fig. 16.13b. Using the nomenclature of Fig. 16.13, it is evident that

$$\begin{Bmatrix} F_{01} \\ F_{04} \end{Bmatrix} = \begin{Bmatrix} P_{01} \\ P_{04} \end{Bmatrix} \qquad (16.45)$$

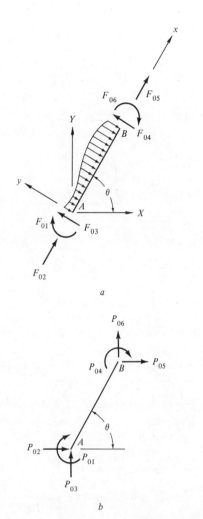

Fig. 16.13. Transformation of restrained-state forces. *a*, general loading and restrained-state forces; *b*, restrained-state forces in structure coordinates.

and

$$\begin{Bmatrix} F_{02} \\ F_{03} \end{Bmatrix} = [T'] \begin{Bmatrix} P_{02} \\ P_{03} \end{Bmatrix} \qquad (16.46)$$

$$\begin{Bmatrix} F_{05} \\ F_{06} \end{Bmatrix} = [T'] \begin{Bmatrix} P_{05} \\ P_{06} \end{Bmatrix} \qquad (16.47)$$

in which $[T']$ is given by Eq. 16.4. By assembly of Eqs. 16.45 through 16.47 and use of the characteristic that $[T']^{-1} = [T']^T$, Eq. 16.48 is obtained.

$$\begin{Bmatrix} P_{01} \\ P_{02} \\ P_{03} \\ P_{04} \\ P_{05} \\ P_{06} \end{Bmatrix}_m = \begin{bmatrix} 1 & 0 & 0 & 0 & 0 & 0 \\ 0 & & & 0 & 0 & 0 \\ 0 & [T']^T & & 0 & 0 & 0 \\ 0 & 0 & 0 & 1 & 0 & 0 \\ 0 & 0 & 0 & 0 & & \\ 0 & 0 & 0 & 0 & [T']^T & \end{bmatrix} \begin{Bmatrix} F_{01} \\ F_{02} \\ F_{03} \\ F_{04} \\ F_{05} \\ F_{06} \end{Bmatrix}_m$$

$$(16.48)$$

The symbolic form of Eq. 16.48 is

$$\{P_o\}_m = [T]^T \{F_o\}_m \qquad (16.49)$$

Execution of Eq. 16.49 for each member of a structure permits direct feeding of the $\{P_o\}_m$ matrices into the appropriate locations of $\{P\}$ in the same manner employed for constructing $[K]$. Actual support conditions are accounted for by a subsequent subdivision of $\{P\}$. This process is described in the following paragraphs.

Figure 16.14a depicts a member that is part of a larger planar frame. In this instance, the member connects to a pin support at A and a free joint at B. The corresponding reactions, R_i and R_j, and possible loads at the degrees of freedom, P_i–P_l, are also indicated. Subscripts are written in general form because the actual numbers depend upon the numbering scheme assigned to the entire framework. In the first stage of matrix analysis, all degrees of freedom are held in restraint. Thus, application of load to member AB produces six internal forces, P_{01} through P_{06}, oriented as shown in Fig. 16.14b. For joint equilibrium to exist, six restraining forces, RF_1–RF_6, must be applied to the joints. The magnitudes of the restraining forces (obtained by applying statics to each joint) are

$$\begin{Bmatrix} RF_1 \\ RF_2 \\ RF_3 \\ RF_4 \\ RF_5 \\ RF_6 \end{Bmatrix} = \begin{Bmatrix} P_{01} \\ P_{02} \\ P_{03} \\ P_{04} \\ P_{05} \\ P_{06} \end{Bmatrix} \qquad (16.50)$$

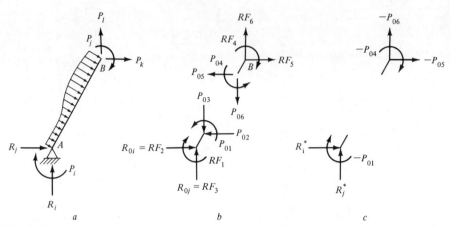

Fig. 16.14. Calculation of equivalent joint loads. *a*, joint conditions; *b*, joint equilibrium in restrained-state; *c*, release state joint forces.

Because joint A is a real pin-support, the two restraining forces RF_2 and RF_3 are true reactions. In recognition of this, initial reaction forces R_{oi} and R_{oj} are shown in Fig. 16.14b, and their magnitudes are given by

$$\begin{Bmatrix} R_{oi} \\ R_{oj} \end{Bmatrix} = \begin{Bmatrix} RF_2 \\ RF_3 \end{Bmatrix} = \begin{Bmatrix} P_{02} \\ P_{03} \end{Bmatrix} \quad (16.51)$$

The remaining restraining forces are fictitious (do not exist in the actual structure) and are subsequently removed by applying equal and opposite *loads* to the released structure (Fig. 16.11c). For the subscripts assigned in Fig. 16.14a and b, the loads required in the second stage of matrix analysis are

$$\begin{Bmatrix} P_i \\ P_j \\ P_k \\ P_l \end{Bmatrix} = \begin{Bmatrix} -RF_1 \\ -RF_4 \\ -RF_5 \\ -RF_6 \end{Bmatrix} = \begin{Bmatrix} -P_{01} \\ -P_{04} \\ -P_{05} \\ -P_{06} \end{Bmatrix}$$

$$(16.52)$$

Equations 16.51 and 16.52 illustrate the role of the six restrained-state forces for a given member in matrix analysis. Depending upon the member's connectivity, some represent contributions to the *restrained state reactions* and some represent contributions to the *released*

state loads. It is essential to realize that the member forces in the latter case *must be changed in sign* prior to assigning them to particular load locations.

The concept demonstrated in Fig. 16.14 can be stated as a general procedure.

1. For each member subjected to transverse load, calculate the member end forces in member coordinates, $\{F_o\}_m$
2. Transform $\{F_o\}_m$ to structure coordinates by the operation

$$\{P_o\}_m = [T]^T \{F_o\}_m$$

3. Forces in $\{P_o\}_m$ that align with the degrees of freedom of the actual structure are extracted, *changed in sign*, and directly fed into the corresponding locations in $\{P\}_E$, the "equivalent load matrix."
4. Forces in $\{P_o\}_m$ that align with reaction forces in the actual structure are extracted and, *without changing signs*, fed into corresponding locations in $\{R_o\}$.
5. Steps 1 through 3 are repeated for all members of the structure. The result is the complete matrices $\{P\}_E$ and $\{R_o\}$.

16.4.1.2 Grids

Transverse loads on a grid member can be treated by the technique described for planar

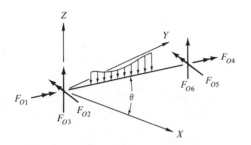

Fig. 16.15. Restrained-state forces for a grid member.

frames. However, the nature of the member end forces and the coordinate transformation differ from those of a planar frame. A transversely loaded grid member, placed at an arbitrary location θ relative to the structure X coordinate axis, is depicted in Fig. 16.15. Torsional moments in the restrained state are obtained from Appendix A, Table A.6. Local end moments and shears in the restrained state have the same magnitude as would exist for a planar frame member, but the local numbering system employed in Fig. 16.15 differs from that of Fig. 16.12. Consequently, the end forces occupy different locations in $\{F_o\}_m$. After assembling the six end forces into $\{F_o\}_m$, Eq. 16.49 is employed to transform the loads to the global coordinate system. The results are then fed into $\{P\}_E$ and $\{R_o\}$. Except for the contents of $[T]$ and the order of numbering of the end forces, the five steps of the general procedure (Section 16.4.1.1) for dealing with transverse loads remains the same.

16.4.2 Treatment of Joint Loads

Transverse loads applied to individual members are not the only type of loads generally applied to structures. Concentrated loads applied directly to the joints are a common occurrence (e.g., see Fig. 16.2) and in trusses are the only loads. Inclusion of these actions in the load matrix is easily accomplished by formulating a *joint load* matrix, $\{P\}_J$. The size of $\{P\}_J$ is $NP \times 1$, where NP is the total number of degrees of freedom for the structure. When concentrated loads (including moments) exist

at the joints, these are entered into the corresponding locations of $\{P\}_J$. The complete load matrix for the analysis of the released state is obtained by combining $\{P\}_J$ with the equivalent joint loads $\{P\}_E$.

16.5 GENERAL SOLUTION PROCEDURE

The concepts and matrix relationships developed in the preceding sections are the basis of a more complete matrix formulation for the direct stiffness method. A step-by-step commentary on the procedure and accompanying matrix operations are assembled in Table 16.1. The sequence and content are general and can be applied to all structural types. The analyst need only employ the particular local member stiffness matrix $[S]_m$ and transformation matrix $[T]$ associated with the type of structure at hand (i.e., planar frame, planar truss, etc.). In addition, the nature of the contents of the initial member force matrix, $\{F_o\}$, is dependent upon the structural type. These matrices have just been discussed for planar structures. Appropriate forms of these matrices for space structures can also be derived, but the topic is outside the scope of this textbook.

The reader should also note that Table 16.1 presumes a two-stage analysis (restrained state and released state). Thus, some of the relationships developed earlier are shown in the table with asterisks if they constitute steps in the released-state analysis.

16.6 EXAMPLE PROBLEMS

Example 16.1. Analyze the given planar frame (Fig. 16.16a) by the matrix stiffness method using transformation matrices. Use $EI = 20,000$ kip-ft^2 and $EA = 200,000$ kips.

Solution: Using Table 16.1 as a guide the solution proceeds as follows

1. Local member stiffness matrices are determined by use of Eq. 16.9. Member AB will be presented as an example.

$$[S]_{AB} = \begin{bmatrix} 4000 & 0 & -300 & 2000 & 0 & 300 \\ 0 & 10{,}000 & 0 & 0 & -10{,}000 & 0 \\ -300 & 0 & 30 & -300 & 0 & -30 \\ 2000 & 0 & -300 & 4000 & 0 & 300 \\ 0 & -10{,}000 & 0 & 0 & 10{,}000 & 0 \\ 300 & 0 & -30 & 300 & 0 & 30 \end{bmatrix}$$

2. The transformation matrix for a planar frame is given in Eq. 16.14. For member AB, $\sin \theta = 0.8$, $\cos \theta = 0.6$, and

$$[T]_{AB} = \begin{bmatrix} 1 & 0 & 0 & 0 & 0 & 0 \\ 0 & 0.6 & 0.8 & 0 & 0 & 0 \\ 0 & -0.8 & 0.6 & 0 & 0 & 0 \\ 0 & 0 & 0 & 1 & 0 & 0 \\ 0 & 0 & 0 & 0 & 0.6 & 0.8 \\ 0 & 0 & 0 & 0 & -0.8 & 0.6 \end{bmatrix}$$

3. The member stiffness matrix in structure coordinates is given by

$$[K]_m = [T]^T[S]_m[T]$$

For member AB,

$$[K]_{AB} = \begin{array}{c} \begin{array}{cccccc} \;①\;\; & \;⑤\;\; & \;⑥\;\; & \;②\;\; & \;③\;\; & \;④\;\; \end{array} \\ \begin{array}{c} ① \\ ⑤ \\ ⑥ \\ ② \\ ③ \\ ④ \end{array} \begin{bmatrix} 4000.0 & 240.0 & -180.0 & 2000.0 & -240.0 & 180.0 \\ 240.0 & 3691.2 & 4785.6 & 240.0 & -3619.2 & -4785.6 \\ -180.0 & 4785.6 & 6410.8 & -180.0 & -4785.6 & -6410.8 \\ 2000.0 & 240.0 & -180.0 & 4000.0 & -240.0 & 180.0 \\ -240.0 & -3619.2 & -4785.6 & -240.0 & 3619.2 & 4785.6 \\ 180.0 & -4785.6 & -6410.8 & 180.0 & 4785.6 & 6410.8 \end{bmatrix} \end{array}$$

in which the connectivity is referenced to the unsupported frame in Fig. 16.16c.

4. The structure stiffness matrix, after $[K]_{AB}$ has been entered, is given in Fig. 16.17. Direct feeding of $[K]_{BC}$ and $[K]_{BD}$ produces the complete stiffness relationship presented in Fig. 16.18. In this figure, the conversion to the actual connectivity (Fig. 16.16b) is also indicated by substitution of the unknown reactions R_1^*–R_8^* and the corresponding zero displacement values.

5. Substructuring of $[K]$ also is indicated in Fig. 16.18.

6. Initial member forces are shown in Fig. 16.16d. For member AB, the 10-kip load is separated into its components, and each is treated by use of Table A.3 in Appendix A. Treatment of the 8-kip axial force is accomplished by use of Table A.7.

$$\{F_o\}_{AB} = \begin{Bmatrix} F_{01} \\ F_{02} \\ F_{03} \\ F_{04} \\ F_{05} \\ F_{06} \end{Bmatrix} = \begin{Bmatrix} -15.0 \text{ kip-ft} \\ 4.0 \text{ kips} \\ 3.0 \text{ kips} \\ 15.0 \text{ kip-ft} \\ 4.0 \text{ kips} \\ 3.0 \text{ kips} \end{Bmatrix}$$

7. The results of step 6 are transformed to structure coordinates

$$\{P_o\}_{AB} = [T]_{AB}^T\{F_o\}_{AB}$$

$$\begin{Bmatrix} P_{01} \\ P_{05} \\ P_{06} \\ P_{02} \\ P_{03} \\ P_{04} \end{Bmatrix} = \begin{Bmatrix} -15.0 \text{ kip-ft} \\ 0.0 \text{ kip} \\ 5.0 \text{ kips} \\ 15.0 \text{ kip-ft} \\ 0.0 \text{ kip} \\ 5.0 \text{ kips} \end{Bmatrix}$$

8. Member forces in structure coordinates are directly fed into the equivalent load matrix and the initial reaction matrix. Forces fed into the equivalent load matrix are first changed in sign! After treating member AB, in which the subscripts are based on the connectivity of member AB shown in Fig. 16.16c, and converting to the symbols shown in Fig. 16.16b gives

Table 16.1.   Procedure for a Semiautomated Matrix Stiffness Analysis.

Description	Matrix Operations
1. Calculate the member stiffness matrices in member coordinates.	$[S]_m$
2. Calculate the transformation matrix for each member.	$[T]$
3. Transform each member stiffness matrix to the structure-oriented coordinate system.	$[K]_m = [T]^T [S]_m [T]$
4. Build the structure stiffness matrix by directly feeding of the member stiffness matrices developed in step 3.	$[K]_m \rightarrow [K]$
5. Substructure $[K]$ into its component parts.	$[K] = \begin{bmatrix} K_{PP} & K_{PR} \\ \hline K_{RP} & K_{RR} \end{bmatrix}$
6. Calculate the initial end forces for each member in member coordinates.	$\{F_0\}_m$
7. Transform the member end forces to structure coordinates.	$\{P_0\}_m = [T]^T \{F_0\}_m$
8. Build the equivalent joint load matrix and the initial reaction matrix by direct feeding of the results of step 7. Elements fed into $\{P\}_E$ are first changed in sign.	$\{P_0\}_m \rightarrow \begin{matrix} \{P\}_E \\ \{R_0\} \end{matrix}$
9. Build the joint load matrix.	$\{P\}_J$
10. Add the existing joint loads $\{P_J\}$ to the equivalent load matrix.	$\{P\} = \{P\}_E + \{P\}_J$
11. Form the governing stiffness relationships.	$\{P\} = [K_{PP}] \{X^*\}$ $\{R^*\} = [K_{RP}] \{X^*\}$
12. Calculate the joint displacements in the released state.	$\{X\} = \{X^*\} = [K_{PP}]^{-1} \{P\}$
13. Calculate the support reactions in the released state.	$\{R^*\} = [K_{RP}] \{X\}$
14. Calculate the support reactions in the actual structure, i.e., combine the results of steps 8 and 13.	$\{R\} = \{R_0\} + \{R^*\}$
15. For each member, extract the joint displacements at its ends from the results of step 11.	$\{X\}_m \leftarrow \{X^*\}$
16. For each member, transform the joint displacements to member coordinates.	$\{\Delta\}_m = \{\Delta^*\} = [T] \{X^*\}_m$
17. Calculate the member end forces in member coordinates.	$\{F^*\}_m = [S]_m \{\Delta^*\}_m$
18. Calculate the member end forces in the actual structure, i.e., combine the results of steps 6 and 17.	$\{F\}_m = \{F_0\} + \{F^*\}_m$

$$\begin{Bmatrix} P_{10} \\ R_{01} \\ R_{02} \\ P_{02} \\ P_{03} \\ P_{04} \end{Bmatrix}_{AB} = \begin{Bmatrix} -15.0 \\ 0.0 \\ 5.0 \\ 15.0 \\ 0.0 \\ 5.0 \end{Bmatrix}$$

and

$$\{R_o\} = \begin{Bmatrix} R_{01} \\ R_{02} \\ R_{03} \\ R_{04} \\ R_{05} \\ R_{06} \\ R_{07} \\ R_{08} \end{Bmatrix} = \begin{Bmatrix} 0 \\ 5.0 \text{ kips} \\ 0 \\ 0.0 \\ 0.0 \\ 40.0 \text{ kip-ft} \\ 0.0 \\ 12.0 \text{ kips} \end{Bmatrix}$$

$$\{P\}_E = \begin{Bmatrix} P_1 \\ P_2 \\ P_3 \\ P_4 \end{Bmatrix} = \begin{Bmatrix} +15.0 \text{ kip-ft} \\ -15.0 \text{ kip-ft} \\ 0 \\ -5.0 \text{ kips} \end{Bmatrix}$$

Similar analysis of members BC and BD yields the final matrices

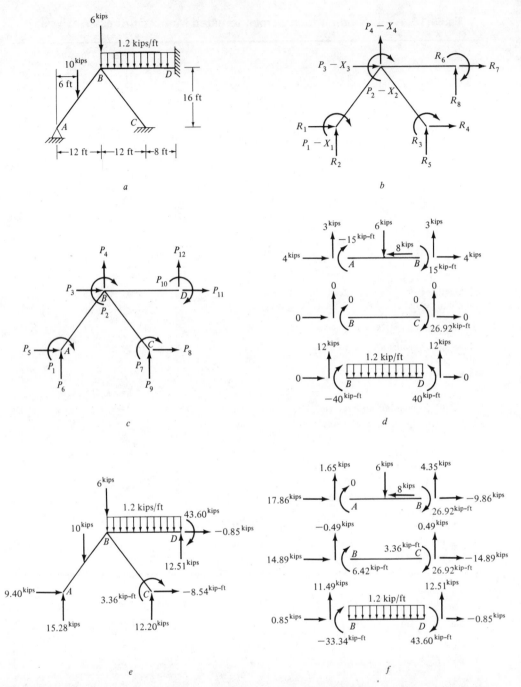

Fig. 16.16.   Planar frame for Example 16.1. *a*, structure and loading; *b*, joint conditions; *c*, unsupported frame; *d*, initial member forces; *e*, final reaction forces; *f*, final member forces.

$$
\begin{array}{c}
\quad\;\; ① \qquad ② \qquad ③ \qquad ④ \qquad ⑤ \qquad ⑥ \qquad ⑦\;\; ⑧\;\; ⑨\;\; ⑩\;\; ⑪\;\; ⑫ \\[4pt]
\begin{array}{c}
① \\ ② \\ ③ \\ ④ \\ ⑤ \\ ⑥ \\ ⑦ \\ ⑧ \\ ⑨ \\ ⑩ \\ ⑪ \\ ⑫
\end{array}
\left[
\begin{array}{cccccccccccc}
4000.0 & 2000.0 & -240.0 & 180.0 & 240.0 & -180.0 & 0 & 0 & 0 & 0 & 0 & 0 \\
2000.0 & 4000.0 & -240.0 & 180.0 & 240.0 & -180.0 & 0 & 0 & .0 & 0 & 0 & 0 \\
-240.0 & -240.0 & 3619.2 & 4785.6 & -3619.2 & -4785.6 & 0 & 0 & 0 & 0 & 0 & 0 \\
180.0 & 180.0 & 4785.6 & 6410.8 & -4785.6 & -6410.8 & 0 & 0 & 0 & 0 & 0 & 0 \\
240.0 & 240.0 & -3619.2 & -4785.6 & 3619.2 & 4785.6 & 0 & 0 & 0 & 0 & 0 & 0 \\
-180.0 & -180.0 & -4785.6 & -6410.8 & 4785.6 & 6410.8 & 0 & 0 & 0 & 0 & 0 & 0 \\
0 & 0 & 0 & 0 & 0 & 0 & 0 & 0 & 0 & 0 & 0 & 0 \\
0 & 0 & 0 & 0 & 0 & 0 & 0 & 0 & 0 & 0 & 0 & 0 \\
0 & 0 & 0 & 0 & 0 & 0 & 0 & 0 & 0 & 0 & 0 & 0 \\
0 & 0 & 0 & 0 & 0 & 0 & 0 & 0 & 0 & 0 & 0 & 0 \\
0 & 0 & 0 & 0 & 0 & 0 & 0 & 0 & 0 & 0 & 0 & 0 \\
0 & 0 & 0 & 0 & 0 & 0 & 0 & 0 & 0 & 0 & 0 & 0
\end{array}
\right]
\end{array}
$$

Fig. 16.17. Structure stiffness matrix after entering $[K]_{AB}$.

$$
\{P\}_E = \left\{\begin{array}{c} +15.0 \\ +25.0 \\ 0 \\ -17.0 \end{array}\right\}
\quad \text{and} \quad
\{R_o\} = \left\{\begin{array}{c} 0.0 \\ 5.0 \\ 0.0 \\ 0.0 \\ 0.0 \\ 40.0 \\ 0.0 \\ 12.0 \end{array}\right\}
$$

$$
\{P\}_J = \left\{\begin{array}{c} P_1 \\ P_2 \\ P_3 \\ P_4 \end{array}\right\} = \left\{\begin{array}{c} 0 \\ 0 \\ 0 \\ -6 \end{array}\right\}
$$

and added to the load matrix

$$
\{P\} = \{P\}_E + \{P\}_J
$$

$$
= \left\{\begin{array}{c} +15.0 \\ 25.0 \\ 0 \\ -17.0 \end{array}\right\} + \left\{\begin{array}{c} 0 \\ 0 \\ 0 \\ -6 \end{array}\right\} = \left\{\begin{array}{c} 15.0 \\ +25.0 \\ 0.0 \\ -23.0 \end{array}\right\}
$$

9, 10. The joint load matrix is established as

$$
\left\{\begin{array}{c} P_1 \\ P_2 \\ P_3 \\ P_4 \\ P_5 \\ P_6 \\ P_7 \\ P_8 \\ P_9 \\ P_{10} \\ P_{11} \\ P_{12} \end{array}\right\} = \left\{\begin{array}{c} P_1 \\ P_2 \\ P_3 \\ P_4 \\ R_1^* \\ R_2^* \\ R_3^* \\ R_4^* \\ R_5^* \\ R_6^* \\ R_7^* \\ R_8^* \end{array}\right\} =
\left[\begin{array}{cccccccccccc}
4000.0 & 2000.0 & -240.0 & 180.0 & 240.0 & -180.0 & 0.0 & 0.0 & 0.0 & 0.0 & 0.0 & 0.0 \\
2000.0 & 12,000.0 & -480.0 & -300.0 & 240.0 & -180.0 & 2000.0 & 240.0 & 180.0 & 2000.0 & 0.0 & 300.0 \\
-240.0 & -480.0 & 17,238.4 & 0.0 & -3619.2 & -4785.6 & -240.0 & -3619.2 & 4785.6 & 0.0 & -10,000.0 & 0.0 \\
180.0 & -300.0 & 0.0 & 12,851.6 & -4785.6 & -6410.8 & -180.0 & 4785.6 & -6410.8 & -300.0 & 0.0 & -30.0 \\
240.0 & 240.0 & -3619.2 & -4785.6 & 3619.2 & 4785.6 & 0.0 & 0.0 & 0.0 & 0.0 & 0.0 & 0.0 \\
-180.0 & -180.0 & -4785.6 & -6410.8 & 4785.6 & 6410.8 & 0.0 & 0.0 & 0.0 & 0.0 & 0.0 & 0.0 \\
0.0 & 2000.0 & -240.0 & -180.0 & 0.0 & 0.0 & 4000.0 & 240.0 & 180.0 & 0.0 & 0.0 & 0.0 \\
0.0 & 240.0 & -3619.2 & 4785.6 & 0.0 & 0.0 & 240.0 & 3619.2 & -4785.6 & 0.0 & 0.0 & 0.0 \\
0.0 & 180.0 & 4785.6 & -6410.8 & 0.0 & 0.0 & 180.0 & -4785.6 & 6410.8 & 4000.0 & 0.0 & 0.0 \\
0.0 & 2000.0 & 0.0 & -300.0 & 0.0 & 0.0 & 0.0 & 0.0 & 0.0 & 0.0 & 0.0 & 300.0 \\
0.0 & 0.0 & -10,000.0 & 0.0 & 0.0 & 0.0 & 0.0 & 0.0 & 0.0 & 0.0 & 10,000.0 & 0.0 \\
0.0 & 300.0 & 0.0 & -300.0 & 0.0 & 0.0 & 0.0 & 0.0 & 0.0 & 300.0 & 0.0 & 30.0
\end{array}\right]
\left\{\begin{array}{c} X_1^* \\ X_2^* \\ X_3^* \\ X_4^* \\ 0 \\ 0 \\ 0 \\ 0 \\ 0 \\ 0 \\ 0 \\ 0 \end{array}\right\}
$$

$$
\left\{\begin{array}{c} \{P\} \\ \{R^*\} \end{array}\right\} = \left[\begin{array}{c|c} [K_{PP}] & [K_{PR}] \\ \hline [K_{RP}] & [K_{RR}] \end{array}\right] \left\{\begin{array}{c} \{X^*\} \\ \{0\} \end{array}\right\}
$$

Fig. 16.18. Complete structure stiffnes matrix.

11, 12, 13. Governing stiffness relationships are extracted from Fig. 16.18 and solved for $\{X^*\}$ and $\{R^*\}$

$$\{X^*\} = \{K_{PP}\}^{-1}\{P\}$$

where

$$[K_{PP}]^{-1} = \begin{bmatrix} 4000.0 & 2000.0 & -240.0 & 180.0 \\ 2000.0 & 12{,}000.0 & -480.0 & -300.0 \\ -240.0 & -480.0 & 17{,}238.4 & 0.0 \\ 180.0 & -300.0 & -300.0 & 12{,}851.6 \end{bmatrix}^{-1}$$

$$\{P\} = \begin{Bmatrix} 15.0 \\ 25.0 \\ 0.0 \\ -23.0 \end{Bmatrix}$$

Thus,

$$\{X^*\} = \begin{Bmatrix} X_1^* \\ X_2^* \\ X_3^* \\ X_4^* \end{Bmatrix} = \begin{Bmatrix} 0.3071 \text{ rad} \\ 0.1530 \text{ rad} \\ 0.008536 \text{ ft} \\ -0.1797 \text{ ft} \end{Bmatrix} 10^{-2}$$

$$\{R^*\} = [K_{RP}]\{X^*\}$$

$$\begin{Bmatrix} R_1^* \\ R_2^* \\ R_3^* \\ R_4^* \\ R_5^* \\ R_6^* \\ R_7^* \\ R_8^* \end{Bmatrix} = \begin{Bmatrix} 9.40 \\ 10.28 \\ 3.36 \\ -8.54 \\ 12.20 \\ 3.60 \\ -0.85 \\ 0.51 \end{Bmatrix}$$

14. Final reactions are obtained as

$$\{R\} = \{R_0\} + \{R^*\} = \begin{Bmatrix} 9.40 \text{ kips} \\ 15.25 \text{ kips} \\ 3.36 \text{ kip-ft} \\ -8.54 \text{ kips} \\ 12.20 \text{ kips} \\ 43.60 \text{ kip-ft} \\ -0.85 \text{ kips} \\ 12.51 \text{ kips} \end{Bmatrix}$$

and are displayed in Fig. 16.16e.

15, 16. Member end displacements in structure coordinates are extracted from $\{X\}$ and transformed to member coordinates, e.g.,

$$\{\Delta^*\}_{AB} = [T]_{AB}\{X^*\}_{AB}$$

$$\begin{Bmatrix} \Delta_1^* \\ \Delta_2^* \\ \Delta_3^* \\ \Delta_4^* \\ \Delta_5^* \\ \Delta_6^* \end{Bmatrix} = [T]_{AB} \begin{Bmatrix} X_1^* \\ 0 \\ 0 \\ X_2^* \\ X_3^* \\ X_4^* \end{Bmatrix}$$

$$= \begin{Bmatrix} 0.3071 \text{ rad} \\ 0.0 \\ 0.0 \\ 0.1530 \text{ rad} \\ -0.1386 \text{ ft} \\ -0.1146 \text{ ft} \end{Bmatrix} \times 10^{-2}$$

17, 18. Member end forces in member coordinates are calculated by employing Eq. 16.7

$$\{F^*\}_{AB} = [S]_{AB}\{\Delta^*\}_{AB} = \begin{Bmatrix} 15.00 \text{ kip-ft} \\ 13.86 \text{ kips} \\ -1.35 \text{ kips} \\ 11.92 \text{ kip-ft} \\ -13.86 \text{ kips} \\ 1.35 \text{ kips} \end{Bmatrix}$$

and combined with the initial member end forces

$$\{F\}_{AB} = \begin{Bmatrix} F_1 \\ F_2 \\ F_3 \\ F_4 \\ F_5 \\ F_6 \end{Bmatrix} = \{F_0\}_{AB} + \{F^*\}_{AB}$$

$$= \begin{Bmatrix} 0 \\ 17.86 \text{ kips} \\ 1.65 \text{ kips} \\ 26.92 \text{ kip-ft} \\ -9.86 \text{ kips} \\ 4.35 \text{ kips} \end{Bmatrix}$$

which completes the structural analysis of member AB. A detailed execution of all the preceding matrix operations for members BC and BD yields the final results given in Fig. 16.16f.

Example 16.2.   Analyze the truss in Fig. 16.19 by the matrix stiffness method using the transformation matrices. Use $EA/L = 5000$ kip/ft for all members.

Solution: Because the only loads in a truss problem are joint loads, the steps associated with transverse loads can be omitted in the general procedure enumerated in Table 16.1. Furthermore, if no actions other than applied joint loads are present, the use of a restrained state is unnecessary. The analysis that follows is performed with these facts in mind and irrelevant steps in Table 16.1 are so noted.

1. Member stiffness matrices (in local coordinates are given by

$$[S]_{AC} = [S]_{AB} = [S]_{CB}$$

$$= \begin{bmatrix} 5000 & 0 & -5000 & 0 \\ 0 & 0 & 0 & 0 \\ -5000 & 0 & 5000 & 0 \\ 0 & 0 & 0 & 0 \end{bmatrix}$$

2. Transformation matrices are formed using Eq. 16.23.

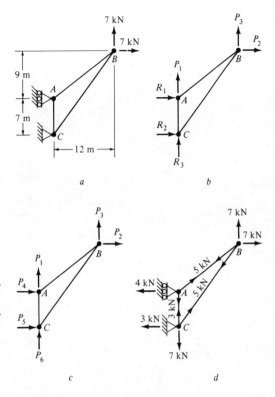

Fig. 16.19.   Planar truss, Example 16.2. *a*, truss and loading; *b*, joint conditions; *c*, unsupported truss; *d*, member forces.

$$[T]_{AC} = \begin{bmatrix} 0 & -1 & 0 & 0 \\ 1 & 0 & 0 & 0 \\ 0 & 0 & 0 & -1 \\ 0 & 0 & 1 & 0 \end{bmatrix}$$

$$[T]_{AB} = \begin{bmatrix} 0.8 & 0.6 & 0 & 0 \\ -0.6 & 0.8 & 0 & 0 \\ 0 & 0 & 0.8 & 0.6 \\ 0 & 0 & -0.6 & 0.8 \end{bmatrix}$$

$$[T]_{CB} = \begin{bmatrix} 0.6 & 0.8 & 0 & 0 \\ -0.8 & 0.6 & 0 & 0 \\ 0 & 0 & 0.6 & 0.8 \\ 0 & 0 & -0.8 & 0.6 \end{bmatrix}$$

3. Member stiffness matrices in structure coordinates are given by

$$[K]_m = [T]^T[S]_m[T]$$

$$[K]_{AC} = \begin{array}{c} \\ \text{④} \\ \text{①} \\ \text{⑤} \\ \text{⑥} \end{array}
\begin{array}{cccc}
\text{④} & \text{①} & \text{⑤} & \text{⑥} \\
\begin{bmatrix} 0 & 0 & 0 & 0 \\ 0 & 5000 & 0 & -5000 \\ 0 & 0 & 0 & 0 \\ 0 & -5000 & 0 & 5000 \end{bmatrix}
\end{array}$$

$$[K]_{CB} = \begin{array}{c} \\ \text{⑤} \\ \text{⑥} \\ \text{②} \\ \text{③} \end{array}
\begin{array}{cccc}
\text{⑤} & \text{⑥} & \text{②} & \text{③} \\
\begin{bmatrix} 1800 & 2400 & -1800 & -2400 \\ 2400 & 3200 & -2400 & -3200 \\ -1800 & -2400 & 1800 & 2400 \\ -2400 & -3200 & 2400 & 3200 \end{bmatrix}
\end{array}$$

$$[K]_{AB} = \begin{array}{c} \\ \text{④} \\ \text{①} \\ \text{②} \\ \text{③} \end{array}
\begin{array}{cccc}
\text{④} & \text{①} & \text{②} & \text{③} \\
\begin{bmatrix} 3200 & 2400 & -3200 & -2400 \\ 2400 & 1800 & -2400 & -1800 \\ -3200 & -2400 & 3200 & 2400 \\ -2400 & -1800 & 2400 & 1800 \end{bmatrix}
\end{array}$$

4. Direct feeding produces the structure stiffness matrix

$$[K] = \begin{array}{c} \\ \text{①} \\ \text{②} \\ \text{③} \\ \text{④} \\ \text{⑤} \\ \text{⑥} \end{array}
\begin{array}{cccccc}
\text{①} & \text{②} & \text{③} & \text{④} & \text{⑤} & \text{⑥} \\
\begin{bmatrix} 6800 & -2400 & -1800 & 2400 & 0 & -5000 \\ -2400 & 5000 & 4800 & -3200 & -1800 & -2400 \\ -1800 & 4800 & 5000 & -2400 & -2400 & -3200 \\ 2400 & -3200 & -2400 & 3200 & 0 & 0 \\ 0 & -1800 & -2400 & 0 & 1800 & 2400 \\ -5000 & -2400 & -3200 & 0 & 2400 & 8200 \end{bmatrix}
\end{array}$$

for the unsupported truss in Fig. 16.19c.

5. By observing the support conditions (restraint at P_4, P_5, and P_6), $[K]$ is substructured (as shown above) to give

$$[K_{PP}] = \begin{bmatrix} 6800 & -2400 & -1800 \\ -2400 & 5000 & 4800 \\ -1800 & 4800 & 5000 \end{bmatrix}$$

$$[K_{RP}] = \begin{bmatrix} 2400 & -3200 & -2400 \\ 0 & -1800 & -2400 \\ -5000 & -2400 & -3200 \end{bmatrix}$$

6, 7, 8. Irrelevant

9. Loads are entered into the joint load matrix.

$$\{P\}_J = \begin{Bmatrix} P_1 \\ P_2 \\ P_3 \end{Bmatrix} = \begin{Bmatrix} 0 \\ 7 \\ 7 \end{Bmatrix} \text{ kN}$$

10. Irrelevant

11, 12, 13. Joint displacements and reactions are calculated. Because there is no restrained state in this case, the usual asterisks are omitted from the nomenclature to emphasize that the values obtained constitute final answers. This is also done in all succeeding steps.

$$\{X\} = [K_{PP}]^{-1}\{P\}$$

$$\begin{Bmatrix} X_1 \\ X_2 \\ X_3 \end{Bmatrix} = \begin{Bmatrix} 6.000 \\ 17.43 \\ -0.5714 \end{Bmatrix} 10^{-4} \text{ m}$$

$$\{R\} = [K_{RP}]\{X\}$$

$$\begin{Bmatrix} R_1 \\ R_2 \\ R_3 \end{Bmatrix} = \begin{Bmatrix} -4.00 \\ -3.00 \\ -7.00 \end{Bmatrix} = \begin{Bmatrix} 4.00 \\ 3.00 \\ 7.00 \end{Bmatrix} kN$$

14. Irrelevant
15, 16. Member end deformations are determined in local coordinates by the relationship
$\{\Delta\} = [T]\{X\}_m$
Member AC:

$$\begin{Bmatrix} \Delta_1 \\ \Delta_2 \\ \Delta_3 \\ \Delta_4 \end{Bmatrix}_{AC} = \begin{bmatrix} 0 & -1 & 0 & 0 \\ 1 & 0 & 0 & 0 \\ 0 & 0 & 0 & -1 \\ 0 & 0 & 1 & 0 \end{bmatrix} \begin{Bmatrix} 0 \\ 6.0 \\ 0 \\ 0 \end{Bmatrix} 10^{-4}$$

$$= \begin{Bmatrix} -6.000 \\ 0 \\ 0 \\ 0 \end{Bmatrix} 10^{-4} \text{ m}$$

Members AB and CB (by similar analysis):

$$\begin{Bmatrix} \Delta_1 \\ \Delta_2 \\ \Delta_3 \\ \Delta_4 \end{Bmatrix}_{AB} = \begin{Bmatrix} 3.6 \\ 4.8 \\ 13.6 \\ -10.92 \end{Bmatrix} 10^{-4}$$

$$\begin{Bmatrix} \Delta_1 \\ \Delta_2 \\ \Delta_3 \\ \Delta_4 \end{Bmatrix}_{CB} = \begin{Bmatrix} 0 \\ 0 \\ 10.0 \\ -14.29 \end{Bmatrix} 10^{-4} \text{ m}$$

17. Member forces follow by Eq. 16.21
Member AC:

$$\begin{Bmatrix} F_1 \\ F_2 \\ F_3 \\ F_4 \end{Bmatrix}_{AB} = \begin{bmatrix} 5000 & 0 & -5000 & 0 \\ 0 & 0 & 0 & 0 \\ -5000 & 0 & 5000 & 0 \\ 0 & 0 & 0 & 0 \end{bmatrix}$$

$$\cdot \begin{Bmatrix} -6.0 \\ 0 \\ 0 \\ 0 \end{Bmatrix} 10^{-4}$$

$$= \begin{Bmatrix} -3 \\ 0 \\ 3 \\ 0 \end{Bmatrix} = \begin{Bmatrix} 3 \text{ kN} \\ 0 \\ 3 \text{ kN} \\ 0 \end{Bmatrix}$$

Members AB and CB: (by similar analysis)

$$\begin{Bmatrix} F_1 \\ F_2 \\ F_3 \\ F_4 \end{Bmatrix}_{AB} = \begin{Bmatrix} -5 \\ 0 \\ 5 \\ 0 \end{Bmatrix}$$

$$\begin{Bmatrix} F_1 \\ F_2 \\ F_3 \\ F_4 \end{Bmatrix}_{CB} = \begin{Bmatrix} -5 \\ 0 \\ 5 \\ 0 \end{Bmatrix} kN$$

18. Irrelevant
Results of the analysis are displayed in Fig. 16.19d.

16.7 SPECIAL CONDITIONS

Numerical problems presented in the preceding sections exemplify a polished approach to matrix analysis of structures subjected to ordinary loads. The only actions considered were transverse loads applied to the members and joints. Support motions, thermal loads, initial strains, internal hinges, and so on, introduce anomalies not taken into account in developing the fundamental analytical procedure. Such irregularities were discussed earlier in the text. In some cases, the technique described could be considered crude. It is useful now to address special structural conditions in light of the improved analysis techniques of the present chapter. Each condition will be discussed in the context of a planar frame.

16.7.1 Support Motions

Support settlements introduce structural effects that are additional to those caused by applied loads. Including such motions is best accomplished by use of Eq. 16.53:

$$\begin{Bmatrix} \{P\} \\ \{R\} \end{Bmatrix} = \begin{bmatrix} [K_{PP}] & [K_{PR}] \\ [K_{RP}] & [K_{RR}] \end{bmatrix} \begin{Bmatrix} \{X\} \\ \{X_0\} \end{Bmatrix}$$

$$(16.53)$$

Equation 16.53 is identical to Eq. 16.40 except for the additional entry, $\{X_0\}$. The subscript zero indicates $\{X_0\}$ contains known values of the support motions. The lines of action of the quantities in $\{X_0\}$ are taken to align with the support reactions, $\{R\}$. Normally, the supports maintain their position, and $\{X_0\}$ is a null matrix. Under this circumstance, Eq. 16.53 degenerates to Eq. 16.40. Otherwise, a matrix solution follows by replacing step 11 in Table 16.1 with the relationships:

$$\{P\} = [K_{PP}]\{X\} + [K_{PR}]\{X_0\} \quad (16.54)$$

$$\{R\} = [K_{RP}]\{X\} + [K_{RR}]\{X_0\} \quad (16.55)$$

which results from the expansion of Eq. 16.53. With $\{P\}$ and $\{X_0\}$ known, Eqs. 16.54 and 16.55 can be evaluated simultaneously. Equation 13.54 can be written as

$$\{P\} - [K_{PR}]\{X_0\} = [K_{PP}]\{X\} \quad (16.56)$$

and solved to give

$$\{X\} = [K_{PP}]^{-1}\{\{P\} - [K_{PR}]\{X_0\}\}$$

$$(16.57)$$

Substitution of Eq. 16.57 into Eq. 16.56 produces the values of the reaction forces. Table 16.1 must be modified in several ways to account for possible support movements. Table 16.2 lists the revised steps. Again, asterisks are used in the table to distinguish the role of these equations in the released-state analysis.

16.7.2 Treatment of Boundary Conditions by Alternative Methods

Section 16.7.1 addressed the issue of the incorporation of known support motions into the matrix stiffness method of analysis. By proper numbering of the degrees of freedom, the arrangement shown in Eq. 16.53 can be guaranteed; i.e., all support restraints should be numbered separately, after the degrees of freedom have been numbered. In this way, no reorganization of the structural equations is necessary to properly partition the matrix $[K]$. However, execution of steps 11 through 13 of Table 16.2 is somewhat awkward when efficient equation-solving routines are to be adopted. To avoid disruption to the orderly storage and retrieval of data in some computer programs, it is desirable to replace step 12, i.e., evaluation of

Table 16.2. Modifications to Table 16.1 to Incorporate Known Support Motions.

Description	Matrix Operations
11. Form the governing stiffness relationships.	$\{P\} = [K_{PP}]\{X^*\} + [K_{PR}]\{X_0\}$ $\{R^*\} = [K_{RP}]\{X^*\} + [K_{RR}]\{X_0\}$
12. Calculate the joint displacements in the released state.	$\{X^*\} = [K_{PP}]^{-1}\{\{P\} - [K_{PR}]\{X_0\}\}$
13. Calculate the support reactions in the released state.	$\{R^*\} = [K_{RP}]\{X^*\} + [K_{RR}]\{X_0\}]$
14. Calculate the support reactions in the actual structure, i.e., combine the results of steps 8 and 13.	$\{R\} = \{R_0\} + \{R^*\}$
15. For each member, extract the joint displacements at its ends from the results of step 11 and the known support motions.	$\{X\}_m \begin{array}{l} \leftarrow \{X^*\} \\ \leftarrow \{X_0\} \end{array}$

$$\{X\} = [K_{PP}]^{-1}[\{P\} - [K_{PR}]\{X_0\}]$$

$$(16.57)$$

with an alternative operation. The alternative computation consists of adjusting the equations

$$[K_{PP}]\{X\} = \{P\} - [K_{PR}]\{X_0\} \quad (16.58)$$

with creation of the trivial relationships,

$$\{X_0\} = \{X_0\} \quad (16.59)$$

prior to solving them as a set.

As an example, consider a structure that has two degrees of freedom and two boundary conditions. The structural equations have the form

$$\begin{Bmatrix} P_1 \\ P_2 \\ P_3 \\ P_4 \end{Bmatrix} = \begin{bmatrix} K_{11} & K_{12} & K_{13} & K_{14} \\ K_{21} & K_{22} & K_{23} & K_{24} \\ K_{31} & K_{32} & K_{33} & K_{34} \\ K_{41} & K_{42} & K_{43} & K_{44} \end{bmatrix} \begin{Bmatrix} X_1 \\ X_2 \\ X_{01} \\ X_{02} \end{Bmatrix}$$

$$(16.60)$$

in which X_{01} and X_{02} are known. The joint displacements for this system would be determined instead of solving the equations

$$\begin{bmatrix} K_{11} & K_{12} & 0 & 0 \\ K_{21} & K_{22} & 0 & 0 \\ 0 & 0 & 1 & 0 \\ 0 & 0 & 0 & 1 \end{bmatrix} \begin{Bmatrix} X_1 \\ X_2 \\ X_{01} \\ X_{02} \end{Bmatrix}$$

$$= \begin{Bmatrix} P_1 - K_{13}X_{01} - K_{14}X_{02} \\ P_2 - K_{23}X_{01} - K_{24}X_{02} \\ X_{01} \\ X_{02} \end{Bmatrix}$$

$$(16.61)$$

As a further simplification of the inferred programming effort, Eq. 16.60 can be replaced by the system

$$\begin{bmatrix} K_{11} & K_{12} & K_{13} & K_{14} \\ K_{21} & K_{22} & K_{23} & K_{24} \\ K_{31} & K_{32} & \alpha K_{33} & K_{34} \\ K_{41} & K_{42} & K_{43} & \alpha K_{44} \end{bmatrix} \begin{Bmatrix} X_1 \\ X_2 \\ X_{01} \\ X_{02} \end{Bmatrix}$$

$$= \begin{Bmatrix} P_1 \\ P_2 \\ \alpha K_{33}X_{01} \\ \alpha K_{44}X_{02} \end{Bmatrix} \quad (16.62)$$

in which α is the largest number available in the computer system. Its effect in a given equation is to make the other stiffness coefficients in the row appear extremely small in comparison to the coefficient on the diagonal. In this way, their presence is essentially meaningless as they have a relative value approaching zero. In effect, the third and fourth equations become

$$\alpha K_{33}X_{01} = \alpha K_{33}X_{01} \quad (16.63)$$

$$\alpha K_{44}X_{02} = \alpha K_{44}X_{02} \quad (16.64)$$

and, after solving the four simultaneous equations, X_{01} and X_{02} assume correct values.

16.7.3 Initial Strains

Strain can be introduced into a structural member by many influences in addition to applied loads. Temperature changes and thermal gradients, shrinkage of concrete, and moisture changes in woods represent *natural* causes of strain. Theoretical member dimensions can only be achieved within a given tolerance. Hot-rolling induces strain (and correspondingly, "residual stress") into a member due to differential cooling. Laminating wood to produce larger structural members induces strain in the unloaded state. These fabrication effects, and construction practices such as prestressing concrete and welding, represent *human* causes of strain.

Natural and human causes of strain are taken into account in the same manner as transverse loads—i.e., by replacing their effect with appropriate initial end forces. Two aspects of this were presented in Chapter 14. These end forces are superimposed upon those caused by transverse loads and entered in the matrix $\{F_0\}$. The operation is performed as part of step 6 in Table 16.1. A simple example is presented to illustrate the general concept.

Example 16.5. A rectangular member is known to have been fabricated too long by an amount Δ_0. In addition, after assembly, the member experiences a linear gradient due to differential temperatures, T_1, and T_2, on opposite faces of the member. Determine the initial end forces for matrix analysis.

Solution:

(a) Figure 16.20 depicts the solution for the fabrication inaccuracy. The initial end forces, F_{02} and F_{05}, must be of sufficient magnitude to compress the member to the theoretical length, L. Equilibrium requires the condition

$$F_{02} = -F_{05}$$

Thus, a constant axial force exists in the member. If the ends were released, the strain at any section would be

$$\epsilon = \frac{\Delta_0}{L}$$

The normal stress required to prevent the strain is

$$\sigma = E\epsilon = \frac{E}{L}\Delta_0$$

and the corresponding force is obtained by integration through the member depth

$$F = \int_0^d \sigma \, dA$$

$$= \int_0^d \frac{E\Delta_0}{L} t \, dy$$

$$F = \frac{EA}{L}\Delta_0$$

where A is the cross-sectional area. This force must be compressive in nature. Therefore,

$$F_{02} = -F_{05} = \frac{EA}{L}\Delta_0$$

It is presumed that Δ_0 is sufficiently small to ignore the difference between L and $L + \Delta_0$.

(b) Figure 16.20b depicts the general forces produced by restraining the thermal effects at each end of the member. The condition

$$F_{01} = -F_{04}$$

is deduced from the symmetry of the structure. Consequently, moment equilibrium requires

$$F_{03} = F_{06} = 0$$

With no external load present, the zero shear and the unknown moment are constant throughout the span. Furthermore, horizontal equilibrium requires

$$F_{02} = -F_{05}$$

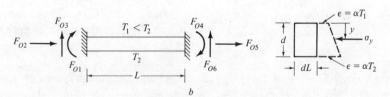

Fig. 16.20. Treatment of initial strains. *a*, fabrication oversize; *b*, thermal gradient.

which implies that the axial force is also constant. Therefore, evaluation of the internal moment and axial force *at any location* establishes their values at the member ends.

The thermal deformation at any location is depicted in Fig. 16.20b. The strain at a distance y from the top of the section is given by

$$\epsilon = \frac{\text{deformation}}{dL} = \alpha T_1 + \alpha(T_2 - T_1)\frac{y}{d}$$

The compressive normal stress required to prevent this strain is given by

$$\sigma_y = \epsilon E = \alpha E T_1 + \alpha E (T_2 - T_1)\frac{y}{d}$$

Integration through the depth produces the corresponding normal force,

$$F = \int_0^d \sigma_y \, dA = \int_0^d \sigma_y t \, dy$$

$$= \frac{\alpha E}{2}(T_1 + T_2) t \, d$$

$$F = \frac{\alpha EA}{2}(T_1 + T_2)$$

where t and d are the thickness and depth of the cross section, respectively. This force must exist at all points, giving

$$F_{02} = -F_{05} = \frac{\alpha EA}{2}(T_1 + T_2)$$

The internal resisting moment is given by

$$M = \int_0^d \sigma_y \left(\frac{d}{2} - y\right) dA$$

$$= -\frac{\alpha EI}{d}(T_2 - T_1)$$

where I is the moment of inertia ($\frac{1}{12}t\,d^3$). The negative sign indicates the internal moment is counterclockwise on the right face; therefore,

$$F_{04} = -F_{01} = \frac{\alpha EI}{d}(T_2 - T_1)$$

(c) The initial force matrix for the combined effect is

$$\begin{Bmatrix} F_{01} \\ F_{02} \\ F_{03} \\ F_{04} \\ F_{05} \\ F_{06} \end{Bmatrix} = \begin{Bmatrix} -\dfrac{\alpha EI}{d}(T_2 - T_1) \\[2mm] \dfrac{EA}{L}\Delta_0 + \alpha\dfrac{EA}{2}(t_1 + T_2) \\[2mm] 0 \\[2mm] \dfrac{\alpha EI}{d}(T_2 - T_1) \\[2mm] -\dfrac{EA}{L}\Delta_0 - \alpha\dfrac{EA}{2}(T_1 + T_2) \\[2mm] 0 \end{Bmatrix}$$

16.7.4 Internal Hinges

Continuity at all joints is a condition presumed in Sections 16.1 through 16.6. The presence of an internal hinge destroys this state at the location of the hinge. Two methods of accounting for such a discontinuity will be described.

The first and most direct approach is based upon a recognition that independent rotational degrees of freedom exist for each member end that frames into the internal hinge. Consider the planar frame shown in Fig. 16.21a. With the degrees of freedom numbered as indicated, the only alteration required in the matrix stiffness analysis occurs in the direct feeding process. Local member stiffness matrices for members BC and CD are constructed in the usual manner and transformed to the global coordinate system. The direct feeding process ensues at this time with a recognition of the separate rotational degrees of freedom at joint C. The member stiffness matrix member BC is fed into locations associated with degrees of freedom numbered 1 through 6. The member stiffness matrix for member CD is fed into locations associated with degrees of freedom Nos. 4, 5, and 7 (i.e., excluding No. 6). In this way, the individual stiffness contributions to the structure are assigned properly. Two additional alterations to the general procedure are necessary. First, forces corresponding to the separate degrees of freedom, No. 6 and 7, must be included in the initial force matrix. These forces are derived separately from members BC and

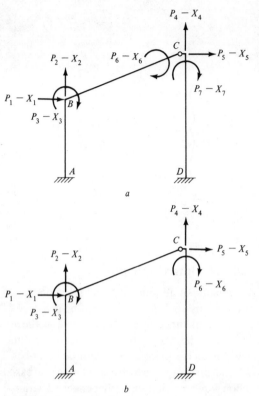

Fig. 16.21. Planar frame with an internal hinge.

CD, respectively. A moment directly applied to joint C must be assigned to either of the rotational directions (the choice is arbitrary). Second, in executing step 15 of Table 16.1 (extracting the member end deformations from $\{X\}$), the two rotations at joint C must be assigned to the appropriate member.

The technique just described is simple in concept but has the undesirable feature of complicating the development of automated computer algorithms. Basically, the additional rotational degree of freedom at each internal hinge destroys the regularity of the joint conditions. In addition to the input necessary to the joint conditions. In addition to the input necessary to note the presence of internal hinges, special routines are needed to accomplish the procedural modifications described in the preceding paragraph. An alternative technique exists and causes less interruption to Table 16.1. Consider the declaration of the degrees of freedom as indicated in Fig. 16.21b. The rotation at joint C of member BC has been omitted in this numbering pattern. With the recognition

that degree of freedom number 6 (associated with member CD) has no influence on member BC, the local member stiffness matrix $[S]_{BC}$ can then be adjusted to account for this fact. The general local stiffness matrix for a planar frame member was first derived by use of Eq. 15.1 and the boundary conditions in Eq. 15.2. With an internal hinge at the right end of the member, the boundary conditions become

$$@x = 0 \qquad y = C_4 = +\delta_L$$

$$@x = 0 \qquad y' = C_3 = -\Phi_L$$

$$@x = L \qquad y = C_1\frac{L^3}{6} + C_2\frac{L^2}{2}$$
$$+ C_3L + C_4 = +\delta_R$$

$$@x = L \qquad \frac{d^2y}{dx^2} = C_1L + C_2 = 0$$

in which Φ_L is measured relative to the original member axis (which is inclined in this case). Solving simultaneously gives

$$C_1 = -\frac{3}{L^2}\Phi_L + \frac{3}{L^3}\delta_L + \frac{3}{L^3}\delta_R$$

$$C_2 = \frac{3}{L}\Phi_L - \frac{3}{L^2}\delta_L + \frac{3}{L^2}\delta_R$$

$$C_3 = -\Phi_L$$

$$C_4 = +\delta_L \qquad (16.66)$$

Substitution into Eqs. 15.4 gives

$$\begin{Bmatrix} M_L \\ V_L \\ V_R \end{Bmatrix} = \begin{bmatrix} \dfrac{3EI}{L} & -\dfrac{3EI}{L^2} & +\dfrac{3EI}{L^2} \\[2mm] -\dfrac{3EI}{L^2} & \dfrac{3EI}{L^3} & -\dfrac{3EI}{L^3} \\[2mm] -\dfrac{3EI}{L^2} & \dfrac{3EI}{L^3} & -\dfrac{3EI}{L^3} \end{bmatrix} \begin{Bmatrix} \Phi_L \\ \delta_L \\ \delta_R \end{Bmatrix}$$

$$(16.67)$$

Equation 15.1 is based upon the quadruple-integration method in which V_R is taken to be positive when acting downward. In the matrix stiffness method, this shear acts upward. The local stiffness relationship for the flexural member depicted in Fig. 16.22 is obtained by changing the signs in the third row of Eq. 16.67 and expanding the size of the matrix to include M_R and ϕ_R, the moment and rotation at end R.

$$
\begin{Bmatrix} M_L \\ V_L \\ M_R \\ V_R \end{Bmatrix} = \begin{bmatrix} \dfrac{3EI}{L} & -\dfrac{3EI}{L^2} & 0 & \dfrac{3EI}{L^2} \\[2mm] -\dfrac{3EI}{L^2} & \dfrac{3EI}{L^3} & 0 & -\dfrac{3EI}{L^3} \\[2mm] 0 & 0 & 0 & 0 \\[2mm] \dfrac{3EI}{L^2} & -\dfrac{3EI}{L^3} & 0 & \dfrac{3EI}{L^3} \end{bmatrix} \begin{Bmatrix} \Phi_L \\ \delta_L \\ \Phi_R \\ \delta_R \end{Bmatrix}
$$

$$(16.68)$$

The explanation for the zeros in the third row is obvious, as M_R should be zero for all displaced states. However, the justification for zeros in the third column is less apparent. Clearly, if Φ_R were to have nonzero value while Φ_L, δ_L,

and δ_R were restrained (set equal to zero), the member would be deformed and the end forces would have nonzero values. Equation 16.68 does not produce such a result. This perplexity is explained by realizing that the four displacement quantities in Eq. 16.68 are not *independent*. Differentiation of Eq. 15.1, substitution of Eqs. 16.66, and setting x equal to L will verify this fact. The zero entries in the third column simply permit expanding the stiffness relationship to 4×4 to facilitate direct feeding, while still forcing the correct value of M_R to result.

The local stiffness matrix for a general planar frame member (with an internal hinge at the right end) is obtained by adding the axial effects to Eq. 16.68 to give

$$
[S]_m = \begin{bmatrix} \dfrac{3EI}{L} & 0 & -\dfrac{3EI}{L^2} & 0 & 0 & \dfrac{3EI}{L^2} \\[2mm] 0 & \dfrac{EA}{L} & 0 & 0 & -\dfrac{EA}{L} & 0 \\[2mm] -\dfrac{3EI}{L^2} & 0 & \dfrac{3EI}{L^3} & 0 & 0 & -\dfrac{3EI}{L^3} \\[2mm] 0 & 0 & 0 & 0 & 0 & 0 \\[2mm] 0 & -\dfrac{EA}{L} & 0 & 0 & \dfrac{EA}{L} & 0 \\[2mm] \dfrac{3EI}{L^2} & 0 & -\dfrac{3EI}{L^3} & 0 & 0 & -\dfrac{3EI}{L^3} \end{bmatrix}
$$

$$(16.69)$$

If the internal hinge is at the left end of the member, a similar derivation produces

$$
[S]_m = \begin{bmatrix} 0 & 0 & 0 & 0 & 0 & 0 \\[2mm] 0 & \dfrac{EA}{L} & 0 & 0 & -\dfrac{EA}{L} & 0 \\[2mm] 0 & 0 & \dfrac{3EI}{L^3} & -\dfrac{3EI}{L^2} & 0 & -\dfrac{3EI}{L^3} \\[2mm] 0 & 0 & -\dfrac{3EI}{L^2} & \dfrac{3EI}{L} & 0 & \dfrac{3EI}{L^2} \\[2mm] 0 & -\dfrac{EA}{L} & 0 & 0 & \dfrac{EA}{L} & 0 \\[2mm] 0 & 0 & -\dfrac{3EI}{L^3} & \dfrac{3EI}{L^2} & 0 & \dfrac{3EI}{L^3} \end{bmatrix}
$$

$$(16.70)$$

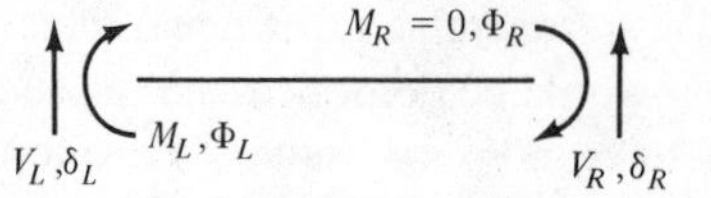

Fig. 16.22. Planar frame member—internal hinge at right end.

If hinges exist at both ends of the member, the local stiffness matrix is obtained by removing all the flexural entries (those involving EI) from either Eq. 16.69 or Eq. 16.70.

Use of Eq. 16.69 for the structure in Fig. 16.21b in lieu of the technique described for Fig. 16.21a is much simpler. The major modifications lie in the use of Eq. 16.69 for member *BC* in place of Eq. 16.9 and in the subsequent feeding into the structure stiffness matrix. The latter step is accomplished by associating the local rotational degree of freedom (after transformation to global coordinates) at the right end with the degree of freedom (after transformation to global coordinates) at the right end with the degree of freedom number 6 of the structure. This is gimmickry, because it has the effect of forcing rotation Φ_R for member *BC* to be set equal to X_6. This is erroneous but harmless because the magnitude of Φ_R has no influence on the final answers for all other quantities. The removal of Φ_R from the formulation (i.e., in Eq. 16.69) is a small price to pay for the convenience of a more regular feeding of the member stiffness matrix. Furthermore, if desired, the true magnitude of Φ_R can be determined subsequently by use of Eq. 15.1 and the known member deformations.

A significant, but subtle, feature that accompanies the use of stiffness matrices modified for the presence of internal hinges is the need to modify the initial end forces due to transverse loads. These forces must align with the independent local degrees of freedom. Consequently, general formulas must be derived for each of the restrained conditions depicted in Fig. 16.23. In all cases, the moment at each hinge is known to be zero. Remaining forces can be calculated by use of integration methods. As a convenience, results for common loadings are listed in Appendix A, Table A.4, for the cases shown in Fig. 16.23a and b. The situation in Fig. 16.23c can be analyzed by

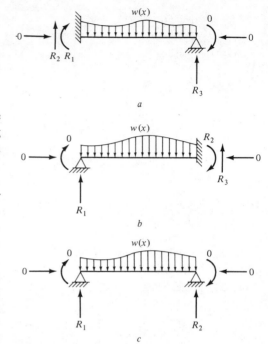

Fig. 16.23. Restraining forces for members containing internal hinges.

simple statics. In all cases, axial forces applied to the member are treated by use of Table A.7 in Appendix A.

16.7.5 Inclined Supports

Occasionally, a roller support is positioned so that its direction of motion does not align with either global coordinate axis. The planar frame in Fig. 16.24 is an illustration of this condition. With the concept of transformation matrices now well understood, it is a straightforward matter to account for an inclined support. After the member stiffnesses have been formed in global coordinates, *each member that frames into an inclined support* must be subject to a secondary transformation. The purpose of the transformation is to rotate the force and displacement vectors from the established global axes to the axes x_s–y_s that specify the position of the inclined roller (see Fig. 16.24). If $[K]_m$ is the global member stiffness matrix, the secondary transformation takes the form

$$[K]'_m = [T_s][K]_m[T_s]^T \quad (16.71)$$

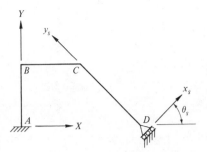

Fig. 16.24. Frame with an inclined support.

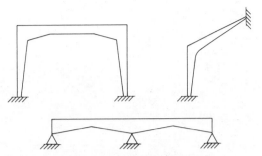

Fig. 16.25. Structure with nonprismatic members.

where

$$[T_s] = \begin{bmatrix} 1 & 0 & 0 & 0 \\ 0 & 1 & 0 & 0 \\ 0 & 0 & \cos\theta_s & \sin\theta_s \\ 0 & 0 & -\sin\theta_s & \cos\theta_s \end{bmatrix}$$

$$(16.72)$$

and θ_s is the inclination of the roller support. Equation 16.72 applies if the roller support is at the far end of the member. When the roller support is at the near end, the transformation matrix has the form

$$[T_s] = \begin{bmatrix} \cos\theta_s & \sin\theta_s & 0 & 0 \\ -\sin\theta_s & \cos\theta_s & 0 & 0 \\ 0 & 0 & 1 & 0 \\ 0 & 0 & 0 & 1 \end{bmatrix} \quad (16.73)$$

Assembly of the structure stiffness matrix and incorporation of support conditions follow the usual procedure. Initial end forces due to transverse loads must be subjected to the matrix operation

$$\{P_0\}'_m = [T_s]\{P_0\}_m \quad (16.74)$$

where $\{P_0\}_m$ are the forces obtained by application of Eq. 16.49. The reader is alerted to the position of the *transposed* transformation matrix in Eq. 16.71 and its absence in Eq. 16.74. This differs from Eq. 16.20 and 16.49, and the reader is encouraged to ponder why.

16.8 NONPRISMATIC MEMBERS

Frequently, structural members are tapered, curved, haunched, or otherwise articulated (see Fig. 16.25) to increase their strength in the regions where the internal forces have their greatest magnitude. Derivation of exact stiffness matrices for these types of members is a prohibitive task and generally not undertaken. Instead, the usual practice is to model the variable cross section by a series of stepped segments such as depicted in Fig. 16.26. In this way, the intersection of each pair of adjoining segments can be declared as individual prismatic members. The transformed stiffness matrix for each segment is then fed into the structure stiffness matrix. Although this technique increases the size of the structure stiffness matrix, it is preferable to attempting an exact formulation of the stiffness matrix.

As an alternative to modeling a nonprismatic member by segmentation, it is possible to determine a numerical approximation to the exact 6 × 6 stiffness matrix. Consider the determination of the elements in the sixth column of the stiffness matrix for the member shown in Fig. 16.27a. By replacing the tapered member with the stepped approximation shown in Fig. 16.26, a 16 × 16 local stiffness matrix can be developed. Using this matrix, the structure is

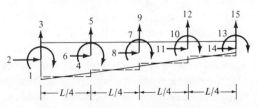

Fig. 16.26. Idealization of a tapered member.

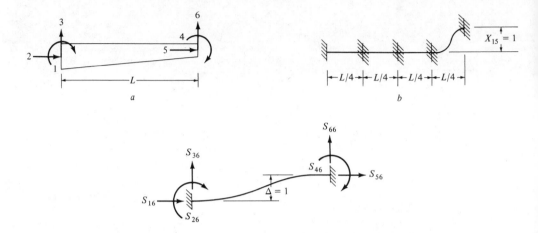

Fig. 16.27.   Approximating the member stiffness matrix of a tapered member.

then analyzed for an imposed unit displacement at the right end. The restrained state for this analysis is depicted in Fig. 16.27b, and it is seen that forces exist only in one segment. The initial end forces for this state are calculated by use of Appendix A, Table A.5, Case f. By releasing the interior supports and completing the matrix stiffness analysis, the end forces shown in Fig. 16.27c result. The forces shown in Fig. 16.27c (obtained by combining the forces in the restrained and released states) constitute the desired stiffness coefficients for the member in Fig. 16.27a. Remaining columns of $[S]_m$ are determined in similar fashion, i.e., by imposing unit displacements at each of the five remaining local degrees of freedom and repeating the analysis for each case. Clearly, this is a time consuming task. A more expedient technique is *matrix condensation*, which is a topic outside the scope of this textbook.

16.9   COMBINED SYSTEMS

Structures constructed of a combination of member types are a frequent occurrence. Suspension bridges, cable-stayed bridges, guyed towers, bracing systems in a frame, and ski lifts are a few common examples. The analysis of combined systems by the matrix stiffness method causes no great difficulty. Recognizing that the structure stiffness matrix is the assemblage of the individual stiffness matrices of the component members is the key to understand-

ing the analytical technique. Initially, the global member stiffness matrix is constructed for each member based upon its structural type. Regardless of the nature of any given member, its presence contributes to the stiffness of the overall system. This is taken into account by the direct feeding of each global stiffness matrix into $[K]$. The size of the member stiffness matrix differs for each type of member, but this is no concern. The contents of any given matrix are fed according to its connectivity to the degrees of freedom and supports of the complete structure. Equivalent member end forces and joint loads are fed into the load matrix by similar considerations. Solution of the resulting structural equations follows the usual steps. The subsequent determination of member deformations and member forces is complicated only by a need to exercise care in extracting displacement quantities from $\{X\}$ that are correct in number, type, and location for each particular member. These are entered in $\{X\}_m$, transformed to local coordinates, and premultiplied by $[S]_m$ to produce the forces. A simple example will serve to illustrate the features just described.

Example 16.6.   Analyze the planar structure illustrated in Fig. 16.28 by the matrix stiffness method. Use $EA/L = 5000$ kip/ft for the cable and $EA/L = 10,000$ kip/ft and $EI/L = 20,000$ kip-ft for the beam.

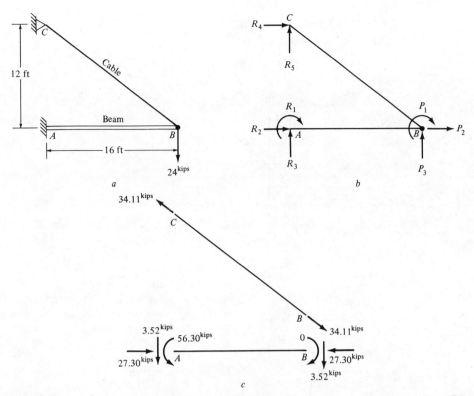

Fig. 16.28. Structure with a combination of member types. *a*, complete structure; *b*, joint conditions; *c*, internal forces.

Solution: The local member stiffness matrices are evaluated by use of Eqs. 16.9 and 16.22. In the case of the horizontal member AB, the global and member stiffness matrices are identical.

$$[K]_{AB} = [S]_{AB} = \begin{array}{c} \\ 1 \\ 2 \\ 3 \\ ① \\ ② \\ ③ \end{array} \begin{array}{cccccc} 1 & 2 & 3 & ① & ② & ③ \\ \begin{bmatrix} 80{,}000 & 0 & 7500 & 40{,}000 & 0 & 7500 \\ 0 & 10{,}000 & 0 & 0 & 10{,}000 & 0 \\ 7500 & 0 & 938 & 7500 & 0 & 938 \\ 40{,}000 & 0 & 7500 & 80{,}000 & 0 & 7500 \\ 0 & 10{,}000 & 0 & 0 & 10{,}000 & 0 \\ 7500 & 0 & 938 & 7500 & 0 & 938 \end{bmatrix} \end{array}$$

$$[S]_{CB} = \begin{bmatrix} 5000 & 0 & 5000 & 0 \\ 0 & 0 & 0 & 0 \\ 5000 & 0 & 5000 & 0 \\ 0 & 0 & 0 & 0 \end{bmatrix}$$

Transformation of $[S]_{CB}$ to global coordinates yields

$$[K]_{CB}$$

$$= \begin{array}{c} \\ 4 \\ 5 \\ ② \\ ③ \end{array} \begin{array}{cccc} 4 & 5 & ② & ③ \\ \begin{bmatrix} 3200 & -2400 & -3200 & 2400 \\ -2400 & 1800 & 2400 & -1800 \\ -3200 & 2400 & 3200 & -2400 \\ 2400 & -1800 & -2400 & 1800 \end{bmatrix} \end{array}$$

Connectivity shown for the above matrices is based upon Fig. 16.28b. Circled and uncircled numbers outside the matrices indicate degrees of freedom and directions of restraint, respectively. With the absence of support motions, it is unnecessary to build the entire structure stiffness matrix. Instead, direct feeding into its partitions produces

$$[K_{PP}] = \begin{array}{c} \\ ① \\ ② \\ ③ \end{array} \begin{array}{ccc} ① & ② & ③ \\ \begin{bmatrix} 80{,}000 & 0 & 7500 \\ 0 & 13{,}200 & -2400 \\ 7500 & -2400 & 2738 \end{bmatrix} \end{array}$$

$$[K_{RP}] = \begin{array}{c} 1 \\ 2 \\ 3 \\ 4 \\ 5 \end{array} \begin{array}{ccc} ① & ② & ③ \\ \begin{bmatrix} 40{,}000 & 0 & 7500 \\ 0 & -10{,}000 & 0 \\ -7500 & 0 & -938 \\ 0 & -3200 & 2400 \\ 0 & 2400 & -1800 \end{bmatrix} \end{array}$$

The load matrix contains the single joint load.

$$\{P\} = \begin{Bmatrix} 0 \\ 0 \\ -24 \end{Bmatrix} \text{ kips}$$

Matrix analysis yields

$$\{X^*\} = [K_{PP}]^{-1}\{P\} = \begin{Bmatrix} 1.408 \\ -2.730 \\ -15.014 \end{Bmatrix} \times 10^{-3} \text{ ft}$$

$$\{R^*\} = [K_{RP}]\{X\} = \begin{Bmatrix} -56.30 \\ 27.30 \\ 3.53 \\ -27.30 \\ 20.47 \end{Bmatrix} \text{ kips}$$

Member deformations are calculated as follows:
    Member AB:

$$\{\Delta^*\}_{AB} = [T]_{AB}\{X^*\}_{AB} = \{X^*_{AB}\}$$

$$= \begin{Bmatrix} 0 \\ 0 \\ 0 \\ X_1 \\ X_2 \\ X_3 \end{Bmatrix} = \begin{Bmatrix} 0 \\ 0 \\ 0 \\ 1.408 \\ -2.730 \\ -15.014 \end{Bmatrix} \times 10^{-3} \text{ ft}$$

$$\{\Delta^*\}_{BC} = [T]_{CB} \begin{Bmatrix} 0 \\ 0 \\ X_2 \\ X_3 \end{Bmatrix}$$

$$= \begin{bmatrix} 0.8 & -0.6 & 0 & 0 \\ 0.6 & 0.8 & 0 & 0 \\ 0 & 0 & 0.8 & -0.6 \\ 0 & 0 & 0.6 & 0.8 \end{bmatrix}$$

$$\cdot \begin{Bmatrix} 0 \\ 0 \\ -2.730 \\ -15.014 \end{Bmatrix} \times 10^{-3}$$

$$= \begin{Bmatrix} 0 \\ 0 \\ -6.82 \\ -13.35 \end{Bmatrix} \times 10^{-3} \text{ ft}$$

and multiplication by the corresponding local member stiffness matrices gives the results shown in Fig. 16.28c. With the absence of transverse loads, the analysis of a restrained state is unnecessary, and all asterisked quantities represent final answers.

16.10   REMARKS

A major goal of this chapter is to extend the reader's understanding of matrix structural analysis beyond simple knowledge of the basic equations. Various concepts have been discussed that allow a more sophisticated approach to the implementation of the matrix stiffness method. The use of transformation matrices and direct feeding of member stiffness matrices introduces a certain degree of automation into the procedure. By combining these concepts with the use of efficient methods for solving simultaneous equations, the fundamental background needed as a prerequisite to producing highly refined computer algorithms would complete. Planar and space structures can be analyzed by use of a common base of mathematical operations and computer-programming techniques. This textbook is intended to prepare the reader for subsequent study of space structures and advanced topics.

PROBLEMS

Note: All structural analysis problems are to be done
    by   the   semi-automated   matrix   analysis

method. Unless other values are stated in the problem, use $EA = 200,000$ kips for U.S. customary units and $EA = 20000$ kN for SI units.

16-1. Find the inverse of $[T'] = \begin{bmatrix} \cos\theta & \sin\theta \\ -\sin\theta & \cos\theta \end{bmatrix}$ and by any method to prove that $[T']^{-1} = [T']^T$.

16-2. Analyze the planar trusses in Problem 14-1.

16-3. Analyze the planar truss in Example 14.4.

16-4. Analyze the planar truss in Example 14.5.

16-5. Analyze the planar truss in Problem 14.2a.

16-6. Analyze the planar truss in Problem 14.2c.

16-7. Analyze the planar truss in Problem 15-18.

16-8. Analyze the planar trusses in Problem 15-26.

16-9. Analyze the continuous beam in Problem 15-8.

16-10. Analyze the continuous beam in Problem 11-7.

16-11. Analyze the continuous beams in Problem 15-7.

16-12. Analyze the continuous beam in Problem 15-10.

16-13. Analyze the continuous beams in Problem 13-7.

16-14. Analyze the continuous beam in Problem 12-5.

16-15. Analyze the continuous beam in Problem 12-11.

16-16. Analyze the planar frame in Problem 15-31.

16-17. Analyze the planar frame in Problem 15-32.

16-18. Analyze the planar frame in Problem 15-33.

16-19. Analyze the planar frames in Problem 15-35.

16-20. Analyze the planar frame in Problem 15-36a.

16-21. Analyze the planar frames in Problem 15-36b.

16-22. Analyze the given planar frame. Use $E = 30,000$ ksi, $I = 100$ in.^{4}, and $A = 2.5$ in.^{2} for all members.

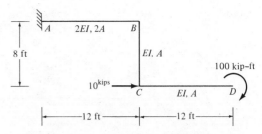

16-23. The given frame is subjected to a wind load of 2 kip/ft on the vertical projection of member AB. Analyze the structure for $EI = 48,000$ kip-ft^2 and $EA = 400,000$ kips.

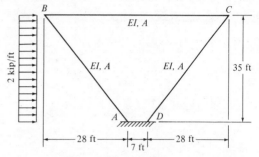

16-24. Analyze the given member for a support settlement of 0.5 in. at point A. See Section 16.7.1.

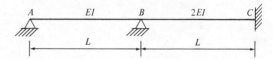

16-25. Analyze the planar truss in Problem 14-11 for the described settlement. See Section 16.7.1.

16-26. Analyze the planar truss in Problem 14-12 for the described settlement. See Section 16.7.1.

16-27. Analyze the planar truss in Problem 14-13 for the described fabrication error. See Section 16.7.3.

16-28. All members of the given truss were fabricated 0.25 in. longer than the dimensions shown. Subsequent to erection, each member was subjected to a temperature increase of 100°F. Determine the resulting member forces. Use $EA = 20,000$ kips and $\alpha = 6.5 \times 10^{-6}/°F$ for all members.

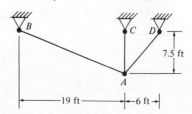

16-29. Analyze the given continuous beam by any of the methods described for the treatment of internal hinges.

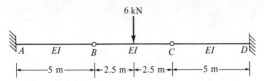

16-30. Repeat Problem 16-29 for the given planar frame. For both members, $EA = 200,000$ kips.

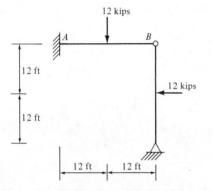

16-31. Build the structure stiffness matrix for the given plane frame. Partition the matrix to reflect boundary conditions.

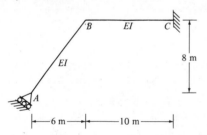

16-32. For each beam derive the member stiffness matrix corresponding to the given degrees of freedom. Use $L = 20$ ft.

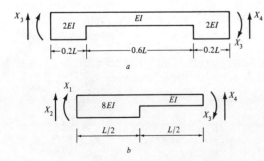

16-33. Build the structure stiffness matrix for the combined system. The cable area is one-tenth of the area of the members. Partition the matrix to reflect boundary conditions.

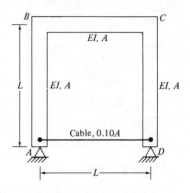

16-34. Repeat Problem 16-33 for the given structure. Both cables have an area equal to one-tenth of the area of the members.

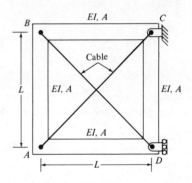

16-35. Build $[K]$ for the given structure. The frame members have $EA = .01EI$. Determine the member forces.

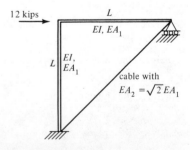

16-36. Execute the matrix operation $[K]_m = [T]^T [S]_m [T]$ for the general grid member described in Section 16.2.3.4.

Appendix A
Tables

Table A.1. Areas and Centroids.

Triangle	$A = \dfrac{1}{2}\,hL$ $\overline{X} = \dfrac{L}{3}$
Triangle	$A = \dfrac{1}{2}\,hL$ $\overline{X}_1 = \dfrac{L+a}{3}$ $\overline{X}_2 = \dfrac{L+b}{3}$
Parabola	$A = \dfrac{1}{3}\,hL$ $\overline{X}_1 = \dfrac{L}{4}$ $\overline{X}_2 = \dfrac{3}{4}\,L$
Parabola	$A = \dfrac{2}{3}\,hL$ $\overline{X}_1 = \dfrac{5}{8}\,L$ $\overline{X}_2 = \dfrac{3}{8}\,L$
Power curve	$A = \dfrac{1}{(n+1)}\,hL$ $\overline{X} = \dfrac{(n+1)}{(n+2)}\,L$
Chord of parabola	$h = \dfrac{1}{8}\,qL^2$ $A = \dfrac{2}{3}\,hL$ $\overline{X} = \dfrac{L}{2}$

Table A.2. Simple Beam Moments.

Loading	Moment Diagram	Moment Values
		$M_{max} = \dfrac{1}{8} qL^2$
		$M_{max} = \dfrac{Pab}{L}$
		$M_{max} = \dfrac{PL}{4}$
		$R_2 = \dfrac{qb}{2L}(2L - b)$ $M_{max} = \dfrac{R_2^2}{2q}$ $M_1 = \dfrac{qab^2}{2L}$
		$M_{max} = M$
		$M_1 = -M\dfrac{a}{L}$ $M_2 = M\dfrac{b}{L}$

Table A.3. Fixed Beam Moments.

Loading	Fixed-End Moments
 a	$M_1 = -\dfrac{PL}{8}$ $M_2 = +\dfrac{PL}{8}$
 b	$M_1 = -\dfrac{Pab^2}{L^2}$ $M_2 = +\dfrac{Pba^2}{L^2}$
 c	$M_1 = -\dfrac{1}{12}\,wL^2$ $M_2 = +\dfrac{1}{12}\,wL^2$
 d	$M_1 = -\dfrac{1}{30}\,wL^2$ $M_2 = +\dfrac{1}{20}\,wL^2$
 e	$M_1 = -\dfrac{1}{L^2}\displaystyle\int_0^L (L-x)^2 x\,w(x)\,dx$ $M_2 = +\dfrac{1}{L^2}\displaystyle\int_0^L (L-x)x^2\,w(x)\,dx$

Table A.4. Propped Beam Moments.

Loading	Fixed-End Moment
a	$M_1 = -\dfrac{3}{16}\,PL$
b	$M_1 = -\dfrac{Pab}{2L^2}\,(b+L)$
c	$M_1 = -\dfrac{1}{8}\,wL^2$
d	$M_1 = -\dfrac{7}{120}\,wL^2$
e	$M_1 = -\dfrac{1}{2L^2}\displaystyle\int_0^L (L-x)(2L-x)\,x\,w(x)\,dx$
f	$M_1 = +\dfrac{1}{2}\,M$

Table A.5. Moments Due to End Displacements.

Beam Condition	Fixed-End Moments
 f	$M_1 = -\dfrac{6EI}{L^2}\,\Delta$ $M_2 = -\dfrac{6EI}{L^2}\,\Delta$
 g	$M_1 = -\dfrac{3EI}{L^2}\,\Delta$

Table A.6. Torsional Fixed-End Moments.

Loading	Fixed-End Moments
	$T_1 = -\dfrac{T}{2}$ $T_2 = -\dfrac{T}{2}$
	$T_1 = -T\dfrac{b}{L}$ $T_2 = -T\dfrac{a}{L}$

Table A.7. Fixed-End Reactions Due to Axial Loads.

Loading	Fixed-End Reactions
	$T_1 = -\dfrac{T}{2}$ $T_2 = -\dfrac{T}{2}$
	$T_1 = -T\dfrac{b}{L}$ $T_2 = -T\dfrac{a}{L}$

Appendix B
Summary of Matrix Algebra

B.1 INTRODUCTION

Matrix algebra allows algebraic information to be presented in an abbreviated form. Applications of matrix algebra range over a wide realm; however, this summary is limited to a discussion of its use in manipulating linear equations. Employment of matrices permits a systematic approach to linear algebra. A matrix, the basic quantity used in matrix algebra, represents the fundamental tool in matrix structural analysis. By expressing the various equations used in structural analysis in a matrix format, widespread use of electronic digital computation is possible.

A solid understanding of the rudimentary matrix operations is an essential prerequisite to undertaking a study of matrix structural analysis. This appendix presents the minimal review necessary to insure an adequate background for comprehending the material in this text. For readers who intend to pursue the subject of matrix structural analysis more extensively, many textbooks on the mathematics of matrix algebra are available.

B.2 DEFINITION AND NOTATION

A rectangular array of numbers arranged in rows and columns constitutes a matrix. The usual symbol for a matrix is either a capital letter within brackets or a capital letter in boldface print. Its contents are displayed as shown in Eq. B.1:

$$[A] = \begin{bmatrix} a_{11} & a_{12} & \cdots & a_{1n} \\ a_{21} & a_{22} & \cdots & a_{2n} \\ \cdot & \cdot & & \cdot \\ \cdot & \cdot & & \cdot \\ \cdot & \cdot & & \cdot \\ a_{m1} & a_{m2} & \cdots & a_{mn} \end{bmatrix} \quad (B.1)$$

The quantities embraced by the brackets, which are called *elements* of the matrix, are subscripted in order to denote their position. The first subscript denotes the row, and the second subscript denotes the column in which the element appears. The *size* or *order* of the matrix is stated by specifying the number of rows and columns it contains. The matrix in Eq. B.1 is $m \times n$ in size.

Certain matrix forms have been given special names and should be mentioned. A matrix that has the same number of rows and columns ($m = n$) is a *square matrix*, such as that shown in Eq. B.2:

$$[A] = \begin{bmatrix} a_{11} & a_{12} & \cdots & a_{1m} \\ a_{21} & a_{22} & \cdots & a_{2m} \\ \cdot & \cdot & & \cdot \\ \cdot & \cdot & & \cdot \\ \cdot & \cdot & & \cdot \\ a_{m1} & a_{m2} & \cdots & a_{mm} \end{bmatrix} \quad (B.2)$$

If for two square matrices $[A]$ and $[B]$, all entries obey the equality $a_{ij} = b_{ji}$, then $[A]$ is termed the *transpose* of $[B]$.

A square matrix in which all the elements except $a_{11}, a_{22}, \cdots a_{mm}$ are zero is called a *diagonal matrix*. Equation B.3 represents such a matrix.

$$[A] = \begin{bmatrix} a_{11} & 0 & \cdots & 0 \\ 0 & a_{22} & & \cdot \\ \cdot & & \cdot & \cdot \\ \cdot & & & \cdot \\ 0 & & \cdots & a_{mm} \end{bmatrix} \quad \text{(B.3)}$$

If the nonzero entries in a diagonal matrix all have the value unity, the matrix is called an *identity* or *unit matrix* and is symbolized by the notation $[I]$.

A *column matrix* contains only a single column of elements, i.e., $(n = 1)$, as shown in Eq. B.4:

$$\{A\} = \begin{Bmatrix} a_{11} \\ a_{21} \\ \cdot \\ \cdot \\ a_{m1} \end{Bmatrix} \quad \text{(B.4)}$$

To distinguish a column matrix, braces are used in place of brackets to enclose both the symbol and the elements.

A *row matrix* contains only a single row of elements, i.e., $(m = 1)$, as shown in Eq. B.5:

$$[A] = [a_{11} \quad a_{12} \quad \cdots \quad a_{1n}] \quad \text{(B.5)}$$

No distinguishing symbol is employed for a row matrix.

Finally, a matrix in which $a_{ij} = a_{ji}$ (i and j being general) is referred to as a *symmetric matrix*. The inverse of a symmetric matrix exhibits a special quality that will be stated later.

B.3 BASIC OPERATIONS

B.3.1 Scalar Multiplication

A given matrix can be multiplied by a scalar number, k, and the result is

$$k[A] = \begin{bmatrix} ka_{11} & ka_{12} & \cdots & ka_{1n} \\ ka_{21} & ka_{22} & & \cdot \\ \cdot & \cdot & & \cdot \\ \cdot & \cdot & & \cdot \\ ka_{m1} & ka_{m2} & \cdots & ka_{mn} \end{bmatrix}$$

For example,

$$\frac{1}{6} \begin{bmatrix} 3 & 12 & 9 \\ 6 & 6 & 3 \\ -12 & -18 & 6 \end{bmatrix} = \begin{bmatrix} 0.5 & 2 & 1.5 \\ 1 & 1 & 0.5 \\ -2 & -3 & 1 \end{bmatrix}$$

B.3.2 Addition and Subtraction

Two matrices, $[A]$ and $[B]$, which must have identical size, can be algebraically combined. The result, is a matrix, $[C]$, of the same size. The symbols associated with matrix addition are

$$[C] = [A] + [B] \quad \text{(B.6)}$$

in which c_{ij}, an element in $[C]$, is given by

$$c_{ij} = a_{ij} + b_{ij} \quad \text{(B.7)}$$

In Eq. B.7, a_{ij} and b_{ij} are elements in matrices $[A]$ and $[B]$, respectively.

The difference between two matrices is expressed as

$$[C] = [A] - [B] \quad \text{(B.8)}$$

in which

$$c_{ij} = a_{ij} - b_{ij} \quad \text{(B.9)}$$

Again, all matrices in Eq. B.8 must be of the same size. As an example, if

$$[A] = \begin{bmatrix} 7 & 4.5 \\ -2 & -6.3 \\ \sqrt{2} & 0 \end{bmatrix} \quad [B] = \begin{bmatrix} 6 & 1.9 \\ 0 & 7.0 \\ \sqrt{2} & \sqrt{3} \end{bmatrix}$$

then

$$[A] + [B] = \begin{bmatrix} (7+6) & (4.5+1.9) \\ (-2+0) & (-6.3+7.0) \\ (\sqrt{2}+\sqrt{2}) & (0+\sqrt{3}) \end{bmatrix} = \begin{bmatrix} 13 & 6.4 \\ -2 & 0.7 \\ 2\sqrt{2} & \sqrt{3} \end{bmatrix}$$

$$[A] - [B] = \begin{bmatrix} (7-6) & (4.5-1.9) \\ (-2-0) & (-6.3-7.0) \\ (\sqrt{2}-\sqrt{2}) & (0-\sqrt{3}) \end{bmatrix} = \begin{bmatrix} 1 & 2.6 \\ -2 & -13.3 \\ 0 & -\sqrt{3} \end{bmatrix}$$

B.3.3 Multiplication

Symbolically, the product of two matrices is indicated by

$$[C] = [A] [B] \qquad (B.10)$$

in which $[A]$ and $[B]$ must be of compatible size. Compatibility of size requires only that the number of *columns* in $[A]$ be equal to the number of *rows* in $[B]$. If the size of $[A]$ is $(m \times n)$ and the size of $[B]$ is $(n \times p)$, the size of the resulting matrix product, $[C]$, is $(m \times p)$, An element in $[C]$ is given by

$$c_{ij} = \sum_{k=1}^{n} a_{ik}b_{kj} \qquad (B.11)$$

For example, if

$$[A] = \begin{bmatrix} 1 & 6 \\ -2 & 5 \\ 3 & 4 \end{bmatrix} \quad [B] = \begin{bmatrix} 7 & 9 \\ 8 & -10 \end{bmatrix}$$

then

$$[C] = \begin{bmatrix} [1(7)+6(8)] & [1(9)+1(-10)] \\ [-2(7)+5(8)] & [-2(9)+5(-10)] \\ [3(7)+4(8)] & [3(9)+4(-10)] \end{bmatrix} = \begin{bmatrix} 55 & -1 \\ 26 & -68 \\ 53 & 13 \end{bmatrix}$$

It is important to note that, in general, $[A][B] \neq [B][A]$. Also, if a matrix is premultiplied or postmultiplied by $[I]$, the result is the original matrix.

B.3.4 Determinant

An operation that has major significance in matrix algebra is the *expansion of the determinant*. Given a square matrix, $[A]$, of size $(n \times m)$, its determinant is symbolically indicated by

$$\det A = \begin{vmatrix} a_{11} & a_{12} & \cdots & a_{1n} \\ a_{21} & a_{22} & \cdots & a_{2n} \\ \cdot & \cdot & & \cdot \\ \cdot & \cdot & & \cdot \\ \cdot & \cdot & & \cdot \\ a_{n1} & a_{n2} & \cdots & a_{nn} \end{vmatrix}$$

$$(B.12)$$

The determinant of a matrix is a number obtained by evaluating either of the expressions in Eq. B.13.

$$\det A = \sum_{k=1}^{n} a_{ik} C_{ik} = \sum_{k=1}^{n} a_{ki} C_{ki} \qquad (i \leqslant n)$$

$$(B.13)$$

in which $C_{ik}(C_{ki})$ is the "cofactor" of $a_{ik}(a_{ki})$. The cofactor of a_{ik} is given by

$$C_{ik} = (-1)^{i+k} M_{ik} \qquad (B.14)$$

In Eq. B.14, the term M_{ik} represents a quantity referred to as the *minor* of element a_{ik}. A minor is a *determinant of reduced order* obtained from the original determinant by removing the elements in the ith row and kth column. Thus, a cofactor (Eq. B.14) is simply a "signed minor," i.e., a positive or negative determinant depending on the values of the subscripts, i and k.

As an example, if

$$[A] = \begin{bmatrix} a_{11} & a_{12} & a_{13} \\ a_{21} & a_{22} & a_{23} \\ a_{31} & a_{32} & a_{33} \end{bmatrix}$$

the minor of element a_{23} is

$$M_{23} = \begin{vmatrix} a_{11} & a_{12} \\ a_{31} & a_{32} \end{vmatrix}$$

and the corresponding cofactor is

$$C_{23} = (-1)^{2+3} M_{23} = - \begin{vmatrix} a_{11} & a_{12} \\ a_{31} & a_{32} \end{vmatrix}$$

The determinant of $[A]$ is calculated by application of Eq. B.13. Choosing the first of the optional summations and selecting $i = 3$ yields

$$\det A = a_{31} C_{31} + a_{32} C_{32} + a_{33} C_{33}$$

Alternatively, choosing the second summation and $i = 3$ yields

$$\det A = a_{13} C_{13} + a_{23} C_{23} + a_{33} C_{33}$$

The former expansion is referred to as an *expansion along the third row* and the latter as an *expansion down the third column*. Both expansions would produce the same final numerical result. More importantly, Eq. B.13 indicates that the choice of the particular row or column for expanding the determinant is totally arbitrary, i.e., all choices give the same answer.

A numerical example will now be given. If

$$[A] = \begin{bmatrix} 6 & -1 & 4 \\ -1 & 8 & 12 \\ 4 & 12 & 64 \end{bmatrix}$$

then expanding the determinant along the second row gives

$$\det A = (-1)C_{21} + (8)C_{22} + (12)C_{23}$$

$$= 1 \begin{vmatrix} -1 & 4 \\ 12 & 64 \end{vmatrix} + 8 \begin{vmatrix} 6 & 4 \\ 4 & 64 \end{vmatrix} - 12 \begin{vmatrix} 6 & -1 \\ 4 & 12 \end{vmatrix}$$

Expanding each of the minors (along the first row in each case) gives

$$\det A = 1\,[(-1)\,(64) - (4)\,(12)]$$
$$+ 8\,[(6)\,(64) - 4(4)]$$
$$- 12\,[6\,(12) - (-1)\,(4)]$$
$$= 1920$$

B.3.5 Inversion

Division of matrices is an undefined operation in matrix algebra. However, an operation similar in meaning to taking the reciprocal of a number exists. This operation is referred to as finding the *inverse of matrix* or, simply, *inversion*. If a is a number, then

$$\left(\frac{1}{a}\right) a = 1$$

in which $(1/a)$ is the *reciprocal* of a. In matrix algebra, if

$$[B]\,[A] = [I]$$

then $[B]$ is the *inverse* of $[A]$ and is denoted as $[A]^{-1}$, i.e.,

$$[A]^{-1}\,[A] = [I] \qquad (B.15)$$

Although the derivation is omitted in this review and summary of matrix algebra, the inverse of $[A]$ is given by

$$[A]^{-1} = \frac{1}{\det A}\ \text{adjoint } A \qquad (B.16)$$

The adjoint A is a matrix expressed as

$$\text{adjoint } A = [CoA]^T = \begin{bmatrix} C_{11} & C_{12} & \cdots & C_{1n} \\ C_{21} & C_{22} & \cdots & C_{2n} \\ \cdot & \cdot & & \cdot \\ \cdot & \cdot & & \cdot \\ \cdot & \cdot & & \cdot \\ C_{n1} & C_{n2} & \cdots & C_{nn} \end{bmatrix}^T$$

$$(B.17)$$

Matrix $[CoA]$ is called the *cofactor matrix* and contains the *cofactors* of $[A]$, i.e., C_{ij} is the cofactor of element a_{ij}. In other words, the adjoint A is the transpose of the cofactor matrix. The inverse of the matrix $[A]$ treated in Section B.3.4 is determined as follows:

$$[CoA] = \begin{bmatrix} +\begin{vmatrix} 8 & 12 \\ 12 & 64 \end{vmatrix} & -\begin{vmatrix} -1 & 12 \\ 4 & 64 \end{vmatrix} & +\begin{vmatrix} -1 & 8 \\ 4 & 12 \end{vmatrix} \\ -\begin{vmatrix} -1 & 4 \\ 12 & 64 \end{vmatrix} & +\begin{vmatrix} 6 & 4 \\ 4 & 64 \end{vmatrix} & -\begin{vmatrix} 6 & -1 \\ 4 & 12 \end{vmatrix} \\ +\begin{vmatrix} -1 & 4 \\ 8 & 12 \end{vmatrix} & -\begin{vmatrix} 6 & 4 \\ 1 & 12 \end{vmatrix} & +\begin{vmatrix} 6 & -1 \\ -1 & 8 \end{vmatrix} \end{bmatrix}$$

$$= \begin{bmatrix} 368 & 112 & -44 \\ 112 & 368 & -76 \\ -44 & -76 & 47 \end{bmatrix}$$

$$[A]^{-1} = \frac{[CoA]^T}{\det A} = \begin{bmatrix} 0.1917 & 0.0583 & -0.0229 \\ 0.0583 & 0.1917 & -0.0396 \\ -0.0229 & -0.0396 & 0.0245 \end{bmatrix}$$

The result demonstrates a useful general property of a symmetric matrix, namely, its inverse is also symmetric. Finally, because

$$[A]^{-1}\,[A] = [I]$$
$$[A]^{-1}\,[A]\,[A]^{-1} = [I]\,[A]^{-1}$$
$$[A]^{-1}\,[A]\,[A]^{-1} = [A]^{-1}$$
$$[A]^{-1}\,[A]\,[A]^{-1} = [A]^{-1}\,[I]$$
$$[A]\,[A^{-1}] = [I]$$

it is apparent that

$$[A]\,[A]^{-1} = [A]^{-1}\,[A] \qquad (B.18)$$

The property—that either premultiplication or postmultiplication of a matrix by its inverse produces the identity matrix—indicates that only square matrices can be inverted. This is evident from the conditions that the identity matrix is square, and any multiplication that produces it must involve compatible matrices. The only way Eq. B.18 can be valid is if $[A]$ is square.

B.4 PARTITIONING OF MATRICES

A partitioned matrix is a matrix subdivided into a number of smaller matrices within the original matrix form. The separation is generally designated by inserting broken lines into the matrix to isolate the submatrices. The matrix

$$[A] = \begin{bmatrix} 1 & 2 & 4 & 4 \\ 3 & 2 & 4 & 4 \\ 4 & 3 & 3 & 2 \\ 4 & 4 & 3 & 4 \end{bmatrix}$$

can be partitioned in numerous ways, e.g.,

$$[A] = \begin{bmatrix} 1 & 2 & 4 & \vdots & 4 \\ 3 & 2 & 2 & \vdots & 4 \\ 4 & 3 & 3 & \vdots & 2 \\ \hdashline 4 & 4 & 3 & \vdots & 4 \end{bmatrix} = \begin{bmatrix} 1 & 2 & 4 & \vdots & 4 \\ 3 & 2 & 2 & \vdots & 4 \\ \hdashline 4 & 3 & 3 & \vdots & 2 \\ 4 & 4 & 3 & \vdots & 4 \end{bmatrix}$$

It is usual to express a partitioned matrix in the form

$$[A] = \begin{bmatrix} [A_{11}] & [A_{12}] & \cdots & [A_{1n}] \\ [A_{21}] & [A_{22}] & \cdots & [A_{2n}] \\ \vdots & \vdots & & \vdots \\ [A_{m1}] & [A_{m2}] & \cdots & [A_{mn}] \end{bmatrix}$$

$$(B.19)$$

The exact configuration of a partitioned matrix depends upon the purpose of the subdivision. As an illustration, multiplication of two partitioned matrices will be discussed.

Let matrices $[A]$ and $[B]$ be partitioned as follows

$$[A] = \begin{bmatrix} [A_{11}] & [A_{12}] \\ [A_{21}] & [A_{22}] \end{bmatrix} \qquad (B.20)$$

$$[B] = \begin{bmatrix} [B_{11}] & [B_{12}] \\ [B_{21}] & [B_{22}] \end{bmatrix} \qquad (B.21)$$

The product, $[C]$, of the two partitioned matrices is given by

$$[C] = \begin{bmatrix} [C_{11}] & [C_{12}] \\ [C_{21}] & [C_{22}] \end{bmatrix} \qquad (B.22)$$

in which

$$[C_{11}] = [A_{11}] [B_{11}] + [A_{12}] [B_{21}]$$
$$(B.23a)$$

$$[C_{12}] = [A_{11}] [B_{12}] + [A_{12}] [B_{22}]$$
$$(B.23b)$$

$$[C_{21}] = [A_{21}] [B_{11}] + [A_{22}] [B_{21}]$$
$$(B.23c)$$

$$[C_{22}] = [A_{21}] [B_{12}] + [A_{22}] [B_{22}]$$
$$(B.23d)$$

The implication of this result is that the original matrices must be partitioned in such a way that each of the multiplications in Eqs. B.23 is possible. This requires that the sizes of the various submatrices be chosen to make each side of Eq. B.23 compatible. This is accomplished if the partitioned matrices are of the following orders

$$\begin{bmatrix} [A_{11}]_{i \times j} & [A_{12}]_{i \times l} \\ [A_{21}]_{k \times j} & [A_{22}]_{k \times l} \end{bmatrix} \quad \text{and} \quad \begin{bmatrix} [B_{11}]_{j \times i} & [B]_{j \times k} \\ [B_{21}]_{l \times i} & [B]_{l \times k} \end{bmatrix}$$

Partitioning matrices reduces the work effort involved in producing a matrix inverse. The procedure for developing the inverse of a partitioned matrix can be found in any standard reference on matrix algebra.

B.5 RELATIONSHIP TO LINEAR EQUATIONS

B.5.1 Representation of Equations

A primary function of matrices is their usefulness in representing a set of linear equations. The set of equations

$$a_{11} x_1 + a_{12} x_2 + \cdots a_{1n} x_n = c_1$$
$$\vdots \qquad\qquad \vdots$$
$$a_{m1} x_1 + a_{12} x_2 + \cdots a_{mn} x_n = c_n$$
$$(B.24)$$

is written in matrix form as

$$[A]\{X\} = \{C\} \qquad (B.25)$$

in which

$$[A] = \begin{bmatrix} a_{11} & \cdots & a_{1n} \\ \vdots & & \vdots \\ a_{m1} & \cdots & a_{mn} \end{bmatrix}$$

$$\{X\} = \begin{Bmatrix} x_1 \\ \vdots \\ x_n \end{Bmatrix}$$

$$\{C\} = \begin{Bmatrix} c_1 \\ \vdots \\ c_n \end{Bmatrix}$$

Algebraic operations commonly performed on linear equations can also be accomplished in matrix form. Multiplying the coefficient matrix, $[A]$, and the matrix of constants, $[C]$, by a scalar is identical to multiplying each of the underlying equations by that same scalar. Addition (subtraction) of two equations is accomplished by simultaneously adding (subtracting) the entries in the corresponding rows of $[A]$ and adding (subtracting) the associated constants in $\{C\}$. Either or both of the equations can be multiplied by a scalar before the addition or subtraction is performed. The matrix equivalent is to multiply the corresponding entries in $[A]$ and $\{C\}$ by the same scalars.

As an example, given

$$6x_1 + 3x_2 = 4$$
$$12x_1 - 12x_2 = 24$$

the matrix equivalent is

$$\begin{bmatrix} 6 & 3 \\ 12 & -12 \end{bmatrix} \begin{Bmatrix} x_1 \\ x_2 \end{Bmatrix} = \begin{Bmatrix} 4 \\ 24 \end{Bmatrix}$$

Consider multiplying the first equation by 2 and subtracting it from the second equation. The result of the first matrix operation is

$$\begin{bmatrix} 12 & 6 \\ 12 & -12 \end{bmatrix} \begin{Bmatrix} x_1 \\ x_2 \end{Bmatrix} = \begin{Bmatrix} 8 \\ 24 \end{Bmatrix}$$

and the result of the second matrix operation is

$$\begin{bmatrix} 12 & 6 \\ 0 & -18 \end{bmatrix} \begin{Bmatrix} x_1 \\ x_2 \end{Bmatrix} = \begin{Bmatrix} 8 \\ 16 \end{Bmatrix}$$

B.5.2 Addition and Subtraction

Consider the two sets of equations presented as Eqs. B.26 and B.27

$$a_{11}x_1 + a_{21}x_2 + a_{13}x_3 = m_1 \quad (B.26a)$$
$$a_{21}x_1 + a_{22}x_2 + a_{23}x_3 = m_2 \quad (B.26b)$$

$$b_{11}x_1 + b_{12}x_2 + b_{13}x_3 = n_1 \quad (B.27a)$$
$$b_{21}x_1 + b_{22}x_2 + b_{23}x_3 = n_2 \quad (B.27b)$$

The pairs of equations can be written in matrix form as

$$\begin{bmatrix} a_{11} & a_{12} & a_{13} \\ a_{21} & a_{22} & a_{23} \end{bmatrix} \begin{Bmatrix} x_1 \\ x_2 \\ x_3 \end{Bmatrix} = \begin{Bmatrix} m_1 \\ m_2 \end{Bmatrix}$$

$$(B.28)$$

$$\begin{bmatrix} b_{11} & b_{12} & b_{13} \\ b_{21} & b_{22} & b_{23} \end{bmatrix} \begin{Bmatrix} x_1 \\ x_2 \\ x_3 \end{Bmatrix} = \begin{Bmatrix} n_1 \\ n_2 \end{Bmatrix}$$

$$(B.29)$$

or in symbolic form

$$[A]\{X\} = \{M\} \qquad (B.30)$$
$$[B]\{X\} = \{N\} \qquad (B.31)$$

If Eq. B.26a and b are added to Eqs. B.27a and b, respectively, the result is

$$(a_{11} + b_{11})x_1 + (a_{12} + b_{12})x_2 + (a_{13} + b_{13})x_3$$
$$= m_1 + n_1 \quad (B.32)$$
$$(a_{21} + b_{21})x_1 + (a_{22} + b_{22})x_2 + (a_{23} + b_{23})x_3$$
$$= m_2 + n_2 \quad (B.33)$$

It is apparent that this algebra can be accomplished in matrix form and expressed as

$$[C]\{X\} = \{P\} \qquad (B.34)$$

in which

$$[C] = [A] + [B]$$
$$\{P\} = \{M\} + \{N\}$$

Clearly, if Eq. B.26a and b are subtracted from Eq. B.27a and b, respectively, the matrix equivalent would be

$$[D]\{X\} = \{Q\} \qquad (B.35)$$

in which

$$[D] = [A] - [B]$$
$$\{Q\} = \{M\} - \{N\}$$

B.5.3 Multiplication

Consider the following two sets of equations:

$$y_1 = a_{11}x_1 + a_{12}x_2 \qquad (B.36a)$$
$$y_2 = a_{21}x_1 + a_{22}x_2 \qquad (B.36b)$$

$$x_1 = b_{11}z_1 + b_{12}z_2 \qquad (B.37a)$$
$$x_2 = b_{21}z_1 + b_{22}z_2 \qquad (B.37b)$$

Substitution of Eq. B.37 into Eq. B.36 produces

$$y_1 = (a_{11}b_{11} + a_{12}b_{21})z_1$$
$$\qquad + (a_{11}b_{12} + a_{12}b_{22})z_2 \qquad (B.38a)$$
$$y_2 = (a_{21}b_{11} + a_{22}b_{21})z_1$$
$$\qquad + (a_{21}b_{12} + a_{22}b_{22})z_2 \qquad (B.38b)$$

In matrix form, Eqs. B.36 and B.37 are written as

$$\{Y\} = [A]\{X\} \qquad (B.39a)$$
$$\{X\} = [B]\{Y\} \qquad (B.39b)$$

in which

$$[A] = \begin{bmatrix} a_{11} & a_{12} \\ a_{21} & a_{22} \end{bmatrix} \quad [B] = \begin{bmatrix} b_{11} & b_{12} \\ b_{21} & b_{22} \end{bmatrix}$$

It is evident now that the coefficients in Eqs. B.38 can be obtained by the matrix operation

$$[C] = [A][B] \qquad (B.40)$$

and the equations themselves as

$$\{Y\} = [C]\{X\} = [A][B]\{X\} \qquad (B.41)$$

B.5.4 Inversion

Frequently, it is necessary to solve a set of simultaneous, linear equations such as those given in Eq. B.36a and b for unknown x-values for a given set of constant coefficients and known values of the variables on the left-hand side of the equations. The matrix form of Eq. B.36 was given in Eq. B.39a. Premultiplying both sides by $[A]^{-1}$ yields

$$[A]^{-1}\{Y\} = [A]^{-1}[A]\{X\}$$
$$\qquad = [I]\{X\} \qquad (B.42)$$

or equivalently,

$$\{X\} = [A]^{-1}\{Y\} \qquad (B.43)$$

Performing the multiplication indicated in Eq. B.43 will produce the desired quantities.
  Consider solving

$$y_1 = 4x_1 + 2x_2$$
$$y_2 = 2x_1 + 4x_2$$

for given values of y_1 and y_2. In matrix form,

$$\begin{Bmatrix} y_1 \\ y_2 \end{Bmatrix} = \begin{bmatrix} 4 & 2 \\ 2 & 4 \end{bmatrix} \begin{Bmatrix} x_1 \\ x_2 \end{Bmatrix}$$

and by inversion

$$\begin{Bmatrix} x_1 \\ x_2 \end{Bmatrix} = \begin{bmatrix} 4 & 2 \\ 2 & 4 \end{bmatrix}^{-1} \begin{Bmatrix} y_1 \\ y_2 \end{Bmatrix}$$

Substitution of known values permits a solution, e.g., for $y_1 = 6$ and $y_2 = 12$:

$$\begin{Bmatrix} x_1 \\ x_2 \end{Bmatrix} = \frac{1}{12} \begin{bmatrix} 4 & -2 \\ -2 & 4 \end{bmatrix} \begin{Bmatrix} 6 \\ 12 \end{Bmatrix}$$

$$= \begin{Bmatrix} 0 \\ 3 \end{Bmatrix}$$

Substitution of the results ($x_1 = 0, x_2 = 3$) into the original matrix equation confirms their correctness.
  Sometimes, it is necessary to solve the same set of linear equations for different sets of known constants, $\{Y\}$. For example, it may be desired to solve both of the following pairs of equations for the values of x_1 and x_2, assuming $y_1 - y_4$ are specified values.

$$y_1 = a_{11}x_1 + a_{12}x_2$$
$$y_2 = a_{21}x_1 + a_{22}x_2 \qquad \text{(a)}$$

and

$$y_3 = a_{11}x_1 + a_{12}x_2$$
$$y_4 = a_{21}x_1 + a_{12}x_2 \qquad \text{(b)}$$

Both (a) and (b) can be expressed in the single matrix form in which $[Y] = [A][X]$

$$\begin{bmatrix} y_1 & y_3 \\ y_2 & y_4 \end{bmatrix} = \begin{bmatrix} a_{11} & a_{12} \\ a_{21} & a_{22} \end{bmatrix} \begin{bmatrix} x_1 & x_1 \\ x_2 & x_2 \end{bmatrix}$$

If done in this way, the operation

$$[X] = [A]^{-1}[Y]$$

produces both sets of answers. The first column of $[X]$ contains the values of x_1 and x_2 that satisfy Eq. (a). The second column contains the values of x_1 and x_2 that satisfy Eq. (b).

B.6 INVERSION BY GAUSS-JORDAN ELIMINATION

Gauss-Jordan elimination is a simple, but powerful, numerical technique for producing a matrix inverse. This method is based upon two realizations.

First, if $[B]$ is the inverse of $[A]$, then

$$[A][B] = [I] \qquad \text{(B.44)}$$

must hold. Premultiplication by $[A]^{-1}$ produces

$$[I][B] = [A]^{-1}$$
$$[B] = [A]^{-1} \qquad \text{(B.45)}$$

In effect, the inverse can be obtained by solving Eq. B.44 for $[B]$.

Second, going from Eq. B.44 to Eq. B.45 represents a simultaneous solution of the linear equations that underlie Eq. B.44. This solution can be obtained by operating on $[B]$ and $[I]$ in the manner described in Section B.5.1.

By forming the partitioned matrix,

$$[[A] \vert [I]] = \begin{bmatrix} a_{11} & a_{12} & \cdots & a_{1n} & 1 & 0 & \cdots & 0 \\ a_{21} & a_{22} & \cdots & a_{2n} & 0 & 1 & \cdots & 0 \\ \cdot & \cdot & & \cdot & \cdot & \cdot & & \cdot \\ \cdot & \cdot & & \cdot & \cdot & \cdot & & \cdot \\ \cdot & \cdot & & \cdot & \cdot & \cdot & & \cdot \\ a_{n1} & a_{n2} & \cdots & a_{nn} & 0 & 0 & \cdots & 1 \end{bmatrix}$$

$$\text{(B.46)}$$

and algebraically operating on the rows, one can produce the form

$$[[I] \vert [B]] = \begin{bmatrix} 1 & 0 & \cdots & 0 & b_{11} & b_{12} & \cdots & b_{1n} \\ 0 & 1 & \cdots & 0 & b_{21} & b_{22} & \cdots & b_{2n} \\ \cdot & \cdot & & \cdot & \cdot & \cdot & & \cdot \\ \cdot & \cdot & & \cdot & \cdot & \cdot & & \cdot \\ \cdot & \cdot & & \cdot & \cdot & \cdot & & \cdot \\ 0 & 0 & \cdots & 1 & b_{n1} & b_{n2} & \cdots & b_{nn} \end{bmatrix}$$

$$\text{(B.47)}$$

The entries to the right of the broken line in Eq. B.47 constitute $[B]$ in Eq. B.45, which is also $[A]^{-1}$.

As an illustration, the matrix inverse obtained in Section B.3.5 by the cofactor method will be obtained by the Gauss-Jordan technique.

Step 1. Form the matrix

$$\begin{bmatrix} 6 & -1 & 4 & 1 & 0 & 0 \\ -1 & 8 & 12 & 0 & 1 & 0 \\ 4 & 12 & 64 & 0 & 0 & 1 \end{bmatrix}$$

Step 2. Divide the first row by 6 to give

Row 1: 1   -0.1667 0.6667 0.1667 0 0

Steps 3 and 4. Multiply row 1 by -1, and subtract the result from row 2 of the original matrix. Multiply row 1 by 4, and subtract the result from row 3 of the original matrix. The outcome is

Matrix 2: $\begin{bmatrix} 1 & -0.16617 & 0.6667 & 0.1667 & 0 & 0 \\ 0 & -7.8333 & 12.6667 & 0.1667 & 1 & 0 \\ 0 & 12.6668 & 61.3332 & -0.6668 & 0 & 1 \end{bmatrix}$

Step 5. Divide the second row of matrix 2 by -7.8333 to produce

Row 2: 0   1   1.6170   0.0213   0.1277   0

Steps 6 and 7. Multiply row 2 by -0.1667, and subtract the result from row 1 of matrix 2. Multiply row 2 by 12.6668, and subtract the result from row 3 of matrix 2. The outcome is

Matrix 3: $\begin{bmatrix} 1 & 0 & 0.9363 & 0.1702 & 0.0213 & 0 \\ 0 & 1 & 1.6170 & 0.0213 & 0.1277 & 0 \\ 0 & 0 & 40.8510 & -0.9366 & -1.6178 & 1 \end{bmatrix}$.

Step 8. Divide the third row of matrix 3 by 40.8510 to produce

Row 3: 0 0 1 -0.0229 -0.0396 0.0245

Steps 9 and 10. Multiply row 3 by 0.9363, and subtract the result from row 1 of matrix 3. Multiply row 3 by 1.6170, and subtract the result from row 2 of the matrix 3. The outcome is

$$
\text{Matrix 4: } \begin{bmatrix} 1 & 0 & 0 & 0.1916 & 0.0584 & -0.0229 \\ 0 & 1 & 0 & 0.0583 & 0.1917 & -0.0396 \\ 0 & 0 & 1 & -0.0229 & -0.0396 & 0.0245 \end{bmatrix}
$$

The (3×3) matrix in the right partition of matrix 4 is $[A]^{-1}$. The precision employed was aimed at minimizing the round-off error. Nevertheless, some slight differences from the earlier solution are evident.

Appendix C
Computer Programs

PROGRAM INVERSE.BAS

```
10  '**********************************************************************
20  '                                                                   *
30  ' PROGRAM INVERSE.BAS                    LANGUAGE: IBM BASICA   3.0  *
40  '                                        SYSTEM  : MICROSOFT DOS 3.3 *
50  '                                                                   *
60  ' PURPOSE: TO PERFORM AN INPLACE INVERSION OF A NPxNP MATRIX [K]     *
70  '                                                                   *
80  ' INPUT  : NP (INTEGER) - NUMBER OF ROWS OR COLUMNS OF MATRIX [K]    *
90  '            [K] (REAL)  - ELEMENTS OF THE [K] MATRIX                *
100 '                                                                   *
110 ' OUTPUT : [K] (REAL)   - ELEMENTS OF THE INVERTED [K] MATRIX        *
120 '                                                                   *
130 ' **********************************************************************
140 CLS
150 '*******
160  READ NP
170 '*******
180 DIM K(NP,NP),IS(NP)
190 ' **************************
200 ' READ THE MATRIX K ROW BY ROW
210 ' **************************
220 FOR I = 1 TO NP
230    FOR J = 1 TO NP
240      READ K(I,J)
250    NEXT J
260 NEXT I
270 '*****************************************
280    PRINT TAB(10) "RESULTS OF INVERSE.BAS"
290    PRINT
300    PRINT TAB(10) "THE ORIGINIAL [K] MATRIX "
310 '*****************************************
320 FOR I = 1 TO NP
330    FOR J = 1 TO NP
340      LOCATE I+4, J*12
350      PRINT USING "+#.###^^^^";K(I,J)
360    NEXT J
370 NEXT I
380 ' *****************************************
390 ' INVERT THE MATRIX K BY GAUSS ELIMINATION
400 ' *****************************************
410 FOR I = 1 TO NP
420    IS(I) = 0
```

492

```
430 NEXT I
440 AMAX=-1
450 FOR I = 1 TO NP
460    IF IS(I) <> O THEN GOTO 510
470    TEMP=ABS(K(I,I))
480    IF (TEMP-AMAX) <= 0 THEN GOTO 510
490    ICOL = I
500    AMAX = TEMP
510 NEXT I
520 IF (AMAX) = 0! THEN GOTO 930
530 IF (AMAX) < 0! THEN GOTO 730
540 IS(ICOL) = 1
550 PIVOT = K(ICOL,ICOL)
560 K(ICOL,ICOL) = 1!
570 PIVOT = 1/PIVOT
580 FOR J = 1 TO NP
590    K(ICOL,J) = K(ICOL,J)*PIVOT
600 NEXT J
610 FOR I = 1 TO NP
620    IF (I-ICOL) = O THEN GOTO 680
630    TEMP = K(I,ICOL)
640    K(I,ICOL) = O.
650    FOR J = 1 TO NP
660       K(I,J) = K(I,J) - K(ICOL,J)*TEMP
670    NEXT J
680 NEXT I
690 GOTO 440
700 ' ****************************
710 ' PRINT THE INVERTED [K] MATRIX
720 ' ****************************
730 LOCATE NP+6,1
740 PRINT TAB(10) "THE INVERTED [K] MATRIX"
750 FOR I = 1 TO NP
760    FOR J = 1 TO NP
770       LOCATE I+NP+7, J*12
780       PRINT USING "+#.###^^^^";K(I,J)
790    NEXT J
800 NEXT I
810 '***********
820 ' INPUT DATA
830 '***********
840 '
850 '     NP
860 DATA 4
870 '
880 ' [K] MATRIX (NPxNP)
890 DATA 4,2,0,0
900 DATA 2,4,0,0
910 DATA 0,0,4,2
920 DATA 0,0,2,4
930 END
```

RESULTS OF INVERSE.BAS

THE ORIGINIAL [K] MATRIX

```
   +4.000E+00   +2.000E+00   +0.000E+00   +0.000E+00
   +2.000E+00   +4.000E+00   +0.000E+00   +0.000E+00
   +0.000E+00   +0.000E+00   +4.000E+00   +2.000E+00
   +0.000E+00   +0.000E+00   +2.000E+00   +4.000E+00
```


THE INVERTED [K] MATRIX

```
   +3.333E-01   -1.667E-01   +0.000E+00   +0.000E+00
   -1.667E-01   +3.333E-01   +0.000E+00   +0.000E+00
   +0.000E+00   +0.000E+00   +3.333E-01   -1.667E-01
   +0.000E+00   +0.000E+00   -1.667E-01   +3.333E-01
```

PROGRAM INVERSE.F

```
C*************************************************************************
C                                                                       *
C     PROGRAM INVERSE.F                         LANGUAGE: FORTRAN 77     *
C                                               SYSTEM   : UNIX V        *
C                                                                       *
C     PURPOSE: TO PERFORM AN INPLACE INVERSION OF A NPxNP MATRIX [K]    *
C                                                                       *
C     INPUT  : NP (INTEGER) - NUMBER OF ROWS OR COLUMNS OF MATRIX [K]   *
C              [K] (REAL)   - ELEMENTS OF THE [K] MATRIX                *
C                                                                       *
C     OUTPUT : [K] (REAL)   - ELEMENTS OF THE INVERTED [K] MATRIX       *
C                                                                       *
C*************************************************************************
C
      PROGRAM INVERSE
      PARAMETER(INP=4)
      REAL K(INP,INP)
      INTEGER IS(INP)
C
      OPEN(UNIT=1,FILE='INPUT', FORM='FORMATTED',STATUS='UNKNOWN')
      OPEN(UNIT=2,FILE='OUTPUT',FORM='FORMATTED',STATUS='UNKNOWN')
C
C     READ THE INPUT FILE HEADER
C
      READ(1,*)
      READ(1,*)
C
      READ(1,*)
      READ(1,*) NP
      READ(1,*)
C
C     READ THE MATRIX K ROW BY ROW
C
      READ(1,*)
      DO 100 I = 1, NP
        READ(1,*)(K(I,J),J=1,NP)
 100  CONTINUE
      READ(1,*)
C
      WRITE(2,*)
      WRITE(2,*) "RESULTS OF INVERSE.F "
      WRITE(2,*)
C
      WRITE(2,*)
      WRITE(2,*) "THE ORIGINIAL [K] MATRIX "
      WRITE(2,*)
      DO 110 I=1,NP
        WRITE(2,10)(K(I,J),J=1,NP)
 10   FORMAT(1P14(E10.3,2X))
 110  CONTINUE
C
C     INVERT THE MATRIX K BY GAUSS ELIMINATION
C
      DO 120 I=1,NP
        IS(I) = 0
 120  CONTINUE
 130  AMAX=-1
      DO 140 I=1,NP
        IF(IS(I).NE.0) GOTO 140
        TEMP=ABS(K(I,I))
        IF((TEMP-AMAX).LE.0) GOTO 140
        ICOL = I
        AMAX = TEMP
 140  CONTINUE
      IF(AMAX.EQ.0) GOTO 200
      IF(AMAX.LT.0) GOTO 180
      IS(ICOL) = 1
      PIVOT = K(ICOL,ICOL)
      K(ICOL,ICOL) = 1.0
      PIVOT = 1/PIVOT
      DO 150 J = 1,NP
        K(ICOL,J) = K(ICOL,J)*PIVOT
```

```
 150    CONTINUE
        DO 160 I=1,NP
           IF((I-ICOL).EQ.0) GOTO 160
           TEMP = K(I,ICOL)
           K(I,ICOL) = 0.0
           DO 170 J = 1,NP
              K(I,J) = K(I,J) - K(ICOL,J)*TEMP
 170       CONTINUE
 160    CONTINUE
        GOTO 130
C
 180    WRITE(2,*)
        WRITE(2,*) "THE INVERTED [K] MATRIX"
        WRITE(2,*)
C
        DO 190 I=1,NP
           WRITE(2,10)(K(I,J),J=1,NP)
 190    CONTINUE
        STOP
 200    END
```

INPUT FILE FOR INVERSE.F
NP
4

THE [K] MATRIX
4.0 2.0 0.0 0.0
2.0 4.0 0.0 0.0
0.0 0.0 4.0 2.0
0.0 0.0 2.0 4.0

  RESULTS OF INVERSE.F

  THE ORIGINIAL [K] MATRIX

4.000e+00	2.000e+00	0.000e+00	0.000e+00
2.000e+00	4.000e+00	0.000e+00	0.000e+00
0.000e+00	0.000e+00	4.000e+00	2.000e+00
0.000e+00	0.000e+00	2.000e+00	4.000e+00

  THE INVERTED [K] MATRIX

3.333e-01	-1.667e-01	0.000e+00	0.000e+00
-1.667e-01	3.333e-01	0.000e+00	0.000e+00
0.000e+00	0.000e+00	3.333e-01	-1.667e-01
0.000e+00	0.000e+00	-1.667e-01	3.333e-01

PROGRAM AMATRIX.BAS

```
10   '**********************************************************************
20   '                                                                    *
30   ' PROGRAM AMATRIX.BAS                     LANGUAGE: IBM BASICA   3.0  *
40   '                                         SYSTEM  : MICROSOFT DOS 3.3 *
50   '                                                                    *
60   ' PURPOSE: TO EXECUTE THE ALTERNATE BASIC MATRIX ANALYSIS METHOD     *
70   '          FOR DETERMINATE TRUSSES.                                  *
80   '                                                                    *
90   ' INPUT  : NP   (INTEGER) - NUMBER OF DEGREES OF FREEDOM             *
100  '          NR   (INTEGER) - NUMBER OF REACTIONS                      *
110  '          [EP] (REAL)    - ELEMENTS OF THE EQUILIBRIUM MATRIX FOR   *
120  '                           DEGREES OF FREEDOM                       *
130  '          [S]  (REAL)    - DIAGONAL ELEMENTS OF THE STIFFNESS MATRIX*
140  '          [C]  (REAL)    - ELEMENTS OF THE COMPATIBILITY MATRIX     *
150  '          [P]  (REAL)    - NEGATIVE OF THE FIXED STATE REACTIONS AT *
160  '                           THE DEGREES OF FREEDOM                   *
170  '          [ER] (REAL)    - ELEMENTS OF THE EQUILIBRIUM MATRIX FOR   *
180  '                           REACTIONS                                *
190  '                                                                    *
200  ' OUTPUT : [X]  (REAL)    - JOINT DISPLACEMENTS                      *
210  '          [D]  (REAL)    - MEMBER DEFORMATIONS                      *
220  '          [F]  (REAL)    - FINAL MEMBER FORCES                      *
230  '          [R]  (REAL)    - FINAL REACTIONS                          *
240  '                                                                    *
250  '**********************************************************************
```

```
260   CLS
270  '*********
280   READ NP,NR
290  '*********
300 DIM EP(NP,NP),S(NP,NP),C(NP,NP),IS(NP),P(NP)
310 DIM X(NP),D(NP),F(NP),ER(NR,NP),R(NR)
320  '**********
330  ' READ   [EP]
340  '**********
350 FOR I = 1 TO NP
360    FOR J = 1 TO NP
370       READ EP(I,J)
380    NEXT J
390 NEXT I
400  ' *********************
410  ' INITIALIZE [S] TO ZERO
420  ' *********************
430 FOR I = 1 TO NP
440    FOR J = 1 TO NP
450       S(I,J) = 0
460    NEXT J
470 NEXT I
480  ' **************************
490  ' READ DIAGONAL ENTRIES IN [S]
500  ' **************************
510 FOR I = 1 TO NP
520    READ S(I,I)
530 NEXT I
540  ' ********
550  ' READ [C]
560  ' ********
570 FOR I = 1 TO NP
580    FOR J = 1 TO NP
590       READ C(I,J)
600    NEXT J
610 NEXT I
620  '*********************
630  ' INITIALIZE {P} TO ZERO
640  '*********************
650 FOR I = 1 TO NP
660    P(I) = 0
670 NEXT I
680  ' ********
690  ' READ [P]
700  ' ********
710 FOR I = 1 TO NP
720    READ P(I)
730 NEXT I
740  ' *********
750  ' READ [ER]
760  ' *********
770 FOR I = 1 TO NR
780    FOR J = 1 TO NP
790       READ ER(I,J)
800    NEXT J
810 NEXT I
820  ' *****************************
830  ' INVERT [EP] BY GAUSS ELIMINATION
840  ' *****************************
850 FOR I = 1 TO NP
860    IS(I) = 0
870 NEXT I
880 AMAX=-1
890 FOR I = 1 TO NP
900    IF IS(I) <> 0 THEN GOTO 950
910    TEMP=ABS(EP(I,I))
920    IF (TEMP-AMAX) <= 0 THEN GOTO 950
930    ICOL = I
940    AMAX = TEMP
950 NEXT I
960 IF (AMAX) = 0! THEN GOTO 2810
970 IF (AMAX) < 0! THEN GOTO 1170
980 IS(ICOL) = 1
990 PIVOT = EP(ICOL,ICOL)
```

```
1000 EP(ICOL,ICOL) = 1!
1010 PIVOT = 1/PIVOT
1020 FOR J = 1 TO NP
1030    EP(ICOL,J) = EP(ICOL,J)*PIVOT
1040 NEXT J
1050 FOR I = 1 TO NP
1060    IF (I-ICOL) = O THEN GOTO 1120
1070    TEMP = EP(I,ICOL)
1080    EP(I,ICOL) = O!
1090    FOR J = 1 TO NP
1100       EP(I,J) = EP(I,J) - EP(ICOL,J)*TEMP
1110    NEXT J
1120 NEXT I
1130 GOTO 880
1140 ' *********************
1150 ' [EP] HAS BEEN INVERTED
1160 ' *********************
1170 ' *****************************
1180 ' OBTAIN [F] = [EP(INVERSE)] * [P]
1190 ' *****************************
1200 FOR I = 1 TO NP
1210    F(I) = 0
1220    FOR L = 1 TO NP
1230       F(I) = F(I) + EP(I,L) * P(L)
1240    NEXT L
1250 NEXT I
1260 ' *****************************
1270 ' INVERT [S] BY GAUSS ELIMINATION
1280 ' *****************************
1290 FOR I = 1 TO NP
1300    IS(I) = 0
1310 NEXT I
1320 AMAX=-1
1330 FOR I = 1 TO NP
1340    IF IS(I) <> O THEN GOTO 1390
1350    TEMP=ABS(S(I,I))
1360    IF (TEMP-AMAX) <= 0 THEN GOTO 1390
1370    ICOL = I
1380    AMAX = TEMP
1390 NEXT I
1400 IF (AMAX) = 0! THEN GOTO 2810
1410 IF (AMAX) < 0! THEN GOTO 1580
1420 IS(ICOL) = 1
1430 PIVOT = S(ICOL,ICOL)
1440 S(ICOL,ICOL) = 1!
1450 PIVOT = 1/PIVOT
1460 FOR J = 1 TO NP
1470    S(ICOL,J) = S(ICOL,J)*PIVOT
1480 NEXT J
1490 FOR I = 1 TO NP
1500    IF (I-ICOL) = O THEN GOTO 1560
1510    TEMP = S(I,ICOL)
1520    S(I,ICOL) = O!
1530    FOR J = 1 TO NP
1540       S(I,J) = S(I,J) - S(ICOL,J)*TEMP
1550    NEXT J
1560 NEXT I
1570 GOTO 1320
1580 ' *********************
1590 ' [K] HAS BEEN INVERTED
1600 ' *********************
1610 ' *****************************
1620 ' SOLVE [D] = [S(INVERSE)] * [F]
1630 ' *****************************
1640 FOR I = 1 TO NP
1650    D(I) = 0
1660    FOR L = 1 TO NP
1670       D(I) = D(I) + S(I,L) * F(L)
1680    NEXT L
1690 NEXT I
1700 ' *****************************
1710 ' INVERT [C] BY GAUSS ELIMINATION
1720 ' *****************************
1730 FOR I = 1 TO NP
```

```
1740    IS(I) = O
1750 NEXT I
1760 AMAX=-1
1770 FOR L = 1 TO NP
1780    IF IS(L) <> O THEN GOTO 1830
1790    TEMP=ABS(C(L,L))
1800    IF (TEMP-AMAX) <= O THEN GOTO 1830
1810    ICOL = L
1820    AMAX = TEMP
1830 NEXT L
1840 IF (AMAX) = 0! THEN GOTO 2810
1850 IF (AMAX) < 0! THEN GOTO 2060
1860 IS(ICOL) = 1
1870 PIVOT = C(ICOL,ICOL)
1880 C(ICOL,ICOL) = 1!
1890 PIVOT = 1/PIVOT
1900 FOR J = 1 TO NP
1910    C(ICOL,J) = C(ICOL,J)*PIVOT
1920 NEXT J
1930 FOR I = 1 TO NP
1940    IF (I-ICOL) = O THEN GOTO 2000
1950    TEMP = C(I,ICOL)
1960    C(I,ICOL) = O!
1970    FOR J = 1 TO NP
1980       C(I,J) = C(I,J) - C(ICOL,J)*TEMP
1990    NEXT J
2000 NEXT I
2010 GOTO 1760
2020 ' ********************
2030 ' [C] HAS BEEN INVERTED
2040 ' ********************
2050 '****************************
2060 ' SOLVE [X] = [C(INVERSE)] * [D]
2070 '****************************
2080 FOR I = 1 TO NP
2090    X(I) = 0
2100    FOR L = 1 TO NP
2110       X(I) = X(I) + C(I,L) * D(L)
2120    NEXT L
2130 NEXT I
2140 '***********************
2150 ' SOLVE [R] = [ER] * [F]
2160 '***********************
2170 FOR I = 1 TO NR
2180    R(I) = 0
2190    FOR L = 1 TO NP
2200       R(I) = R(I) + ER(I,L) * F(L)
2210    NEXT   L
2220 NEXT I
2230 '********************************
2240    PRINT "RESULTS OF AMATRIX.BAS"
2250    PRINT "FOR EXMAPLE 14.6 (2nd ed.)"
2260    PRINT
2270 '********************************
2280 '**************************
2290    PRINT "JOINT DISPLACEMENTS"
2300 '**************************
2310 FOR I = 1 TO NP
2320    PRINT "X(";I;") = ";USING"+#.###^^^^";X(I)
2330 NEXT I
2340 PRINT
2350 '**************************
2360    PRINT "MEMBER DEFORMATIONS"
2370 '**************************
2380 FOR I = 1 TO NP
2390    PRINT "D(";I;") = ";USING"+#.###^^^^";D(I)
2400 NEXT I
2410 PRINT
2420 '********************
2430    PRINT "MEMBER FORCES"
2440 '********************
2450 FOR I = 1 TO NP
2460    PRINT "F(";I;") = ";USING"+#.###^^^^";F(I)
```

```
2470 NEXT I
2480 PRINT
2490 '*******************
2500   PRINT "REACTIONS"
2510 '*******************
2520 FOR I = 1 TO NR
2530   PRINT "R(";I;") = ";USING"+#.###^^^^";R(I)
2540 NEXT I
2550 ' ********************************
2560 ' DATA FOR EXAMPLE 14.6 (2nd EDITION)
2570 ' ********************************
2580 '
2590 '    NP,NR
2600 DATA 2, 4
2610 '
2620 ' [EP] MATRIX (NPxNP)
2630 DATA -1.000, -0.530
2640 DATA  0.000, -0.848
2650 '
2660 ' [S] MATRIX (NPxNP) (DIAGONAL ELEMENTS)
2670 DATA 750, 397.499
2680 '
2690 ' [C] MATRIX (NPxNP)
2700 DATA -1.000,  0.000
2710 DATA -0.530, -0.848
2720 '
2730 ' [P] MATRIX (NP)
2740 DATA  7, -9
2750 '
2760 ' [ER] MATRIX (NRxNP)
2770 DATA  0.000,  0.000
2780 DATA  1.000,  0.000
2790 DATA  0.000,  0.848
2800 DATA  0.000,  0.530
2810 END
```

```
RESULTS OF AMATRIX.BAS
FOR EXMAPLE 14.6 (2nd ed.)

JOINT DISPLACEMENTS
X( 1 ) = +1.683E-02
X( 2 ) = -4.201E-02

MEMBER DEFORMATIONS
D( 1 ) = -1.683E-02
D( 2 ) = +2.670E-02

MEMBER FORCES
F( 1 ) = -1.263E+01
F( 2 ) = +1.061E+01

REACTIONS
R( 1 ) = +0.000E+00
R( 2 ) = -1.263E+01
R( 3 ) = +9.000E+00
R( 4 ) = +5.625E+00
```

PROGRAM AMATRIX.F

```
C********************************************************************
C                                                                  *
C   PROGRAM AMATRIX.F                        LANGUAGE: FORTRAN 77   *
C                                            SYSTEM  : UNIX V       *
C                                                                  *
C   PURPOSE: TO EXECUTE THE ALTERNATE BASIC MATRIX ANALYSIS METHOD  *
C            FOR DETERMINATE TRUSSES.                              *
C                                                                  *
C   INPUT  : NP  (INTEGER) - NUMBER OF DEGREES OF FREEDOM          *
C            NR  (INTEGER) - NUMBER OF REACTIONS                   *
C            [EP]  (REAL)  - ELEMENTS OF THE EQUILIBRIUM MATRIX FOR *
C                            DEGREES OF FREEDOM                    *
```

```
C               [S]    (REAL)  - DIAGONAL ELEMENTS OF THE STIFFNESS MATRIX *
C               [C]    (REAL)  - ELEMENTS OF THE COMPATIBILITY MATRIX       *
C               [P]    (REAL)  - NEGATIVE OF THE FIXED STATE REACTIONS AT   *
C                                THE DEGREES OF FREEDOM                      *
C               [ER]   (REAL)  - ELEMENTS OF THE EQUILIBRIUM MATRIX FOR     *
C                                REACTIONS                                   *
C                                                                           *
C  OUTPUT : [X]        (REAL)  - JOINT DISPLACEMENTS                        *
C               [D]    (REAL)  - MEMBER DEFORMATIONS                        *
C               [F]    (REAL)  - MEMBER FORCES                              *
C               [R]    (REAL)  - REACTIONS                                  *
C                                                                           *
C***************************************************************************
C
      PROGRAM AMATRIX
      PARAMETER (INP=2,INR=4)
      REAL P(INP),EP(INP,INP),S(INP,INP),C(INP,INP),ER(INR,INP)
      REAL X(INP),D(INP),F(INP),R(INR)
      INTEGER NP,NR
C
      OPEN(UNIT=1,FILE='INPUT', FORM='FORMATTED',STATUS='UNKNOWN')
      OPEN(UNIT=2,FILE='OUTPUT',FORM='FORMATTED',STATUS='UNKNOWN')
C
C     READ INPUT HEADER
C
      READ(1,*)
      READ(1,*)
      READ(1,*)
C
      READ(1,*)
      READ(1,*) NP,NR
      READ(1,*)
C
C     READ [EP]
C
      READ(1,*)
      DO 100 I=1,NP
        READ(1,*)(EP(I,J),J=1,NP)
 100  CONTINUE
      READ(1,*)
C
C     INITIALIZE [S] TO ZERO
C
      DO 110 I=1,NP
        DO 120 J=1,NP
          S(I,J)=0
 120    CONTINUE
 110  CONTINUE
C
C     READ DIAGONAL ENTRIES IN [S]
C
      READ(1,*)
      READ(1,*)(S(I,I),I=1,NP)
      READ(1,*)
C
C     READ [C]
C
      READ(1,*)
      DO 130 I=1,NP
        READ(1,*)(C(I,J),J=1,NP)
 130  CONTINUE
      READ(1,*)
C
C     INITIALIZE {P} TO ZERO
C
      DO 140 I=1,NP
        P(I)=0
 140  CONTINUE
C
C     READ {P}
C
      READ(1,*)
      READ(1,*)(P(I),I=1,NP)
      READ(1,*)
```

```
C
C       READ [ER]
C
        READ(1,*)
        DO 150 I=1,NR
          READ(1,*)(ER(I,J),J=1,NP)
 150    CONTINUE
        READ(1,*)
C
        CALL INVERT(EP,NP)
C
C       OBTAIN {F} = [EP(INVERSE)] * {P}
C
        DO 160 I=1,NP
          F(I)=0
          DO 170 L=1,NP
            F(I)=F(I)+EP(I,L)*P(L)
 170      CONTINUE
 160    CONTINUE
C
        CALL INVERT(S,NP)
C
C       OBTAIN {D} = [S(INVERSE)] * {F}
C
        DO 180 I=1,NP
          D(I)=0
          DO 190 L=1,NP
            D(I)=D(I)+ S(I,L)*F(L)
 190      CONTINUE
 180    CONTINUE
C
        CALL INVERT(C,NP)
C
C       OBTAIN {X} = [C(INVERSE)] * {D}
C
        DO 200 I=1,NP
          X(I)=0
          DO 210 L=1,NP
            X(I)=X(I)+C(I,L)*D(L)
 210      CONTINUE
 200    CONTINUE
C
C       OBTAIN {R} = [ER] * {F}
C
        DO 220 I=1,NR
          R(I)=0
          DO 230 L=1,NP
            R(I)=R(I)+ER(I,L)*F(L)
 230      CONTINUE
 220    CONTINUE
C
        WRITE(2,*)
        WRITE(2,*) "RESULTS OF AMATRIX.F"
        WRITE(2,*) "FOR EXAMPLE 14.6 (2nd ed.)"
C
        WRITE(2,*)
        WRITE(2,*) "JOINT DISPLACEMENTS"
        DO 240 I=1,NP
          WRITE(2,10) I,X(I)
 10     FORMAT(3X,'X(',I2,') = ',1PE10.3)
 240    CONTINUE
C
        WRITE(2,*)
        WRITE(2,*) "MEMBER DEFORMATIONS"
        DO 250 I=1,NP
          WRITE(2,11) I,D(I)
 11     FORMAT(3X,'D(',I2,') = ',1PE10.3)
 250    CONTINUE
C
        WRITE(2,*)
        WRITE(2,*) "MEMBER FORCES"
        DO 260 I=1,NP
          WRITE(2,12) I,F(I)
 12     FORMAT(3X,'F(',I2,') = ',1PE10.3)
```

```
 260   CONTINUE
C
       WRITE(2,*)
       WRITE(2,*) "REACTIONS"
       DO 270 I=1,NR
         WRITE(2,13)  I,R(I)
  13   FORMAT(3X,'R(',I2,') = ',1PE10.3)
 270   CONTINUE
C
       STOP
       END
C
C
C
       SUBROUTINE INVERT(k,np)
       PARAMETER(INP=2)
       REAL K(INP,INP)
       INTEGER IS(INP)
C
C      INVERT THE MATRIX K BY GAUSS ELIMINATION
C
       DO 100 I=1,NP
         IS(I) = 0
 100   CONTINUE
 110   AMAX=-1
       DO 120 I=1,NP
         IF(IS(I).NE.0) GOTO 120
         TEMP=ABS(K(I,I))
         IF((TEMP-AMAX).LE.0) GOTO 120
         ICOL = I
         AMAX = TEMP
 120   CONTINUE
       IF(AMAX.EQ.0) GOTO 170
       IF(AMAX.LT.0) GOTO 160
       IS(ICOL) = 1
       PIVOT = K(ICOL,ICOL)
       K(ICOL,ICOL) = 1.0
       PIVOT = 1/PIVOT
       DO 130 J = 1,NP
         K(ICOL,J) = K(ICOL,J)*PIVOT
 130   CONTINUE
       DO 140 I=1,NP
         IF((I-ICOL).EQ.0) GOTO 140
         TEMP = K(I,ICOL)
         K(I,ICOL) = 0.0
         DO 150 J = 1,NP
           K(I,J) = K(I,J) - K(ICOL,J)*TEMP
 150     CONTINUE
 140   CONTINUE
       GOTO 110
 160   RETURN
 170   STOP
       END


INPUT FILE FOR AMATRIX.F
DATA FOR EXAMPLE 14.6 (2nd EDITION)

NP NR
2  4

[EP] MATRIX (NPxNP)
-1.000 -0.530
 0.000 -0.848

[S] MATRIX (NPxNP) (DIAGONAL ELEMENTS)
750 397.499

[C] MATRIX (NPxNP)
-1.000  0.000
-0.530 -0.848

[P] MATRIX (NP)
7 -9
```

[ER] MATRIX (NRxNP)
```
0.000   0.000
1.000   0.000
0.000   0.848
0.000   0.530
```

RESULTS OF AMATRIX.F
FOR EXAMPLE 14.6 (2nd ed.)

JOINT DISPLACEMENTS
```
 X( 1) =   1.683e-02
 X( 2) = -4.201e-02
```

MEMBER DEFORMATIONS
```
 D( 1) = -1.683e-02
 D( 2) =  2.670e-02
```

MEMBER FORCES
```
 F( 1) = -1.263e+01
 F( 2) =  1.061e+01
```

REACTIONS
```
 R( 1) =   0.000e+00
 R( 2) = -1.263e+01
 R( 3) =  9.000e+00
 R( 4) =  5.625e+00
```

PROGRAM BMATRIX.BAS

```
10   '***********************************************************************
20   '                                                                     *
30   ' PROGRAM BMATRIX.BAS                      LANGUAGE: IBM BASICA   3.0  *
40   '                                          SYSTEM  : MICROSOFT DOS 3.3 *
50   '                                                                     *
60   ' PURPOSE: TO EXECUTE THE BASIC MATRIX ANALYSIS METHOD INCLUDING       *
70   '          SPECIAL "ACTIONS" AND APPLIED LOADS.                        *
80   '                                                                     *
90   ' INPUT   : NP   (INTEGER) - NUMBER OF DEGREES OF FREEDOM              *
100  '           NF   (INTEGER) - NUMBER OF MEMBERS                         *
110  '           NR   (INTEGER) - NUMBER OF REACTIONS                       *
120  '           [EP] (REAL)    - ELEMENTS OF THE EQUILIBRIUM MATRIX FOR    *
130  '                            DEGREES OF FREEDOM                        *
140  '           [S]  (REAL)    - DIAGONAL ELEMENTS OF THE STIFFNESS MATRIX *
150  '           [P]  (REAL)    - NEGATIVE OF THE FIXED STATE REACTIONS AT  *
160  '                            THE DEGREES OF FREEDOM                    *
170  '           [ER] (REAL)    - ELEMENTS OF THE EQUILIBRIUM MATRIX FOR    *
180  '                            REACTIONS                                 *
190  '           [FO] (REAL)    - INITIAL FIXED STATE MEMBER FORCES FOR     *
200  '                            THE DEGREES OF FREEDOM                    *
210  '           [RO] (REAL)    - INITIAL FIXED STATE MEMBER FORCES FOR     *
220  '                            THE REACTIONS                             *
230  '                                                                     *
240  ' OUTPUT  : [X]  (REAL)    - JOINT DISPLACEMENTS                       *
250  '           [D]  (REAL)    - MEMBER DEFORMATIONS                       *
260  '           [F]  (REAL)    - FINAL MEMBER FORCES                       *
270  '           [R]  (REAL)    - FINAL REACTIONS                           *
280  '                                                                     *
290  '***********************************************************************
300  CLS
310  '*************
320  READ NP,NF,NR
330  '*************
340  DIM EP(NP,NF),S(NF,NF),C(NF,NP),SC(NF,NP),K(NP,NP),IS(NP),P(NP)
350  DIM X(NP),D(NF),F(NF),FO(NF),ER(NR,NF),R(NR),RO(NR)
360  '***********
370  ' READ  [EP]
380  '***********
390  FOR I = 1 TO NP
400     FOR J = 1 TO NF
410        READ EP(I,J)
```

```
420    NEXT J
430 NEXT I
440 ' **************************
450 ' INITIALIZE S TO NULL MATRIX
460 ' **************************
470 FOR I = 1 TO NF
480    FOR J = 1 TO NF
490      S(I,J) = 0
500    NEXT J
510 NEXT I
520 ' **************************
530 ' READ DIAGONAL ENTRIES IN [S]
540 ' **************************
550 FOR I = 1 TO NF
560    READ S(I,I)
570 NEXT I
580 ' ********
590 ' READ [P]
600 ' ********
610 FOR I = 1 TO NP
620      READ P(I)
630 NEXT I
640 ' *********
650 ' READ [ER]
660 ' *********
670 FOR I = 1 TO NR
680    FOR J = 1 TO NF
690      READ ER(I,J)
700    NEXT J
710 NEXT I
720 ' ***********************
730 ' DO [C] = [EP(TRANSPOSE)]
740 ' ***********************
750 FOR I = 1 TO NF
760    FOR J = 1 TO NP
770      C(I,J) = EP(J,I)
780    NEXT J
790 NEXT I
800 ' *****************
810 ' MULTIPLY [S] * [C]
820 ' *****************
830 FOR I = 1 TO NF
840    FOR J = 1 TO NP
850      SC(I,J) = 0
860      FOR L = 1 TO NF
870        SC(I,J) = SC(I,J) + S(I,L) * C(L,J)
880      NEXT L
890    NEXT J
900 NEXT I
910 ' ***********************
920 ' OBTAIN [K] = [EP] * [SC]
930 ' ***********************
940 FOR I = 1 TO NP
950   FOR J = 1 TO NP
960     K(I,J) = 0
970     FOR L = 1 TO NF
980        K(I,J) = K(I,J) + EP(I,L) * SC(L,J)
990      NEXT L
1000    NEXT J
1010 NEXT I
1020 CLS
1030 ' *****************************
1040 ' INVERT [K] BY GAUSS ELIMINATION
1050 ' *****************************
1060 FOR I = 1 TO NP
1070    IS(I) = 0
1080 NEXT I
1090 AMAX=-1
1100 FOR I = 1 TO NP
1110    IF IS(I) <> 0 THEN GOTO 1160
1120    TEMP=ABS(K(I,I))
1130    IF (TEMP-AMAX) <= 0 THEN GOTO 1160
1140    ICOL = I
1150    AMAX = TEMP
```

```
1160 NEXT I
1170 IF (AMAX) = 0! THEN GOTO 1290
1180 IF (AMAX) < 0! THEN GOTO 1380
1190 IS(ICOL) = 1
1200 PIVOT = K(ICOL,ICOL)
1210 K(ICOL,ICOL) = 1!
1220 PIVOT = 1/PIVOT
1230 FOR J = 1 TO NP
1240    K(ICOL,J) = K(ICOL,J)*PIVOT
1250 NEXT J
1260 FOR I = 1 TO NP
1270    IF (I-ICOL) = O THEN GOTO 1330
1280    TEMP = K(I,ICOL)
1290    K(I,ICOL) = O!
1300    FOR J = 1 TO NP
1310       K(I,J) = K(I,J) - K(ICOL,J)*TEMP
1320    NEXT J
1330 NEXT I
1340 GOTO 1090
1350 ' ********************
1360 ' [K] HAS BEEN INVERTED
1370 ' ********************
1380 LOCATE NP+4,1
1390 '****************************
1400 ' OBTAIN [X] = [K(INVERSE)] * [P]
1410 '****************************
1420 FOR I = 1 TO NP
1430    X(I) = 0
1440    FOR L = 1 TO NP
1450       X(I) = X(I) + K(I,L) * P(L)
1460    NEXT L
1470 NEXT I
1480 '****************************
1490   PRINT "RESULTS OF BMATRIX.BAS"
1500   PRINT "FOR EXAMPLE 14.7 (2nd ed.)"
1510   PRINT
1520 '****************************
1530 '****************************
1540   PRINT "JOINT DISPLACEMENTS"
1550 '****************************
1560 INCR = 10
1570 FOR I = 1 TO NP
1580      PRINT "X(";I;") = ";USING"+#.###^^^^";X(I)
1590 NEXT I
1600 '********************
1610 'OBTAIN [D] = [C] * [X]
1620 '********************
1630 FOR I = 1 TO NF
1640    D(I) = 0!
1650    FOR L = 1 TO NP
1660       D(I) = D(I) +  C(I,L) * X(L)
1670    NEXT L
1680 NEXT I
1690 PRINT
1700 '****************************
1710   PRINT "MEMEBER DEFORMATIONS"
1720 '****************************
1730 FOR I = 1 TO NF
1740    PRINT "D(";I;") = ";USING"+#.###^^^^";D(I)
1750 NEXT I
1760 '********************
1770 ' OBTAIN [F] = [S] * [D]
1780 '********************
1790 FOR I = 1 TO NF
1800    F(I) = 0
1810    FOR L = 1 TO NF
1820       F(I) = F(I) + S(I,L) * D(L)
1830    NEXT L
1840 NEXT I
1850 '********************
1860 ' OBTAIN [R] = [ER] * [F]
1870 '********************
1880 FOR I = 1 TO NR
1890    R(I) = 0
```

```
1900    FOR L = 1 TO NF
1910      R(I) = R(I) + ER(I,L) * F(L)
1920    NEXT   L
1930 NEXT I
1940 ' *****************************
1950 ' READ [FO] AND COMBINE WITH [F]
1960 ' *****************************
1970 FOR I = 1 TO NF
1980      READ FO(I)
1990      F(I) = F(I) + FO(I)
2000 NEXT I
2010 PRINT
2020 '*********************
2030   PRINT "MEMBER FORCES"
2040 '*********************
2050 FOR I = 1 TO NF
2060    PRINT "F(";I;") = ";USING"+#.###^^^^";F(I)
2070 NEXT I
2080 ' *****************************
2090 ' READ [RO] AND COMBINE WITH [R]
2100 ' *****************************
2110 FOR I = 1 TO NR
2120    READ RO(I)
2130    R(I) = R(I) + RO(I)
2140 NEXT I
2150 PRINT
2160 '******************
2170    PRINT "REACTIONS"
2180 '******************
2190 FOR I = 1 TO NR
2200    PRINT "R(";I;") = ";USING"+#.###^^^^";R(I)
2210 NEXT I
2220 ' *********************************
2230 ' DATA FOR EXAMPLE 14.7 (2nd EDITION)
2240 ' *********************************
2250 '
2260 '    NP,NF,NR
2270 DATA 3,  5, 5
2280 '
2290 ' [EP] MATRIX (NPxNF)
2300 DATA 0.000, -1.000,  0.000, -0.707,  0.000
2310 DATA 1.000,  0.000,  0.000,  0.707,  0.000
2320 DATA 0.000,  0.000,  1.000,  0.000,  0.707
2330 '
2340 ' [S] MATRIX (NFxNF) (DIAGONAL ELEMENTS)
2350 DATA 0.01, 0.01, 0.01, 0.00707, 0.00707
2360 '
2370 ' [P] MATRIX (NP)
2380 DATA 35.36, -35.36, -100.0
2390 '
2400 ' [ER] MATRIX (NRxNF)
2410 DATA  0.000,  0.000,  0.000,  0.000, -0.707
2420 DATA -1.000,  0.000,  0.000,  0.000, -0.707
2430 DATA  0.000,  0.000,  0.000,  0.707,  0.000
2440 DATA  0.000,  0.000, -1.000, -0.707,  0.000
2450 DATA  0.000,  1.000,  0.000,  0.000,  0.707
2460 '
2470 ' [FO] MATRIX (NF)
2480 DATA 0.000,  0.000,  100.0,  50.00,  0.000
2490 '
2500 ' [RO] MATRIX (NR)
2510 DATA 0.000,  0.000,  35.36, -135.36, 0.000


RESULTS OF BMATRIX.BAS
FOR EXAMPLE 14.7 (2nd ed.)

JOINT DISPLACEMENTS
X( 1 ) = +2.072E+03
X( 2 ) = -2.072E+03
X( 3 ) = -7.389E+03

MEMEBER DEFORMATIONS
D( 1 ) = -2.072E+03
D( 2 ) = -2.072E+03
```

```
D( 3 ) = -7.389E+03
D( 4 ) = -2.929E+03
D( 5 ) = -5.224E+03

MEMBER FORCES
F( 1 ) = -2.072E+01
F( 2 ) = -2.072E+01
F( 3 ) = +2.611E+01
F( 4 ) = +2.929E+01
F( 5 ) = -3.693E+01

REACTIONS
R( 1 ) = +2.611E+01
R( 2 ) = +4.683E+01
R( 3 ) = +2.072E+01
R( 4 ) = -4.683E+01
R( 5 ) = -4.683E+01
Ok
```

PROGRAM BMATRIX.F

```fortran
C*****************************************************************************
C                                                                          *
C        PROGRAM BMATRIX.F                        LANGUAGE: FORTRAN 77      *
C                                                 SYSTEM  : UNIX V          *
C                                                                          *
C        PURPOSE: TO EXECUTE THE BASIC MATRIX ANALYSIS METHOD INCLUDING    *
C                 SPECIAL "ACTIONS" AND APPLIED LOADS.                      *
C                                                                          *
C        INPUT  : NP   (INTEGER) - NUMBER OF DEGREES OF FREEDOM            *
C                 NF   (INTEGER) - NUMBER OF MEMBERS                        *
C                 NR   (INTEGER) - NUMBER OF REACTIONS                      *
C                 [EP] (REAL)    - ELEMENTS OF THE EQUILIBRIUM MATRIX FOR   *
C                                  DEGREES OF FREEDOM                       *
C                 [S]  (REAL)    - DIAGONAL ELEMENTS OF THE STIFFNESS MATRIX*
C                 [P]  (REAL)    - NEGATIVE OF THE FIXED STATE REACTIONS AT *
C                                  THE DEGREES OF FREEDOM                   *
C                 [ER] (REAL)    - ELEMENTS OF THE EQUILIBRIUM MATRIX FOR   *
C                                  REACTIONS                                *
C                 [FO] (REAL)    - INITIAL FIXED STATE MEMBER FORCES FOR    *
C                                  THE DEGREES OF FREEDOM                   *
C                 [RO] (REAL)    - INITIAL FIXED STATE MEMBER FORCES FOR    *
C                                  THE REACTIONS                            *
C                                                                          *
C        OUTPUT : [X]  (REAL)    - JOINT DISPLACEMENTS                      *
C                 [D]  (REAL)    - MEMBER DEFORMATIONS                      *
C                 [F]  (REAL)    - FINAL MEMBER FORCES                      *
C                 [R]  (REAL)    - FINAL REACTIONS                          *
C                                                                          *
C*****************************************************************************
C
      PROGRAM BMATRIX
      PARAMETER(INP=3, INF=5, INR=5 )
      REAL EP(INP,INF),S(INF,INF),C(INF,INP),SC(INF,INP),K(INP,INP)
      REAL P(INP),X(INP),D(INF),F(INF),FO(INF),ER(INR,INF),R(INR)
      REAL RO(INR)
      INTEGER NP,NF,NR
C
      OPEN(UNIT=1,FILE='INPUT' ,FORM='FORMATTED',STATUS='UNKNOWN')
      OPEN(UNIT=2,FILE='OUTPUT',FORM='FORMATTED',STATUS='UNKNOWN')
C
C     READ INPUT HEADER
C
      READ(1,*)
      READ(1,*)
      READ(1,*)
C
      READ(1,*)
      READ(1,*)NP,NF,NR
      READ(1,*)
C
C     READ [EP]
C
```

```
      READ(1,*)
      DO 100 I = 1,NP
        READ(1,*)(EP(I,J),J=1,NF)
100   CONTINUE
      READ(1,*)
C
C     INITIALIZE S TO ZERO
C
      DO 110 I = 1,NF
        DO 120 J = 1,NF
          S(I,J) = 0
120     CONTINUE
110   CONTINUE
C
C     READ DIAGONAL ENTRIES IN [S]
C
      READ(1,*)
      READ(1,*)(S(I,I),I=1,NF)
      READ(1,*)
C
C     READ [P]
C
      READ(1,*)
      READ(1,*)(P(I),I=1,NP)
      READ(1,*)
C
C     READ [ER]
C
      READ(1,*)
      DO 130 I=1,NR
        READ(1,*)(ER(I,J),J=1,NF)
130   CONTINUE
      READ(1,*)
C
C     DO [C] = [EP(TRANSPOSE)]
C
      DO 140 I = 1,NF
        DO 150 J = 1,NP
          C(I,J) = EP(J,I)
150     CONTINUE
140   CONTINUE
C
C     MULTIPLY [S] * [C]
C
      DO 160 I = 1,NF
        DO 170 J = 1,NP
          SC(I,J) = 0
          DO 180 L = 1,NF
            SC(I,J) = SC(I,J) + S(I,L) * C(L,J)
180       CONTINUE
170     CONTINUE
160   CONTINUE
C
C     OBTAIN [K] = [EP] * [SC]
C
      DO 190 I = 1,NP
        DO 200 J = 1,NP
          K(I,J) = 0
          DO 210 L = 1,NF
            K(I,J) = K(I,J) + EP(I,L) * SC(L,J)
210       CONTINUE
200     CONTINUE
190   CONTINUE
C
      CALL INVERT(K,NP)
C
C     OBTAIN [X] = [K(INVERSE)] * [P]
C
      DO 220 I = 1,NP
        X(I) = 0
        DO 230 L = 1,NP
          X(I) = X(I) + K(I,L) * P(L)
230     CONTINUE
220   CONTINUE
```

```
C
      WRITE(2,*)
      WRITE(2,*)  "RESULTS OF  BMATRIX.F"
      WRITE(2,*)  "FOR EXAMPLE 14.7 (2nd ed.)"
C
      WRITE(2,*)
      WRITE(2,*)  "JOINT DISPLACEMENTS"
      DO 240 I = 1,NP
         WRITE(2,10) I,X(I)
  10  FORMAT(3X,'X(',I2,') = ',1PE10.3)
 240  CONTINUE
C
C     OBTAIN [D] = [C] * [X]
C
      DO 250 I = 1,NF
         D(I) =0.0
         DO 260 L = 1,NP
            D(I) = D(I) +  C(I,L) * X(L)
 260     CONTINUE
 250  CONTINUE
C
      WRITE(2,*)
      WRITE(2,*)  "MEMBER DEFORMATIONS"
      DO 270 I = 1,NF
         WRITE(2,11) I,D(I)
  11  FORMAT(3X,'D(',I2,') = ',1PE10.3)
 270  CONTINUE
C
C     OBTAIN [F] = [S] * [D]
C
      DO 280 I = 1,NF
         F(I) = 0
         DO 290 L = 1,NF
            F(I)=F(I) + S(I,L) * D(L)
 290     CONTINUE
 280  CONTINUE
C
C     OBTAIN [R] = [ER] * [F]
C
      DO 300 I=1,NR
         R(I)=0
         DO 310 L=1,NF
            R(I) = R(I) + ER(I,L) * F(L)
 310     CONTINUE
 300  CONTINUE
C
C     READ [FO] AND COMBINE WITH [F]
C
      READ(1,*)
      READ(1,*)(FO(I),I=1,NF)
      READ(1,*)
      DO 320 I = 1,NF
         F(I) = F(I) + FO(I)
 320  CONTINUE
C
      WRITE(2,*)
      WRITE(2,*)  "MEMBER FORCES"
      DO 330 I = 1,NF
         WRITE(2,12)I,F(I)
  12  FORMAT(3X,'F(',I2') = ',1PE10.3)
 330  CONTINUE
C
C     READ [RO] AND COMBINE WITH [R]
C
      READ(1,*)
      READ(1,*)(RO(I),I=1,NR)
      READ(1,*)
      DO 340 I=1,NR
         R(I)=R(I)+RO(I)
 340  CONTINUE
C
      WRITE(2,*)
      WRITE(2,*)  "REACTIONS"
      DO 350  I = 1,NR
```

```
       WRITE(2,13) I,R(I)
 13    FORMAT(3X,'R(',I2,') = ',1PE10.3)
350    CONTINUE
       STOP
       END
C
C
C
       SUBROUTINE INVERT(k,np)
       PARAMETER(INP=3)
       REAL K(INP,INP)
       INTEGER IS(INP)
C
C      INVERT THE MATRIX K BY GAUSS ELIMINATION
C
       DO 100 I=1,NP
         IS(I) = 0
100    CONTINUE
110    AMAX=-1
       DO 120 I=1,NP
         IF(IS(I).NE.0) GOTO 120
         TEMP=ABS(K(I,I))
         IF((TEMP-AMAX).LE.0) GOTO 120
         ICOL = I
         AMAX = TEMP
120    CONTINUE
       IF(AMAX.EQ.0) GOTO 170
       IF(AMAX.LT.0) GOTO 160
       IS(ICOL) = 1
       PIVOT = K(ICOL,ICOL)
       K(ICOL,ICOL) = 1.0
       PIVOT = 1/PIVOT
       DO 130 J = 1,NP
         K(ICOL,J) = K(ICOL,J)*PIVOT
130    CONTINUE
       DO 140 I=1,NP
         IF((I-ICOL).EQ.0) GOTO 140
         TEMP = K(I,ICOL)
         K(I,ICOL) = 0.0
         DO 150 J = 1,NP
           K(I,J) = K(I,J) - K(ICOL,J)*TEMP
150      CONTINUE
140    CONTINUE
       GOTO 110
160    RETURN
170    STOP
       END
```

```
INPUT FILE FOR BMATRIX.F
DATA FOR EXAMPLE 14.7 (2nd EDITION)

NP  NF  NR
3   5   5


[EP] MATRIX (NPxNF)
0.000 -1.000  0.000 -0.707  0.000
1.000  0.000  0.000  0.707  0.000
0.000  0.000  1.000  0.000  0.707


[S] MATRIX (NFxNF) (DIAGONAL ELEMENTS)
0.01  0.01  0.01  0.00707  0.00707


[P] MATRIX (NP)
35.36 -35.36 -100.0


[ER] MATRIX (NRxNF)
 0.000  0.000  0.000  0.000 -0.707
-1.000  0.000  0.000  0.000 -0.707
 0.000  0.000  0.000  0.707  0.000
 0.000  0.000 -1.000 -0.707  0.000
 0.000  1.000  0.000  0.000  0.707
```

```
[F0] MATRIX (NF)
0.000   0.000   100.0   50.00   0.000

[R0] MATRIX (NR)
0.000   0.000   35.36   -135.36   0.000
```

RESULTS OF BMATRIX.F
FOR EXAMPLE 14.7 (2nd ed.)

```
JOINT DISPLACEMENTS
 X( 1) =   2.072e+03
 X( 2) =  -2.072e+03
 X( 3) =  -7.389e+03

MEMBER DEFORMATIONS
 D( 1) =  -2.072e+03
 D( 2) =  -2.072e+03
 D( 3) =  -7.389e+03
 D( 4) =  -2.929e+03
 D( 5) =  -5.224e+03

MEMBER FORCES
 F( 1) =  -2.072e+01
 F( 2) =  -2.072e+01
 F( 3) =   2.611e+01
 F( 4) =   2.929e+01
 F( 5) =  -3.693e+01

REACTIONS
 R( 1) =   2.611e+01
 R( 2) =   4.683e+01
 R( 3) =   2.072e+01
 R( 4) =  -4.683e+01
 R( 5) =  -4.683e+01
```

PROGRAM BEAM.BAS

```
10  '************************************************************************
20  '                                                                      *
30  '  PROGRAM BEAM.BAS                        LANGUAGE: IBM BASICA    3.0  *
40  '                                          SYSTEM  : MICROSOFT DOS 3.3  *
50  '                                                                      *
60  '  PURPOSE: TO EXECUTE THE BASIC MATRIX ANALYSIS METHOD FOR BEAMS      *
70  '                                                                      *
80  '  SIGN CONVENTION:    DEGREE OF FREEDOM        POSITIVE IS:           *
90  '                         ROTATION                 CLOCKWISE           *
100 '                         VERTICAL                 UPWARD              *
110 '                                                                      *
120 '  INPUT  : NP     (INTEGER) - NUMBER OF DEGREES OF FREEDOM            *
130 '           NM     (INTEGER) - NUMBER OF MEMBERS                       *
140 '           E       (REAL)   - MODULUS OF ELASTICITY                   *
150 '           MEMB   (INTEGER) - MEMBER IDENTIFICATION NUMBER            *
160 '           PN(1)  (INTEGER) - LEFT  ROTATIONAL DEGREE OF FREEDOM      *
170 '           PN(2)  (INTEGER) - LEFT  VERTICAL   DEGREE OF FREEDOM      *
180 '           PN(3)  (INTEGER) - RIGHT ROTATIONAL DEGREE OF FREEDOM      *
190 '           PN(4)  (INTEGER) - RIGHT VERTICAL   DEGREE OF FREEDOM      *
200 '                  NOTE: FOR RESTRAINT USE PN( ) = NP + 1              *
210 '           ML      (REAL)   - MEMBER LENGTH                           *
220 '           INERTIA (REAL)   - MOMENT OF INERTIA                       *
230 '           I      (INTEGER) - DEGREE OF FREEDOM                       *
240 '           PMAG    (REAL)   - NEGATIVE OF THE SUM OF THE FIXED STATE  *
250 '                              REACTIONS AT THE DEGREE OF FREEDOM I    *
260 '           {F0}    (REAL)   - FIXED STATE MEMBER FORCES GIVEN AS      *
270 '                              ROTATIONAL @ PN(1), VERTICAL @ PN(2)    *
280 '                              ROTATIONAL @ PN(3), VERTICAL @ PN(4)    *
290 '                              FOR ALL MEMBERS IN SEQUENTIAL ORDER     *
300 '                                                                      *
310 '  OUTPUT : {X}     (REAL)   - JOINT DISPLACEMENTS                     *
320 '           {F}     (REAL)   - MEMBER END FORCES                       *
330 '                                                                      *
340 '************************************************************************
```

```
350 CLS
360 '***********
370   READ NP,NM,E
380 '***********
390 NP1 = NP + 1
400 DIM P(NP1),X(NP1),PM(6),IS(NP1),K(NP1,NP1),PN(4),F0(4)
410 '**********************
420 ' INITIALIZE [K] TO ZERO
430 '**********************
440 FOR I = 1 TO NP1
450   FOR J = 1 TO NP1
460     K(I,J) = 0
470   NEXT J
480 NEXT I
490 FOR II = 1 TO NM
500 ' ****************************************
510   GOSUB 1910 'OBTAIN MEMBER STIFFNESS [KM]
520 '
530 '   ADD [KM] TO STRUCTURE STIFFNESS [K]
540 ' ****************************************
550   FOR I = 1 TO 4
560     N1 = PN(I)
570     FOR J = 1 TO 4
580       N2 = PN(J)
590       K(N1,N2)=K(N1,N2) + KM(I,J)
600     NEXT J
610   NEXT I
620 NEXT II
630 INCR = 10
640 ' ****************************************
650 ' INVERT THE MATRIX K BY GAUSS ELIMINATION
660 ' ****************************************
670 FOR I = 1 TO NP
680   IS(I) = 0
690 NEXT I
700 AMAX=-1
710 FOR I = 1 TO NP
720   IF IS(I) <> O THEN GOTO 770
730   TEMP=ABS(K(I,I))
740   IF (TEMP-AMAX) <= 0 THEN GOTO 770
750   ICOL = I
760   AMAX = TEMP
770 NEXT I
780 IF (AMAX) = 0! THEN GOTO 1160
790 IF (AMAX) < 0! THEN GOTO 1010
800 IS(ICOL) = 1
810 PIVOT = K(ICOL,ICOL)
820 K(ICOL,ICOL) = 1!
830 PIVOT = 1/PIVOT
840 FOR J = 1 TO NP
850   K(ICOL,J) = K(ICOL,J)*PIVOT
860 NEXT J
870 FOR I = 1 TO NP
880   IF (I-ICOL) = O THEN GOTO 940
890   TEMP = K(I,ICOL)
900   K(I,ICOL) = O.
910   FOR J = 1 TO NP
920     K(I,J) = K(I,J) - K(ICOL,J)*TEMP
930   NEXT J
940 NEXT I
950 GOTO 700
960 ' **********************************
970 '     MATRIX K HAS BEEN INVERTED
980 '
990 ' INITIALIZE LOAD MATRIX {P} TO ZERO
1000 ' **********************************
1010 FOR I = 1 TO NP
1020   P(I) = 0
1030 NEXT I
1040 '************
1050   READ I,PMAG
1060 '************
1070 IF I = O THEN GOTO 1100
1080 P(I) = PMAG
```

```
1090 GOTO 1050
1100 ' *****************************
1110 ' OBTAIN {X} = [K(INVERSE)] * {P}
1120 ' *****************************
1130 FOR I = 1 TO NP
1140     X(I) = O
1150     FOR L = 1 TO NP
1160        X(I) = X(I) + K(I,L) * P(L)
1170     NEXT L
1180 NEXT I
1190 '**************************
1200  PRINT "RESULTS OF BEAM.BAS"
1210  PRINT "FOR EXAMPLE 15.5 (2nd ed.)"
1220  PRINT
1230 '**************************
1240 '*******************************
1250 PRINT TAB(1);"JOINT DISPLACEMENTS"
1260 '*******************************
1270 FOR I = 1 TO NP
1280     PRINT "X(";I;") = ";USING"+#.###^^^^";X(I)
1290 NEXT I
1300 X(NP1) = O
1310 '****************************************************************
1320 ' CREATE FILE #1 FOR STORING FIXED STATE MEMBER FORCES [F0]
1330 '****************************************************************
1340   OPEN "F0MATRIX" FOR OUTPUT AS #1
1350 '
1360 ' *****************************
1370 ' STORE [F0] MATRICES ON FILE #1
1380 ' *****************************
1390 FOR II = 1 TO NM
1400     FOR LL = 1 TO 4
1410        READ F0(LL)
1420        PRINT #1, F0(LL)
1430     NEXT LL
1440 NEXT II
1450 CLOSE #1
1460 '****************************************************************
1470  PRINT " ": '        HEADER FOR MEMBER END FORCES
1480  PRINT"MEMB   L. MOMENT    L. SHEAR     R. MOMENT    R. SHEAR   "
1490 ' ****************************************************************
1500 '*****************************
1510 ' RESTORE DATA AND REREAD VALUES
1520 '*****************************
1530 RESTORE
1540 READ NP,NM,E
1550 OPEN "F0MATRIX" FOR INPUT AS #1
1560 FOR II = 1 TO NM
1570 ' *************************************************
1580   GOSUB 1910 :'OBTAIN MEMBER STIFFNESS MATRIX, AGAIN
1590 '
1600 '     CALCULATE RELEASED STATE MEMBER FORCES [F]
1610 ' *************************************************
1620     FOR I = 1 TO 4
1630        PM(I) = 0
1640        FOR J = 1 TO 4
1650           N3 = PN(J)
1660           PM(I) = PM(I) + KM(I,J) * X(N3)
1670        NEXT J
1680     NEXT I
1690 ' ****************************************************************************
1700 ' OBTAIN [F0] FROM FILE #1 AND COMBINE WITH RELEASED STATE MEMBER FORCES
1710 ' ****************************************************************************
1720     FOR LL = 1 TO 4
1730        INPUT #1, F0(LL)
1740     NEXT LL
1750     F1 = PM(1) + F0(1)
1760     F2 = PM(2) + F0(2)
1770     F3 = PM(3) + F0(3)
1780     F4 = PM(4) + F0(4)
1790 '     ******************
1800 '     PRINT MEMBER FORCES
1810 '     ******************
```

```
1820      PRINT USING "###_     ";MEMB;
1830      PRINT USING "+#.###^^^^_ _ ";F1,F2,F3,F4
1840 NEXT II
1850 CLOSE #1
1860 GOTO 2190
1870 ' ***********************************************
1880 ' SUBROUTINE FOR BUILDING MEMBER STIFFNESS MATRIX [K]
1890 ' ***********************************************
1900 '*******************************************
1910   READ MEMB,PN(1),PN(2),PN(3),PN(4),ML,INERTIA
1920 '*******************************************
1930 C1 =  4 * E * INERTIA/ML
1940 C2 =  6 * E * INERTIA/ML/ML
1950 C3 = 12 * E * INERTIA/ML/ML/ML
1960 ' ******************************************
1970 ' BUILD UPPER RIGHT TRIANGLE INCLUDING DIAGONAL
1980 ' ******************************************
1990 KM(1,1) = C1
2000 KM(1,2) =-C2
2010 KM(1,3) = C1/2!
2020 KM(1,4) = C2
2030 KM(2,2) = C3
2040 KM(2,3) =-C2
2050 KM(2,4) =-C3
2060 KM(3,3) = C1
2070 KM(3,4) = C2
2080 KM(4,4) = C3
2090 ' ************************************
2100 ' BUILD LOWER RIGHT TRIANGLE BY SYMMETRY
2110 ' ************************************
2120 FOR J = 1 TO 4
2130   JP1 = J + 1
2140   FOR L = JP1 TO 4
2150     KM(L,J) = KM(J,L)
2160   NEXT L
2170 NEXT J
2180 RETURN
2190 END
2200 '*****************************************
2210 'DATA FOR EXAMPLE 15.5 IN TEXT (2ND EDITION)
2220 '*****************************************
2230 '
2240 '    NP,NM, E
2250 DATA  3, 2, 1
2260 '
2270 ' MEMB | PN(1) PN(2)|| PN(3) PN(4) | ML,INERTIA
2280 ' LEFT>|  ROT. VERT ||  ROT.  VERT |<RIGHT
2290 DATA  1,    1,    3,      2,     4,   16,  1
2300 DATA  2,    2,    4,      4,     4,   16,  1
2310 '
2320 '     I,PMAG
2330  DATA 1, 64
2340  DATA 2,-32
2350  DATA 3,-24
2360 '
2370 ' STOP CARD (END OF PMAG)
2380 DATA 0,0
2390 '
2400 '{F0} MATRIX FOR PN(1) PN(2) PN(3) PN(4) FOR
2410 '              ALL MEMBERS IN SEQUENTIAL ORDER
2420 DATA -64,24,64,24
2430 DATA -32, 8,32, 8
```

```
RESULTS OF BEAM.BAS
FOR EXAMPLE 15.5 (2nd ed.)

JOINT DISPLACEMENTS
X( 1 ) = -3.456E+03
X( 2 ) = -1.408E+03
X( 3 ) = -4.710E+04

MEMB   L. MOMENT    L. SHEAR     R. MOMENT    R. SHEAR
  1    -3.052E-05  +7.629E-06   +3.840E+02   +4.800E+01
  2    -3.840E+02  +4.100E+01   -1.440E+02   -2.500E+01
```

PROGRAM BEAM.F

```
C************************************************************************
C                                                                     *
C       PROGRAM BEAM.F                        LANGUAGE: FORTRAN 77     *
C                                             SYSTEM  : UNIX V         *
C                                                                     *
C       PURPOSE: TO EXECUTE THE BASIC MATRIX ANALYSIS METHOD FOR BEAMS *
C                                                                     *
C       SIGN CONVENTION:     DEGREE OF FREEDOM        POSITIVE IS:     *
C                            ROTATION                 CLOCKWISE        *
C                            VERTICAL                 UPWARD           *
C                                                                     *
C       INPUT  : NP      (INTEGER) - NUMBER OF DEGREES OF FREEDOM      *
C                NM      (INTEGER) - NUMBER OF MEMBERS                 *
C                E       (REAL)    - MODULUS OF ELASTICITY             *
C                MEMB    (INTEGER) - MEMBER IDENTIFICATION NUMBER      *
C                PN(1)   (INTEGER) - LEFT  ROTATIONAL DEGREE OF FREEDOM *
C                PN(2)   (INTEGER) - LEFT  VERTICAL   DEGREE OF FREEDOM *
C                PN(3)   (INTEGER) - RIGHT ROTATIONAL DEGREE OF FREEDOM *
C                PN(4)   (INTEGER) - RIGHT VERTICAL   DEGREE OF FREEDOM *
C                    NOTE: FOR RESTRAINT USE PN( ) = NP + 1            *
C                ML      (REAL)    - MEMBER LENGTH                     *
C                INERTIA (REAL)    - MOMENT OF INERTIA                 *
C                I       (INTEGER) - DEGREE OF FREEDOM                 *
C                PMAG    (REAL)    - NEGATIVE OF THE SUM OF THE FIXED STATE *
C                                    REACTIONS AT THE DEGREE OF FREEDOM I *
C                {F0}    (REAL)    - FIXED STATE MEMBER FORCES GIVEN AS *
C                                    ROTATIONAL @ PN(1), VERTICAL @ PN(2) *
C                                    ROTATIONAL @ PN(3), VERTICAL @ PN(4) *
C                                    FOR ALL MEMBERS IN SEQUENTIAL ORDER *
C                                                                     *
C       OUTPUT : {X}     (REAL)    - JOINT DISPLACEMENTS               *
C                {F}     (REAL)    - MEMBER END FORCES                 *
C                                                                     *
C************************************************************************
C
        PROGRAM BEAM
        PARAMETER(INP=3, INM=2, INP1=4)
        REAL E,PMAG,F0(4),X(INP1),PM(6),K(INP1,INP1)
        REAL KM(4,4),P(INP)
        INTEGER NP,NM,PN(4)
C
        OPEN(UNIT=1,FILE='INPUT' ,FORM='FORMATTED',STATUS='UNKNOWN')
        OPEN(UNIT=2,FILE='OUTPUT',FORM='FORMATTED',STATUS='UNKNOWN')
C
C       READ INPUT HEADER
C
        READ(1,*)
        READ(1,*)
        READ(1,*)
C
        READ(1,*)
        READ(1,*) NP,NM,E
        READ(1,*)
C
        NP1 = NP + 1
C
C       INITIALIZE [K] TO ZERO
C
        DO 100 I=1,NP1
          DO 110 J=1,NP1
            K(I,J) = 0
 110      CONTINUE
 100    CONTINUE
C
C       READ MEMBER STIFFNESS HEADER
C
        READ(1,*)
        READ(1,*)
C
        DO 120 II = 1,NM
C
        CALL MEMBSTIF(KM,MEMB,PN,E)
```

```
C
C       ADD [KM] TO STRUCTURE STIFFNESS [K]
C
          DO 130 I = 1,4
            N1 = PN(I)
            DO 140 J = 1,4
              N2 = PN(J)
              K(N1,N2)=K(N1,N2) + KM(I,J)
 140        CONTINUE
 130      CONTINUE
 120    CONTINUE
C
        CALL INVERT(K,NP)
C
C       INITIALIZE {P} TO ZERO
C
        DO 150 I = 1,NP
          P(I) = O
 150    CONTINUE
C
        READ(1,*)
        READ(1,*)
 160    READ(1,*) I,PMAG
        IF(I.EQ.0)GOTO 170
        P(I)=PMAG
        GOTO 160
 170    READ(1,*)
C
C       OBTAIN {X} = [K(INVERSE)] * {P}
C
        DO 180 I=1,NP
          X(I) = 0
          DO 190 L=1,NP
            X(I) = X(I) + K(I,L) * P(L)
 190      CONTINUE
 180    CONTINUE
C
        WRITE(2,*)
        WRITE(2,*) "RESULTS OF BEAM.F"
        WRITE(2,*) "FOR EXAMPLE 15.5 (2nd ed.)"
C
        WRITE(2,*)
        WRITE(2,*) "JOINT DISPLACEMENTS"
        DO 200 I=1,NP
          WRITE(2,10) I,X(I)
 10     FORMAT(3X,'X(',I2,') = ',1PE10.3)
 200    CONTINUE
        X(NP1)=0
C
C       READ HEADER FOR FIXED STATE MEMBER FORCES
C
        READ(1,*)
        READ(1,*)
C
C       CREATE FILE #3 TO STORE FIXED STATE MEMBER FORCES [F0]
C
        OPEN(UNIT=3,FILE='F0MATRIX',FORM='FORMATTED',STATUS='UNKNOWN')
C
C       STORE [F0] MATRIX ON FILE #3
C
        DO 210 II= 1,NM
          READ (1,*)(F0(LL),LL=1,4)
          WRITE(3,*)(F0(LL),LL=1,4)
 210    CONTINUE
C
        WRITE(2,*)
        WRITE(2,*) "MEMB    L. MOMENT    L. SHEAR",
       *              " R. MOMENT    L. SHEAR"
C
C       REWIND DATA AND REREAD VALUES
C
        REWIND 1
        REWIND 3
C
```

```
C       READ INPUT HEADER
C
        READ(1,*)
        READ(1,*)
        READ(1,*)
C
        READ(1,*)
        READ(1,*) NP,NM,E
        READ(1,*)
C
C       READ MEMBER STIFFNESS HEADER
C
        READ(1,*)
        READ(1,*)
C
        DO 220 II=1,NM
C
          CALL MEMBSTIF(KM,MEMB,PN,E)
C
C         CALCULATE RELEASED STATE MEMBER FORCES [F]
C
          DO 230 I = 1,4
            PM(I) = 0
            DO 240 J = 1,4
              N3 = PN(J)
              PM(I) = PM(I) + KM(I,J) * X(N3)
  240       CONTINUE
  230     CONTINUE
C
C         OBTAIN [F0] FROM FILE #3 AND COMBINE WITH RELEASED STATE MEMBER FORCES
C
          READ(3,*)(F0(LL),LL=1,4)
          F1 = PM(1) + F0(1)
          F2 = PM(2) + F0(2)
          F3 = PM(3) + F0(3)
          F4 = PM(4) + F0(4)
C
C         WRITE MEMBER FORCES
C
          WRITE(2,11)MEMB,F1,F2,F3,F4
   11     FORMAT(3X,I3,3X,4(1PE10.3,2x))
C
  220   CONTINUE
        STOP
        END
C
C
C
        SUBROUTINE MEMBSTIF(KM,MEMB,PN,E)
C
        PARAMETER(INP=3, INM=2, INP1=4)
        REAL E,ML,INERTIA,KM(4,4)
        INTEGER PN(4)
C
        READ(1,*)MEMB,PN(1),PN(2),PN(3),PN(4),ML,INERTIA
C
        C1 =  4.0 * E * INERTIA/ML
        C2 =  6.0 * E * INERTIA/ML/ML
        C3 = 12.0 * E * INERTIA/ML/ML/ML
C
C       BUILD UPPER RIGHT TRIANGLE INCLUDING DIAGONAL
C
        KM(1,1) = C1
        KM(1,2) =-C2
        KM(1,3) = C1/2.0
        KM(1,4) = C2
        KM(2,2) = C3
        KM(2,3) =-C2
        KM(2,4) =-C3
        KM(3,3) = C1
        KM(3,4) = C2
        KM(4,4) = C3
C
C       BUILD LOWER RIGHT TRIANGLE BY SYMMETRY
```

```
C
      DO 100 J = 1,4
        JP1 = J + 1
        DO 110 L = JP1,4
          KM(L,J) = KM(J,L)
 110    CONTINUE
 100  CONTINUE
      RETURN
      END
C
C
C
      SUBROUTINE INVERT(K,NP)
      PARAMETER(INP=3, INM=2, INP1=4)
      REAL K(INP1,INP1)
      INTEGER IS(INP1)
C
C     INVERT THE MATRIX K BY GAUSS ELIMINATION
C
      DO 100 I=1,NP
        IS(I) = 0
 100  CONTINUE
 110  AMAX=-1
      DO 120 I=1,NP
        IF(IS(I).NE.0) GOTO 120
        TEMP=ABS(K(I,I))
        IF((TEMP-AMAX).LE.0) GOTO 120
        ICOL = I
        AMAX = TEMP
 120  CONTINUE
      IF(AMAX.EQ.0) GOTO 170
      IF(AMAX.LT.0) GOTO 160
      IS(ICOL) = 1
      PIVOT = K(ICOL,ICOL)
      K(ICOL,ICOL) = 1.0
      PIVOT = 1/PIVOT
      DO 130 J = 1,NP
        K(ICOL,J) = K(ICOL,J)*PIVOT
 130  CONTINUE
      DO 140 I=1,NP
        IF((I-ICOL).EQ.0) GOTO 140
        TEMP = K(I,ICOL)
        K(I,ICOL) = 0.0
        DO 150 J = 1,NP
          K(I,J) = K(I,J) - K(ICOL,J)*TEMP
 150    CONTINUE
 140  CONTINUE
      GOTO 110
 160  RETURN
 170  STOP
      END


INPUT FILE FOR BEAM.F
DATA FOR EXAMPLE 15.5 (2ND EDITION)

NP NM  E
 3  2  1

MEMB | PN(1) PN(2)|| PN(3) PN(4) | ML,INERTIA
LEFT>|  ROT.  VERT ||  ROT.  VERT |<RIGHT
  1  |   1     3   ||   2     4   | 16  1
  2  |   2     4   ||   4     4   | 16  1

I PMAG
1  64
2 -32
3 -24
0   0   < STOP CARD (END OF PMAG)

{F0} MATRIX FOR PN(1) PN(2) PN(3) PN(4) FOR
ALL MEMBERS IN SEQUENTIAL ORDER
-64  24  64  24
-32   8  32   8
```

```
RESULTS OF BEAM.F
FOR EXAMPLE 15.5 (2nd ed.)

JOINT DISPLACEMENTS
  X( 1) = -3.456e+03
  X( 2) = -1.408e+03
  X( 3) = -4.710e+04

MEMB    L. MOMENT    L. SHEAR    R. MOMENT    L. SHEAR
  1     3.052e-05   -7.629e-06   3.840e+02    4.800e+01
  2    -3.840e+02    4.100e+01  -1.440e+02   -2.500e+01
```

PROGRAM TRUSS.BAS

```
10  '********************************************************************
20  '                                                                  *
30  '    PROGRAM TRUSS.BAS                     LANGUAGE: IBM BASICA   3.0 *
40  '                                          SYSTEM  : MICROSOFT DOS 3.3 *
50  '                                                                  *
60  '  PURPOSE: TO EXECUTE THE BASIC MATRIX ANALYSIS METHOD FOR TRUSSES *
70  '                                                                  *
80  '  SIGN CONVENTION:    DEGREE OF FREEDOM         POSITIVE IS:      *
90  '                      VERTICAL                  UPWARDS           *
100 '                      HORIZONTAL                TO THE RIGHT      *
110 '                                                                  *
120 '   INPUT  : NP    (INTEGER) - NUMBER OF DEGREES OF FREEDOM        *
130 '            NF    (INTEGER) - NUMBER OF MEMBERS                   *
140 '            E     (REAL)    - MODULUS OF ELASTICITY               *
150 '            MEMB  (INTEGER) - MEMBER NUMBER                       *
160 '            P1    (INTEGER) - HORZ. DEGREE OF FREEDOM (MEMBER ORIGIN) *
170 '            P2    (INTEGER) - VERT. DEGREE OF FREEDOM (MEMBER ORIGIN) *
180 '            P3    (INTEGER) - HORZ. DEGREE OF FREEDOM (MEMBER END) *
190 '            P4    (INTEGER) - VERT. DEGREE OF FREEDOM (MEMBER END) *
200 '            H     (REAL)    - HORZ. DISTANCE BETWEEN THE ENDS OF  *
210 '                             THE MEMBER.                          *
220 '            V     (REAL)    - VERT. DISTANCE BETWEEN THE ENDS OF  *
230 '                             THE MEMBER.                          *
240 '            A     (REAL)    - AREA OF THE MEMBER                  *
250 '            I     (INTEGER) - DEGREE OF FREEDOM                   *
260 '            PMAG  (REAL)    - NEGATIVE OF THE SUM OF THE FIXED STATE *
270 '                             REACTIONS AT THE DEGREE OF FREEDOM I *
280 '                                                                  *
290 '   OUTPUT : [X]   (REAL)    - JOINT DISPLACEMENTS                 *
300 '            [F]   (REAL)    - MEMBER FORCES                       *
310 '                                                                  *
320 '********************************************************************
330 CLS
340 '************
350   READ NP,NF,E
360 '************
370 NP1 = NP + 1
380 NP4 = NP + 4
390 DIM P(NP4),X(NP4),F(NP4),IS(NP4),K(NP4,NP4)
400 '*********************
410 'INITIALIZE [K] TO ZERO
420 '*********************
430 FOR I = 1 TO NP1
440   FOR J = 1 TO NP1
450     K(I,J) = 0
460   NEXT J
470 NEXT I
480 '
490 FOR I = 1 TO NF
500 ' *************************
510   READ MEMB,P1,P2,P3,P4,H,V,A
520 ' *************************
530   ML = SQR(H*H+V*V)
```

```
540    MSIN = V/ ML
550    MCOS = H/ ML
560    C1 = E * A * MCOS * MCOS / ML
570    C2 = E * A * MSIN * MCOS / ML
580    C3 = E * A * MSIN * MSIN / ML
590    K(P1,P1) = K(P1,P1) + C1
600    K(P1,P2) = K(P1,P2) + C2
610    K(P1,P3) = K(P1,P3) - C1
620    K(P1,P4) = K(P1,P4) - C2
630    K(P2,P1) = K(P2,P1) + C2
640    K(P2,P2) = K(P2,P2) + C3
650    K(P2,P3) = K(P2,P3) - C2
660    K(P2,P4) = K(P2,P4) - C3
670    K(P3,P1) = K(P3,P1) - C1
680    K(P3,P2) = K(P3,P2) - C2
690    K(P3,P3) = K(P3,P3) + C1
700    K(P3,P4) = K(P3,P4) + C2
710    K(P4,P1) = K(P4,P1) - C2
720    K(P4,P2) = K(P4,P2) - C3
730    K(P4,P3) = K(P4,P3) + C2
740    K(P4,P4) = K(P4,P4) + C3
750 NEXT I
760 '*****************************************
770 ' INVERT THE MATRIX K BY GAUSS ELIMINATION
780 '*****************************************
790 FOR I = 1 TO NP
800    IS(I) = 0
810 NEXT I
820 AMAX=-1
830 FOR I = 1 TO NP
840    IF IS(I) <> 0 THEN GOTO 890
850    TEMP=ABS(K(I,I))
860    IF (TEMP-AMAX) <= 0 THEN GOTO 890
870    ICOL = I
880    AMAX = TEMP
890 NEXT I
900 IF (AMAX) = 0! THEN GOTO 1270
910 IF (AMAX) < 0! THEN GOTO 1120
920 IS(ICOL) = 1
930 PIVOT = K(ICOL,ICOL)
940 K(ICOL,ICOL) = 1!
950 PIVOT = 1/PIVOT
960 FOR J = 1 TO NP
970    K(ICOL,J) = K(ICOL,J)*PIVOT
980 NEXT J
990 FOR I = 1 TO NP
1000    IF (I-ICOL) = 0 THEN GOTO 1060
1010    TEMP = K(I,ICOL)
1020    K(I,ICOL) = 0.
1030    FOR J = 1 TO NP
1040      K(I,J) = K(I,J) - K(ICOL,J)*TEMP
1050    NEXT J
1060 NEXT I
1070 GOTO 820
1080 '*****************************
1090 '    MATRIX K HAS BEEN INVERTED
1100 ' INITIALIZE {P} MATRIX TO ZERO
1110 '*****************************
1120 FOR I = 1 TO NP
1130    P(I) = 0
1140 NEXT I
1150 '**********
1160    READ I,PMAG
1170 '**********
1180 IF I = 0 THEN GOTO 1210
1190 P(I) = PMAG
1200 GOTO 1160
1210 '***********************
1220 ' CALCULATE {X} = [K] * {P}
1230 '***********************
1240 FOR I = 1 TO NP
1250    X(I) = 0
```

```
1260    FOR L = 1 TO NP
1270      X(I) = X(I) + K(I,L) * P(L)
1280    NEXT L
1290 NEXT I
1300 '***************************
1310  PRINT "RESULTS OF TRUSS.BAS"
1320  PRINT "FOR EXAMPLE 14.5 (2nd ed.)
1330  PRINT
1340 '***************************
1350 '*********************************
1360  PRINT TAB(1);"JOINT DISPLACEMENTS"
1370 '*********************************
1380 FOR I = 1 TO NP
1390    PRINT "X(";I;") = ";USING"+#.###^^^^";X(I)
1400 NEXT I
1410 X(NP1) = O
1420 '***************************
1430 ' RESTORE DATA & REREAD VALUES
1440 '***************************
1450 RESTORE
1460 READ NP,NF,E
1470 FOR I = 1 TO NF
1480 '***************************
1490    READ MEMB,P1,P2,P3,P4,H,V,A
1500 '***************************
1510    ML = SQR(H*H + V*V)
1520    MSIN = V / ML
1530    MCOS = H / ML
1540    F(I) = (E*A/ML) *((X(P3)-X(P1))*MCOS + (X(P4)-X(P2))*MSIN)
1550 NEXT I
1560 PRINT " "
1570 '***************************
1580  PRINT TAB(1);"MEMBER FORCES"
1590 '***************************
1600 FOR I = 1 TO NF
1610    PRINT "F(";I;") = ";USING"+#.###^^^^";F(I)
1620 NEXT I
1630 END
1640 '*********************
1650 ' DATA FOR EXAMPLE 14.5
1660 '*********************
1670 '
1680 '    NP NF  E
1690  DATA 2,4,30000
1700 '
1710 '    MEMB | PN(1) PN(2) || PN(3) PN(4) | H     V     A
1720 ' ORIGIN>| HORZ  VERT  || HORZ  VERT  |<END
1730  DATA  1,    2,    1,     3,    3,    -60, 144, 1.5
1740  DATA  2,    2,    1,     3,    3,      0, 144, 1.5
1750  DATA  3,    2,    1,     3,    3,    108, 144, 1.5
1760  DATA  4,    2,    1,     3,    3,    192, 144, 1.5
1770 '
1780 '      I,PMAG
1790  DATA 1,-10
1800 '
1810 ' STOP CARD (END OF PMAG)
1820  DATA 0,0


RESULTS OF TRUSS.BAS
FOR EXAMPLE 14.5 (2nd ed.)


JOINT DISPLACEMENTS
X( 1 ) = -1.351E-02
X( 2 ) = +5.754E-03


MEMBER FORCES
F( 1 ) = +4.237E+00
F( 2 ) = +4.223E+00
F( 3 ) = +1.840E+00
F( 4 ) = +6.572E-01
```

PROGRAM TRUSS.F

```
C************************************************************************
C                                                                      *
C       PROGRAM TRUSS.F                         LANGUAGE: FORTRAN 77    *
C                                               SYSTEM : UNIX V         *
C                                                                      *
C       PURPOSE: TO EXECUTE THE BASIC MATRIX ANALYSIS METHOD FOR TRUSSES *
C                                                                      *
C       SIGN CONVENTION:    DEGREE OF FREEDOM          POSITIVE IS:     *
C                           VERTICAL                   UPWARDS          *
C                           HORIZONTAL                 TO THE RIGHT     *
C                                                                      *
C       INPUT  : NP    (INTEGER) - NUMBER OF DEGREES OF FREEDOM         *
C                NF    (INTEGER) - NUMBER OF MEMBERS                    *
C                E     (REAL)    - MODULUS OF ELASTICITY                *
C                MEMB  (INTEGER) - MEMBER NUMBER                        *
C                P1    (INTEGER) - HORZ. DEGREE OF FREEDOM (MEMBER ORIGIN) *
C                P2    (INTEGER) - VERT. DEGREE OF FREEDOM (MEMBER ORIGIN) *
C                P3    (INTEGER) - HORZ. DEGREE OF FREEDOM (MEMBER END)  *
C                P4    (INTEGER) - VERT. DEGREE OF FREEDOM (MEMBER END)  *
C                H     (REAL)    - HORZ. DISTANCE BETWEEN THE ENDS OF    *
C                                  THE MEMBER.                          *
C                V     (REAL)    - VERT. DISTANCE BETWEEN THE ENDS OF    *
C                                  THE MEMBER.                          *
C                A     (REAL)    - AREA OF THE MEMBER                   *
C                I     (INTEGER) - DEGREE OF FREEDOM                    *
C                PMAG  (REAL)    - NEGATIVE OF THE SUM OF THE FIXED STATE *
C                                  REACTIONS AT THE DEGREE OF FREEDOM I  *
C                                                                      *
C       OUTPUT : [X]   (REAL)    - JOINT DISPLACEMENTS                  *
C                [F]   (REAL)    - MEMBER FORCES                        *
C                                                                      *
C************************************************************************
C
        PROGRAM TRUSS
        PARAMETER (INP=2, INF=4, INP4=6)
        REAL E,H,V,A,PMAG,X(INP4),F(INP4),P(INP4),K(INP4,INP4)
        REAL ML,MSIN,MCOS
        INTEGER NP,NF,MEMB,P1,P2,P3,P4
C
        OPEN(UNIT=1,FILE='INPUT', FORM='FORMATTED',STATUS='UNKNOWN')
        OPEN(UNIT=2,FILE='OUTPUT',FORM='FORMATTED',STATUS='UNKNOWN')
C
C       READ INPUT HEADER
C
        READ(1,*)
        READ(1,*)
        READ(1,*)
C
        READ(1,*)
        READ(1,*) NP,NF,E
        READ(1,*)
C
        NP1 = NP + 1
        NP4 = NP + 4
C
C       INITIALIZE [K] TO ZERO
C
        DO 100 I = 1 ,NP1
          DO 110 J = 1 ,NP1
          K(I,J) = 0
  110     CONTINUE
  100   CONTINUE
C
C       READ MEMBER STIFFNESS HEADER
C
        READ(1,*)
        READ(1,*)
C
        DO 120 I = 1 ,NF
C
        READ(1,*)MEMB,P1,P2,P3,P4,H,V,A
C
        ML = SQRT(H*H+V*V)
```

```
       MSIN = V/ ML
       MCOS = H/ ML
       C1 = E * A * MCOS * MCOS / ML
       C2 = E * A * MSIN * MCOS / ML
       C3 = E * A * MSIN * MSIN / ML
       K(P1,P1) = K(P1,P1) + C1
       K(P1,P2) = K(P1,P2) + C2
       K(P1,P3) = K(P1,P3) - C1
       K(P1,P4) = K(P1,P4) - C2
       K(P2,P1) = K(P2,P1) + C2
       K(P2,P2) = K(P2,P2) + C3
       K(P2,P3) = K(P2,P3) - C2
       K(P2,P4) = K(P2,P4) - C3
       K(P3,P1) = K(P3,P1) - C1
       K(P3,P2) = K(P3,P2) - C2
       K(P3,P3) = K(P3,P3) + C1
       K(P3,P4) = K(P3,P4) + C2
       K(P4,P1) = K(P4,P1) - C2
       K(P4,P2) = K(P4,P2) - C3
       K(P4,P3) = K(P4,P3) + C2
       K(P4,P4) = K(P4,P4) + C3
 120   CONTINUE
C
C      INVERT THE MATRIX K BY GAUSS ELIMINATION
C
       CALL INVERT(K,NP)
C
C      INITIALIZE {P} MATRIX TO ZERO
C
       DO 130 I = 1 ,NP
         P(I) = 0
 130   CONTINUE
C
C      READ {P}
C
       READ(1,*)
       READ(1,*)
 140   READ(1,*) I,PMAG
       IF(I.EQ.0) GOTO 150
       P(I) = PMAG
       GOTO 140
C
C      CALCULATE {X} = [K] * {P}
C
 150   DO 160 I = 1 ,NP
         X(I) = 0
         DO 170 L = 1 ,NP
           X(I) = X(I) + K(I,L) * P(L)
 170     CONTINUE
 160   CONTINUE
C
       WRITE(2,*)
       WRITE(2,*)"RESULTS OF TRUSS.F"
       WRITE(2,*)"FOR EXAMPLE 14.5 (2nd ed.)"
C
       WRITE(2,*)
       WRITE(2,*)"JOINT DISPLACEMENTS"
       DO 180 I = 1 ,NP
         WRITE(2,10) I,X(I)
 10    FORMAT(3X,'X(',I2,') = ',1PE10.3)
 180   CONTINUE
       X(NP1) = 0
C
C      REWIND DATA & REREAD VALUES
C
       REWIND 1
C
C      READ INPUT HEADER
C
       READ(1,*)
       READ(1,*)
       READ(1,*)
C
       READ(1,*)
```

```
      READ(1,*) NP,NF,E
      READ(1,*)
C
C     READ MEMBER STIFFNESS HEADER
C
      READ(1,*)
      READ(1,*)
C
      DO 190 I = 1 ,NF
C
         READ(1,*) MEMB,P1,P2,P3,P4,H,V,A
C
         ML = SQRT(H*H + V*V)
         MSIN = V / ML
         MCOS = H / ML
         F(I) = (E*A/ML) *((X(P3)-X(P1))*MCOS + (X(P4)-X(P2))*MSIN)
 190  CONTINUE
C
      WRITE(2,*)
      WRITE(2,*) "MEMBER FORCES"
      DO 200 I = 1 ,NF
         WRITE(2,11) I,F(I)
  11  FORMAT(3X,'F(',I2,') = ',1PE10.3)
 200  CONTINUE
      STOP
      END
C
C
C
      SUBROUTINE INVERT(K,NP)
      PARAMETER (INP=2, INF=4, INP4=6)
      REAL K(INP4,INP4)
      INTEGER IS(INP4)
C
C     INVERT THE MATRIX K BY GAUSS ELIMINATION
C
      DO 100 I=1,NP
         IS(I) = 0
 100  CONTINUE
 110  AMAX=-1
      DO 120 I=1,NP
         IF(IS(I).NE.0) GOTO 120
         TEMP=ABS(K(I,I))
         IF((TEMP-AMAX).LE.0) GOTO 120
         ICOL = I
         AMAX = TEMP
 120  CONTINUE
      IF(AMAX.EQ.0) GOTO 170
      IF(AMAX.LT.0) GOTO 160
      IS(ICOL) = 1
      PIVOT = K(ICOL,ICOL)
      K(ICOL,ICOL) = 1.0
      PIVOT = 1/PIVOT
      DO 130 J = 1,NP
         K(ICOL,J) = K(ICOL,J)*PIVOT
 130  CONTINUE
      DO 140 I=1,NP
         IF((I-ICOL).EQ.0) GOTO 140
         TEMP = K(I,ICOL)
         K(I,ICOL) = 0.0
         DO 150 J = 1,NP
            K(I,J) = K(I,J) - K(ICOL,J)*TEMP
 150     CONTINUE
 140  CONTINUE
      GOTO 110
 160  RETURN
 170  STOP
      END


INPUT FILE FOR TRUSS.F
DATA FOR EXAMPLE 14.5 (2ND EDITION)

NP NF    E
 2  4  30000
```

MEMB |	PN(1)	PN(2) ||	PN(3)	PN(4) |	H	V	A
ORIGIN>|	HORZ	VERT ||	HORZ	VERT |	<END		
1	2	1	3	3	-60	144	1.5
2	2	1	3	3	0	144	1.5
3	2	1	3	3	108	144	1.5
4	2	1	3	3	192	144	1.5

```
I PMAG
1 -10
0    0  < STOP CARD (END OF PMAG)
```

```
RESULTS OF TRUSS.F
FOR EXAMPLE 14.5 (2nd ed.)

JOINT DISPLACEMENTS
 X( 1) = -1.351e-02
 X( 2) =  5.754e-03

MEMBER FORCES
 F( 1) =  4.237e+00
 F( 2) =  4.223e+00
 F( 3) =  1.840e+00
 F( 4) =  6.572e-01
```

PROGRAM FRAME.BAS

```
10  '*************************************************************************
20  '                                                                        *
30  '   PROGRAM FRAME.BAS                        LANGUAGE: IBM BASICA    3.0  *
40  '                                            SYSTEM  : MICROSOFT DOS 3.3  *
50  '                                                                        *
60  '   PURPOSE: TO EXECUTE THE BASIC MATRIX ANALYSIS METHOD FOR FRAMES      *
70  '                                                                        *
80  '   SIGN CONVENTION:    DEGREE OF FREEDOM          POSITIVE IS:          *
90  '                       ROTATION                   CLOCKWISE             *
100 '                       HORIZONTAL                 TO THE RIGHT          *
110 '                       VERTICAL                   UPWARD                *
120 '                                                                        *
130 '   INPUT  : NP     (INTEGER) - NUMBER OF DEGREES OF FREEDOM             *
140 '            NM     (INTEGER) - NUMBER OF MEMBERS                        *
150 '            E       (REAL)   - MODULUS OF ELASTICITY                    *
160 '            MEMB   (INTEGER) - MEMBER IDENTIFICATION NUMBER             *
170 '            PN(1)  (INTEGER) - LEFT  ROTATIONAL DEGREE OF FREEDOM       *
180 '            PN(2)  (INTEGER) - LEFT  HORIZONTAL DEGREE OF FREEDOM       *
190 '            PN(3)  (INTEGER) - LEFT  VERTICAL   DEGREE OF FREEDOM       *
200 '            PN(4)  (INTEGER) - RIGHT ROTATIONAL DEGREE OF FREEDOM       *
210 '            PN(5)  (INTEGER) - RIGHT HORIZONTAL DEGREE OF FREEDOM       *
220 '            PN(6)  (INTEGER) - RIGHT VERTICAL   DEGREE OF FREEDOM       *
230 '              NOTE: FOR RESTRAINT USE PN( ) = NP + 1                    *
240 '            H       (REAL)   - HORIZ. DIST. BETWEEN MEMBER ENDS         *
250 '            V       (REAL)   - VERT.  DIST. BETWEEN MEMBER ENDS         *
260 '            AREA    (REAL)   - AREA OF MEMBER                           *
270 '            INERTIA (REAL)   - MOMENT OF INERTIA OF MEMBER              *
280 '            I      (INTEGER) - DEGREE OF FREEDOM                        *
290 '            PMAG    (REAL)   - NEGATIVE OF THE SUM OF THE FIXED STATE   *
300 '                               REACTIONS AT THE DEGREE OF FREEDOM I     *
310 '            {FO}    (REAL)   - FIXED STATE MEMBER FORCES GIVEN AS       *
320 '                               ROT @PN(1), HORZ @PN(2), VERT @PN(3)     *
330 '                               ROT @PN(4), HORZ @PN(5), VERT @PN(6)     *
340 '                               FOR EACH MEMBER IN SEQUENTIAL ORDER      *
350 '                                                                        *
360 '   OUTPUT : {X}     (REAL)   - JOINT DISPLACEMENTS                      *
370 '            {F}     (REAL)   - MEMBER END FORCES                        *
380 '                                                                        *
390 '*************************************************************************
400 CLS
410 '************
420  READ NP,NM,E
430 '************
440 NP1 = NP + 1
450 DIM P(NP),X(NP1),PM(6),IS(NP1),K(NP1,NP1),KM(6,6),PN(6),F0(6)
460 '*********************
470 'INITIALIZE [K] TO ZERO
480 '*********************
```

```
490 FOR I = 1 TO NP1
500    FOR J = 1 TO NP1
510       K(I,J) = 0
520    NEXT J
530 NEXT I
540 FOR II = 1 TO NM
550 ' ****************************************
560    GOSUB 1960 'OBTAIN MEMBER STIFFNESS [KM]
570 ' ****************************************
580 ' ADD [KM] TO STRUCTURE STIFFNESS [K]
590 ' ****************************************
600    FOR I = 1 TO 6
610      N1 = PN(I)
620      FOR J = 1 TO 6
630        N2 = PN(J)
640          K(N1,N2) = K(N1,N2) + KM(I,J)
650      NEXT J
660    NEXT I
670 NEXT II
680 ' ****************************************
690 ' INVERT THE MATRIX K BY GAUSS ELIMINATION
700 ' ****************************************
710 FOR I = 1 TO NP
720    IS(I) = 0
730 NEXT I
740 AMAX=-1
750 FOR I = 1 TO NP
760    IF IS(I) <> 0 THEN GOTO 810
770    TEMP=ABS(K(I,I))
780    IF (TEMP-AMAX) <= 0 THEN GOTO 810
790    ICOL = I
800    AMAX = TEMP
810 NEXT I
820 IF (AMAX) = 0! THEN GOTO 1200
830 IF (AMAX) < 0! THEN GOTO 1050
840 IS(ICOL) = 1
850 PIVOT = K(ICOL,ICOL)
860 K(ICOL,ICOL) = 1!
870 PIVOT = 1/PIVOT
880 FOR J = 1 TO NP
890    K(ICOL,J) = K(ICOL,J)*PIVOT
900 NEXT J
910 FOR I = 1 TO NP
920    IF (I-ICOL) = 0 THEN GOTO 980
930    TEMP = K(I,ICOL)
940    K(I,ICOL) = 0.
950    FOR J = 1 TO NP
960      K(I,J) = K(I,J) - K(ICOL,J)*TEMP
970    NEXT J
980 NEXT I
990 GOTO 740
1000 '***********************************
1010 ' MATRIX K HAS BEEN INVERTED
1020 '
1030 ' INITIALIZE LOAD MATRIX {P} TO ZERO
1040 '***********************************
1050 FOR I = 1 TO NP
1060    P(I) = 0
1070 NEXT I
1080 '**********
1090    READ I,PMAG
1100 '**********
1110 IF I = 0 THEN GOTO 1140
1120 P(I) = PMAG
1130 GOTO 1090
1140 ' ****************************
1150 ' OBTAIN {X} = [K(INVERSE)] * {P}
1160 ' ****************************
1170 FOR I = 1 TO NP
1180    X(I) = 0
1190    FOR L = 1 TO NP
1200      X(I) = X(I) + K(I,L) * P(L)
1210    NEXT L
1220 NEXT I
```

```
1230 '***************************
1240  PRINT "RESULTS OF FRAME.BAS"
1250  PRINT "FOR EXAMPLES 15.10 & 15.11 (2nd ed.)"
1260  PRINT
1270 '***************************
1280 '***********************************
1290  PRINT TAB(1);"JOINT DISPLACEMENTS"
1300 '***********************************
1310 FOR I = 1 TO NP
1320     PRINT "X(";I;") = ";USING"+#.###^^^^";X(I)
1330 NEXT I
1340   X(NP1) = O
1350 '*****************************************************************
1360 ' CREATE FILE #1 FOR STORING FIXED STATE MEMBER FORCES [F0]
1370 '*****************************************************************
1380   OPEN "F0MATRIX" FOR OUTPUT AS #1
1390 '*****************************
1400 ' STORE [F0] MATRICES ON FILE #1
1410 '*****************************
1420 FOR II = 1 TO NM
1430     FOR LL = 1 TO 6
1440        READ F0(LL)
1450        PRINT #1, F0(LL)
1460     NEXT LL
1470 NEXT II
1480 CLOSE #1
1490 '***********************************************
1500 PRINT " ": '    HEADER FOR MEMBER END FORCES
1510 PRINT"MEMB    L. MOMENT      L. SHEAR      L. AXIAL";
1520 PRINT     "    R. MOMENT      R. SHEAR      R. AXIAL"
1530 '***********************************************
1540 '***************************
1550 ' RESTORE DATA & REREAD VALUES
1560 '***************************
1570 RESTORE
1580 READ NP,NM,E
1590 OPEN "F0MATRIX" FOR INPUT AS #1
1600 FOR II = 1 TO NM
1610 ' ****************************************************
1620    GOSUB 1960  "OBTAIN MEMBER STIFFNESS MATRIX, AGAIN
1630 '
1640 '     CALCULATE RELEASED STATE MEMBER FORCES [F]
1650 ' ****************************************************
1660     FOR I = 1 TO 6
1670     PM(I) = 0
1680       FOR J = 1 TO 6
1690          N3 = PN(J)
1700          PM(I) = PM(I) + KM(I,J) * X(N3)
1710       NEXT J
1720     NEXT I
1730 ' ******************************************************************
1740 ' OBTAIN [F0] FROM FILE #1 AND COMBINE WITH RELEASED STATE MEMBER FORCES
1750 ' ******************************************************************
1760     FOR LL = 1 TO 6
1770        INPUT #1, F0(LL)
1780     NEXT LL
1790     F1 =   PM(1) + F0(1)
1800     F2 = -PM(2) * MSIN + PM(3) * MCOS + F0(2)
1810     F3 =  PM(2) * MCOS + PM(3) * MSIN + F0(3)
1820     F4 =   PM(4) + F0(4)
1830     F5 = -PM(5) * MSIN + PM(6) * MCOS + F0(5)
1840     F6 =  PM(5) * MCOS + PM(6) * MSIN + F0(6)
1850 '     ******************
1860 '    PRINT MEMBER FORCES
1870 '     ******************
1880     PRINT USING "###_    ";MEMB;
1890     PRINT USING "+#.###^^^^_ _ ";F1,F2,F3,F4,F5,F6
1900 NEXT II
1910 CLOSE #1
1920 GOTO 2430
1930 '***********************************************
1940 ' SUBROUTINE FOR BUILDING MEMBER STIFFNESS MATRIX [K]
1950 '***********************************************
1960 READ MEMB,PN(1),PN(2),PN(3),PN(4),PN(5),PN(6),H,V,AREA,INERTIA
```

```
1970 ML = SQR(H*H+V*V)
1980 MSIN = V/ ML
1990 MCOS = H/ ML
2000 C1 =   4 * E * INERTIA/ML
2010 C2 =   6 * E * INERTIA/ML/ML * MSIN
2020 C3 =   6 * E * INERTIA/ML/ML * MCOS
2030 C4 = 12 * E * INERTIA/ML/ML/ML * MSIN * MSIN
2040 C5 = 12 * E * INERTIA/ML/ML/ML * MCOS * MCOS
2050 C6 = 12 * E * INERTIA/ML/ML/ML * MSIN * MCOS
2060 C7 = E * AREA/ML * MSIN * MSIN
2070 C8 = E * AREA/ML * MCOS * MCOS
2080 C9 = E * AREA/ML * MSIN * MCOS
2090 ' **********************************************
2100 ' BUILD UPPER RIGHT TRIANGLE INCLUDING DIAGONAL
2110 ' **********************************************
2120 KM(1,1) =    C1
2130 KM(1,2) =    C2
2140 KM(1,3) = -C3
2150 KM(1,4) =    C1 / 2
2160 KM(1,5) = -C2
2170 KM(1,6) =    C3
2180 KM(2,2) =    C8 + C4
2190 KM(2,3) =    C9 - C6
2200 KM(2,4) =    C2
2210 KM(2,5) = -C8 - C4
2220 KM(2,6) = -C9 + C6
2230 KM(3,3) =    C7 + C5
2240 KM(3,4) = -C3
2250 KM(3,5) = -C9 + C6
2260 KM(3,6) = -C7 - C5
2270 KM(4,4) =    C1
2280 KM(4,5) = -C2
2290 KM(4,6) =    C3
2300 KM(5,5) =    C8 + C4
2310 KM(5,6) =    C9 - C6
2320 KM(6,6) =    C7 + C5
2330 ' **********************************
2340 ' BUILD LOWER RIGHT TRIANGLE BY SYMMETRY
2350 ' **********************************
2360    FOR J = 1 TO 6
2370      JP1 = J+ 1
2380      FOR L = JP1 TO 6
2390        KM(L,J) = KM(J,L)
2400      NEXT L
2410    NEXT J
2420 RETURN
2430 END
2440 '*********************************************************
2450 ' DATA FOR EXAMPLES 15.10 & 15.11 IN TEXT (2ND EDITION)
2460 '*********************************************************
2470 '   NP, NM,  E
2480 DATA 7,  3,  20000
2490 '
2500 'MEMB, | NP(1) NP(2) NP(3) || NP(4) NP(5) NP(6) | H, V, AREA,INERTIA
2510 ' LEFT>|  ROT. HORZ  VERT  ||  ROT. HORZ  VERT  |<RIGHT
2520 DATA 1,   1,    8,    8,      2,    3,    4,    8, 6, 10, 1
2530 DATA 2,   2,    3,    4,      5,    6,    7,   10, 0, 10, 2
2540 DATA 3,   5,    6,    7,      8,    8,    8,    6,-8, 10, 1
2550 '
2560 '    I, PMAG
2570 DATA 2, 28.80
2580 DATA 4,-12.96
2590 DATA 5,-19.20
2600 DATA 7, -7.04
2610 '
2620 ' STOP CARD (END OF PMAG)
2630 DATA 0,0
2640 '
2650 ' {F0} MATRIX AT NP(1),NP(2),NP(3),NP(4),NP(5),NP(6)
2660 '              FOR ALL MEMBERS IN SEQUENTIAL ORDER
2670 DATA     0,     0,    0,    0,     0,    0
2680 DATA -28.8,     0,12.96, 19.2,    0, 7.04
2690 DATA     0,     0,    0,    0,     0,    0
```

```
RESULTS OF FRAME.BAS
FOR EXAMPLES 15.10 & 15.11 (2nd ed.)

JOINT DISPLACEMENTS
X( 1 ) = +1.130E-03
X( 2 ) = +1.025E-03
X( 3 ) = +5.934E-03
X( 4 ) = -9.238E-03
X( 5 ) = -2.062E-03
X( 6 ) = +5.298E-03
X( 7 ) = +2.976E-03

MEMB   L. MOMENT    L. SHEAR    L. AXIAL    R. MOMENT    R. SHEAR    R. AXIAL
  1   -3.815E-06   +4.199E-02  +1.592E+01  -4.199E-01  -4.199E-02  -1.592E+01
  2   +4.199E-01   -3.374E+00  +2.567E+01  +2.372E+01  +3.374E+00  -5.671E+00
  3   -2.372E+01   +3.920E+00  +1.596E+01  -1.548E+01  -3.920E+00  -1.596E+01
```

PROGRAM FRAME.F

```
C**********************************************************************
C                                                                    *
C   PROGRAM FRAME.F                          LANGUAGE: FORTRAN 77     *
C                                            SYSTEM  : UNIX V         *
C                                                                    *
C   PURPOSE: TO EXECUTE THE BASIC MATRIX ANALYSIS METHOD FOR FRAMES   *
C                                                                    *
C   SIGN CONVENTION:     DEGREE OF FREEDOM          POSITIVE IS:      *
C                        ROTATION                   CLOCKWISE         *
C                        HORIZONTAL                 TO THE RIGHT      *
C                        VERTICAL                   UPWARD            *
C                                                                    *
C   INPUT  : NP     (INTEGER) - NUMBER OF DEGREES OF FREEDOM          *
C            NM     (INTEGER) - NUMBER OF MEMBERS                     *
C            E      (REAL)    - MODULUS OF ELASTICITY                 *
C            MEMB   (INTEGER) - MEMBER IDENTIFICATION NUMBER          *
C            PN(1)  (INTEGER) - LEFT  ROTATIONAL DEGREE OF FREEDOM    *
C            PN(2)  (INTEGER) - LEFT  HORIZONTAL DEGREE OF FREEDOM    *
C            PN(3)  (INTEGER) - LEFT  VERTICAL   DEGREE OF FREEDOM    *
C            PN(4)  (INTEGER) - RIGHT ROTATIONAL DEGREE OF FREEDOM    *
C            PN(5)  (INTEGER) - RIGHT HORIZONTAL DEGREE OF FREEDOM    *
C            PN(6)  (INTEGER) - RIGHT VERTICAL   DEGREE OF FREEDOM    *
C                 NOTE: FOR RESTRAINT USE PN( ) = NP + 1             *
C            H      (REAL)    - HORIZ. DIST. BETWEEN MEMBER ENDS      *
C            V      (REAL)    - VERT.  DIST. BETWEEN MEMBER ENDS      *
C            AREA   (REAL)    - AREA OF MEMBER                        *
C            INERTIA (REAL)   - MOMENT OF INERTIA OF MEMBER           *
C            I      (INTEGER) - DEGREE OF FREEDOM                     *
C            PMAG   (REAL)    - NEGATIVE OF THE SUM OF THE FIXED STATE *
C                              REACTIONS AT THE DEGREE OF FREEDOM I   *
C            {F0}   (REAL)    - FIXED STATE MEMBER FORCES GIVEN AS    *
C                              ROT @PN(1), HORZ @PN(2), VERT @PN(3)   *
C                              ROT @PN(4), HORZ @PN(5), VERT @PN(6)   *
C                              FOR EACH MEMBER IN SEQUENTIAL ORDER    *
C                                                                    *
C   OUTPUT : {X}    (REAL)    - JOINT DISPLACEMENTS                   *
C            {F}    (REAL)    - MEMBER END FORCES                     *
C                                                                    *
C**********************************************************************
C
        PROGRAM FRAME
        PARAMETER (INP=7,INM=3,INP1=8)
        REAL E,PMAG,F0(6),X(INP1),P(INP1),PM(6)
        REAL K(INP1,INP1),KM(6,6),MSIN,MCOS
        INTEGER NP,NM,MEMB,PN(6)
C
        OPEN(UNIT=1,FILE='INPUT' ,FORM='FORMATTED',STATUS='UNKNOWN')
        OPEN(UNIT=2,FILE='OUTPUT',FORM='FORMATTED',STATUS='UNKNOWN')
C
C       READ INPUT HEADER
C
        READ(1,*)
```

```
        READ(1,*)
        READ(1,*)
C
        READ(1,*)
        READ(1,*) NP,NM,E
        READ(1,*)
C
        NP1 = NP + 1
C
C       INITIALIZE [K] TO ZERO
C
        DO 100 I=1,NP1
          DO 110 J=1,NP1
            K(I,J) = 0
 110      CONTINUE
 100    CONTINUE
C
C       READ MEMBER STIFFNESS HEADER
C
        READ(1,*)
        READ(1,*)
C
        DO 120 II=1,NM
C
           CALL MEMBSTIF(KM,MEMB,PN,E,MCOS,MSIN)
C
C          ADD [KM] TO STRUCTURE STIFFNESS [K]
C
           DO 130 I=1,6
             N1 = PN(I)
             DO 140 J=1,6
               N2 = PN(J)
               K(N1,N2) = K(N1,N2) + KM(I,J)
 140         CONTINUE
 130       CONTINUE
 120    CONTINUE
C
        CALL INVERT(K,NP)
C
C       INITIALIZE {P} TO ZERO
C
        DO 150 I=1,NP
          P(I)=0
 150    CONTINUE
C
        READ(1,*)
        READ(1,*)
 160    READ(1,*) I,PMAG
        IF(I.EQ.0) GOTO 170
        P(I)=PMAG
        GOTO 160
 170    READ(1,*)
C
C       OBTAIN {X} = [K(INVERSE)] * {P}
C
        DO 180 I=1,NP
          X(I)=0
          DO 190 L=1,NP
            X(I)=X(I)+K(I,L)*P(L)
 190      CONTINUE
 180    CONTINUE
C
        WRITE(2,*)
        WRITE(2,*) "RESULTS OF FRAME.F"
        WRITE(2,*) "FOR EXAMPLES 15.10 & 15.11 (2nd ed.)"
C
        WRITE(2,*)
        WRITE(2,*) "JOINT DISPLACEMENTS"
        DO 200 I=1,NP
          WRITE(2,10) I,X(I)
 10     FORMAT(3X,'X(',I2,') = ',1PE10.3)
 200    CONTINUE
        X(NP1)=0
```

```
C
C       READ HEADER FOR FIXED STATE MEMBER FORCES
C
        READ(1,*)
        READ(1,*)
C
C       CREATE FILE #3 TO STORE FIXED STATE MEMBER FORCES [F0]
C
        OPEN(UNIT=3,FILE='F0MATRIX',FORM='FORMATTED',STATUS='UNKNOWN')
C
C       STORE [F0] MATRIX ON FILE #3
C
        DO 210 II=1,NM
            READ (1,*)(F0(LL),LL=1,6)
            WRITE(3,*)(F0(LL),LL=1,6)
  210   CONTINUE
C
        WRITE(2,*)
        WRITE(2,*)"MEMB    L. MOMENT    L. SHEAR    L. AXIAL",
       *              "  R. MOMENT    R. SHEAR    R. AXIAL"
C
C       REWIND DATA AND REREAD VALUES
C
        REWIND 1
        REWIND 3
C
C       READ INPUT HEADER
C
        READ(1,*)
        READ(1,*)
        READ(1,*)
C
        READ(1,*)
        READ(1,*) NP,NM,E
        READ(1,*)
C
C       READ MEMBER STIFFNESS HEADER
C
        READ(1,*)
        READ(1,*)
C
        DO 220 II=1,NM
C
          CALL MEMBSTIF(KM,MEMB,PN,E,MCOS,MSIN)
C
C         CALCULATE RELEASED STATE MEMBER FORCES [F]
C
          DO 230 I=1,6
            PM(I)=0
            DO 240 J=1,6
              N3 = PN(J)
              PM(I) = PM(I)+KM(I,J)*X(N3)
  240       CONTINUE
  230     CONTINUE
C
C         OBTAIN [F0] FROM FILE #3 AND COMBINE WITH RELEASED STATE MEMBER FORCES
C
          READ(3,*)(F0(LL),LL=1,6)
          F1 = PM(1) + F0(1)
          F2 =-PM(2) * MSIN + PM(3) * MCOS + F0(2)
          F3 = PM(2) * MCOS + PM(3) * MSIN + F0(3)
          F4 = PM(4) + F0(4)
          F5 =-PM(5) * MSIN + PM(6) * MCOS + F0(5)
          F6 = PM(5) * MCOS + PM(6) * MSIN + F0(6)
C
C         WRITE MEMBER FORCES
C
          WRITE(2,11) MEMB,F1,F2,F3,F4,F5,F6
   11     FORMAT(3X,I3,3X,6(1PE10.3,2X))
C
  220   CONTINUE
        STOP
        END
```

```
C
C
      SUBROUTINE MEMBSTIF(KM,MEMB,PN,E,MCOS,MSIN)
      PARAMETER (INP=7,INM=3,INP1=8)
      REAL E,ML,INERTIA,KM(6,6),ML,MSIN,MCOS
      INTEGER PN(4)
C
      READ(1,*)MEMB,PN(1),PN(2),PN(3),PN(4),PN(5),PN(6),H,V,AREA,INERTIA
C
      ML = SQRT(H*H + V*V)
      MSIN = V/ML
      MCOS = H/ML
      C1 =  4 * E * INERTIA/ML
      C2 =  6 * E * INERTIA/ML/ML * MSIN
      C3 =  6 * E * INERTIA/ML/ML * MCOS
      C4 = 12 * E * INERTIA/ML/ML/ML * MSIN * MSIN
      C5 = 12 * E * INERTIA/ML/ML/ML * MCOS * MCOS
      C6 = 12 * E * INERTIA/ML/ML/ML * MSIN * MCOS
      C7 = E * AREA/ML * MSIN * MSIN
      C8 = E * AREA/ML * MCOS * MCOS
      C9 = E * AREA/ML * MSIN * MCOS
C
C     BUILD UPPER RIGHT TRIANGLE INCLUDING DIAGONAL
C
      KM(1,1) =   C1
      KM(1,2) =   C2
      KM(1,3) =  -C3
      KM(1,4) =   C1 / 2
      KM(1,5) =  -C2
      KM(1,6) =   C3
C
      KM(2,2) =   C8 + C4
      KM(2,3) =   C9 - C6
      KM(2,4) =   C2
      KM(2,5) =  -C8 - C4
      KM(2,6) =  -C9 + C6
C
      KM(3,3) =   C7 + C5
      KM(3,4) =  -C3
      KM(3,5) =  -C9 + C6
      KM(3,6) =  -C7 - C5
C
      KM(4,4) =   C1
      KM(4,5) =  -C2
      KM(4,6) =   C3
C
      KM(5,5) =   C8 + C4
      KM(5,6) =   C9 - C6
C
      KM(6,6) =   C7 + C5
C
C     BUILD LOWER TRIANGULAR BY SYMMETRY
C
      DO 100 J = 1,6
        LP1 = J + 1
        DO 110 L = LP1,6
          KM(L,J) = KM(J,L)
110     CONTINUE
100   CONTINUE
      RETURN
      END
C
C
C
      SUBROUTINE INVERT(K,NP)
      PARAMETER (INP=7,INM=3,INP1=8)
      REAL K(INP1,INP1)
      INTEGER IS(INP1)
C
C     INVERT THE MATRIX K BY GAUSS ELIMINATION
C
      DO 100 I=1,NP
        IS(I) = 0
```

```
100    CONTINUE
110    AMAX=-1
       DO 120 I=1,NP
         IF(IS(I).NE.0) GOTO 120
         TEMP=ABS(K(I,I))
         IF((TEMP-AMAX).LE.0) GOTO 120
         ICOL = I
         AMAX = TEMP
120    CONTINUE
       IF(AMAX.EQ.0) GOTO 170
       IF(AMAX.LT.0) GOTO 160
       IS(ICOL) = 1
       PIVOT = K(ICOL,ICOL)
       K(ICOL,ICOL) = 1.0
       PIVOT = 1/PIVOT
       DO 130 J = 1,NP
         K(ICOL,J) = K(ICOL,J)*PIVOT
130    CONTINUE
       DO 140 I=1,NP
         IF((I-ICOL).EQ.0) GOTO 140
         TEMP = K(I,ICOL)
         K(I,ICOL) = 0.0
         DO 150 J = 1,NP
           K(I,J) = K(I,J) - K(ICOL,J)*TEMP
150      CONTINUE
140    CONTINUE
       GOTO 110
160    RETURN
170    STOP
       END
```

```
INPUT FILE FOR FRAME.F
DATA FOR EXAMPLES 15.10 & 15.11 (2ND EDITION)

NP   NM    E
 7    3   20000
```

MEMB,	NP(1)	NP(2)	NP(3)		NP(4)	NP(5)	NP(6)	H	V	AREA	INERTIA
LEFT>	ROT.	HORZ	VERT		ROT.	HORZ	VERT	<RIGHT			
1	1	8	8		2	3	4	8	6	10	1
2	2	3	4		5	6	7	10	0	10	2
3	5	6	7		8	8	8	6	-8	10	1

```
I    PMAG
2    28.80
4   -12.96
5   -19.20
7    -7.04
0    0   < STOP CARD (END OF PMAG)
```

```
{F0} MATRIX AT NP(1),NP(2),NP(3),NP(4),NP(5),NP(6)
FOR ALL MEMBERS IN SEQUENTIAL ORDER
   0      0      0       0      0      0
-28.8     0    12.96   19.2     0     7.04
   0      0      0       0      0      0
```

```
RESULTS OF FRAME.F
FOR EXAMPLES 15.10 & 15.11 (2nd ed.)

JOINT DISPLACEMENTS
X( 1) =   1.130e-03
X( 2) =   1.025e-03
X( 3) =   5.933e-03
X( 4) =  -9.238e-03
X( 5) =  -2.062e-03
X( 6) =   5.298e-03
X( 7) =   2.976e-03
```

MEMB	L. MOMENT	L. SHEAR	L. AXIAL	R. MOMENT	R. SHEAR	R. AXIAL
1	0.000e+00	4.199e-02	1.592e+01	-4.198e-01	-4.199e-02	-1.592e+01
2	4.198e-01	-3.374e+00	2.567e+01	2.372e+01	3.374e+00	-5.671e+00
3	-2.372e+01	3.920e+00	1.596e+01	-1.548e+01	-3.920e+00	-1.596e+01

Index

Analysis of Structures, methods. *See* Structures; Structural analysis
Angle change
 definition, 199
 concentrated, 202
Angle-change principle, 199
 concepts, 199–203, 206
 corrective rigid body rotation, 205–207
 use in computing frame displacements, 201–204, 207–209
 cantilevered frames, 202
Areas of geometrical shapes, 478
Axial deformation
 neglect of, 95, 190, 249, 282, 316, 419
 in unit-load method. *See* Unit-load method
Axial forces, in fixed beams, 482
Axial stress. *See* Stress

Basic matrix method
 beams and planar frames, 382
 compatibility, 362–365
 equilibrium, 361
 material conditions, 366
 matrices. *See* Matrices, in structural analysis planar trusses
 computer programs, 310, 378
 initial member deformations, 376
 procedure, 367
 released state, 375
 restrained state, 373
 statically determinate, 372
 statically indeterminate, 360
 structure compatibility modes, 369, 371
 support settlements, 372
Beams
 analysis by conjugate beam method. *See* Conjugate beam method
 by direct stiffness method. *See* Direct stiffness method

 by double integration. *See* Double integration method
 by flexibility method. *See* Fexibility method
 by moment-distribution method. *See* Moment-distribution method
 by quadruple integration. *See* Quadruple integration method
 by slope-deflection method. *See* Slope-deflection method
 by unit-load method. *See* Unit-load method
 definition, 5–6
 forces in, 68
 shear and moment diagrams. *See* Moment diagrams
Bending
 biaxial, 42
 uniaxial, 42
 unsymmetrical, 43
Bracing 55–58
 stability. *See* Stability
 lateral. *See* Stability

Centroids of areas, 478
Compatibility
 conditions in flexibility method, 279
 definition. *See* Criteria of structural analysis
 in planar trusses, 362–366
 matrix. *See* Basic matrix analysis
 modes. *See* Direct stiffness method
Complementary strain energy, 229
Complementary strain energy density, 229
Complementary work, 229
Composite behavior, 19
Compound structures
 analysis by direct stiffness method, 472
 by unit-load method. *See* Unit-load method for general structures
 by flexibility method. *See* Flexibility method
Computer programs, 310, 378, 431, 492

Conjugate beam method
  example problems, 218–221
  relationship to conjugate grid method, 215
  transforming support conditions, 216
    rules for, 217
  treatment of internal hinges, 217–218
Conjugate grid
  loading on, 209
  rules for constructing, 209
Conjugate grid method, 209–215
  procedure, 209–212
  relationship angle change principle, 209
  sign convention, 209–210
Constitutive laws, 28
Continua
  definition, 2, 6
  flexibility method. *See* Flexibility method
  unit-load method. *See* Unit-load method
Coordinates. *See* Transformation matrices
  global, 409, 442
  local, 177, 262, 442
Criteria for structural analysis, 26, 361
  compatibility, 24
  concepts, 23–24
  continua, 23
  equilibrium, 24
  framed structures, 23
  material laws, See Material
  sources of error, 24

Deformation. *See* Axial deformation
  axial, 40, 232–242, 254
  flexural, 43, 78–90, 94–95, 242–246
  shear, 90–92, 253
  torsional, 44–45, 253
Degrees of freedom, 260
  at internal hinges, 265–269
  definition, 262
  determination of
    including axial deformation, 262–265
    neglecting axial deformation, 269–275
  general equations
    beams, 264
    grids, 264
    planar frames, 263–272
    planar trusses, 264
    space frames, 263–272
    space trusses, 263
  sidesway motions, 272,
Design, definition, 1
Direct stiffness method, 382–431
  assembly of load matrix, 452–455
  assembly of structure stiffness matrix
    beams, 396–400
    planar trusses, 410–412
  boundary conditions, 464–465
  computer programs, 431. *See* Computer programs
  inclined supports. *See* Transformation matrices
  initial strains, 465–467

  internal hinges, 467–470
  local member stiffness matrix
    overchanging member, 385
    referenced to displaced chord, 385
    supported member, 384,
    unsupported member, 383
  member (basic) compatibility modes
    beams, 388
    planar frames, 420–423
    planar trusses, 409–410
    procedure
      beams, 400–405
        joint loads, 402, 404
        released state, 402
        restraining moments, 400–401
        restrained state, 401
        sign conventions, 403
    compound structures, 472–474
    planar frames, 418–431
      assembly of structure stiffness matrix, 423–426
      calculation of internal forces, 428–431
      equivalent joint loads, 426–428
      global member stiffness matrix, 420–423
      released state, 418
      restrained state, 418
    planar trusses, 408–415
      assembly of structure stiffness matrix, 410–412
      calculation of member forces, 413
      global member stiffness matrix, 408–415
  structure compatibility modes
    beams, 389–392
    general definition, 387
    planar frames, 415–418
    planar trusses, 406–408
  support motions, 463–464
  using transformation matrices, 455–463, see also
    Transformation matrices
Displacements, analysis of
  by Angle-Change Principle. *See* Principle(s), Angle
    Change
  by basic matrix analysis. *See* basic matrix analysis
  by conjugate beam method. *See* Conjugate beam
    method
  by direct stiffness method. *See* Direct stiffness method
  by double integration. *See* Double integration method
  by quadruple integration. *See* Quadruple integration
    method
  by slope-deflection method. *See* Slope-deflection
    method
Double integration method
  for beams, 80–90
  for planar frames, 178
  use of binomial functions, 85–88

Elastic, strain energy, 20
Elastic curve, equation of
  beams, 80
  equations of, 36–37
  equilibrium. *See* Criteria for structural analysis

planar frames, 177–179
  of planar trusses. *See* Planar trusses

First order structural analysis, 92
Fixed beam moments, 318, 336–355
  end displacements, due to, 482
  torsion, due to, 482
Flat plates, 6
Flexibility coefficients, 233
  symmetry of, 303
  by unit load method, 303–305
Flexibility method
  beam analysis by, 286–293
  compound structure analysis by, 308–310
  fundamentals of, 279
  general equations, 301
  including a continuum, 310
  initial deformations, 307
  planar frame analysis by, 293–297
  planar truss analysis by, 298–301, 305
  selecting redundants, 302
  support settlements, 306
Flexural stress. *See* Stress
Framed structures
  definition, 4
  equilibrium, 36–37
  modeling. *See* Idealized model
  types, 4–6
Free body diagrams, 37–39

Gauss Jordan elimination, 490
Generalized deformations, 261
Generalized displacements, 260–316
Geometrically unstable
  definition. *See* Stability
  determination of, 51–53
Global member stiffness matrix. *See* Direct stiffness
        method
Governing differential equation
  definition, 23
  for flexural displacement, 78
Grids, 5–6
  loads, 34
  modeling. *See* Idealized model

Idealized model, 2
  grid, 48
  individual member, 49
  joints. *See* Joints
  material laws, 2
  mathematical model, 2
  schematic diagram, 2
  selecting an, 50–51
  sources of error, 24
  substructuring, 45–51
  supports. *See* Supports
  three-dimensional, 45–47
  two-dimensional, 47–48
Internal forces, 33, 39
Internal hinge, 4, 63

Joints, types, 4

Law of Conservation of Work and Energy, 2, 226, 238
  formulation for, 227
Load matrix, assembly of
Load paths, 16–19
  concept, 16
  primary and secondary, 17–19
  structural foundation, 16
Loads
  applied, 34–35
  dead, 9
  dynamic, 22
  earthquake, 13
  environmental, 9
  occupancy live, 9
  snow, 13
  static, 22
  sources of magnitude, 9–16
  thermal, 13–16
  wind, 10–13

Material
  constitutive laws, 28–30
  elastic, 22
  inelastic, 22, 32
  isotropic, 28
  laws, 2
  modulus of elasticity, 27
  nonlinear, 31
  Poisson's ratio, 27
  shearing modulus of elasticity, 27
Matrices
  general
    adjoint, 486
    cofactor, 486
    column, 484
    diagonal, 484
    identity, 484
    notation, 483
    partitioned, 487
    row, 484
    square, 483
    symmetric, 484
    unit, 484
  in structural analysis. *See also* Direct stiffness method;
        Transformation matrices
    compatibility, 231, 362
    equilibrium, 231, 361
    global member stiffness, 444–451
    initial force, 375, 401
    initial reaction, 375
    joint displacement, 231, 365
    load, 231
    local member stiffness, 366
    member deformation, 231, 365
    member force, 231, 361
    reaction, 361
    structure stiffness, 234, 367, 382

Matrix algebra. *See* Transformation matrices
  addition, 484
  cofactors, 492
  element, 483
  expansion of a determinant, 491
  linear equations
    addition of, 488
    representation of, 487
    solution by inversion, 489
    subtraction of, 488
  multiplication, 487
  order of matrix, 484
  scalar multiplication, 484
  size of a matrix, 484
  transpose, 484
Member stiffness
  in basic matrix analysis method. *See* Basic matrix
    analysis method
  in direct stiffness method. *See* Direct stiffness method
  in moment distribution method. *See* Moment-distribu-
    tion method
Member stiffness matrix
  assembly of, 451–452. *See* Direct stiffness method
  by transformation matrices, 451
Membrane, 9
Moment-area method
  cantilever beams, 107
  derivation. *See* Moment-area theorems
  effect of chord displacement, 128
  frames, 181
    cantilevered, 183–187
    general frames, 187–190
  overhangs, 111
  reference tangents, 107
  sign convention, 113
  simply supported beams, 108
  with internal hinges, 131
Moment-area theorems
  derivation, 103–108
  use of, 105–116
Moment diagrams
  by parts
    cantilever method, 117
    purpose, 116–117
    simple beam method, 120
    use in moment area method, 127
  for fixed beams, 480
  for propped beams, 481
  for simple beams, 479
  plotting
    by mathematical integration, 71–78
    by moment equation, 68–71
    by statics, 68–71
    for beams, 68–78
    for frames, 180
Moment-distribution method
  carryover factor, 332–334
  carryover moment, 332–336
  distributed moment, 334–346
  distribution factor, 334–336

  fixed-end moments, 336
  for beams, 336–344
  for planar frames, 344–356
    with multiple translational degrees of freedom, 353–
     356
    with one translational degree of freedom, 347–353
    with rotational degrees of freedom, 344–347
  member stiffness, 332–336
  sidesway corrections, 348, 353–356
  support settlements, 341
  tables for, 338
  torsional member stiffness, 356–357
  unbalanced moment, 334–336
Moment equations, 68, 72
  simplified, 246
  use in unit load method. *See* Unit-load method

Non-prismatic members, 471

P-delta effect, 93
Planar frames
  analysis
    by Angle-Change principle. *See* Angle-change
     principle
    by conjugate grid method. *See* Conjugate grid
     method
    by direct stiffness method. *See* Direct stiffness
     method
    by flexibility method. *See* Flexibility method
    by integration methods. *See* Double integration,
     Quadruple integration
    by moment distribution method. *See* Moment-distri-
     bution method
    by slope-deflection method. *See* Slope-deflection
     method
    by unit-load method. *See* Unit-load method
  definition, 5–6
  independent unknown forces, 172
  loads, 34
  modeling. *See* Idealized model
Planar trusses
  analysis
    by direct stiffness method. *See* Direct stiffness
     method
    by equilibrium equations, 141–145
    by flexibility method. *See* Flexibility method
    by method of joints, 145–149
    by method of sections, 149–154
    by unit load method. *See* Unit-load method
    by visual method of joints, 147–149
  definition, 6
  equilibrium, 138–141
  geometric concepts, 154
  loads, 34
  member inclination diagram, 142
  models. *See* Idealized models
  stability. *See* Stability
  types, 136–138
Plane strain, 30
Plane-stress, 29

Principle(s)
  Angle-Change, 199–200
  of Virtual Work, 231

Quadruple integration method
  for beams, 80–90
  use of binomial functions, 88–90

Reactions
  at common supports, 35
  for beams, 61–65, 68–71
Released state
  in matrix analysis. *See* Direct stiffness method
  in moment distribution. *See* Moment distribution
Restrained state, 400–402, 418, 426–428, 452–455, 472

Shear deformation. *See* Deformation, shear
Shear diagram, 70, 74–76, 180
Shear equations
  beams 68, 72
  frames, 177
Shear force
  in beams. *See* Beams, forces in
  relationship to load and moment, 71
Shearing stress. *See* Stress
Shell, 7
Simple beam moment diagrams. *See* Moment diagrams
        for simple beams
Slope-deflection method, 316–328
  equilibrium conditions, 320–322
  general equations
    non-prismatic members, 328
    prismatic members, 317–320
Small-deflection theory assumptions, 80, 94
Space frame, 5
  loads, 34
  models. *See* Idealized models, three dimensional
Space truss, 6
  loads, 34
Stability
  bracing. *See* Bracing
  buckling, 54
  geometrical, 51, 169
  of trusses
    compound trusses, 159
    example problems, 161–162
    general criteria, 155
    simple trusses, 157
  statical, 51
Statical indeterminacy
  definition, 61, 141, 190
  determination for beams, 65–66, 279–289
    general equation, 65, 282
  determination for planar frames, 190–194, 279–286
    general equations, 190–192, 282–284
    method of trees, 193
    neglecting axial deformation, 282–283
  determination for trusses, 155–162
    general equation, 157

Statically determinate
  beams, 61
  definition, 61, 141, 169
  planar frames, 169, 190
  trusses, 136
Statically unstable
  definition. *See* Stability
  determination, 51–53
Stiffness coefficient, 20
Strain, 2
  axial, 26
  shearing, 27
Strain energy density, 227
  complementary, 229
  concept of, 227
Strain energy, internal, 226
  complementary, 229
  due to axial force, 228
  due to flexure, 228
  due to torsion, 228
  formulation for, 227
  total, 227
Stress, 2
  axial, 26, 40–42
  flexural stress, 42–43
  resultants, 33
  shearing, 27, 43
  torsional, 44
Stress-strain diagram, 27
Structural analysis
  assumptions in, 80, 92–95
  classical methods, 22
  computer methods, 22–23
  criteria for. *See* Criteria for structural analysis
  definition, 1
  first-order, 93
  linear, elastic, 92
  matrix. *See* Basic matrix analysis method, Direct stiff-
        ness method
  nonlinear methods, 23
  objective, 21
  of planar frames, 176
  second order, 94
Structures
  categories, 1–2
  continua, 2
  framed, 2
Structure stiffness matrix. *See* Matrices, in structural
        analysis
  by expanding member stiffness matrix
    beams, 392–396
    planar frames, 418
    planar trusses, 408
  by physical meaning
    beams, 388–392
    planar frames, 415–418
    planar trusses, 405–408
  stiffness coefficients, 387
Substructuring. *See* Idealized models
Superposition, principle of, 95

Supports
  idealized, 3, 35
  types, 4, 35
Support settlements
  by basic matrix method. *See* Basic matrix method
  by direct stiffness method. *See* Direct stiffness method
  by flexibility method. *See* Flexibility method
  by moment-distribution method. *See* Moment-distribution method

Theorem(s)
  Betti's Reciprocal, 233–235
  Maxwell's Reciprocal, 232–233
Theory of elasticity, 22, 28, 256
  field problems, 28, 31
  plane-strain state. *See* Plane strain
  plane-stress state. *See* Plane stress
  strain-displacement relationships, 30
  stress-strain relationships, 28–29
Thermal effects, joint loads for, 465–467
Transformation matrices
  for inclined supports, 470–471
  for load matrix
    grid member, 454–455
    planar frame member, 452–454
  for member stiffness matrix
    grid member, 449–451
    planar frame member, 444–447
    planar truss member, 447–448
  member coordinates, 441–442
  structure coordinates, 441–442

Unit-load method
  axial deformation, 238–242, 254
  by use of simplified moment equations, 246–249

flexural deformation, 242–246
  for beams, 242–246
  for continua, 256
  for general structures, 252
  for grid structures, 255
  for planar trusses, 238–242
  for planar frames
    including axial deformation, 251–252
    neglecting axial deformation, 249–251
  for space frames, 255
  general equation, 254
  shear deformation, 253
  torsional deformation, 253

Virtual displacements, 232
  pseudo-, 232
Virtual strain energy, 231
  pseudo-, 232
  unit-load method, 238–240, 242–244, 253–257
Virtual strain energy density, 231
Virtual work
  concept, 20, 230
  pseudo-, 232
  unit-load method, 238–240, 242–244, 253–257

Work
  complementary, 229
  concept, 19–21
  elastic strain, 20
  real, 20, 226
  unit-load method, 238–240, 242–244,
  virtual. *See* Virtual work
Working lines, 3